Introduction to Development Engineering

Temina Madon • Ashok J. Gadgil
Editors

Richard Anderson • Lorenzo Casaburi •
Kenneth Lee • Arman Rezaee
Co-Editors

Introduction to Development Engineering

A Framework with Applications from the Field

 Springer

Editors
Temina Madon
Center for Effective Global Action
University of California, Berkeley
Berkeley, CA, USA

Ashok J. Gadgil
Department of Civil and Environmental
Engineering
University of California, Berkeley
Berkeley, CA, USA

Co-Editors
Richard Anderson
Department of Computer Science
and Engineering
University of Washington
Seattle, WA, USA

Lorenzo Casaburi
Department of Economics
University of Zurich
Zurich, Switzerland

Kenneth Lee
Chief Research and Evaluation Officer
The Pharo Foundation
Nairobi, Kenya

Arman Rezaee
Department of Economics
University of California, Davis
Davis, CA, USA

ISBN 978-3-030-86067-7   ISBN 978-3-030-86065-3   (eBook)
https://doi.org/10.1007/978-3-030-86065-3

For Aditi, Arya, and Soma (T.M.)
For Anjali, Meghana,
and Madhurima (A.J.G.)

Preface

A defining question for every future generation is how to improve human welfare, and reduce inequality, while respecting the boundaries of our planet. Indeed, sustainable development is perhaps the greatest challenge we face as a species. It will require continuous innovation and adaptation, across massive scales.

Development Engineering explores the design of technological solutions that can rapidly accelerate life outcomes for people in poverty – particularly for those who struggle most – without disruptive impacts on biodiversity and the environment. This textbook serves as an introduction to the nascent field, which sits at an intersection of development economics and engineering.

The scope of the textbook is broad, ranging from mobile services for low-literacy users, to hardware solutions that bring clean water and electricity to remote environments. It is also highly interdisciplinary, drawing on methods and theory from political science and psychology as well as engineering and computer science.

The book's diverse subject matter is woven together by a set of common challenges that engineers and scientists face in designing technologies to accelerate the development of disadvantaged communities. These challenges include market failures (like high transport costs and financial exclusion) as well as institutional weaknesses (like poor regulatory capacity or corruption). These forces commonly prevent even promising technologies from improving welfare at any meaningful scale. The book also considers the behavioral and social constraints facing communities, including the heavy burden of stress and anxiety among those living in poverty. How can researchers learn to design around these constraints?

The opening part of this book offers a history of "technology-for-development" and an overview of the channels through which technological innovations can influence economic development. This is followed by a practical framework for research that helps transform a vast multi-disciplinary field into a tractable practice. This framework consists of four interlinked activities: innovation, iterative implementation, evaluation, and adaptation for scale.

The remainder of the textbook consists of six thematic parts, each focused on a different sector: energy and resources; agricultural markets; education and jobs; water, sanitation, and health; governance; and connectivity. Each part contains multiple case studies describing landmark research that has influenced the field of development engineering. The case studies demonstrate the practical framework for research, analyze the researchers' pivots and failures, and incorporate theories and concepts relevant to each specific sector.

This book has prioritized research that directly integrates engineering innovation with technically rigorous methods from the social sciences, such as randomized evaluation. This is the essence of the new field of "development engineering," which connects insights from development economics and related disciplines with engineering in order to promote impact at scale.

The intended audience for this book includes students, faculty, and other researchers involved in the design, use, and evaluation of technologies that sustainably accelerate development. It is also written for development practitioners (e.g., engineers, service providers, and technocrats) who work in low- and middle-income countries. Our motivation in writing this book is to help define a discipline for those working on technology-for-development. For several decades, the field has been dominated by practice (e.g., programs like "Engineers without Borders"), emphasizing service learning in the absence of theory and empirical research. This text attempts to provide a coherent intellectual framework for addressing the challenges of poverty through the design and scaling of better technologies.

The book is inspired by the lessons learned from a multi-year, $30 million investment in development engineering by the U.S. Agency for International Development. This investment, made through an award to the University of California, has supported more than 100 pilot and large-scale research projects in 30 countries, all focused on technologies for low-income communities.

We feel that the text is particularly timely given the increasing emphasis in universities (and in business) on social innovation and social entrepreneurship. New generations of young scientists and engineers are interested in careers that advance social welfare, tackle inequality, and directly address climate change. They are passionate about improving living conditions for people at the base of the economic pyramid. However, there is a need for a rigorous discipline to help channel this energy and ground it in an understanding of complex economic, political, and social systems. We aim to fill this gap with *An Introduction to Development Engineering*.

Berkeley, CA, USA   Temina Madon

Berkeley, CA, USA   Ashok J. Gadgil

Seattle, WA, USA   Richard Anderson

Zurich, Switzerland   Lorenzo Casaburi

Nairobi, Kenya   Kenneth Lee

Davis, CA, USA   Arman Rezaee

Acknowledgments

The impetus for this textbook first emerged from a conversation with three dear colleagues who have helped forge this new field of development engineering: Eric Brewer, Edward Miguel, and Catherine Wolfram. We are indebted to these pioneers for their individual intellectual contributions, and for their help in disciplining a complex but vital area of research.

In truth, this book (and our field) has been shaped by decades of research published by a diverse set of colleagues across the world. Art Rosenfeld, Rob Socolow, and Amulya (A.K.N.) Reddy are three outstanding researchers who have impacted how we deal with the real world. Alain de Janvry, Rachel Glennerster, and Elisabeth Sadoulet have shaped our thinking, in pragmatic ways, about how markets and institutions operate. Many others have contributed through their participation in the Development Impact Lab, including Gaetano Borriello, Silvia Hostetler, Craig McIntosh, Joyashree Roy, Evan Thomas, and Amos Winter as well as Engineer Bainomugisha, Michael Callen, Edward Kirumira, and Paul Niehaus.

Along the way, a number of colleagues at UC Berkeley have contributed to our thinking. We are grateful, above all, to Anustubh Agnihotri for insights and research into the political and disruptive nature of technology; we are grateful also for his gracious and persistent management of the textbook's drafting. We acknowledge valuable input from faculty and students across the campus, including Joshua Blumenstock, Jennifer Bussell, Prabal Dutta, Daniel Fletcher, Meredith Fowlie, Ken Goldberg, Angeli Kirk, Jeremy Magruder, Kara Nelson, Robert On, Kweku Opoku-Agyemang, Matthew Podolsky, and Javier Rosa.

We are particularly grateful to Maryanne McCormick, Shankar Sastry, and the team at the Blum Center for Developing Economies for spearheading this initiative in partnership with the Center for Effective Global Action. We are also indebted to past members of the Development Impact Lab staff, including Carson Christiano, Guillaume Kroll, Heather Lofthouse, Anh-Thi Le, Sophi Martin, Kevin McCarthy, Lina Nilsson, and Sarah White, for driving this learning odyssey.

The contours of this textbook have been profoundly shaped by early instructors in development engineering at Berkeley, including Alice Agogino, Paul Gertler, David Levine, Matthew Potts, and Jennifer Walske. We are also grateful to participants in the 2020 Development Engineering graduate courses at UC Berkeley and the University of Washington, for providing helpful insights and feedback on early versions of the text. In particular, we thank doctoral students Casey Finnerty and Isa Ferrall, who have drafted the companion instructor guide for the textbook.

This endeavor would not have been possible without the support of the U.S. Agency for International Development, which has supported an impressive body of research in development engineering and has enabled us to publish the textbook as an open access title. Colleagues at the agency, including Ticora Jones, Tara Hill, and David Ferguson, have been valuable partners, supporters, and mentors. We would also like to acknowledge Berber Kremer, Amy Bilton, Susan Amrose, and the editorial board of the journal *Development Engineering*, which has advanced the field immeasurably over the last 5 years.

Finally, much of the research described in this textbook has been supported by dedicated individuals at non-profit organizations like BRAC, Innovations for Poverty Action, the Abu Latif Jameel Poverty Action Lab, Evidence Action, and GiveDirectly. We are grateful for their meticulous, creative, and tireless work in the domain of economic development. We are also grateful to all of the research participants, individual contributors, chapter authors, and communities whose life experiences are represented in this book. Your efforts are helping to put innovation to work for the resilience of people and planet.

Contents

Contributors

Kwame Abrokwah Brixels Company Limited, Accra, Ghana

Achyuta Adhvaryu William Davidson Institute, Ross School of Business, University of Michigan, Michigan, MI, USA
Good Business Lab, New Delhi, Delhi, India

Joshua Adkins Electrical Engineering and Computer Science, University of California, CA, USA
nLine, Inc, Berkeley, CA, USA

Anustubh Agnihotri University of California Berkeley, Berkeley, CA, USA

Jenny C. Aker Teffs University, Medford, MA, USA

Susan Amrose Department of Mechanical Engineering, Massachusetts Institute of Technology, Cambridge, MA, USA

Richard Anderson Department of Computer Science and Engineering, University of Washington, Seattle, WA, USA

Mary Claire Barela University of the Philippines Diliman, Diliman, Philippines

Susanna Berkouwer The Wharton School, University of Pennsylvania, Philadelphia, PA, USA

Menna Bishop London School of Economics, London, UK

Waylon Brunette Paul G Allen School of Computer Science & Engineering, University of Washington, Seattle, WA, USA

Robin Burgess London School of Economics, London, UK

Lorenzo Casaburi Department of Economics, University of Zurich, Zurich, Switzerland

Brandon D. Clark Department of Chemical Engineering, Stanford University, Stanford, CA, USA

Maria Theresa Cunanan University of the Philippines Diliman, Diliman, Philippines

Prabal Dutta Electrical Engineering and Computer Science, University of California, CA, USA
nLine, Inc, Berkeley, CA, USA

Burak Eskici University of California at San Diego, San Diego, CA, USA

Raissa Fabregas Lyndon B. Johnson School of Public Affairs, The University of Texas at Austin, Austin, TX, USA

Cedric Angelo Festin University of the Philippines Diliman, Diliman, Philippines

Smit Gade Good Business Lab, New Delhi, Delhi, India

Ashok J. Gadgil Department of Civil and Environmental Engineering, University of California, Berkeley, Berkeley, CA, USA

Piyush Gandhi Good Business Lab, New Delhi, Delhi, India

Lavanya Garg Good Business Lab, New Delhi, Delhi, India

Tomoko Harigaya Senior Researcher, Precision Agriculture for Development (PAD), Boston, MA, USA

Carl Hartung Paul G Allen School of Computer Science & Engineering, University of Washington, Seattle, WA, USA

Kurtis Heimerl Paul G. Allen School of Computer Science and Engineering, Seattle, WA, USA

Dana Hernandez Department of Civil and Environmental Engineering, University of California at Berkeley, Berkeley, CA, USA

Christopher Hyun University of California, Berkeley, CA, USA

Esther Jang Paul G. Allen School of Computer Science and Engineering, Seattle, WA, USA

Matthew William Johnson Paul G. Allen School of Computer Science and Engineering, Seattle, WA, USA

Mansi Kabra Good Business Lab, New Delhi, Delhi, India

Noah Klugman Electrical Engineering and Computer Science, University of California, CA, USA
nLine, Inc, Berkeley, CA, USA

Michael Kremer University Professor in Economics and the College and the Harris School of Public Policy (Recipient of 2019 Nobel Prize for Economics), University of Chicago, Chicago, IL, USA

Tanu Kumar William & Mary, Williamsburg, VA, USA

Kenneth Lee Chief Research and Evaluation Officer, The Pharo Foundation, Nairobi, Kenya

James D. Long University of Washington, Seattle, WA, USA

Temina Madon Center for Effective Global Action, University of California, Berkeley, Berkeley, CA, USA

Philip Martinez University of the Philippines Diliman, Diliman, Philippines

Ashwin Nair University of Virginia, Charlottesville, VA, USA

Ankita Nanda Good Business Lab, New Delhi, Delhi, India

Kara L. Nelson Department of Civil & Environmental Engineering, University of California Berkeley, Davis Hall, Berkeley, CA, USA

Keren Neza Center for Technology and Economic Development, New York University, New York, NY, USA

Yaw Nyarko Division of Social Science, New York University Abu Dhabi, New York University, New York, NY, USA

Anant Nyshadham Ross School of Business, University of Michigan, Michigan, MI, USA

Kevin D. Orner Department of Civil & Environmental Engineering, University of California Berkeley, Davis Hall, Berkeley, CA, USA

Angela Orozco Center for Technology and Economic Development, New York University, New York, NY, USA

Pat Pannuto Computer Science and Engineering, University of California, San Diego, CA, USA

Good Business Lab, New Delhi, Delhi, India

Mamta Pimoli Good Business Lab, New Delhi, Delhi, India

Nicola Pitchford School of Psychology, University of Nottingham, Nottingham, UK

Matthew Podolsky Electrical Engineering and Computer Science, University of California, CA, USA

Alison E. Post University of California, Berkeley, CA, USA

Ravindra Ramrattan *(deceased)* Ravindra Ramrattan wrote this book while at Kenya

Isha Ray University of California, Berkeley, CA, USA

Agha Ali Raza Lahore University of Management Sciences (LUMS), Lahore, Punjab, Pakistan

Arman Rezaee Department of Economics, University of California, Davis, CA, USA

Patrick Shaw Brown University, Providence, RI, USAResearch Triangle International, Durham, NC, USA

Anant Sudarshan Ross School of Business, University of Michigan, Michigan, MI, USA

Jay Taneja Electrical and Computer Engineering, University of Massachusetts, Amherst, MA, USA

William A. Tarpeh Stanford University, Stanford, CA, USA

Kentaro Toyama University of Michigan School of Information, Ann Arbor, MI, USA

Aditya Vashistha Cornell University, Ithaca, NY, USA

Ronel Vincent Vistal University of the Philippines Diliman, Diliman, Philippines

Laurel Wheeler Department of Economics, University of Alberta, Edmonton, AB, Canada

Daniel L. Wilson Geocene Inc., Berkeley, CA, USA

Catherine Wolfram Haas School of Business, University of California, Berkeley, CA, USA

Céline Zipfel London School of Economics, London, UK

About the Editors

Temina Madon is on the professional faculty of the University of California, Berkeley, Haas School of Business, and a member of South Park Commons, where she advises early-stage startups. Previously, she led business development at machine learning startup Atlas AI. Earlier, Madon was the founding executive director of the Center for Effective Global Action (CEGA), a global research network focused on human and economic development, with headquarters at UC Berkeley. In this role, she managed the Development Impact Lab, a USAID-funded consortium of universities advancing the field of "development engineering." Madon worked with Ashok Gadgil to conceptualize a framework for this book and recruit contributors. Madon began her career in science and technology policy, working first in the US Senate (as an AAAS Science and Technology Policy Fellow) and later at the US National Institutes of Health (NIH). She holds a PhD from UC Berkeley in computational neuroscience and a BS from MIT.

Ashok J. Gadgil is a faculty senior scientist and former director of the Energy and Environmental Technologies Division at Lawrence Berkeley National Laboratory, and a Distinguished Professor of Civil and Environmental Engineering at the University of California, Berkeley. Since 2012, Gadgil is the faculty director of a large multi-campus, multi-disciplinary USAID-funded project, Development Impact Lab, with headquarters at UC Berkeley. Since 2006, he has taught graduate courses at UC Berkeley on inventing, implementing, and scaling up technologies for development. His research and technology inventions have been recognized with several significant awards and honors. Along with Dr. Temina Madon, Gadgil led the conceptualization of this book, shaping its content, framework, and selection of chapter authors. Gadgil holds a PhD in physics from UC Berkeley and an MSc in physics from the Indian Institute of Technology Kanpur.

Richard Anderson is a Professor in the Paul G. Allen School of Computer Science and Engineering at the University of Washington. His research interest is in Computing for the Developing World, with work spanning educational technology, mobile data management tools, global health information systems, and digital financial services. He has conducted research at the Mathematical Sciences Research Institute, the Indian Institute of Science, Microsoft Research, and PATH, a Seattle based NGO working on health technologies for low resource environments. He has been recognized with the NSF Presidential Young Investigator award, the University of Washington College of Engineering Faculty Innovator for Teaching Award, and the 2020 ACM Eugene L. Lawler Award for Humanitarian Contributions within Computer Science and Informatics. Previously, he worked in the theory and implementation of algorithms, including parallel algorithms, computational geometry, and scientific applications. He graduated with a B.A. in Mathematics from Reed College and a Ph.D. in Computer Science from Stanford University.

Lorenzo Casaburi is an associate professor in the Department of Economics at the University of Zurich. His main line of research focuses on agricultural markets in Sub-Saharan Africa, with an emphasis on market structure, behavioral insights, and agricultural finance. His research has received funding from the European Research Council, the Swiss National Foundation, the U.S. Agency for International Development, UK Aid, and others. He is a research fellow at the Centre for Economic Policy Research and Bureau for Research and Economic Analysis of Development, and a research affiliate at the International Growth Centre, Innovations for Poverty Action, and the Abdul Latif Jameel Poverty Action Lab. Lorenzo holds a BA from the University of Bologna and a PhD in economics from Harvard. Before joining the University of Zurich, he was a postdoctoral fellow at Stanford University.

Kenneth Lee is the director of Air Quality Life Index® and a senior research associate in the Department of Economics at the University of Chicago. He researches questions in the areas of development economics, environmental and energy economics, and environmental health. He has designed and published field experiments in both Kenya and India. Prior to this, he was the executive director of the Energy Policy Institute at the University of Chicago (EPIC) in India. He also held research fellow positions at the Center for Effective Global Action and the Energy Institute at Haas. He holds a PhD from the University of California, Berkeley, a master's in international affairs from Columbia University, and a bachelor's from McGill University. Earlier in his career, he worked as an investment banker in Toronto and London, covering media and telecoms companies in Africa, Europe, and Canada.

Arman Rezaee is Assistant Professor of Economics at the University of California, Davis. His research focuses on intersections of service delivery, political economy, and technology. He makes use of large-scale field experiments that leverage cellular technology, as well as natural experiments using historical archival data. Much of his work focuses on Pakistan. He also has active projects in Uganda and the Philippines. His work has been supported by the Center for Effective Global Action, the Bill & Melinda Gates Foundation, the International Growth Centre, the Abu Latif

Jameel Poverty Action Lab, the Policy Design Evaluation Lab, Private Enterprise Development in Low-Income Countries, and the University of California Labs. Before obtaining his PhD in economics from UC San Diego, he earned his master's in public policy from the Harvard Kennedy School of Government, where he was a public policy and international affairs fellow.

Chapter 1
Introduction to Development Engineering

Anustubh Agnihotri, Temina Madon ⓘ, and Ashok J. Gadgil ⓘ

1 What is Development Engineering?

Technological change has always played a role in shaping human progress. From the power loom to the mobile phone, new technologies have continuously influenced how social and economic activities are organized—sometimes for better and sometimes for worse. Agricultural technologies, for example, have increased the efficiency of agricultural production and catalyzed the restructuring of economies (Bustos et al., 2016). At the same time, these innovations have degraded the environment and, in some cases, fueled inequality (Foster and Rosenzweig, 2008; Pingali, 2012). Information technology has played a catalytic role in social development, enabling collective action and inclusive political movements (Enikolopov et al., 2020; Manacorda & Tesei, 2020); yet it has also fueled political violence and perhaps even genocide (Pierskalla & Hollenbach, 2013; Fink, 2018).

Nevertheless, the United Nations (UN) has recognized technology as key to achieving the Sustainable Development Goals (SDGs), a set of global policy targets adopted by 193 national governments for implementation by 2030.[1] An outstanding question is how to *systematically* harness technology for sustainable development?

---

[1] United Nations Sustainable Development Goals at https://sdgs.un.org/goals

A. Agnihotri (✉)
University of California Berkeley, Berkeley, CA, USA
e-mail: anustubh@berkeley.edu

T. Madon
Center for Effective Global Action, University of California, Berkeley, Berkeley, CA, USA

A. J. Gadgil
Department of Civil and Environmental Engineering, University of California, Berkeley, Berkeley, CA, USA

© The Author(s) 2023
T. Madon et al. (eds.), *Introduction to Development Engineering*,
https://doi.org/10.1007/978-3-030-86065-3_1

Fortunately, the research community has begun to offer paths forward. In this textbook, we introduce the nascent field of *development engineering*, an area of research focused on discovering generalizable technological solutions that can improve development outcomes in poverty-constrained settings. It integrates the theory and methods of development economics (and other social sciences) with the practice of *engineering*, promoting the co-design of engineering advances alongside the social and economic innovations required for impact in the "real world." The resulting solutions—whether they focus on intensifying agricultural production, enhancing early child development, or expanding access to sanitation—are well positioned to succeed at scale, and within planetary boundaries.

As a field, development engineering is closely aligned with the recent movement to scientifically validate different approaches to poverty reduction, exemplified in the 2019 Nobel Prize in Economic Sciences (awarded to development economists Abhijit Banerjee, Esther Duflo, and Michael Kremer)[2]. These researchers and their co-authors have helped pioneer the use of randomized controlled trials in public policy, bringing a precise and incremental approach to solving the problems of poverty. Development engineering follows in this tradition, yet is distinct in its focus on technological innovation as a tool for achieving sustainable development.

For all the promise of technology to accelerate sustainable development, we must also recognize the potential for new tools to harm people and the environment. Indeed the motivation in launching this new field has been, in part, the long string of failures in the area of "technology for good." There is a rich history of engineering projects that have been technically sophisticated but have failed to achieve social impact in the real world—or worse, have rolled back the frontiers of human development. Examples include costly but ineffective attempts to improve educational outcomes through low-cost laptops (Cristia et al., 2017; Kraemer et al., 2009); water rollers[3] that were intended to facilitate water transport but failed to gain adoption within targeted communities (Borland, 2014; Crabbe, 2012; Stellar, 2010); and large-scale irrigation systems that failed to deliver promised benefits (Higginbottom et al., 2021).

These failures have a number of elements in common. First, it is not obvious, *ex ante*, that such projects should fail, and the causes of failure are not always clear. They are often well intentioned efforts, employing human-centered design to better meet the needs of individual users. Yet they often overlook the top-down view of development: the politics, institutions, and social norms that surround any user. These conditions can doom the most well-intentioned efforts to fail.

Second, engineers operating in the context of poverty often lack information about users' habits. Take this as a thought experiment: as a consumer in a well-functioning market, you benefit from a vast infrastructure for data collection that reveals the economic behavior of you and people like you. The firms that service your needs have access to your web traffic logs, digital payments, utility meters, and mobile location data—not to mention household economic surveys, government economic indicators, and industry analyst reports. But what about the homeless

---

[2] See https://www.nobelprize.org/uploads/2019/10/advanced-economicsciencesprize2019.pdf

[3] See https://www.hipporoller.org

consumer who lives in urban poverty, subsisting on free meals and donations? Or the rural subsistence farmer who uses cash to operate in informal markets?

The most disadvantaged households are rarely reached by business analysts and government enumerators. Just 10 percent of households in rural India have access to formal sources of credit; the vast majority leave no trace in the credit market (Demirguc-Kunt et al., 2018). Fewer than half of all nations in sub-Saharan Africa have conducted a nationally representative household economic survey in the last decade (Yeh et al., 2020). People living in poverty, by definition, are excluded from participating actively in formal markets. As a result, their preferences are rarely captured in market price signals or routine consumer data. They may provide feedback to researchers in the form of self-reported preferences (e.g., through focus groups or interviews), but these inputs may be biased and unreliable. Without reliable insights to guide technology design, it is unsurprising that so many engineers have failed to achieve impact.

In recent years, we have developed better techniques to observe the preferences and behaviors of underserved communities. These include low-cost sensors for monitoring product use, automated digitization of administrative records, and even behavioral experiments conducted outside the lab, in "the field." Some of these tools will be discussed in future chapters; they are increasingly being used by engineers to design for people excluded from conventional markets.

A third challenge is the paucity of research identifying the long-term economic and social impacts of new technologies (largely for a lack of investment in rigorous evaluation). Rarely have the developers of "pro-poor" technologies had the resources to evaluate the downstream social and economic impacts of their inventions. We are all familiar with the use of randomized, controlled trials (RCTs) in medicine; these methods are used to rigorously measure the effects of a novel medical treatment or prophylactic, across large populations of patients. More recently, software developers have adopted this approach to test the effects of different product features, using rapid experimentation to generate user feedback in a process known as A/B testing. Yet the tools of rigorous evaluation have only slowly diffused into the broader engineering community. This is despite the fact that engineers are interventionists at heart, seeking to make changes to markets, the environment, and people's lives.

Through collaboration with economists, political scientists, and public health researchers, engineers are now investigating the impacts of their inventions. Adapting the experimental methods used in medical trials (and more recently in public policy), we can now ask: How does the use of tablets in classrooms affect learning outcomes, both for the highest-performing students and those in the bottom quantile (Chap. 11)? How does the introduction of improved cookstove technology affect household consumption and nutrition (Chap. 15)? What is the impact of mobile telephony on local economies (Chap. 11), and what is the development impact of access to grid electricity (Chap. 5)?

Rigorous evaluation can help explain the causal relationships between a technology and its downstream impacts, including impacts on the climate and the environment (Alpízar & Ferraro, 2020). It allows us to learn how technologies effect change, and it teaches us about the economic and social constraints that any successful solution must address. Experiments in real-world settings have also led to a better understanding of how technologies get adopted in disadvantaged communities. These insights can be used to weave novel behavioral, economic, and social interventions into the design of technological solutions.

What does a "development engineering" innovation look like? One of the earliest examples is a community-scale water chlorination technology for rural households, designed by a team of engineers and economists. For user convenience and perceptual salience, it is a brightly colored device placed at high-traffic points of water collection, like springs. It dispenses just the right amount of chlorine to fill the typical household's container, and it is provided free of charge. Its design is based on rigorous studies of users' willingness to pay, their consumption habits, and an understanding of how social pressure influences hygiene practices (Kremer et al., 2011; Null et al., 2012). The system is now being scaled to millions of households across sub-Saharan Africa, with appropriate adaptations; and it is widely viewed as one of the most sustainable modern solutions for providing clean water to rural communities (Ahuja et al., 2015).

Technologies like these leverage important recent insights from economics—for example, the finding that poverty-constrained households do not use preventive health technologies (like insecticide treated bednets) when pricing is non-zero[4] (Dupas, 2014). They are built for specific social, behavioral, environmental, and economic contexts. This means that when markets cannot deliver the desired development impact, the public sector (or civil society) is leveraged as the channel for delivery.

In some sense, development engineering is similar to other problem-focused fields, like environmental engineering and bioengineering, in that it combines two or more disparate disciplines to holistically address a defined set of problems. By definition it is highly interdisciplinary, combining insights from development economics and political science as well as computer science, environmental science, and of course engineering. Similarly, it is applied: there is a limited focus on basic research and an emphasis on identifying innovations that solve problems reliably (and at scale) within complex "real-world" environments. It is unique in its emphasis on the challenges faced by individuals and communities subjected to poverty and marginalization.

Defining Terms: Technology, Invention, Intervention
In this textbook, we refer to a "technological solution" as a technology integrated with the social and economic interventions required to achieve impact at scale. When brought together, these two elements solve a development problem that neither could have achieved independently. In some cases, we will use the word "innovation" in place of the word "solution." To help

(continued)

---

[4] See https://www.povertyactionlab.org/case-study/free-bednets-fight-malaria.

navigate the jargon-rich world of development engineering, here we define a set of common terms that you will find throughout the textbook.

Technology is the body of scientific and engineering knowledge *and* its application to improve the production of goods, the delivery of services, and the accomplishment of societal objectives. Technology can take the form of novel systems, practices, or processes.

An **invention** is a unique device, method, process, or composition that is technically novel, nonobvious, and often patentable. An invention is the result of a creative process that involves the discovery of something new. It may not require new technology. For example, invention of the lightbulb brought together multiple existing technologies in a new arrangement, yielding a useful and novel product.

An **intervention** is an action taken to effect or modify the outcomes of individuals, populations, and systems. In the context of development engineering, an intervention may be a social or economic strategy designed to change the behaviors of markets, institutions, and households. Interventions can be innovative, and they may involve technologies or inventions, but these are not required.

Development engineering is a practice, but it is also a field of research, with a research agenda that explores how technological solutions (and their design) can be optimized and applied for sustainable development. While the design of technology has been well studied in developed markets, it is less clear how innovations should be designed to solve development challenges. The field aims to generate technological solutions that can be rigorously evaluated, can perform reliably at scale, and can improve millions of lives.

The authors of the various case studies in this textbook speak from experience. They have engaged in research and collaboration across disciplines and over many years. Electrical engineers studying power grids have learned in the field alongside development economists exploring the demand for electricity in rural communities. Political scientists interested in post-conflict state capacity have collaborated with computer scientists on the design of digital governance technologies. They have also advanced the measurement of social and economic outcomes, leveraging tools like remote sensing, mobile data, and networked sensors to observe and understand the process of sustainable development. By learning each other's languages—and defining this new discipline—we are able to form a more coherent, systematic approach to global development challenges.

While we attempt to define development engineering in the opening chapters of this book, the research community has offered several diverse definitions of the field (Nilsson et al., 2014; Agogino & Levine, 2016). Taken together, these perspectives are beginning to shape an important dialogue about technology and its role in sustainable development. We value these contributions, and we aim for this textbook to offer a comprehensible (if not comprehensive) synthesis of research to date.

2 Intellectual History of the Field

The concepts of "engineering for development" and "technology for development" have taken many forms over the last few decades. This section sketches an intellectual history of the field, tracing the different paradigms that have dominated our thinking about technology in resource-constrained settings. We start with research on the broad relationship between technological change and human development and then review the various movements employing technology as a solution for societal challenges. We conclude by explaining how this new field differs from earlier paradigms.

It is well established that technological innovation is central to economic growth. Technological advances, with an enduring consistency, have led to increases in the productive capacity of societies, allowing them to move from scarcity to surplus (Landes, 2003; Nelson & Nelson, 2005). Economic historians have studied this process in great detail, starting with the industrial revolution (Mokyr, 2018; Landes, 2003; Polanyi & MacIver, 1944; Piketty, 2014). Propelled by technological innovation, the industrial revolution had a profound impact on the thinking of philosophers and economists. It introduced the idea that technological transformations can make persistent improvements in economic conditions; it also established the centrality of markets in shaping the economic life of individuals and societies. It introduced the notion that human intervention can actually shift the course of our development (Smith, 2010).[5]

However, the idea that human development could be achieved through policy intervention did not take root until the end of the second World War and the so-called Marshall Plan. Postwar policy initiatives focused on economic growth across war-torn Europe, with the underlying assumption that technological progress would increase productivity and create economic surplus (Landes, 2003, Keynes, 2018). Such progress was "engineered" through large-scale industrialization that was managed by corporations and guided by governments through economic policy. The success in spurring postwar economic growth led to a Western concept of *development* that had well-defined stages of growth, with all societies passing through distinct phases and eventually converging through the diffusion of technology (Rostow, 1960).

In the postwar era, Europe's success in using large-scale industrial technology to solve the challenges of production led to the transfer of these technologies to less developed countries, with the aim of rapidly transforming their economies. However, this effort to transplant technology was riddled with failures. Not only did many of these technologies (like synthetic fertilizers and large-scale dams) create

---

[5] Prior to the industrial revolution, economic growth was seen as cyclical. Scholars embraced the notion of a Malthusian trap, an argument that views technical progress as linear and population growth as exponential. Malthus (in his famous essay published in 1798) argued that for a fixed technical growth in resources, small populations will experience greater per capita income, resulting in population growth that overshoots the available resource base. This, in turn, will reduce per capita income, inevitably driving a contraction of the population.

unforeseen environmental harm; they also failed to be widely adopted or fell into disuse (e.g., handpumps to access groundwater).

2.1 Appropriate Technology Movement

The movement for appropriate technology emerged, in part, as a reaction to the frustrations stemming from attempts to rapidly replicate "Western" models of technology-driven growth in lower-income settings. The Western model often excluded community input, treating people as recipients of intervention rather than participants in development.

Peaking in the 1970s and 1980s, the appropriate technology movement argued for small-scale technological solutions that were based on local needs and "appropriate" for the nature of local endowments, rather than implemented by central authorities (Schumacher, 2011, Dunn, 1979). The movement borrowed heavily from the Gandhian ideal of self-reliant village communities. It also viewed the adoption of technology, and its consequences, through the lens of equality, by focusing on *who* adopts a technology, and how the gains from a technology are distributed. As a consequence, the approach has focused on local and indigenous production of (appropriate) technology, so that communities benefit from wider-scale adoption in multiple ways.

Impact on the environment is also a central tenet of the movement, with a strong emphasis on sustainability and the use of renewable sources. An example of a widely adopted appropriate technology is the treadle-pump for irrigation, which is easily constructed at the village level and sustainably enables the farmer to provide water to his or her fields (Adeoti et al., 2007). In reality, this innovation has been delivered through a centralized nongovernmental organization (NGO) to enable product quality certification ("KrishiBandhu"), signaling some of the shortcomings of this approach.

The appropriate technology movement has had a deep impact on how the development community thinks about the role of technology in shaping lives of people in poor communities. It has highlighted the need to pay closer attention to the negative environmental externalities of industrial technology. However, appropriate technologies have not seen widespread and sustained adoption over the medium to the long run. Critiques have suggested that the lack of attention to the role of markets and scalability has limited the success of "appropriate" technologies (Rybczynski, 1980; Willoughby, 1990).

2.2 Market-Oriented Approaches

In parallel to the appropriate technology movement is a long history of leveraging market-based incentives to stimulate innovation for resource-poor settings. The idea of profit at the "bottom of the pyramid," popularized by CK Prahalad, asserts that there are large, untapped market opportunities in low-resource communities that can be exposed by making technologies more affordable for the poor (Prahalad, 2009). Rather than viewing people who live under $2 a day as passive recipients of development aid, this approach views them as consumers of profitable goods and services. Given the very large number of people living in resource-poor environments, even a small profit margin can yield substantial profits at scale. While the poorest households cannot afford a bottle of shampoo or a box of tea, they do desire, and can afford, a small sachet that is cheaply priced. This approach has encouraged corporations to pursue profit while ensuring that people with limited resources can access the products they need. This approach too has its limitations, since it focuses exclusively on needs that can be addressed through market expansion. Large "public goods" requirements—like education and health—are not always effectively met by this approach.

A different market-oriented approach has focused on the productive and creative capacity of people living in resource-poor settings. Challenging the often held assumptions that associate technological innovation with high levels of formal education, this approach emphasizes the entrepreneurial and generative capabilities of the poor as "frugal innovators." The idea is that within resource-constrained settings, local innovators can develop technologies with unique forms and functionalities, tailored to local problems and environments. Anil Gupta's Honey Bee network leverages the traditional knowledge created by grassroot innovators to identify and screen new technologies for scale up (Gupta, 2006). An example of this is the biosand filter, an adaptation of centuries-old indigenous technology that was refined for scale-up in 1990. It is now estimated to serve more than 4 million people in 55 countries.

Like Prahalad's market-oriented approach, the view of people in resource-poor environments as technology creators leads to technologies that are adapted to local contexts and preferences. This can have spillover benefits for wealthier consumers, when products optimized for low-income communities move into developed markets. Indeed the unique nature of innovations from resource-constrained settings has led to a so-called "boomerang" effect, with products designed for scarcity benefiting users in more prosperous economies (Immelt et al., 2009; Winter & Govindarajan, 2015). For example, the leveraged freedom chair which provided users navigating uneven terrain in rural India with added control and flexibility was also successfully marketed in the United States as GRIT Freedom Chair, at a higher cost (Judge et al., 2015). Thus, market-oriented approaches have focused on people in under-resourced conditions as both consumers and producers of technological innovation for solving development problems.

2.3 Humanitarian Engineering

Humanitarian engineering is a paradigm that explores how engineering solutions can be used to provide access to basic human needs—like water, sanitation, energy, and shelter—in response to disasters, emergencies, and other resource-challenged environments. Unlike market-oriented approaches, humanitarian engineering takes a rights-based view, placing the needs of communities as the central motivation behind intervening. It often relies on researchers and innovators contributing their time to develop a technological solution that solves a well-identified problem within a community.

While the field of humanitarian engineering has begun to embrace market-based solutions, for example, through the distribution of cash transfers to households recovering from economic shocks, it is unclear whether private sector approaches actually work, particularly when it comes to provision of goods like water and sanitation (Martin-Simpson et al., 2018). Alongside recent exploration of market-based programming, there has been an emphasis on the design of "dual-use" solutions that operate in an emergency and also enhance community resilience by building preparedness for future emergencies. For example, a project to provide clean drinking water within a refugee tent camp might be taken up by a voluntary organization like engineers without borders but designed to support sustained use as the camp evolves into a longer-term settlement.

Humanitarian engineering has been especially effective when applied to disaster mitigation, a process that prepares disaster-prone communities to rebuild using resilient technologies. For example, the Berkeley-Darfur Stove, developed initially for Darfur refugees, now serves more than 60,000 families in different settings across Africa (see PotentialEnergy.org). UVWaterworks, a water purification technology initially developed in response to a cholera epidemic in India, now serves 26 million customers across 5 different countries (see WaterHealth.com).

2.4 ICTD

The proliferation of information and communication technology (ICT) across the world has fundamentally altered how individuals access and receive information, search for jobs, obtain government services, engage with financial institutions, and communicate with others. With more than 3 billion Internet users worldwide, ICT plays a central role in how under-resourced communities experience social and economic development (WDR, 2016). Gains from access to ICT can be significant for people who previously lacked access to the technology: for example, fish markets in Kerala saw dramatic reduction in spatial price variation after the introduction of cell phones, which allowed fishermen and wholesalers to more easily exchange information (Jensen, 2007). Similarly, M-pesa, a mobile-based money transfer application introduced in Kenya, has allowed millions of people to easily access remittance flows (Mbiti & Weil, 2015). However, the adoption and benefits of ICTs depend heavily on social and economic factors. For example, more educated people living in urban areas are more likely to have access to smartphones (World Development Report, 2016, Pg 167).

The field of ICT for Development (ICTD or ICT4D) has focused on understanding how this digital divide can be bridged, by making access to ICTs more equitable. One thrust of the field is how to reduce information asymmetries, so that remote and disconnected populations can connect to markets. For example, modifications to communication services like interactive voice response (IVR) enable those with low literacy to access relevant digital information (Chu et al., 2009; Mudliar et al., 2012).

ICTD researchers have also partnered with governments to change how states deliver services to their citizens. The most common innovation is the deployment of "helplines" that enable citizens to register their grievances through web-based or IVR platforms. Thoughtful design of these systems can empower marginalized citizens, providing new channels for reporting their grievances (Chakraborty et al., 2017). This approach has also been adopted by civil society, enabling individuals and communities to act collectively and voice their grievances (World Development Report, 2016, Chap. 3). For example, IVR platforms are being used to help smallholder farmers to raise concerns and grievances with local authorities (Patel et al., 2010).

A corrective critical perspective for the field of ICTD explores the inability of technology, by itself, to improve welfare and the need for institutional arrangements that support technological solutions and their effective adoption (Toyama, 2015; Johri and Pal, 2012). Indeed in the private sector, deployment of ICTs often focuses on the end-user and the product, without close attention to institutional arrangements, power dynamics, and the cultural environment of targeted users. For example, the one-laptop per child (OLPC) program aimed to transform learning by providing every child with an affordable laptop. However, it failed to achieve the impact at scale by failing to account for local cultures and preferences within the educational system (Kraemer et al., 2009).

2.5   Human-Centered and Participatory Design

A persistent challenge in "technology for development" is that products are designed by people who are far removed from the end-user's context. Human-centered design (HCD) advocates for a product design strategy that explicitly centers around the daily experiences of people in their native environments. The hypothesis of HCD is that failing to understand and empathize with the user's needs and requirements can lead to failure in adoption when the technology finally arrives at the user's doorstep. As discussed earlier, the water-roller was designed to help women in rural low-income settings access large quantities of water. Yet it fell into disuse as a result of severe design flaws, including failures on uneven terrain and the size of the product, which failed to meet women's needs (Crabbe, 2012). HCD emphasizes the perspective of the user and her environment, focusing on the complete product cycle from interface to manufacturing, distribution, and repair (Donaldson, 2009). A successful example of HCD is the wheelchair by the Gear Lab at MIT, which serves people with disabilities. The specific needs of disabled people living in low-income settings were incorporated into a redesign of the traditional wheelchair model, allowing users to traverse more rugged terrain with greater maneuverability (Winter and Govindarajan, 2015).

A related effort has been that of participatory design (or co-design), which actively involves end-users and other stakeholders in the design process (Spinuzzi, 2005; Steen, 2013). Thus, the consumers of the new technology provide their inputs from initial ideation to finalization and production. The active involvement of the end user ensures that the design of a new product does not leave out needs of the consumers. However, the deep involvement of a small number of end-users can limit the effort taken to get feedback from a larger, more representative sample of customers. It remains unclear whether human-centered design and co-design result in innovations that achieve superior development outcomes at scale. However, they are a promising complement to approaches that focus on market constraints, institutional failures, and social and behavioral norms.

2.6 Development Engineering

Development engineering borrows from many of the intellectual paradigms mentioned above but also differentiates itself in key ways. Like appropriate technology and frugal innovation, it pursues the well-being of people living in resource-constrained environments (as opposed to targeting rapid industrialization, or macroeconomic growth). Yet unlike these movements, development engineering brings attention to the importance of markets and political institutions in shaping human development. As with humanitarian engineering, we focus on sustainability and resilience, yet we also seek to discover the causal mechanisms through which technology shapes sustainable development over the long term. By studying the *mechanisms* of development, development engineering aims for generalizable lessons that extend beyond any one context, population, or environment.

In many ways, this new field follows in the tradition of ICTD, particularly its emphasis on interdisciplinary collaboration. It seeks to bring insights from the rapid adoption and positive impact of ICTs to other important areas of engineering, including some with great economic promise (like off-grid energy and precision agriculture) and some with importance for health (such as wastewater treatment and sanitation). As such, development engineering extends beyond ICTD's focus on information and computing to include civil and environmental engineering, mechanical engineering, electrical and power systems engineering, materials science, chemical engineering, and related disciplines. And unlike market-oriented approaches, development engineering does not rely on one particular strategy for the implementation of a technological innovation: if markets are the appropriate channel, they are leveraged—while not ruling out the option of delivering a technology through government agencies, nongovernmental organizations (NGOs), or communities.

Indeed development engineering has emerged in the absence of a profit motive, driven by university researchers focused on efficiently meeting the unmet demands of disadvantaged people. These university actors have worked alongside international development agencies, governments, social enterprises, and for-profit ventures to create "testbeds" for innovations that can advance progress toward the SDGs. This team-based architecture has allowed for the accumulation of knowledge and the discovery of generalizable solutions, while also facilitating the transition to scale of effective solutions.

On that note, we should point out that development engineering focuses explicitly on the *scalability* of technological solutions. It does not emphasize "boutique" or bespoke solutions to niche problems nor does it rely exclusively on the participatory approaches that some technical groups (e.g., MIT D-Lab) have developed. The scalability and generalizability of research findings are viewed as critically essential and important features of development engineering, while recognizing that scale-up of any innovation will require localization, customization, and adaptation to local conditions.

References

Adeoti, A., et al. (2007). *Treadle pump irrigation and poverty in Ghana* (Vol. 117). IWMI.

Aghion, P., & Howitt, P. (1990). *A model of growth through creative destruction*. National Bureau of Economic Research.

Ahuja, A., Gratadour, C., Hoffmann, V., Jakiela, P., Lapeyre, R., Null, C., Rostapshova, O., & Sheely, R. (2015). *Chlorine dispensers in Kenya: scaling for results, 3ie Grantee Final Report*. International Initiative for Impact Evaluation (3ie).

Alpízar, F., & Ferraro, P. J. (2020). The environmental effects of poverty programs and the poverty effects of environmental programs: The missing RCTs. *World Development, 127*, 104783.

Banerjee, A. V., Banerjee, A., & Duflo, E. (2011). *Poor economics: A radical rethinking of the way to fight global poverty*. Public Affairs.

Basu, P. (2006). *Improving Access to Finance for India's Rural Poor*. The World Bank. https://doi.org/10.1596/978-0-8213-6146-7

Beaman, L., BenYishay, A., Magruder, J., & Mobarak, A. M. (2018). *Can network theory-based targeting increase technology adoption?*. National Bureau of Economic Research.

Borland, R. (2014). The PlayPump. *The gameful world: Approaches, issues, applications*, 323–338.

Brewer, E., et al. (2005). The case for technology in developing regions. *Computer, 38*(6), 25–38.

Bustos, P., Caprettini, B., & Ponticelli, J. (2016). Agricultural productivity and structural transformation: Evidence from Brazil. *American Economic Review, 106*(6), 1320–1365.

Chakraborty, D., Ahmad, M. S., & Seth, A. (2017). Findings from a civil society mediated and technology assisted grievance redressal model in rural India. In *Proceedings of the Ninth International Conference on Information and Communication Technologies and Development*.

Chu, G., Satpathy, S., Toyama, K., Gandhi, R., Balakrishnan, R., & Menon, S. R. (2009, April). Featherweight multimedia for information dissemination. In *2009 International Conference on Information and Communication Technologies and Development (ICTD)* (pp. 337–347). IEEE.

Crabbe, A. (2012). Three strategies for sustainable design in the developing world. *Design Issues, 28*(2), 6–15.

Cristia, J., Ibarrarán, P., Cueto, S., Santiago, A., & Severín, E. (2017). Technology and child development: Evidence from the one laptop per child program. *American Economic Journal: Applied Economics, 9*(3), 295–320.

Demirguc-Kunt, A., Klapper, L., Singer, D., Ansar, S., & Hess, J. (2018). The Global Findex Database 2017: Measuring financial inclusion and the fintech revolution.. The World Bank.

Donaldson, K. (2009). The future of design for development: three questions. *Information Technologies & International Development, 5*(4), 97.

Dunn, P. D. (1979). *Appropriate technology: Technology with a human face.* Macmillan International Higher Education.

Dupas, P. (2014). Getting essential health products to their end users: Subsidize, but how much? *Science, 345*(6202), 1279–1281.

Dym, C. L., et al. (2005). Engineering design thinking, teaching, and learning. *Journal of engineering education, 94*(1), 103–120.

Enikolopov, R., Makarin, A., & Petrova, M. (2020). Social media and protest participation: Evidence from Russia. *Econometrica, 88*(4), 1479–1514.

Fink, C. (2018). Dangerous speech, anti-muslim violence, and facebook in Myanmar. *Journal of International Affairs, 71*(1.5), 43–52.

Foster, A., & Rosenzweig, M. (2008). Inequality and the sustainability of agricultural productivity growth: Groundwater and the Green Revolution in rural India. In *Prepared for the India Policy Conference at Stanford University* (Vol. 5).

Gupta, A. K. (2006). From sink to source: The Honey Bee Network documents indigenous knowledge and innovations in India. *Innovations: Technology, Governance, Globalization, 1*(3), 49–66.

Higginbottom, T. P., Adhikari, R., Dimova, R., Redicker, S., & Foster, T. (2021). Performance of large-scale irrigation projects in sub-Saharan Africa. *Nature Sustainability*, 1–8.

Immelt, J. R., Govindarajan, V., & Trimble, C. (2009). How GE is disrupting itself. *Harvard Business Review, 87*(10), 56–65.

Jensen, R. (2007). The digital provide: Information (technology), market performance, and welfare in the South Indian fisheries sector. *The Quarterly Journal of Economics, 122*(3), 879–924.

Johri, A., & Pal, J. (2012). Capable and convivial design (CCD): a framework for designing information and communication technologies for human development. *Information Technology for Development, 18*(1), 61–75.

Judge, B. M., Hölttä-Otto, K., & Winter, A. G. (2015). Developing world users as lead users: a case study in engineering reverse innovation. *Journal of Mechanical Design, 137*(7).

Keynes, J. M. (2018). *The General Theory of Employment, Interest, and Money.* Springer.

Kraemer, K. L., Dedrick, J., & Sharma, P. (2009). One laptop per child: vision vs. reality. *Communications of the ACM, 52*(6), 66–73.

Kremer, M., Miguel, E., Mullainathan, S., Null, C., & Zwane, A. P. (2011). *Social engineering: Evidence from a suite of take-up experiments in Kenya.* Unpublished Working Paper.

Landes, D. S. (2003). *The unbound Prometheus: technological change and industrial development in Western Europe from 1750 to the present.* Cambridge University Press.

Levine, D. I., Agogino, A. M., & Lesniewski, M. A. (2016). Design thinking in development engineering. *International Journal of Engineering Education, 32*(3), 1396–1406.

Madon, T., et al. (2007). *Implementation science*, 1728–1729.

Manacorda, M., & Tesei, A. (2020). Liberation technology: Mobile phones and political mobilization in Africa. *Econometrica, 88*(2), 533–567.

Mani, A., et al. (2013). Poverty impedes cognitive function. *Science, 341*(6149), 976–980.

Martin-Simpson, S., Parkinson, J., & Katsou, E. (2018). Measuring the benefits of using market based approaches to provide water and sanitation in humanitarian contexts. *Journal of Environmental Management, 216*, 263–269.

Mbiti, Isaac, and David N. Weil. "Mobile banking: The impact of M-Pesa in Kenya." *African Successes, Volume III: Modernization and Development.* University of Chicago Press, 2015. 247-293.

Mokyr, J. (2018). *The British industrial revolution: an economic perspective.* Routledge.

Mudliar, P., Donner, J., & Thies, W. (2012). Emergent practices around CGNet Swara, voice forum for citizen journalism in rural India. *Proceedings of the Fifth International Conference on Information and Communication Technologies and Development.*

Mullainathan, S., & Shafir, E. (2013). *Scarcity: Why having too little means so much.* Macmillan.

Nelson, R. R., & Nelson, R. R. (2005). *Technology, institutions, and economic growth.* Harvard University Press.

Nilsson, L., Madon, T., & Shankar Sastry, S. (2014). Toward a new field of development engineering: linking technology design to the demands of the poor. *Procedia Engineering, 78,* 3–9.

Null, C., Kremer, M., Miguel, E., Hombrados, J. G., Meeks, R., & Zwane, A. P. (2012). Willingness to pay for cleaner water in less developed countries: Systematic review of experimental evidence. In *The International Initiative for Impact Evaluation (3iE).*

Patel, N., Chittamuru, D., Jain, A., Dave, P., & Parikh, T. S. (2010). Avaaj otalo: a field study of an interactive voice forum for small farmers in rural india. In *Proceedings of the SIGCHI Conference on Human Factors in Computing Systems* (pp. 733–742).

Pierskalla, J. H., & Hollenbach, F. M. (2013). Technology and collective action: The effect of cell phone coverage on political violence in Africa. *American Political Science Review,* 207–224.

Piketty, Thomas. "Capital in the 21st Century." (2014).

Pingali, P. L. (2012). Green revolution: impacts, limits, and the path ahead. *Proceedings of the National Academy of Sciences, 109*(31), 12302–12308.

Polak, Paul. *Out of poverty: What works when traditional approaches fail..* ReadHowYouWant. com, 2009.

Polanyi, K., & MacIver, R. M. (1944). *The great transformation* (Vol. 2). Beacon press.

Postel, S., et al. (2001). Drip irrigation for small farmers: A new initiative to alleviate hunger and poverty. *Water International, 26*(1), 3–13.

Prahalad, Coimbatore K. *The fortune at the bottom of the pyramid, revised and updated 5th anniversary edition: Eradicating poverty through profits.* FT Press, 2009.

Radjou, N., Prabhu, J., & Ahuja, S. (2012). *Jugaad innovation: Think frugal, be flexible, generate breakthrough growth.* Wiley.

Rostow, W. W. (1960). *The Five Stages of Growth–A Summary* (pp. 4–16).

Roy, B., & Hartigan, J. (2008). Empowering the rural poor to develop themselves: The barefoot approach (Innovations case narrative: barefoot college of Tilonia). *Innovations: Technology, Governance, Globalization, 3*(2), 67–93.

Rybczynski, Witold. "Paper heroes; a review of appropriate technology." (1980).

Schumacher, E. F. (2011). *Small is beautiful: A study of economics as if people mattered.* Random House.

Smith, A. (2010). *The Wealth of Nations: An inquiry into the nature and causes of the Wealth of Nations..* Harriman House Limited.

Solow, R. M. (1957). Technical change and the aggregate production function. *The Review of Economics and Statistics,* 312–320.

Spinuzzi, C. (2005). The methodology of participatory design. *Technical Communication, 52*(2), 163–174.

Steen, M. (2013). Co-design as a process of joint inquiry and imagination. *Design Issues, 29*(2), 16–28.

Stellar, D. (2010). The PlayPump: what went wrong? In *State of the Planet The PlayPump What Went Wrong Comments, Np 1.*

Sterling, S. R., & Bennett, J. K. (2012). Three Bad Assumptions: Why Technologies for Social Impact Fail. *International Journal of Computational Engineering Research (IJCER), 2*(4), 1012–1015.

Toyama, K. (2015). *Geek heresy: Rescuing social change from the cult of technology.* Public Affairs.

Willoughby, K. W. (1990). Technology choice: A critique of the appropriate technology movement.. Dr Kelvin Wayne Willoughby.

Winter, A., & Govindarajan, V. (2015). Engineering reverse innovations principles for creating successful products for emerging markets. *Harvard Business Review, 93*(7-8), 80–89.

World Bank Group. (2016). *World development report 2016: digital dividends*. World Bank Publications.

Yeh, C., Perez, A., Driscoll, A., Azzari, G., Tang, Z., Lobell, D., ... Burke, M. (2020). Using publicly available satellite imagery and deep learning to understand economic well-being in Africa. *Nature communications, 11*(1), 1–11.

Chapter 2
Technology and Development

Menna Bishop, Robin Burgess, and Céline Zipfel

1 Introduction

Two major challenges face humanity in the coming century. The first is to generate the innovations and productivity improvements that will keep people on a path to higher standards of living. The second is to ensure that expanding human activity does not generate negative environmental externalities that block this path to progress.[1] In short, our future is about balancing the need for growth with the externalities that arise from that growth.

---

[1] Both these challenges are enshrined in the sustainable development goals (SDGs): SDG1 – "End poverty in all its forms everywhere"; SDG2 – "End hunger, achieve food security and improved nutrition, and promote sustainable agriculture"; SDG3 – "Ensure healthy lives and promote well-being for all at all ages"; SDG4 – "Ensure inclusive and equitable quality education and promote lifelong learning opportunities for all"; SDG6 – "Ensure availability and sustainable management of water and sanitation for all"; SDG7 – "Ensure access to affordable, reliable, sustainable and modern energy for all"; SDG8 – "Promote sustained, inclusive and sustainable economic growth, full and productive employment and decent work for all", SDG9 – "Build resilient infrastructure, promote inclusive and sustainable industrialization, and foster innovation"; SDG12 – "Ensure sustainable consumption and production patterns"; SDG13 – "Take urgent action to combat climate change and its impacts"; SDG14 – "Conserve and sustainably use the oceans, seas and marine resources for sustainable development."

M. Bishop
University of Warwick, Coventry, UK
e-mail: menna.bishop@warwick.ac.uk

R. Burgess (✉)
London School of Economics, London, UK
e-mail: r.burgess@lse.ac.uk

C. Zipfel
Stockholm School of Economics, Stockholm, Sweden
e-mail: Celine.Zipfel@hhs.se

© The Author(s) 2023
T. Madon et al. (eds.), *Introduction to Development Engineering*,
https://doi.org/10.1007/978-3-030-86065-3_2

How both these challenges play out will be determined in large part by what happens in developing countries. It is here that the need to banish poverty is greatest. Indeed, 196 countries have signed up to the goal of eliminating extreme poverty by 2030. It is also in developing countries that environmental externalities from growth are increasing at the most rapid rate and where populations stand to be most affected.

These challenges are both complex and multifaceted and so require not just the development of new innovations but also their careful adaption to developing country contexts. Ensuring that new technologies work when deployed at scale in this way will require the coming together of engineers, technologists, economists, and policymakers. It is this collaborative approach to addressing development challenges which is at the core of development engineering.

This chapter provides an overview of some of the research areas where technology and development can be brought together in a fruitful manner. As with any nascent field, this is a preliminary and incomplete set of topics which is intended to foster and encourage further research in this exciting and important area of work.

We begin with the role of technology in shaping productivity and economic growth. A focus on productivity and growth is inescapable when one is considering poor populations and poor countries. This makes it natural for us to devote considerable attention to the role of technology in the production side of the economy. Hence, we consider how technological innovations can make firms in agriculture, manufacturing, and services more productive. Here, it will be made clear that information and communication technologies (ICTs) such as mobile phones have a critical role to play. The chapter also examines how trade can be made to flow more freely, both internally within countries and externally between countries, and the development benefits that follow from this. Furthermore, we explore how investments in communication and transportation infrastructures can improve the functioning of markets and accelerate structural change and the movement of people from less productive to more productive jobs.

Structural change is ultimately what drives poverty downwards as workers become more productive.[2] For most poor people in the world, labor is their only asset. This means that much of development is about getting people into better jobs, with people's incomes largely being determined by the returns on their labor. We therefore consider how technology can be harnessed in this vital process, both in improving the efficiency of job search and matching and in expanding the size of the markets within which people can sell their labor. We also comment on how the automation of labor market activities brings both challenges and opportunities as the nature of work changes across the world. Finally, this section will summarize evidence on how technology can be used in schools to widen access to quality education.

Also central to production, particularly for populations with little capital, is access to financial technologies such as mobile money. These are beginning to

---

[2] Structural change in the economy is a process involving the movement of people from less productive jobs (e.g., subsistence agriculture) to more productive jobs.

transform the landscape of financial services in the developing world by bringing the ability to borrow and save to populations who have traditionally been excluded from formal financial institutions, thus allowing them to fund new productive activities. These financial technologies may also expedite the movement of money and make households more resilient when hit with different types of shocks. Finally, digital payment technologies are increasingly being used by governments and firms to reduce leakages in payment systems, whether this be payments for workers or welfare transfers to poor populations. Together, these can bring improvements in efficiency and changes in household financial behavior that stand to have long-run implications for development.

It is clear that different inputs are needed for people and firms to take on the modern production activities that drive structural change. One key input is electricity. Our chapter will outline how innovations in renewables such as solar are being combined with technological improvements in national grids to universalize access to electricity. This is revolutionizing the choice of electricity source for firms and households across the developing world, many of whom have been entirely without access to electricity until now. Given that the bulk of greenhouse gas emissions come from the combustion of fossil fuels, one major issue that needs to be confronted in the energy space is how electricity is generated and how the associated environmental externalities are reflected in policies and pricing. Equally critical is the removal of electricity theft and subsidies which are often regressive but are endemic to public utilities in several developing countries. Improvements in generation technologies, energy efficiency, and technologies for monitoring consumption can all be harnessed to ensure that rapid growth in electricity demand in developing countries will be met in a manner that is as sustainable as possible.

This concern with sustainability extends beyond the issue of energy generation and is discussed throughout the chapter. Both growth and the externalities from growth are now deemed central to the study of development economics, which was not the case a decade ago. As such, we will consider how human activity will affect the environment and how technology can be used to minimize the negative environmental externalities that emanate from this activity in a range of different areas, including agriculture and transport. This will involve considering what type of growth is occurring, including the kinds of firms involved and the types of work people engage in but also how technology can be harnessed to counter processes like climate change.

Strong state capacity is required to steer a path towards sustainable growth and development. Here, there is increasing recognition that technology can play an important role, both in the form of platforms that promote political engagement and in applications that serve to increase the accessibility, monitoring, and integrity of elections. Together, these serve to improve the accountability of governments to citizens and help close the gap between the design of policy and its implementation that often exists in developing countries. Closing this gap can lead to improvements in the functioning of the state, for example, in ensuring that populations have access to a higher quality of public services.

In the final part of the chapter, we consider how technology can play a role in development by shaping access to healthcare. Exciting new work is documenting how, in doing so, technology can help to bridge the gap in health between developing and developed countries. This also loops back into what types of work people can do, the returns to their labor, and how well firms function.

The complexity of the development challenges that humanity is facing in each of these areas demands that technological innovations and ideas be brought to bear on them. This is what makes the study of development engineering so important. It is really about bringing in frontier thinking from technology into development, to guide interventions and policy decisions. Only in doing so will we have a chance at achieving the difficult balance between ending extreme poverty by 2030 and doing so in a way that does not block the path of prosperity for future generations.

2 Firms, Trade, and Infrastructure

The relationship between technology and productivity is at the core of many economic models, with theories that stress the importance of technological innovations for sustained growth making up a significant share of the growth economics literature (e.g., Solow, 1957; Aghion & Howitt, 1992; Romer, 1990). In this section, we review the existing literature on two channels linking technology to development, an area of research that remains comparatively small. First, technology can improve productivity in firms and agriculture. Second, technology can facilitate trade, enhancing the integration of markets both within and across countries (Donaldson, 2015). These in turn have the propensity to accelerate the structural change process which underpins economic development.

2.1 Boosting Productivity in Firms and Agriculture

2.1.1 Agriculture

To date, agricultural productivity remains relatively low in the world's poorest regions, with subsistence farming still the most widespread occupation in many developing countries (Lowder et al., 2016). An important body of research in development economics has been dedicated to understanding the role technology can play in rectifying these vast disparities in productivity across countries. This may take the form of agricultural biotechnologies, including improved seeds and fertilizers promising higher yields, or ICTs used to receive and share agriculture-relevant information. The hope is that resulting shifts in productivity can eventually enable developing countries to jumpstart their structural change processes, facilitating the flow of labor from agriculture into the manufacturing and service sectors.

The adoption of agricultural biotechnologies such as improved fertilizers and seeds has been a hallmark of production in many developed countries, given their proven impact on yields and cost (Brookes & Barfoot, 2018). Many have studied how these benefits have been extended to developing countries, who in 2018 accounted for 54% of global biotech crop area (ISAA, 2018). For example, in a study in Kenya, Duflo et al. (2008) estimated annualized rates of return of 70% to the use of chemical fertilizer. Similarly, Bustos et al. (2016) study the introduction of genetically engineered soybean seeds in Brazil and its positive impact on agricultural productivity. They show that such technologies can in turn trigger industrial growth, releasing labor from agriculture and allowing it to shift towards industry and services.

ICTs represent another avenue for boosting agricultural productivity in developing countries. This encompasses the use of mobile phones, Internet, television, and radio to receive and share information on prices, weather, and farming techniques, both within private networks and as part of government initiatives. By improving information circulation and connectivity, ICTs stand to help farmers optimize production decisions. Rosenzweig and Udry (2019) shed further light on this in their study of weather forecasts in India. They argue that accurate forecasts increase profits by allowing farmers to allocate resources to exploit rainfall conditions. For example, in areas where forecasts are accurate, they show that a pessimistic forecast lowers planting-stage investment and use of rainfall-sensitive crops. In settings such as India (Jensen, 2007) and Niger (Aker, 2010), the use of mobile phones for sharing price information was furthermore shown to promote arbitrage in fish and grain markets.

ICTs can also reduce the cost of agricultural extension services, particularly in low population density regions. Fabregas et al. (2019) discuss how this may be vital for increasing exposure to science-based agriculture advice. They furthermore note that GPS-enabled devices may facilitate the tailoring of information, for example, notifying farmers of local pest outbreaks, as well as two-way communication, where farmers are able to ask specific questions. Digital Green is one development organization harnessing ICTs in this way. It connects farmers with experts and disseminates information via video. In India, it was shown to increase adoption of certain practices sevenfold relative to traditional training and visit-based extension approaches and to be 10 times more effective per dollar spent (Ghandi et al., 2009). Overall, in sub-Saharan Africa and India, meta-analyses indicate that information transmission via mobiles has increased yields by 4% and the likelihood of adopting recommended agrochemical inputs by as much as 22% (Fabregas et al., 2019).

Despite significant enthusiasm for biotech crops, development initiatives have sometimes been confronted with poor adoption rates, presenting an important puzzle for economists and policymakers. Where returns have proven heterogeneous across farmers, this may simply reflect optimal decision-making. For example, in rural Kenya, Suri (2011) showed that at least some farmers are better off not adopting technologies such as hybrid maze. In other cases, however, returns to a technology have been shown to significantly exceed the cost of the investment, implying other constraints might be at play (Duflo et al., 2011).

Here, one important strand of the literature has emphasized the role of market inefficiencies, including poor infrastructure, insecure land rights, and missing markets, particularly in the area of financial services (Jack, 2011). For example, where agriculture is rain-fed and farmers lack access to formal insurance, incentives to invest in inputs and technologies may be diminished. This has been tested experimentally in Ghana by Karlan et al. (2014), who find that when provided with insurance against the primary risks they face, farmers could find the resources to increase expenditures on their farms. Similarly, even where the benefits of ICTs are well known, farmers may be unable to access or act on information due to lack of telecommunications, electricity, and transportation infrastructure or illiteracy.

Behavioral factors can also inhibit the diffusion of agricultural technologies. In the case of ICTs, for example, information disseminated via these interfaces may not be afforded the same level of trust as that shared via social networks and traditional extension services. Here, some have pointed to the role of social learning and network effects in the diffusion of agricultural technologies. Studies such as Foster and Rosenzweig (1995), Conley and Udry (2010), Bandiera and Rasul (2006), and Duflo et al. (2010) suggest that their adoption might pose less of a risk for poor farmers when the cost of experimentation is shared with others.

Interlinked with the desire to increase agricultural efficiency are sustainability concerns. It has been estimated that the global population will reach 9 billion by 2050, which will demand a 60% increase in agricultural production (Alexandratos & Bruinsma 2012). However, this intensification risks accelerating climate change through its contribution to greenhouse gas emissions and resource degradation. Moreover, our ability to meet the growing global demand is in itself threatened by climate change and the damaging impact of changing weather patterns on yields, increasing the risk of food insecurity in developing regions (Ignaciuk & Mason-D'Croz, 2014). As such, it is imperative that agricultural technologies enable farmers' adaptation to a changing climate while minimizing any further environmental degradation. Evidence suggests that biotechnologies have contributed significantly to agriculture's sustainability, for example, by reducing the need for pesticide spray and facilitating cuts in fuel use and tillage changes (Brookes & Barfoot, 2018). There has also been much enthusiasm surrounding conservation agriculture (CA) practices. Centered on diversified crop rotation, minimum soil tillage, and maintenance of permanent soil cover, these intend to prevent soil erosion and degradation and enhance biodiversity whilst improving yields (FAO, 2001).

One notable application of CA is the Kenya Cereal Enhancement Program – Climate Resilience Agricultural Livelihoods Window (KCEP-CRALW), co-funded by the European Union and the International Fund for Agricultural Development (IFAD). This provides smallholders with access to training, improved inputs, and CA services through an electronic voucher system, hoping to reduce poverty and promote food security in Kenya's vast arid lands (IFAD, 2015). IFAD reports that the initiative has already reached 83,000 famers and is currently being scaled up throughout Kenya (IFAD, 2020). However, evidence regarding the success of CA in practice has been mixed. For example, in a study of rural Zimbabwe, Michler et al. (2019) find that CA produces no yield gains and sometimes yield losses in years of average rainfall, but does mitigate the negative impacts of deviations in rainfall. Other studies have shown that yield gains are heavily context-dependent and may only be observable after several years (Stevenson et al., 2014). Ultimately, rigorous evidence on the effectiveness of these technologies and the potential trade-offs involved in their uptake remains limited, as are adoption rates. As such, this must be an important focus for research in coming years.

Overall, much remains unknown as to the constraints to technology adoption in agriculture. Solutions might lie in packaged interventions that relax multiple constraints simultaneously, for example, making complementary infrastructure investments. Another area worthy of further investigation is the unequal adoption of biotech crops across the developing world. While adoption of biotech crops has reached over 90% in countries such as Brazil, Argentina, and India, other regions have been lagging behind (ISAAA, 2018). This links to persisting political opposition to genetically modified crops in certain regions, which must be met with dialogue between plant scientists, economists, and policymakers across countries (Elliott & Keller, 2016). Simultaneously, research must continue to seek paths to achieve sustainable intensification of our agricultural production.

2.1.2 Firms

Another fundamental link between technology and development relates to its ability to improve the productivity of firms in manufacturing and services. Neoclassical economic theory has long accepted technological change as the sole driver of long-term growth since the seminal work of Solow (1957). This prompted a vast body of research aimed at understanding the drivers of productivity (Syverson, 2011; Bloom & Van Reenen, 2010) and testing the impact of technological innovations.

In the context of developed countries, various studies have documented a positive impact of new technologies on industry-wide productivity. One notable example is the minimill's introduction to the US steel manufacturing sector, to which Collard-Wexler and De Loecker (2015) attribute a significant increase in productivity in the second half of the twentieth century. Here, they identify a reallocation of output among incumbents, where the least productive firms that failed to adopt the new technology were driven out of the industry. Other evidence has robustly linked the 1990s acceleration in US productivity to the development of ICTs.[3] For example, focusing on the valve manufacturing industry, Bartel et al. (2007) find that new IT investments improved the efficiency of all stages of the production process and increased the skill requirements of machine operators.

There also exists a small but growing literature on technology's role in firm and sector productivity in the developing world. Exploiting firm-level survey data, Commander et al. (2011) find evidence of a large, positive productivity effect of ICT

---

[3] See Draca, Sadun, and Van Reenen (2007) for a review of the literature on ICTs and firm productivity.

adoption in both Brazil and India. In Brazil, this effect was largest for firms simultaneously investing in flattening organizational structures and, in India, in areas with better infrastructure. Moreover, ICT capital intensity was negatively correlated with poor infrastructure and pro-worker labor regulation. A more recent study by Atkin et al. (2017a) investigates the introduction of a new cutting technology in Pakistan's soccer-ball industry. While this was found to reduce waste and increase technical efficiency for nearly all firms in the sector, adoption rates were low when the technology was offered free of charge to a random subset of these. The authors attribute this to misaligned incentives in piece-rate contracts, where employees were initially slowed by the new technology and faced no private incentive to reduce waste, resulting in resistance to adoption. Indeed, when employees were offered financial incentives conditional on demonstrating competence in using the technology, adoption increased significantly.

Ultimately, the literature on technology's impact on firm productivity in developing countries remains in its nascent stages. However, the examples above provide an important cautionary tale for policymakers – though new technologies can be highly effective at raising productivity in firms, we must not lose sight of the conditions that encourage adoption and high returns. This should involve consideration of infrastructure investments and the regulatory environment.

2.2   Facilitating Trade

Firms and households in developing countries face relatively restricted access to markets. Whether a result of policy barriers (e.g., regulations) or poor transportation infrastructure, this prevents firms from engaging in trade and inhibits the transfer of technology. This also stops households from accessing cheaper goods and seeking more productive work opportunities which are key drivers of growth and development. As such, many experts have argued for measures to facilitate trade, such as investments in transportation infrastructure. The use of ICTs can also help to overcome transport costs and facilitate arbitrage, improving market efficiency and firm performance and reducing waste.

A small but growing empirical literature emphasizes the positive welfare effects of trade for both producers and consumers in developing countries. For example, Atkin et al. (2017b) randomize access to export markets for small carpet-making firms in Egypt. They find evidence of "learning by exporting," where exporting firms witnessed improvements in technical efficiency, resulting in higher output quality and profits. On the consumer side, Atkin et al. (2018) study retail FDI (foreign supermarkets' entry into the local retail sector) in Mexico. They detect large welfare gains for the average household, primarily driven by lower living costs. Finally, Redding and Sturm (2008) offer rigorous evidence on the importance of market access for economic development using Germany's division and reunification as a natural experiment.

One important factor that inhibits trade and its associated welfare benefits is high transportation costs. In recent years, a number of studies have demonstrated how combatting this via infrastructure investments can accelerate economic development. For example, railroads – a major technological advancement of the nineteenth century – were shown to facilitate internal trade and market access and produce lasting growth and welfare effects in settings such as colonial India (Donaldson, 2018) and nineteenth-century US (Donaldson & Hornbeck, 2016). Similarly, focusing on 15 countries in sub-Saharan Africa who have a port as their largest city, Storeygard (2016) finds that incomes in secondary cities are highly sensitive to the cost of transport to the largest city.

By reducing transport costs, transportation infrastructure investments also allow households to access external work opportunities, which may improve the allocation of human capital. For example, Morten and Olivera (2018) study the road networks connecting Brasilia and Brazil's state capitals. These are shown to have decreased both trade and migration costs, resulting in important welfare gains. Adukia et al. (2020) focus instead on transport networks to rural areas in their study of India's $40 billion program to construct all-weather roads to nearly 200,000 villages. They detect a positive impact on adolescent schooling outcomes, particularly in areas where the relative return to high-skill work increased the most. This is consistent with market access having raised the return to human capital investment, suggesting the potential for long-run gains to infrastructure investments. In a later study of the same intervention, Asher and Novosad (2020) detect a large reallocation of workers out of agriculture. This is reinforced by a model developed by Gollin and Rogerson (2014) in which reducing transportation costs generates agricultural productivity gains, in turn decreasing the fraction of people working in subsistence farming.

On the other hand, another strand of the literature has cautioned that gains from infrastructure and trade may be unevenly distributed. This may be the case where certain regions are disadvantaged in ways that prevent them from realizing the benefits of improved market access and instead are confronted with issues such as capital flight. Faber (2014) explores this question in his study of China's National Trunk Highway system, as a by-product of which many peripheral counties were connected to major production centers. He argues that declining trade costs produced adverse growth effects for peripheral regions as economic activity and employment were displaced to urban regions. In another study of China's transportation networks, Banerjee et al. (2020a) detect only a small positive effect on sectoral GDP per capita and no effect on growth. Tsivanidis (2018) sheds further light on the distribution of infrastructure gains, focusing on urban public transport. He studies the construction of the world's largest bus rapid transit system in Bogotá, estimating large aggregate output and welfare gains. However, he finds that these were accrued slightly more by high-skilled workers, despite this being a service relied upon more by the low skilled. His model points to knock-on effects on commuting costs, commuting decisions, and wages. Together, these studies underline the emphasis that must be placed on general equilibrium effects in the study of transportation infrastructure and provide a cautionary note to policymakers attempting to use such investments as means of targeting certain groups or regions.

Another consideration which must factor importantly into infrastructure investments relates to environmental concerns. For example, Balboni (2019) documents the high coastal concentration of populations and infrastructure and demonstrates the negative impact of rising sea levels on the profitability of this allocation. She argues that, under a central sea-level rise scenario, 72% higher welfare gains could have been achieved by a foresighted allocation avoiding the most vulnerable regions. Others have cautioned that improved transportation infrastructure risks deforestation and biodiversity loss (Damania & Wheeler, 2015; Dasgupta & Wheeler, 2016). These highlight the importance of accounting for future environmental change when making contemporary infrastructure investment decisions. Sustainability considerations must also steer the nature of the transportation investments being made and how these are powered. Here, an encouraging example is India's Dedicated Freight Corridor, a major infrastructure investment which stands to increase India's share of rail in freight transportation and generate significant reductions to CO_2 emissions (Pangotra & Shukla, 2012).

Alongside transportation infrastructure, another force which may act to facilitate trade is the improved circulation of market information. The fast expansion of ICTs, in particular mobile phones, has been shown to generate major efficiency gains. For example, Jensen (2007) documents how the 1997–2000 expansion of mobile phone coverage along India's Keralan coast allowed fishermen to call sellers at different markets in search of the best price for their output. This resulted in a dramatic fall in price variation across local markets, higher profits, and the elimination of waste, generating welfare gains for producers and consumers alike. Similarly, Aker (2010) finds that the 2000–2006 introduction of mobile phones in Niger reduced price dispersion by 10–16% and increased profits, which she attributes to reduced search costs for farmers.

3 Labor Markets and Structural Change

Development is all about structural change, that is the movement of people from less productive to more productive jobs. In many cases, this involves movement across geography, for example, from the countryside to cities or from poor to rich countries. How technology can encourage this process of development is an area of significant interest in which research in economics is beginning to make some inroads. These technologies, however, also imply that the nature of work is changing, with the enormous expansion of platform-type employment as well as the disruption arising from the mechanization of many occupations. Technology thus has the ability to create new employment opportunities as well as to destroy them. Navigating the path to understanding how technology influences labor markets will require careful investigation into how precisely occupational structures are being affected, rather than assuming that technology will be either a good or a bad thing for employment.

3.1 Bridging the Employer-Employee Information Gap

Research has shown labor market frictions to be an important impediment to firm growth (Greenwald, 1986; Gibbons & Katz, 1991; Abebe et al., 2017a, 2017b), suggesting that improving the efficiency of the job-matching process could be crucial for developing countries. This may be especially the case given their large youth populations, with over ten million Africans entering the labor force annually (Mohammed, 2015), coupled with the high youth unemployment rates characterizing many poorer regions (McKenzie, 2017). Here, technology offers immense opportunities, particularly in labor markets where information may be scarce and for workers considering migration.

In their seminal paper, Bryan et al. (2014) show that encouraging workers to seasonally migrate to cities improves the welfare of rural households in Bangladesh. This is predicated on workers being able to find jobs. Technologies which allow them to do so and indeed allow employers to contract workers for specific tasks are becoming ever more prevalent across the developing world. For example, many of India's migrant workers are contracted online by recruitment companies to fill positions carrying out a huge range of services in Indian cities. Babajob is the country's largest marketplace for informal and entry-level formal jobs, having registered 6.1 million jobseekers and over 370,000 employers across India as of 2018. To promote its accessibility, the platform offers a range of online and offline access options, including Internet, text messaging, and interactive voice response (GIE, 2018).

The availability of contracting online has also meant that large numbers of workers from low- and middle-income countries can secure employment across international borders before migrating to these jobs. Some obvious examples of this are the huge numbers of workers who move from India, Pakistan, and Bangladesh to the Gulf countries to work in construction and other activities.[4] But online contracting is also central to the movement of health workers and domestic maids, for example, from the Philippines to a whole range of countries (Calenda, 2016). The advent of these technologies implies that labor markets have become much broader, in effect expanding job opportunities from the national to the international.

While the benefits in theory could be large, existing research on technology's impact on the efficiency of the job-matching process is extremely limited. Dammert et al. (2013) provide the first experimental evidence in a study of a public labor market intermediation service (LMI) in Peru, randomly assigning registered job seekers to be contacted via SMS or traditional methods (in person or by phone). They detect a positive impact on employment which was larger for SMS intermediation, though not statistically significantly so. Moreover, the employment effect dissipated after 3 months and was not accompanied by an effect on matching efficiency. They furthermore note that those with less labor market experience

---

[4] See, for example, the Musaned electronic platform used in Saudi Arabia to hire Bangaldeshi domestic workers.

seemed to benefit less from the service and were less likely to search for jobs through digital means. Ultimately, digitization of the intermediation service proved both viable and cost-effective but was not in any way transformative.

Overall, the body of research on these technologies is too small to reach a conclusion regarding their impact on job search behavior, migration, unemployment duration, the quality of matches, and labor market efficiency in developing countries. Moving forward, attention should also be paid to their inclusionary potential, for example, the effectiveness of Babajob's use of different access options in combatting potential barriers such as illiteracy. Ultimately, it is also clear that technologies that facilitate job search will be restricted to regions with sufficient telecommunications and electricity infrastructures. Complementary investments in these will be vital to ensure the success of any such platforms. However, it is clear that by increasing the size of the markets over which workers can search and by improving matches between employers and workers, these technologies undoubtedly hold promise.

3.2   Transforming the World of Work

Technological advances prompt questions about the future of work, with automation and digitization threatening the existence of jobs that can be carried out by robots. Prominent examples include assembly-line work and clerical jobs involving "routine tasks" (Autor et al. 2003). Indeed, various studies have attributed declines in employment in these positions in the US across the twentieth century to automation and digitization (Autor et al., 2003; Autor, 2015; Levy & Murnane, 2004). Conversely, others have emphasized complementarities between automation and human labor, painting a more optimistic picture for the future of work. This relates to a conjecture by Autor (2014) that some tasks may be inherently "uncodifiable." In this way, automation may enhance the value of labor that humans can uniquely supply and protect certain occupations from substitution altogether. This is consistent with the phenomenon of job polarization that has been well-documented in developed countries, i.e., the simultaneous growth of high-skill, high-wage and low-skill, low-wage occupations.[5]

What can these debates, which have largely revolved around developed countries, tell us about what lies in store for countries at earlier stages in the structural change process? On the one hand, the "alarmist" view argues that larger employment shares in low-skill work in developing countries imply that more of their jobs could feasibly be taken over by technological change that is skill-biased. Indeed, the World Bank (2016) estimates that two-thirds of all jobs in developing countries are susceptible to automation. Others have furthermore warned of potential "reshoring,"

---

[5] See Goos & Manning 2007; Goos et al., 2014; Autor & Dorn 2013; Autor et al., 2015; Autor et al., 2006, 2008; Autor 2014; Michaels et al., 2014; Graetz & Michaels 2015.

where labor in poorer countries gradually loses its cost advantage to robots in richer ones, ceasing the outsourcing of manufacturing that has hitherto been an important source of growth (Schlogl & Sumner, 2018; Maloney & Molina, 2016). For these reasons, Schlogl and Sumner (2018) project that, in developing countries, automation will cause wage stagnation and "premature deindustrialization," as proposed by Rodrik (2016). On the other hand, by breaking existing barriers to entry and efficiency, Maloney and Molina (2016) argue that technology may support new ICT-intensive industries. For example, Kenya's M-PESA mobile money program works with nearly 400,000 agents across the Democratic Republic of Congo, Egypt, Ghana, Kenya, Lesotho, Mozambique, and Tanzania (Vodafone, 2020), and India's IT sector created 200,000 new jobs in 2019 alone (Bhattacharya, 2020). Further dampening concerns surrounding automation, others have pointed out that technologies typically diffuse at an altogether slower rate in developing countries. Indeed, many of the jobs being carried out by humans in poorer regions have long-since been automated in developed countries (World Bank, 2016).

Several notable attempts have been made to quantify technology's impact on labor markets in developing countries to date. Maloney and Molina (2016) use census data to test for job polarization in the developing world, of which they find no strong evidence. However, they point to a relative decline in employment in the "plant and machine operators and assemblers" category in Indonesia, Brazil, and Mexico since around 2000 as potential evidence of incipient deindustrialization. In another study, Hjort and Poulsen (2019) provide the first direct evidence on the causal relationship between ICTs and labor markets in a developing country context, exploiting the gradual arrival of fast Internet in a range of African countries. They detect a large increase in employment rates and argue that this is driven by job creation in higher-skill occupations. For example, they find evidence of firm entry in South Africa, primarily in ICT-intensive sectors such as finance, and increased productivity of existing firms in Ethiopia. Moreover, the authors show that, after obtaining a fast Internet connection, firms in Ghana, Kenya, Mauritania, Nigeria, Senegal, and Tanzania seemed to engage in more exporting, online communication with clients and training. These benefits appear to have also extended to workers with lower skill levels. In particular, the increase in the probability of employment that the authors observe is of comparable magnitude for those with primary, secondary, and tertiary education. Together, these results paint a different picture to that described by the more alarmist view of technological change's labor market impact.

Overall, we are yet to find rigorous evidence of labor market polarization and displacement in developing countries of the kind experienced in developed countries to date. However, this is an understudied area far from the point of consensus. Additional uncertainty for workers in developed and developing countries alike is introduced by future technological advances that will inevitably extend the range of automatable jobs. Here, machine learning and artificial intelligence have raised particular concerns, given their potential to master "nonroutine" tasks (Autor, 2015; Webb, 2019).

With vast uncertainty ahead, countries must enact forward-looking policy change to equip their workforces as best possible for skill biased technology change. One option is to focus on education investment. Encouragingly, the World Bank (2019) report that, between 2000 and 2014, the share of employment in high-skill occupations increased by 8 percentage points in Bolivia and 13 percentage points in Ethiopia. However, Schlogl and Sumner (2018) caution that education is no panacea. Indeed, even developed countries with far higher average skill levels are struggling to insulate their labor forces from competition with new technologies. Rather, the authors underline the likely future importance of a social safety net. How exactly developing countries might fund this remains an extremely pressing open question.

3.3   Widening Access to Quality Education

Education represents a critical input to labor markets and development, and it is the focus of SDG 4 to "ensure inclusive and equitable quality education and promote lifelong learning opportunities for all" (UN, 2020). As described above, educational investments may also help equip the labor force for future skill-biased technological change. It is estimated that there were 750 million illiterate adults in 2016, largely concentrated in South Asia and sub-Saharan Africa, and 262 million children aged 6–17 out of school in 2017 (UN, 2019). However, measures to increase enrolment will do little to improve attainment unless the education provided in schools is of high quality (Banerjee et al., 2008). Whether technological applications can help guarantee this has been the focus of an increasing number of studies.

One important strand of the literature has focused on computer-assisted learning (CAL). This consists of educational software programs which often adjust to students' achievement levels. Optimists have highlighted how these can target some of the issues endemic to education provision in developing countries, such as underqualified teachers and large classrooms with significant heterogeneity in student learning levels. Indeed, in their review of the literature on education and technology, Bulman and Fairlie (2016) note that effects are generally stronger in developing country contexts, perhaps due to lower levels of human capital. However, this overall positive result masks some important nuances.

In particular, Banerjee et al. (2008) conduct a randomized evaluation of a CAL program in Vadodara, India. Here, grade 2 students were given weekly access to a computer for playing math games which tailored themselves to students' achievement levels. The authors detect substantial test score gains, though these dissipated over time. In another study of India, Muralidharan et al. (2019) evaluate the "Mindspark" CAL program for after-school instruction of middle-school students, which also offers customized content. They find significant improvements in test scores which were particularly large for weaker students, implying the program effectively catered to a wide range of learning levels.

On the other hand, in a randomized evaluation in Gujarat, Linden (2008) found that CAL caused students to learn significantly less when computers were used as a substitute for the normal curriculum. However, when used as a complement, i.e., as an out-of-school program, CAL had a positive effect on learning, especially for weaker students. This the author attributes to the program's design, which reviewed material in the existing curriculum. Outside of India, Carrillo et al. (2010) evaluate a mathematics and language CAL program for primary school students in Ecuador. They detect positive impacts on mathematics test scores and an insignificant (negative) effect on language test scores. Together, these results suggest that CAL is able to promote learning for students of a wide range of abilities but also reinforce the importance of careful format design and considerations of the context into which CAL is being implemented.

In contrast to generally positive results in evaluations of CAL, programs focused on improving access to computers have shown limited success. Such was the case for Beuermann et al. (2015), who analyze an experiment in Peru in which students were given a laptop for home use and fail to detect an impact on academic achievement. In another study, Barrera-Osorio and Linden (2009) evaluate a national program in Colombia to install computers in public schools, also training teachers on how to use them in specific subjects. Here again, the authors detect no impact on student outcomes, which they attribute to their limited incorporation into classroom teaching. These results underline the potential ineffectiveness of interventions that do not tailor technologies to a specific need.

Throughout this chapter, we have documented how successful applications of mobile technologies can have a vital impact on living standards and economic development. In addition to their use in school instruction, these encompass digital agricultural extension programs, mobile money, online job portals, social media, and simple SMS communication. However, the persistent problem of illiteracy in developing countries serves as an important barrier to their use, preventing people from accessing the associated benefits. While development initiatives have taken measures to help overcome this, for example, harnessing interactive voice response technologies, another solution is to tackle the problems of illiteracy and digital illiteracy head-on via targeted initiatives. Research has also shown that this can in turn enhance learning outcomes in other areas.

In particular, Aker et al. (2012) evaluate the experimental incorporation of mobile phone instruction into a standard adult education program in Niger. This was found to substantially increase writing and math test scores compared to the original model, with a relative improvement in math scores still visible after 7 months. The authors attribute this to improved motivation and effort in the classroom, which links to their conjecture that being able to use mobile phones for other services increases the returns to education. Treated students also used mobile phones more actively beyond the classroom, which may have served as means for them to practice their skills. In a later study of the same program, Aker and Ksoll (2020) document that immediate gains in reading scores persisted after 2 years and also reveal a wide range of other socioeconomic effects. In particular, individuals who received additional mobile phone instruction had more diverse income-generating activities, improved food security and asset ownership, were more likely to sell a cash crop, and were more likely to save. The authors were unable to disentangle whether these effects are due to improved learning outcomes or use of the mobile technology itself, an important question for future research.

Overall, the limited success of programs focusing on the distribution of hardware suggests that access to technology may not be the binding constraint to its successful use in education (Bulman & Fairlie, 2016; Escueta et al., 2017). Rather, this seems to depend on the careful design of the CAL programs that students use them for and their thoughtful integration into existing curriculums. In these cases, it has been proven that technology may enhance the learning of students across the achievement spectrum. Moreover, it appears that equipping students with digital literacy can improve learning outcomes in traditional areas of academic study and generate further welfare and economic benefits. However, the success of technology-enabled education broadly rests on the quality of telecommunications and electricity infrastructure, which remains dire in many developing contexts. In sub-Saharan Africa in particular, under 50% of all primary and lower-secondary schools have access to electricity, the Internet, computers, and basic drinking water (UN, 2019). Investing in these crucial amenities must be a first order priority for education initiatives.

It is also vital that we do not understate the importance of human capital in the education workforce. Improvements in educational outcomes will hinge on the skill, motivation, and efforts of teachers in the delivery of both ICT-enhanced and more traditional lessons. Indeed, Aker et al. (2012) note that teachers in their sample with higher education were "better able to harness mobile phones to improve students' educational experiences." Here, technology can also play its own role, for example, in enabling the monitoring of teacher attendance. In an experiment in India, Duflo et al. (2012) show that such initiatives can provoke substantial increases in attendance with positive knock-on effects on student outcomes.

There is also hope that technology can be exploited to educate children in settings where teachers are unavailable altogether. This has become particularly salient in light of the COVID-19 pandemic. For example, the Vodafone foundation reports that over 1.1 million young people in Africa are accessing educational materials online via their e-school programs (Vodafone, 2020). Here, future research must also consider whether and how technology can be used to tackle the gender gaps in educational attainment that remain in developing countries.

4   Financial Technologies

Throughout this chapter, we have documented the range of new opportunities for communication and service provision that have accompanied mobile technology's fast expansion throughout the developing world. One such case is financial technologies, namely, mobile money. This generally refers to the application of mobile phones for sending and receiving money. Transfers take place via SMS for a small fee, and do not require ownership of a formal bank account. Deposits and withdrawals are made by visiting a mobile money agent, a process akin to exchanging cash with "e-money." In this way, mobile money is far removed from the mobile banking applications used by many in developed countries.

Demirgüç-Kunt et al. (2020) estimate that there are 1.7 billion unbanked adults globally, 1.1 billion of whom have a mobile phone. This has sparked enthusiasm surrounding mobile money as a way of bringing formal financial institutions to populations hitherto reliant exclusively on cash. In doing so, it has the propensity to significantly reduce transaction costs and facilitate saving, with important implications for welfare and development. For firms and governments harnessing mobile money for the transfer of wages and welfare payments, there is also scope for significant efficiency gains and reduced leakage. However, whether this technology actually constitutes a development "leapfrog" for low-income countries, whereby mobile banking facilitates universal financial inclusion while bypassing the formal banking sector, remains unclear. This may crucially depend on countries' regulatory frameworks, among other important factors. These issues are discussed in this section.

4.1 Increasing Financial Resilience

In the absence of access to financial services, households revert to inefficient, risky, and costly methods for making transfers to friends and relatives. These include informal practices, such as asking friends or bus drivers to pass on cash, and use of money transfer services, such as Western Union (Aker, 2018). Mobile money stands to significantly facilitate these processes – senders can transfer funds through a simple SMS, requiring only that both parties own a mobile phone and the payment of a transaction fee. A new body of research is beginning to show that, in doing so, mobile money is having an important impact on household financial behavior and welfare.

This question is examined by Jack and Suri (2014) in their seminal study of M-PESA, the developing world's most renowned mobile money success story. The authors document that users experience a significant reduction in the transaction costs associated with sending remittances, which in turn facilitated inter-household risk sharing. In particular, following a negative income shock, M-PESA users received more remittances and from a wider range of sources, allowing them to avoid cutting their consumption. In another study of the same program, the authors reveal that improved smoothing also extended to negative health shocks (Suri et al., 2012). Batista and Vicente (2018) confirm that these effects are not limited to Kenya's M-PESA, offering some of the first experimental evidence in a study of rural Mozambique. Here again, they detect evidence of improved smoothing and reductions in hunger episodes, which they attribute to increased receipt of remittances. These studies speak to mobile money's ability to bolster the financial resilience of poor households in the face of shocks that would have otherwise cut into their consumption and education spending, suggesting significant welfare gains.

Beyond facilitating consumption smoothing, there is evidence of wider socioeconomic effects following mobile money's introduction. For example, in rural Mozambique, Batista and Vicente (2018) detect a fall in agricultural activity in treatment areas as well as an increase in out-migration, which can be a route out of poverty (Bryan et al., 2014). Similarly, in a long-run evaluation of M-PESA's economic impacts, Jack and Suri (2014) estimate that access to mobile money lifted almost 194,000 households, equivalent to 2% of Kenyan households overall, out of extreme poverty, which they attribute to improved financial resilience and higher savings. They also document a shift of women out of subsistence agriculture and into business and retail occupations. Together, these studies indicate a causal link between mobile money and development via improvements to the economic lives of the poor.

Recent years have seen mobile money extending beyond simple transfer and savings facilities to include other traditional financial services, such as loans. For example, Safaricom's M-Shwari enables consumers to open a bank account and make deposits and withdrawals via M-PESA. They can also request a loan, the decision for which is based on financial history data (Suri, 2017). In their evaluation, Bharadwaj et al. (2019) document that M-Shwari had high take-up and effectively improved access to credit and resilience to income shocks. However, the potential for transformative socioeconomic impacts was limited by the small and short-term nature of the loans. In another study of Indonesia, Harigaya (2016) evaluates the experimental digitization of a group microfinance program via mobile money, which meant deposits and withdrawals were more convenient and could be carried out in the absence of peers. The author detects a decline in savings, which he attributes to a weakening of the peer effects that underline the motivation for group banking, as well as sensitivity to fees, which, though small, appeared to increase the salience of transaction costs. Blumenstock et al. (2016) study a separate mobile money innovation in which this was used for wage payments in an Afghan firm, and half of employees were experimentally assigned a 5% default savings contribution. This was found to significantly increase savings, which the authors attribute to the overcoming of behavioral barriers. Together, these studies highlight the potential for innovative applications of mobile money but caution that behavioral factors can work to both their advantage and disadvantage.

4.2 Facilitating Firm and Government Transactions

In recent years, applications of mobile money have been extending from person-to-person payments to also include person-to-business and government-to-person (or NGO-to-person) payments (Suri, 2017). This can include the payment of wages, as described above, as well as the transfer of welfare payments to households, transactions which would otherwise have to take place via cash. In this way, mobile money may produce important efficiency gains for firms, governments, and NGOs. This question is evaluated by Blumenstock et al. (2015) in another

RCT in Afghanistan wherein a subset of employees was transitioned from cash to mobile money payments. They detect significant benefits to the organization, which included significant savings in salary disbursement activities. In an experiment in Niger, Aker et al. (2016) study an NGO application of mobile money to send emergency cash transfers to households following a drought, of which women were the primary beneficiaries. Here, households in which women received electronic rather than cash payments had improved diet diversity and their children consumed one third of an extra meal per day. In addition to time savings in obtaining the transfer, the authors present evidence of improved female bargaining power to explain their results. Overall, these studies show that the efficiency gains enabled by mobile money can benefit both firms and households.

Applications of mobile money are also coming to form an important part of the COVID-19 response, partly to prevent disease transmission via cash exchanges. For example, telecommunication operators across Africa, including M-PESA, have removed fees on small mobile money transactions (Flood, 2020). Others have lowered the barriers to opening accounts, such as by waiving additional documentation requirements (Peyton, 2020). In a move to support struggling small businesses, M-PESA has also raised daily transaction limits (Finextra Research, 2020). Similarly, governments have been using mobile money to scale up emergency welfare programs. For example, Togolese informal workers have been able to register for and receive a state grant via their mobile phones (Financial Times, 2020). The UNHCR has also been distributing mobile phones and SIM cards to displaced families and making assistance payments into these (Faivre, 2020). This positive shock to the use of mobile money stands to kick-start its diffusion in areas where adoption has been hitherto limited.

Dampening enthusiasm surrounding mobile money are concerns surrounding its inclusionary potential, i.e., whether mobile money can effectively extend access to financial services throughout the developing world. This relates to the finding that initial adopters of mobile money may be positively selected in terms of education and income (Batista & Vicente, 2020; Suri et al., 2012). Here, researchers have pointed to potential barriers such as lack of trust, illiteracy, affordability, and possession of official documents required to sign up for accounts, issues which may disproportionately affect women. Indeed, the GSM Association (2019) report that women in low- and middle-income countries are 10% less likely to own a mobile phone and 23% less likely to use mobile Internet in the first place. This ownership gap extends to 28% in South Asia. These figures underline the need for more research into what drives female adoption of mobile phones and mobile money in particular.

Another important barrier to the widespread diffusion of mobile money is the strength of the agent infrastructure, to which Suri (2017) attributes M-PESA's success relative to other mobile money initiatives. For example, in her study of Niger, Aker (2018) discusses how, in one region, households lived an average of 15 km away from the nearest agent. In such contexts, adoption of mobile money was low, despite high costs of alternative methods for sending remittances and high ownership of mobile phones.

Overall, it is clear that mobile money is already beginning to transform the landscape of financial services in the developing world. Its use has been proven to facilitate transactions for individuals, firms, governments, and NGOs, translating to improvements in efficiency and changes in household financial behavior that stand to have long-run implications for development. Recent years have also seen mobile money technology applied to a range of other financial services characteristic of more traditional bank accounts. However, whether these can have the transformative impact that appears to accompany mobile money's basic transfer and savings functions is yet unknown. There are also many remaining questions surrounding the factors that impact the widespread diffusion of mobile money, including behavioral biases, the agent infrastructure, illiteracy, and affordability. Moving forward, there is high demand for further rigorous economic evidence to inform these debates.

5   Energy and Environment

To this day, nearly 1 billion people remain without an electricity connection, and many others receive only partial and intermittent supply (IEA, 2019). Lack of access to this vital technology constrains the set of productive activities households can engage in and inhibits firm performance. The desire to scale energy generation to close this gap has inspired investments into both national grid capacity and alternative off-grid technologies. However, as with any other form of technology, availability may not be the binding constraint to its successful use. Rather, barriers such as electricity theft and poorly functioning utilities continue to inhibit the uninterrupted flow of power to households and firms and its associated benefits for growth and development. Simultaneously, it is vital that growing energy demand in developing countries is met with sustainable sources. These challenges can only be met by harnessing technological innovation.

5.1   Scaling Energy Access

It has been well-documented that reliable energy access is an essential ingredient to economic development (Moneke, 2019; Lipscomb et al., 2013), with many emphasizing an important firm performance channel. For example, both Kassem (2020) and Rud (2012) detect positive impacts of electrification on the entry and performance of manufacturing firms in their respective studies of Indonesia and India. In their evaluation of the Indian textiles industry, Allcott et al. (2016) furthermore document that electricity shortages reduced average output by about 5–10%, underlining the importance of the grid's reliability.

Another dimension of electrification's promise lies in its potential to generate behavioral change, namely, by allowing households prolonged access to light and powering time-saving appliances such as fridges and microwaves. These extend the time available for studying and productive tasks, which may in turn generate longer-run economic benefits to further motivate electrification efforts. Several studies have attempted to capture these effects. Most notably, Dinkelman (2011) studies South Africa's mass rollout of the grid to rural households. She finds that electrification significantly boosted female employment by releasing women from home production and fostering micro-entrepreneurship. This positive impact on female employment is also detected by Grogan (2018) in his study of rural, indigenous households in Guatemala. Interestingly, Fujii and Shonchoy (2020) furthermore find evidence of a negative impact of electrification on fertility. Ultimately, effects of this nature are harder to study, and so much remains unknown as to the longer-run social and economic impact of electrification. However, it is vital that these remain an important focus of research, lest we understate the benefits of and therefore underinvest in electricity.

In line with the well-documented benefits of electrification, infrastructure investments to scale up national energy generation and transmission have become a mainstay of development policy. Here, many African countries are turning to hydropower, exploiting the abundant water supply provided by their rivers. A landmark example is the Grand Ethiopian Renaissance Dam, set to become Africa's largest hydropower project with a 6000 MW capacity (Power Technology, 2020). Overall, hydropower is set to provide 90% of Ethiopia's electricity and also constitutes a promising avenue for growth as it intends to export surplus to neighboring countries (IHA, 2017). In South America, Brazil houses the continent's largest installed hydropower capacity (IHA, 2018), which accounts for around 80% of domestic electricity generation (IEA, 2020). Solar has become another important avenue for scaling national energy generation. 2016 saw the completion of Rwanda's Rwamagana Solar Power Plant, whose 8.5 MW capacity made it East Africa's first utility-scale solar power plant (Mininfra, 2020). Another standout case is Kenya, quickly approaching 100% renewable energy generation thanks to investment in geothermal, wind, and hydropower sources. In 2019, it opened the Lake Turkana Wind Power farm, the largest in Africa with 365 turbines and a 310 MW capacity (Dahir, 2019).

In spite of the aforementioned investments in national energy capacity, a significant share of the developing world is expected to remain off-grid due to the high cost of extensions to remote, rural areas. In particular, of the 315 million set to gain access to electricity in Africa's rural regions by 2040, it is estimated that only 30% will be connected to national grids (African Progress Panel, 2017). Instead, many are anticipated to be electrified via off-grid and micro-grid technologies. These can consist of systems powering individual households or larger-scale ones serving several at a time. They are able to reach remote regions at relatively low cost and can be powered by renewable sources. Smaller devices are insufficient for powering large appliances such as fans, fridges, and TVs, instead primarily intended for phone charging and lighting. These technologies have been met with much enthusiasm and are inspiring a generation of tech entrepreneurs. For example, in Côte d'Ivoire, Evariste Akoumian's Solarpak makes backpacks with built-in solar panels that collect energy while children walk to school (Capron, 2016). These absorb enough energy during the day to power a lamp for 4 to 5 h – enough to allow children to do their homework at night.

In practice, however, evidence from India suggests that the theoretical benefits of microgrids have not always manifested. In a study targeting non-electrified households in Uttar Pradesh, Aklin et al. (2015) installed solar microgrids in 81 randomly selected villages that were previously reliant on kerosene lamps to light their homes. They document an increase in electricity supply, as indicated by reduced kerosene expenditure, but detect no broader socioeconomic impacts. The authors suggest that the power supply supported by the microgrids had been insufficient to encourage business activities or the accumulation of social and human capital.

Others have assessed the microgrid's viability in contexts where it is available alongside the grid. In a study of Bihar, Burgess et al. (2020a) conduct an experiment in which microgrids were offered to a sample of villages at different prices. In this setting, households were also faced with the choices of individual household solar panels (own solar), diesel generators, and the grid, which was being rolled out over the course of the study. They found that, at market price, just 6% of households purchased microgrids, increasing to 19% under a 50% subsidy. Moreover, they saw microgrid demand collapse following grid extensions and improvements in own solar quality. Overall, they argue that richer households have a strong preference for the grid's higher capacity and predict that future income growth will drive electrification primarily via the grid. Simultaneously, off-grid solar will play a key role for poorer households in more remote areas. In this sense, off-grid solar, though highly valuable in situations where the grid alternative is not available, may ultimately be supplanted by the grid. Interestingly, however, the grid itself may increasingly be powered by renewables, including solar. In another study in India, Fowlie et al. (2019) document one company's experience deploying microgrids in Rajasthan. These are also met with low demand, eventually forcing the company to cease its operations. Here, the authors point to competition with the grid, with politicians' promises of imminent local extensions potentially deterring microgrid purchases.

Ultimately, in the face of large government subsidies for grid connections, microgrids may find it impossible to compete. In India in particular, the government has several schemes offering free grid connections to households below the poverty line. Here, Fowlie et al. (2019) argue for transparency in state expansion schedules and measures to ensure microgrids can be technologically integrated into the grid if and when this does arrive. Demand for microgrids also seems to be constrained by their relatively low capacity, which in turn may inhibit the socioeconomic benefits of electrification from manifesting. In this regard, there is certainly room for tech professionals to provide valuable improvements to these energy provision systems.

A natural conclusion of the above discussion on the development benefits of the grid, combined with the proven limitations of off-grid alternatives, would be to pursue rapid universal grid extension. However, a number of recent studies have documented that such efforts may too be alone insufficient to secure the economic benefits of reliable access. In particular, in a study of Western Kenya,

Lee et al. (2016) reveal that, despite high population density and extensive grid coverage, electrification rates remain at 5% on average for rural households and 22% for rural businesses. Moreover, half of unconnected households are "under grid," meaning they could be connected to a low-voltage line at a relatively low cost. In another study, Lee et al. (2020) experimentally offer households in this setting the opportunity to connect to the grid at different subsidized prices. They detect surprisingly low demand even at high subsidy rates, which, combined with costs of supplying connections, imply rural electrification may even have reduced welfare.

Researchers have proposed different explanations for the advent of low-grid demand. Lee et al. (2020) point to credit constraints and the overall low quality of grid provision, with evidence of excess costs from leakage during construction, bureaucratic red tape, low grid reliability, and unaccounted for spillovers. For example, in addition to short-term blackouts lasting minutes or hours, households in rural Kenya face long-term blackouts which can extend for months. Others have argued that issues of grid quality are themselves a function of poor enforcement of payments and informal connections, i.e., individuals illegally connecting for free, a pervasive problem in developing countries. This is documented by Burgess et al. (2020b) in their study of Bihar, whose state electricity utility recovers just 34% of its costs. They argue that this is overwhelmingly the result of electricity having been treated as a right regardless of payment, producing tolerance for subsidies, theft, and nonpayment. The end result is an insolvent utility dependent on government bailouts and tightly rationed supply. McRae (2015) studies a similar paradox in Colombia. He demonstrates that quality upgrades are unprofitable for utility firms, as the state subsidizes their losses and low-income households prefer to receive a low-quality service for which they do not pay.

Ultimately, many developing countries are stuck in a low-quality electricity equilibrium in which households pay little for poor supply and governments battle the competing objectives of sustaining utility companies and retaining the political support of low-income households. More research must focus on finding paths to enhancing the reliability of grid supply and improving the organizational performance and financial sustainability of utilities. Without doing so, it may be impossible to generate the level of demand for and supply of reliable electricity to achieve the development benefits of electrification.

5.2   Guaranteeing Sustainability

World energy demand is on the rise, with the majority of growth set to stem from low-income countries (Wolfram et al., 2012). While this has the potential to spur welcome growth and development, higher energy consumption will also increase levels of pollution and other externalities. Though some national grid investment projects involve large-scale exploitation of renewable energies, others raise sustainability concerns. For example, the China-Pakistan Economic Corridor consists of $60 billion worth of infrastructure projects, including $35 billion for the scaling-up of Pakistan's energy supply (Stacey, 2018). Almost 75% of the new generation capacity will be coal-fired, contributing to an expected rise in the coal share of Pakistan's energy mix from 3% in 2017 to 20% in 2025 (Downs, 2019). Efficiency issues also arise on the consumer-end, with nontechnical losses such as electricity theft prevalent across the developing world, as discussed above. In sub-Saharan Africa in particular, Kojima and Trimble (2016) estimate the total annual value of uncollected electricity bills at 0.17% of national GDP on average. In this section, we discuss the potential for technological innovations to promote both sustainability and the efficient production and consumption of energy.

Electronic meters represent one such technology hailed as a potential solution to energy losses. One variety is the smart meter, which records electricity consumption and communicates this to the distribution company. These were a feature of the microgrids deployed by Fowlie et al. (2019) in Rajasthan, India. Here, however, the authors found that interpersonal relations and inter-caste dynamics rendered operators unwilling to enforce penalties in practice, making cost recovery impossible and the scheme ultimately unviable. Moreover, in his study of Colombia, McRae (2015) argues that grid upgrades involving the installation of meters are unprofitable for utilities, given that many households would be unwilling to pay a nonzero marginal cost even for a higher quality supply and that their existing losses were covered by state subsidies.

Perhaps a more promising technological solution is that of prepaid metering, which has grown in popularity across both the developed and developing world. Here, consumers must credit their accounts in order to access electricity, thus transferring the enforcement burden away from the utility. Jack and Smith (2020) evaluate the viability of this technology in a study of Cape Town, South Africa, in collaboration with the local utility. They detect a 14% decrease in consumption but a net increase in revenue thanks to improved cost recovery, mostly driven by a subset of poor customers who were delinquent on bills. The authors argue that the prevalence of this group in other developing countries suggests the technology could be effective in other settings. However, they caution that they are unable to assess the effect of the prepaid meters on theft, a large increase in which could undo positive revenue effects.

Another category of technologies that may encourage sustainability are energy-saving appliances, such as LED lighting. Evidence suggests that this can also have positive productivity co-benefits, as was recently shown in a study of Indian factories by Adhvaryu and Nyshadham (2020). The authors found that LED lighting, which emits approximately seven times less heat, reduced factory floor temperatures by several degrees, in turn leading to increased productivity. Another notable example is that of cooking appliances. An estimated 3 billion people across low- and middle-income countries continue to rely on traditional stoves and solid fuels such as firewood, biomass, or charcoal for heating and cooking (CCA, 2020). These contribute to deforestation and household air pollution, the death rates for which are highly concentrated in developing countries (Landrigan et al., 2018).

Various initiatives have attempted to encourage transitions away from these harmful methods, with varying success. Kar et al. (2019) evaluate one such program in Karnataka, India, which uses loans and subsidies to promote the use of liquefied petroleum gas. Despite increases in enrolment, they detect no impact on fuel sales, suggesting beneficiaries were not fully transitioning away from solid fuels. This underlines the potential barriers governments may face in encouraging households to replace traditional appliances with cleaner ones, especially in cases where individual gains are only felt once widespread adoption has been achieved.

Another major question facing developed and developing countries alike in their attempts to shift towards renewables relates to energy storage. This is especially the case for wind and solar as variable renewable energy (VRE) sources. For instance, wind generation can be high at night, generating power at a time where demand is low. Similarly, extra energy from peak sunlight generation cannot be easily stored for use at a later period. Even more concerning is the prospect of increasingly unpredictable weather patterns, which are only likely to heighten these issues. Existing scientific efforts to overcome these challenges are primarily taking place in developed countries. As such, solutions proposed are unlikely to be well-attuned to the electricity infrastructures characterizing developing countries, which for example often suffer from insufficient capacity (De Sisternes et al., 2020). In recognition of this disparity, the World Bank has committed $1 billion to a new program focused on accelerating investments in battery storage designed for developing and middle-income countries (World Bank, 2018d).

The extent of the challenge facing developing countries must not be understated. In order to guarantee growth and development, it is necessary that they find paths to scale up energy generation in a way that is sustainable for the planet and financially viable. Off-grid alternatives have thus far seen limited success and, without further technological adaptations, are unlikely to be able to meet the growing energy needs of populations in developing countries. However, without improvements to their organizational performance, national grids will also fail in this respect. There is thus a vital need for the economics and engineering communities to convene to design energy policies and programs that are suited to the needs and constraints of developing countries.

6 State Capacity and Public Sector Delivery

The growth of low-income countries may be constrained by limited state capacity, lack of transparency and accountability, and poor public service delivery. Technology can help address the agency problem in governments, both inside layers of government, by improving government transparency, public service monitoring and state effectiveness, and between government and citizens, by improving accountability and expanding political participation.

6.1   Bolstering State Effectiveness and Accountability

An important theme throughout this chapter has been the importance of infrastructure for growth and development. In addition to transport, telecommunications, and energy infrastructures, this extends to the administrative and fiscal infrastructures used by the government to raise and spend tax revenue. These are critical for the effective provision of the welfare programs and public goods that firms and households rely upon. However, these infrastructures tend to be much weaker in developing countries. Here, tax shares in GDP resemble those of now developed countries from 100 years ago (Besley & Persson, 2013) and administrations are grappling with corruption, evasion, and leakage (World Bank, 2003; Olken, 2006; Olken & Pande, 2012). This prevents the secure flow of funds and services to intended beneficiaries and disproportionately hurts the most vulnerable (World Bank, 2018a).

The question of how to tackle these issues is a vital one for development. Here again, policymakers and economists have been hoping that at least part of the answer lies in technological innovation. In particular, government administrations are increasingly harnessing ICTs and other forms of technology, ranging from electronic procurement platforms to biometric authentication. These share the general benefits of limiting informal interactions between officials and households and firms and streamlining administrative processes. However, each can be faced with its own limitations. These are discussed in the following section.

One essential faculty of the state is the procurement of goods and services from firms, accounting for an estimated $820 billion of annual spending in developing countries alone (World Bank, 2018b). However, this process is often marred by corruption and collusion, for example, where contracts are restricted to select insiders and firms collude to raise prices. In the attempt to overcome this important barrier to effective provision, governments have been transitioning towards electronic procurement platforms (e-procurement) that provide a standardized online mechanism for advertising bids and awarding and pricing contracts. By increasing the transparency of the bidding process, reducing the cost of submitting a bid and restricting interaction between officials and firms, e-procurement intends to facilitate the entry of non-favored firms and those beyond the local area. This enhanced competition may in turn improve project costs and quality. On the other hand, a more sophisticated system risks excluding those without Internet access. Thus, whether e-procurement is ultimately to the benefit of competition or the quality of public provision is an empirical question.

Lewis-Faupel et al. (2016) provide some of the first rigorous evidence on e-procurement, exploiting its gradual roll-out for public works programs in India and Indonesia. This was found to improve project timeliness in Indonesia and road quality in India, which the authors contend was driven by entry of higher-quality firms beyond the home region. In another study, Abdallah (2015) documents preliminary evidence on a similar e-procurement scheme in Bangladesh. He detects a 12% decrease in the price-to-cost ratio of procurement packages, amounting to over $10 million in estimated savings for 2013 alone. He suggests this may have been driven by reduced political influence rather than by increased national competition. Conversely, using cross-country evidence, Kochanova et al. (2018) find that e-procurement is only associated with higher public procurement competitiveness in developed countries. Ultimately, despite a few promising results, much remains unknown about the success of these initiatives and what might drive the observed effects.

Another prerequisite for the efficient provision of public services is the effectiveness of civil servants. To this end, various studies have evaluated the application of ICTs to monitor employee performance and improve work incentives. One notable example is a study by Debnath and Sekhri (2017) of school meal provision in Bihar, in which intermediate officials had become a source of leakage. In the attempt to tackle this problem, the state rolled out an Interactive Voice Response System (IVRS) that made daily calls to schools to collect reliable data with which to hold officials accountable. This reform was found to improve the likelihood of lunch provision as well as the quality and quantity of meals, suggesting a significant reduction in leakage. Other applications have targeted absenteeism among frontline providers. For example, Duflo et al. (2012) conduct an experiment wherein tamper-proof cameras were used to record teacher attendance in India's NGO schools, data which was then used to determine wages. The authors detect an impressive 21 percentage point decrease in absenteeism, which furthermore translated into higher student test scores. In a similar experiment with nurses, Banerjee et al. (2008) also document an initial drastic improvement in attendance. However, this began to dissipate after just 6 months, at which point the health authority began to undermine the system through granting exemptions. Together, these studies underline both technology-enabled monitoring's potential for motivating service providers and the necessity of political will for guaranteeing successful implementation.

In developing countries, many households are acutely reliant on social safety nets to fund basic expenditures, programs on which 1.5% of GDP is spent on average by their governments (World Bank, 2018c). However, the effectiveness of these well-intentioned initiatives can be seriously inhibited by leakage and identity fraud, issues that another set of technology-enabled initiatives are attempting to target. These include transitioning from cash to electronic disbursal of benefits, with accounts linked to recipients' biometric information. The Indian government in particular has rested much hope on such technologies. Its Aadhaar program has become the world's largest biometric digital identification system, having registered over one billion users in just 6 years (Nilekani, 2018). Research in economics is beginning to respond with evaluations of these efforts.

Muralidharan et al. (2016) assess biometric authentication at scale, experimentally introducing biometric smartcards for the receipt of pension and workfare payments in collaboration with the Andhra Pradesh state government. They detect large, positive returns that far exceeded costs for both programs, including a reduction in leakages and "ghost beneficiaries," i.e., false benefit claims on behalf of other individuals, as well as improved public satisfaction. In another study, Barnwal (2019) examines the Indian government's deployment of biometric authentication for household receipt of subsidized fuel, attempting to stem its diversion to a black market for firms (who otherwise had to pay a higher, taxed price). He argues that biometric authentication was initially effective in curtailing this illegal activity. However, when the reform was unexpectedly reversed following political lobbying by local officials, black market supply was quickly re-established. Overall, these seem to suggest that programs utilizing biometric authentication can effectively prevent the leakage of vital transfers but may be susceptible to loss of political will.

A related technology that has been harnessed by the Indian government is e-invoicing, whose application to their flagship workfare program is evaluated by Banerjee et al. (2020a, 2020b). These digital platforms enable "just-in-time" financing, where funds are dispensed only in response to specific invoices, rather than in the form of advances to be justified down the line. Drawing upon an experiment in collaboration with the Bihar state government, the authors detect a fall in leakage in the form of fewer ghost workers and a decrease in local officials' wealth, culminating in a 24% fall in expenditure. However, this was accompanied by only a small increase in beneficiaries and no change in wages or projects completed, as well as longer delays in household payments, consistent with the higher administrative burden placed on local officials. Echoing the experience of Barnwal (2019), the Bihar experiment was unexpectedly ceased after a few months. However, a comparable scheme was subsequently rolled out across India, providing the authors with a longer-run quasi-experiment. Here, they find a similar, persistent fall in expenditure. Overall, e-invoicing seems to have yielded long-run fiscal savings without generating direct benefits to households of the kind witnessed by Mularidharan et al. (2016). This relates to the broader question of the potential co-existence of winners and losers to technological applications.

Many studies have documented significant savings following the deployment of technology-enabled initiatives by governments in developing countries, sometimes also translating into a higher quality of service for firms and households. However, these have unsurprisingly sometimes been met with efforts by local officials with vested interests to thwart their implementation. Overcoming these political barriers is a vital obstacle to progress. The success of these initiatives will also rely upon levels of technical proficiency and complementary infrastructure that may be hard to come by in poorer regions and demand significant investment to overcome inevitable logistical hurdles. Such was the experience of Muralidharan et al. (2016) in India, where after 2 years, only around 50% of subdistricts had been converted to smartcard payments. Furthermore, there remain many open questions surrounding general equilibrium effects, in particular whether these initiatives simply displace corruption to other areas.

Another area of increasing importance relates to the privacy concerns that accompany large-scale state data collection efforts, particularly regarding their potential use as part of wider state surveillance efforts (Drèze & Tiwari, 2016). These have become highly salient in light of the COVID-19 pandemic, which has seen several nations request the universal downloading of contact-tracing apps. In many cases, there is limited transparency as to who can access the data and how it might be used post-pandemic (O'Neill et al., 2020). Given the speed at which administrations in developing countries have been harnessing technologies, it is vital that research continues to respond with their rigorous evaluation.

6.2 Improving Political Participation and Electoral Integrity

Another channel through which technology stands to impact the functioning of government is by increasing political participation. This view, described by Diamond (2010) as the "liberation technology" argument, emphasizes that ICTs such as Internet, mobile phones, and social media can facilitate communication and information transmission and promote political mobilization. These technologies are also being harnessed to increase the accessibility and integrity of elections, namely, by helping to overcome illiteracy and prevent the tampering of vote counts. This is vital for the effective functioning of democracy and ensuring that people's preferences are reflected in policy, which too often fall short in developing countries. These have in turn been shown to have implications for the extent to which policies are efficient and support the needs of the most vulnerable (Becker, 1983; Besley & Burgess, 2002).

Following the rapid diffusion of ICTs across the developing world, the question of their impact on political mobilization has been the focus of an increasing number of studies. One notable example is a study by Manacorda and Tesei (2020) exploiting detailed georeferenced data on protest incidence and participation across the African continent. They estimate that mobile phone coverage does increase the probability of a protest but only during times of economic downturn – periods where grievances are high and the opportunity cost to participation is low. The authors also find evidence of "strategic complementarities" in protesting, whereby fellow community members' participation reduces the costs and increases the returns to one's own participation. These appear to be enhanced by mobile technology, which can easily publicize information on protest attendance.

Others have focused on the role of social media, a relatively new technology whose impact on democracy has been the subject of much debate, both in developed and developing countries. Here, Diamond (2010) cautions that technology is "open to both noble and nefarious purposes." For example, alongside potential reductions of communication costs and equalizing effects, there has been much concern surrounding their control by powerful actors and the proliferation of "fake news" (Allcott & Gentzkow, 2017). Here, existing evidence, especially in the context of developing countries, remains limited, though a number of studies are beginning to uncover social media's impact on political mobilization. For example, Enikolopov et al. (2020) study the effect of one popular social network on a wave of political protests in 2011 Russia, using quasi-random variation in the platform's penetration. They detect evidence of a positive causal impact on both the incidence and size of protests, which they argue is driven by reduction in the cost of collective action. In another study, Acemoglu et al. (2018) examine the impact of street protests in Egypt's Arab spring. They find that Twitter activity data is a predictor of protest incidence, suggesting a role for social media in facilitating coordination. Finally, in China, Qin et al. (2017) document a surprisingly large number of posts on China's Sina Weibo microblogging platform discussing politically sensitive topics and corruption allegations and find that the latter is predictive of future corruption charges of specific individuals. Building further on these initial results presents an important area for future research.

When citizens lack an effective mechanism for expressing their political preferences, governments face weakened incentives to deliver policies that will benefit them. This underlines the importance of both the accessibility and integrity of elections, an area to which ICTs have also been applied. For example, Fujiwara (2015) documents an important barrier to political participation that disproportionately affects the most vulnerable: illiteracy. In Brazil, he describes how this meant ballots were often erroneously completed and had to be discarded. As a result, when an electronic system offering guidance and visual aids was implemented in the late 1990s, the author reveals that millions of citizens, particularly those with lower levels of education, were de facto enfranchised. The implications for both political and health outcomes are striking: the vote-share of left-wing parties increased, which in turn translated into higher government expenditure on healthcare and improved service utilization and outcomes. More specifically, uneducated mothers received more prenatal visits, and there was a lower incidence of low-weight births.

In 2012, it was estimated that under 40% of elections in low- and middle-income countries were free and fair, a problem often attributed to lack of information (World Bank, 2016). Here, it has been shown that technology-enabled monitoring stands to make an important contribution. For example, Callen and Long (2015) study the issue of aggregation fraud, where votes are altered in the process of being added up across polling stations. In an Afghani election, they experimentally announce a "photo quick count," where provisional results posted at individual stations are photographed and compared to post-aggregation figures. This was found to provoke a 60% reduction in theft of election materials by candidate representatives and a 25% reduction in votes for politically powerful candidates. In Callen et al. (2016), the authors conduct an experiment to evaluate a similar photo technology in a Ugandan election. They detect similar evidence of reduced illegal practices and a fall in the vote share for the incumbent. Together, these studies make clear that photo quick counts can be a simple yet effective method of improving electoral integrity, which the authors furthermore emphasize are well-suited to being scaled via citizen-based implementation as mobile access expands. Relatedly, Aker et al. (2017) study the use of SMS messaging for the reporting of electoral irregularities in Mozambique, detecting a 5% increase in political participation as a result. These results underline the potential of ICTs to empower citizens in reinforcing the accountability of vital democratic processes.

Political participation and electoral integrity are the backbone of a functioning democracy. Research in economics has documented how these have often fallen short in developing countries, particularly in ways that disproportionately hurt the most vulnerable. The studies discussed above have highlighted some important successes in terms of how technologies have been harnessed to tackle these problems, which in the case of Fujiwara (2015) translated into tangible impacts on political outcomes and household welfare. However, we are left with a number of important unanswered questions, especially as regards their long-run effectiveness.

In particular, the use of ICTs to enhance political engagement and improve election monitoring will only be effective insofar as these cannot be subverted by powerful interests.

7 Health

In developing countries, access to healthcare is still a major problem faced by a large share of the poor, especially in rural areas. This constrains both living standards and productive capacity. Technology may offer solutions to both sides of the problem by expanding access to quality healthcare.

7.1 Improving the Quality and Delivery of Healthcare Services

For centuries, landmark innovations in health technology have generated drastic improvements in living standards and life expectancies, which in turn enhance the stock of human capital. In this way, they are vital to economic development. However, there remain vast global inequalities in health, with at least half the world's population still lacking access to essential health services (WHO, 2017). In developing countries, households experience worse access to immunization, professional birth attendants and contraception and suffer disproportionately from HIV, malaria, pollution-related diseases, and unsafe drinking water (Landrigan et al., 2018; UN, 2019). Overcoming these disparities is at the heart of Sustainable Development Goal (SDG) 3, to "ensure healthy lives and promote well-being for all at all ages", as well as SDG6, to "ensure availability and sustainable management of water and sanitation for all" (UN, 2020).

Health technologies targeting issues from which poor populations suffer disproportionately are vital in the fight against global health inequality. One such example is soil-transmitted helminth infections or worms, with a quarter of the world's population estimated to be at risk (Landrigan et al., 2018). In their landmark paper, Miguel & Kremer (2004) evaluate a program in which Kenyan primary school students were experimentally administered deworming drugs. They detect substantial health benefits and reductions in school absenteeism, effects which also spilled over to the non-treated. A decade later, Baird et al. (2016) estimate significant long-run impacts. In particular, men that had been treated as boys were found to work an additional 2.5 hours per week and spend more time in nonagricultural self-employment and were more likely to work in manufacturing. Similarly, women were more likely to have attended secondary school and worked more in nonagricultural self-employment and growing cash crops, rather than in traditional agriculture. Together, these studies indicate a link between healthcare technologies and the welfare improvements and transition out of agriculture that are intrinsic to our understanding of development.

Other studies have focused on health technologies for combatting malaria, a leading cause of death in the developing world that is both curable and preventable (WHO, 2019). For example, Bleakley (2010) studies anti-malaria campaigns in the United States in the early twentieth century and in Brazil, Colombia, and Mexico in the mid-twentieth century, made possible by critical scientific discoveries. These provoked significant declines in the incidence of the disease. Moreover, the author finds that cohorts more exposed to eradication efforts as children had higher literacy and incomes in their adult lives, which he attributes to a positive effect on labor productivity. Lucas (2010) also examines malaria eradication campaigns in his study of Paraguay and Sri Lanka, which effectively eliminated malaria in both countries. The author estimates that this in turn had a positive effect on educational attainment and literacy in both countries.

Despite many successful efforts to eradicate malaria and worms, these and other critical health problems remain endemic in certain parts of the world. In the case of malaria, 2018 saw an estimated 228 million cases worldwide, with almost 85% of the global burden concentrated in just 19 countries in sub-Saharan Africa and India (WHO, 2019). In order to generate further progress, it is clear that more investment in the dissemination of medications and insecticides, public health, and hygiene and health education will be required. What remains unclear, however, is whether applications of new technologies such as ICTs can support these vital functions.

Optimism regarding the application of ICTs to improve the efficiency and quality of health provision has been particularly strong in the case of rural areas, where existing services are most limited. For example, Lemay et al. (2012) study a small-scale intervention in which an SMS-based mobile network was established between community health workers (CHWs) in rural Malawi. They report improved information sharing between CHWs and district staff, for example, allowing them to report stockouts, ask medical questions, and manage emergency cases. In another study of rural Guatemala, Martinez et al. (2018) evaluate a smartphone application designed for supporting traditional birth attendants as they examined patients. These care providers are heavily relied upon by local communities but typically have limited support and linkage with public hospitals. The application was found to increase referral rates for pregnancy and childbirth complications. These positive initial results suggest that ICTs' proven benefits for enhancing connectivity and information in markets might be equally crucial for healthcare services.

The use of drones for the delivery of medical products, such as blood, vaccines, and insulin, represents another flagship example of technology's potential for improving healthcare provision. The company spearheading these efforts is Zipline, who in 2016 established the world's first and only national scale commercial medical drone delivery service in Rwanda, a country whose geography is notoriously hard to navigate (McCall, 2019). In 2019, Zipline delivered over 65% of Rwanda's blood supply outside of the capital (Robotics and Automation News, 2019).

Health technologies are also being used to deal with emerging challenges related to the COVID-19 outbreak. This includes the use of drones in the distribution of medical supplies, a service Zipline plans to offer (Zipline, 2020), as well as to broadcast social distancing regulations and monitor people's compliance, as has been carried out in Rwanda and India (Uwiringiyimana, 2020; Jamkhandikar, 2020). Furthermore, ICT applications are enabling virtual doctors' appointments and the dissemination of COVID-related information, for example, in mass government text messages (Flood, 2020). ICTs will also be vital in global efforts to track the virus's spread, echoing their use in Sierra Leone during the Ebola outbreak (O'Donovan & Bersin, 2015).

Overall, we have much to thank health technology for in terms of the life expectancies we enjoy today. Their harnessing to tackle the health problems suffered disproportionately in developing regions is also helping to slowly chip away at the vast inequalities in health outcomes that persist between countries. Research in economics has furthermore demonstrated the presence of knock-on effects these can have on education, incomes, and welfare. It is widely accepted that further progress will hinge on basic investments into public health and sanitation infrastructures and proven health technologies. However, much less is known about the potential contribution that applications of new technologies, such as ICTs, can make to these efforts. Indeed, Sundin et al. (2016) document that most of these initiatives fail to progress beyond the pilot stage but emphasize the importance of social and economic factors over technological ones. For example, they describe how health-related mobile phone applications have been limited by patients' inability to charge their phones. As such, they argue that health technologies must be supported by thorough knowledge of sociocultural dynamics and business practices in order to be able to effectively scale.

8 Conclusion

Development and growth are fundamentally about the spread of innovations and ideas. In this chapter, we reviewed a range of areas of work which can help to enhance this spread. A key insight that we have gleaned is that the same ingenuity that has driven human progress since time immemorial will also be required to tackle the externalities that have been engendered by that progress.

Development engineering is fundamentally about harnessing innovations and bringing them to bear on development challenges, but that is not where the subject stops. Indeed, much of the value of recent work is concerned with how alliances between technologists, engineers, economists, and policymakers can enable the design of interventions that work at scale in developing countries in particular. This is easier said than done as there is often a tougher and more complex set of constraints to overcome in scaling technological innovation in developing versus developed countries. These are not just related to the way markets work but also to some fundamental barriers associated with the design of regulations and the way that political and other institutions work. In this sense, development engineering is truly a field that bridges between science and social science.

The types of challenges faced by developing countries are not only large in magnitude, but also extremely urgent. There are close to a billion people in extreme poverty and close to a billion without electricity, with nearly all growth in energy demand over the next few decades expected to stem from developing countries (Wolfram et al., 2012). With extremely young populations, there is the challenge of how millions of young women and men will find meaningful work (Alfonsi et al. 2020). There also exists a wealth of unanswered questions regarding the relationship between growth and the environmental damages that accompany it, with evidence that these may be more acute in the world's poorer regions (Burgess et al., 2017; Greenstone & Jack, 2015). If we are serious about eliminating extreme poverty by 2030 and about shielding more vulnerable populations from the effects of climate change, and pollution, then we cannot sit on our hands.

This chapter has begun to point to some areas where technology can play a critical role in addressing the major development challenges which are enshrined in the SDGs, but it is still just a start. In many ways, it is more of a call to arms for a diverse set of actors from the private sector, civil society, academia, and government to come together to maximize the positive role that different technologies can play in the process of development. Only in this way can we generate the innovations and productivity improvements needed to keep humans on a trajectory to higher living standards while ensuring that negative externalities generated by this growth do not block the path to progress.

References

Abdallah, W. (2015). *Effect of electronic public procurement: Evidence from Bangladesh*. Working paper.

Abebe, G., Caria, S., Fafchamps, M., Falco, P., Franklin, S., & Quinn, S. (2017a). *Anonymity or distance? Job search and labor market exclusion in a growing African city*. LSE.

Abebe, G., Caria, S., Fafchamps, M., Falco, P., Franklin, S., Quinn, S., & Shilpi, F. (2017b). *Job fairs: Matching firms and workers in a field experiment in Ethiopia*. Working paper.

Acemoglu, D., Hassan, T., & Tahoun, A. (2018). The power of the street: Evidence from Egypt's Arab spring. *Review of Financial Studies, 31*, 1–42.

Adhvaryu, A., Kala, N., & Nyshadham, A. (2020). The light and the heat: Productivity co-benefits of energy-saving technology. *The Review of Economics and Statistics, 102*(4), 779–792.

Adukia, A., Asher, S., & Novosad, P. (2020). Educational investment responses to economic opportunity: Evidence from Indian road construction. *American Economic Journal: Applied Economics, 12*(1), 348–376.

African Progress Panel (APP). (2017). *Lights power action: Electrifying Africa*. February 2017.

Aghion, P., & Howitt, P. (1992). A model of growth through creative destruction. *Econometrica, 60*(2), 323–351.

Aker, J. C. (2010). Information from markets near and far: Mobile phones and agricultural markets in Niger. *American Economic Journal: Applied Economics, 2*(3), 46–59.

Aker, J. (2018). *Migration, money transfers and mobile money: Evidence from Niger*. Pathways for prosperity commission background paper series.

Aker, J. C., & Ksoll, C. (2020). Can ABC lead to sustained 123? The medium-term effects of a technology-enhanced adult education program. *Economic Development and Cultural Change, 68*(3), 1081–1102.

Aker, J. C., Ksoll, C., & Lybbert, T. J. (2012). Can mobile phones improve learning? Evidence from a field experiment in Niger. *American Economic Journal: Applied Economics, 4*(4), 94–120.

Aker, J. C., Boumnijel, R., McClelland, A., & Tierney, N. (2016). Payment mechanisms and antipoverty programs: Evidence from a mobile money cash transfer experiment in Niger. *Economic Development and Cultural Change, 65*(1), 1–37.

Aker, J. C., Collier, P., & Vicente, P. C. (2017). Is information power? Using mobile phones and free newspapers during an election in Mozambique. *The Review of Economics and Statistics, 99*(2), 185–200.

Aklin, M., Bayer, B., Harish, S. P., & Urpelainen, J. (2015). *Rural electrification with off grid community microgrids: An impact evaluation in Uttar Pradesh, India*. IGC Report.

Alfonsi, L., Bandiera, O., Bassi, V., Burgess, R., Rasul, I., Sulaiman, M., & Vitali, A. (2020). Tackling youth unemployment: Evidence from a labor market experiment in Uganda. *Econometrica, 88*(6), 2369–2414.

Allcott, H., & Gentzkow, M. (2017). Social media and fake news in the 2016 election. *Journal of Economic Perspectives, 31*(2), 211–236.

Allcott, H., Collard-Wexler, A., & O'Connell, S. D. (2016). How do electricity shortages affect industry? Evidence from India. *American Economic Review, 106*(3), 587–624.

Alexandratos, N. & Bruinsma, J. (2012). World agriculture towards 2030/2050: the 2012 revision. ESA Working Paper No. 12-03. FAO.

Asher, S., & Novosad, P. (2020). Rural roads and local economic development. *American Economic Review, 110*(3), 797–823.

Atkin, D., Chaudhry, A., Chaudry, S., Khandelwal, A. K., & Verhoogen, E. (2017a). Organizational barriers to Technology adoption: Evidence from soccer-ball producers in Pakistan*. *The Quarterly Journal of Economics, 132*(3), 1101–1164.

Atkin, D., Khandelwal, A. K., & Osman, A. (2017b). Exporting and firm performance: Evidence from a randomized experiment*. *The Quarterly Journal of Economics, 132*(2), 551–615.

Atkin, D., Faber, B., & Gonzalez-Navarro, M. (2018). Retail globalization and household welfare: Evidence from Mexico. *Journal of Political Economy, 126*(1), 1–73.

Autor, D. H. (2014). *Polanyi's paradox and the shape of employment growth*.

Autor, D. H. (2015). Why are there still so many jobs? The history and future of workplace automation. *Journal of Economic Perspectives, 29*(3), 3–30.

Autor, D. H., & Dorn, D. (2013). The growth of low-skill service jobs and the polarization of the us labor market. *American Economic Review, 103*(5), 1553–1597.

Autor, D. H., Levy, F., & Murnane, R. J. (2003). The skill content of recent technological change: An empirical exploration*. *The Quarterly Journal of Economics, 118*(4), 1279–1333.

Autor, D. H., Katz, L. F., & Kearney, M. S. (2006). The polarization of the U.S. labor market. *American Economic Review, 96*(2), 189–194.

Autor, D. H., Katz, L. F., & Kearney, M. S. (2008). Trends in US wage inequality: Revising the revisionists. *The Review of Economics and Statistics, 90*(2), 300–323.

Autor, D. H., Dorn, D., & Hanson, G. H. (2015). Untangling trade and technology: Evidence from local labor markets. *The Economic Journal, 125*(584), 621–646.

Baird, S., Hicks, J. H., Kremer, M., & Miguel, E. (2016). Worms at work: Long-run impacts of a child health investment*. *The Quarterly Journal of Economics, 131*(4), 1637–1680.

Balboni, C. (2019). *In harm's way? Infrastructure investments and the persistence of coastal cities*.

Bandiera, O., & Rasul, I. (2006). Social networks and technology adoption in northern Mozambique*. *The Economic Journal, 116*(514), 869–902.

Banerjee, A., Duflo, E., & Glennerster, R. (2008). Putting a band-aid on a corpse: Incentives for nurses in the Indian public health care system. *Journal of the European Economic Association, 6*(2–3), 487–500.

Banerjee, A., Duflo, E., Imbert, C., Mathew, S., & Pande, R. (2020a). E-governance, accountability, and leakage in public programs: Experimental evidence from a financial management reform in India. *American Economic Journal: Applied Economics, 12*(4), 39–72.

Banerjee, A., Duflo, E., & Qian, N. (2020b). On the road: Access to transportation infrastructure and economic growth in China. *Journal of Development Economics, 145*, 102442.

Barnwal, P. (2019). *Curbing leakage in public programs: Evidence from India's direct benefit transfer policy for LPG subsidies*.

Barrera-Osorio, F. & Linden, L. L. (2009). *The use and misuse of computers in education: evidence from a randomized experiment in Colombia.* Policy Research Working Paper, (1).

Bartel, A., Ichniowski, C., & Shaw, K. (2007). How does information Technology affect productivity? Plant-level comparisons of product innovation, process improvement, and worker skills*. *The Quarterly Journal of Economics, 122*(4), 1721–1758.

Batista, C. & Vicente, P. (2018). *Is mobile money changing rural Africa? Evidence from a field experiment.*

Batista, C., & Vicente, P. (2020). Adopting mobile money: Evidence from an experiment in rural Africa. *AEA Papers and Proceedings, 110*, 594–598.

Becker, G. S. (1983). A theory of competition among pressure groups for political influence. *The Quarterly Journal of Economics, 98*(3), 371–400.

Besley, T., & Burgess, R. (2002). The political economy of government responsiveness: Theory and evidence from India. *The Quarterly Journal of Economics, 117*(4), 1415–1451.

Besley, T., & Persson, T. (2013). Chapter 2 - taxation and development. In A. J. Auerbach, R. Chetty, M. Feldstein, & E. Saez (Eds.), *Handbook of public economics* (Vol. 5, pp. 51–110). Elsevier.

Beuermann, D. W., Cristia, J., Cueto, S., Malamud, O., & Cruz Aguayo, Y. (2015). One laptop per child at home: Short-term impacts from a randomized experiment in Peru. *American Economic Journal: Applied Economics, 7*(2), 53–80.

Bharadwaj, P., Jack, W., & Suri, T. (2019). *Fintech and household resilience to shocks: Evidence from digital loans in Kenya.*

Bhattacharya, A. (2020). *India's IT industry created more jobs in 2019 than a year ago.* Quartz India.

Bleakley, H. (2010). Malaria eradication in the Americas: A retrospective analysis of childhood exposure. *American Economic Journal: Applied Economics, 2*(2), 1–45.

Bloom, N., & Van Reenen, J. (2010). Why do management practices differ across firms and countries? *Journal of Economic Perspectives, 24*(1), 203–224.

Blumenstock, J., Callen, M., Ghani, T., & Koepke, L. (2015). *Promises and pitfalls of mobile money in Afghanistan: Evidence from a randomized control trial.*

Blumenstock, J., Callen, M., & Ghani, T. (2016). *Mobile-izing savings with automatic contributions: Experimental evidence on present bias and default effects in Afghanistan.*

Brookes, G., & Barfoot, P. (2018). Farm income and production impacts of using gm crop technology 1996–2016. *GM Crops & Food, 9*(2), 59–89.

Bryan, G., Chowdhury, S., & Mobarak, A. M. (2014). Underinvestment in a profitable technology: The case of seasonal migration in Bangladesh. *Econometrica, 82*(5), 1671–1748.

Bulman, G., & Fairlie, R. (2016). Chapter 5 - technology and education: Computers, software, and the internet. In *Volume 5 of handbook of the economics of education* (pp. 239–280). Elsevier.

Burgess, R., Deschenes, O., Donaldson, D., & Greenstone, M. (2017). *Weather, climate change and death in India.* LSE.

Burgess, R., Greenstone, M., Ryan, N., & Sudarshan, A. (2020a). The consequences of treating electricity as a right. *Journal of Economic Perspectives, 34*(1), 145–169.

Burgess, R., Greenstone, M., Ryan, N., & Sudarshan, A. (2020b). *Demand for electricity on the global electrification frontier.*

Bustos, P., Caprettini, B., & Ponticelli, J. (2016). Agricultural productivity and structural transformation: Evidence from Brazil. *American Economic Review, 106*(6), 1320–1365.

Calenda, D. (2016). *Case studies in the international recruitment of nurses: Promising practices in recruitment among agencies in the United Kingdom, India, and the Philippines.*

Callen, M., & Long, J. D. (2015). Institutional corruption and election fraud: Evidence from a field experiment in Afghanistan. *American Economic Review, 105*(1), 354–381.

Callen, M., Gibson, C. C., Jung, D. F., & Long, J. D. (2016). Improving electoral integrity with information and communications technology. *Journal of Experimental Political Science, 3*(1), 4–17.

Capron, A. (2016). Pas de lumière pour faire ses devoirs? *Voici le cartable solaire ivoirien. Les Observateurs de France, 24.*

Carrillo, P. E., Onofa, M., & Ponce, J.. (2010). *Information Technology and student achievement: Evidence from a randomized experiment in Ecuador.* IDB working paper series.

Clean Cooking Alliance (CCA). (2020). *Clean cooking Alliance 2019 annual report.*

Collard-Wexler, A., & De Loecker, J. (2015). Reallocation and technology: Evidence from the US steel industry. *American Economic Review, 105*(1), 131–171.

Commander, S., Harrison, R., & Menezes-Filho, N. (2011). ICT and productivity in developing countries: New firm-level evidence from Brazil and India. *The Review of Economics and Statistics, 93*(2), 528–541.

Conley, T. G., & Udry, C. R. (2010). Learning about a new technology: Pineapple in Ghana. *American Economic Review, 100*(1), 35–69.

Dahir, A. L. (2019). *Africa's largest wind power project is now open in Kenya.* Quartz Africa.

Damania, R. & Wheeler, D. (2015). *Road improvement and deforestation in the Congo basin countries.* Policy Research Working Paper; No. 7274. World Bank.

Dammert, A. C., Galdo, J. C., & Galdo, V. (2013). *Digital labor-market intermediation and job expectations: Evidence from a field experiment.* IZA Discussion Paper No. 7395.

Dasgupta, S. & Wheeler, D. (2016). *Minimizing ecological damage from road improvement in tropical forests.* Policy Research Working Paper; No. 7826. World Bank, 2016.

De Sisternes, F. J., Worley, H., Mueller, S., & Jenkin, T. (2020). Scaling-up sustainable energy storage in developing countries. *Journal of Sustainability Research, 2*(1), e200002. https://doi.org/10.20900/jsr20200002

Debnath, S. & Sekhri, S. (2017). *No free lunch: Using technology to improve the efficacy of school feeding programs.*

Demirguc-Kunt, A., Klapper, L., Singer, D., Ansar, S., & Hess, J. (2020). The global findex database 2017: Measuring financial inclusion and opportunities to expand access to and use of financial services. *The World Bank Economic Review, 34*(Supplement1), S2–S8.

Diamond, L. (2010). Liberation Technology. *Journal of Democracy, 3*, 69–83.

Dinkelman, T. (2011). The effects of rural electrification on employment: New evidence from South Africa. *American Economic Review, 101*(7), 3078–3108.

Donaldson, D. (2015). The gains from market integration. *Annual Review of Economics, 7*(1), 619–647.

Donaldson, D. (2018). Railroads of the raj: Estimating the impact of transportation infrastructure. *American Economic Review, 108*(4–5), 899–934.

Donaldson, D., & Hornbeck, R. (2016). Railroads and American economic growth: A "market access" approach *. *The Quarterly Journal of Economics, 131*(2), 799–858.

Downs, E. (2019). *The China-Pakistan economic corridor power projects: Insights into environmental and debt sustainability.* Columbia SIPA.

Drèze, J., & Tiwari, S. (2016). *Making Aadhaar compulsory jeopardizes privacy rights.* IGC.

Duflo, E., Kremer, M., & Robinson, J. (2008). How high are rates of return to fertilizer? Evidence from field experiments in Kenya. *American Economic Review, 98*(2), 482–488.

Duflo, E., Michael K., and Jonathan, R. (2010). Understanding Technology Adoption: Fertilizer in Western Kenya, Evidence from Field Experiments. Unpublished.

Duflo, E., Kremer, M., & Robinson, J. (2011). Nudging farmers to use fertilizer: Theory and experimental evidence from Kenya. *American Economic Review, 101*(6), 2350–2390.

Duflo, E., Hanna, R., & Ryan, S. P. (2012). Incentives work: Getting teachers to come to school. *American Economic Review, 102*(4), 1241–1278.

Elliott, K. A., & Keller, J. M. (2016). *Can GMOs play a role in a new green revolution for Africa?* Center for Global Development.

Enikolopov, R., Makarin, A., & Petrova, M. (2020). Social media and protest participation: Evidence from Russia. *Econometrica, 88*, 1479–1514.

Escueta, M., Quan, V., Nickow, A. J., & Oreopoulos, P. (2017). *Education technology: An evidence-based review.* (23744).

Faber, B. (2014). Trade integration, market size, and industrialization: Evidence from China's National Trunk Highway System. *The Review of Economic Studies, 81*(3), 1046–1070.

Fabregas, R., Kremer, M., & Schilbach, F. (2019). Realizing the potential of digital development: The case of agricultural advice. *Science, 366*(6471).

Faivre, F. (2020). *Mobile money helps displaced Congolese survive amid coronavirus threat*. UNHCR.

Finextra Research. (2020). *Covid-19: M-PESA waives fees to discourage cash usage*.

Flood, Z. (2020). *How Africa's tech innovators respond to the coronavirus pandemic*. Al Jazeera.

Food and Agriculture Organizaton of the United Nations (FAO) (2001). Land, and W. D. Division. The Economics of Conservation Agriculture. FAO, 2001.

Foster, A. D., & Rosenzweig, M. R. (1995). Learning by doing and learning from others: Human capital and technical change in agriculture. *Journal of Political Economy, 103*(6), 1176–1209.

Fowlie, M., Khaitan, Y., Wolfram, C., & Wolfson, D. (2019). Solar microgrids and remote energy access: How weak incentives can undermine smart technology. *Economics of Energy Environmental Policy, 8*.

Fujii, T., & Shonchoy, A. S. (2020). Fertility and rural electrification in Bangladesh. *Journal of Development Economics, 3*, 102430.

Fujiwara, T. (2015). Voting technology, political responsiveness, and infant health: Evidence from Brazil. *Econometrica, 83*(2), 423–464.

Gandhi, R., Veeraraghavan, R., Toyama, K., & Ramprasad, V. (2009). Digital green: Participatory video and mediated instruction for agricultural extension, *5*(1), 1–15.

Gibbons, R., & Katz, L. F. (1991). Layoffs and lemons. *Journal of Labor Economics, 4*, 351–380.

Gollin, D., & Rogerson, R. (2014). Productivity, transport costs and subsistence agriculture. *Journal of Development Economics, 107*(C), 38–48.

Goos, M., & Manning, A. (2007). Lousy and lovely jobs: The rising polarization of work in Britain. *The Review of Economics and Statistics, 89*(1), 118–133.

Goos, M., Manning, A., & Salomons, A. (2014). Explaining job polarization: Routine-biased technological change and offshoring. *American Economic Review, 104*(8), 2509–2526.

Graetz, G., & Michaels, G. (2015). Robots at work. *The Review of Economics and Statistics, 100*(5), 753–768.

Greenstone, M., & Jack, B. K. (2015). Envirodevonomics: A research agenda for an emerging field. *Journal of Economic Literature, 53*(1), 5–42.

Greenwald, B. C. (1986). Adverse selection in the labour market. *The Review of Economic Studies, 53*(3), 325–347.

Grogan, L. (2018). Time use impacts of rural electrification: Longitudinal evidence from Guatemala. *Journal of Development Economics, 135*, 304–317.

GSM Association (GSMA). (2019). *Connected women: The mobile gender gap report 2019*.

Harigaya, T. (2016). *Effects of digitization on financial behaviors: Experimental evidence from the Philippines*. Working Paper.

Hjort, J., & Poulsen, J. (2019). The arrival of fast internet and employment in Africa. *American Economic Review, 109*(3), 1032–1079.

IEA. (2019). World Energy Outlook 2019. IEA.

IEA. (2020). Brazil. IEA.

IFAD. (2015). *Kenya cereal enhancement programme climate resilient agricultural livelihoods window (KCEP-CRAL)*. International Fund for Agricultural Development (IFAD), 2015.

IFAD. (2020). *Before and during covid-19, an e-voucher initiative makes a difference for Kenyan farmers*. International Fund for Agricultural Development (IFAD), 2020.

Ignaciuk, A., & Mason-D'Croz, D. (2014). Modelling adaptation to climate change in agriculture, *70*.

International Hydropower Association (IHA). (2017). Ethiopia.

International Hydropower Association (IHA). (2018). Brazil.

ISAAA. (2018). Global status of commercialized biotech/gm crops in 2018: Biotech crops continue to help meet the challenges of increased population and climate change. 54, 2018.

Jack, K. (2011). *Market inefficiencies and the adoption of agricultural technologies in developing countries*. Center for Effective Global Action.

Jack, K., & Smith, G. (2020). Charging ahead: Prepaid metering, electricity use, and utility revenue. *American Economic Journal: Applied Economics, 2*, 134–168.

Jack, W., & Suri, T. (2014). Risk sharing and transactions costs: Evidence from Kenya's mobile money revolution. *American Economic Review, 104*(1), 183–123.

Jamkhandikar, S. (2020). *With drones and tests, India battles to keep virus out of Mumbai's slums.* Reuters.

Jensen, R. (2007). The digital provide: Information (Technology), market performance, and welfare in the south Indian fisheries sector*. *The Quarterly Journal of Economics, 122*(3), 879–924.

Kar, A., Pachauri, S., Bailis, R., & Zerriffi, H. (2019). Using sales data to assess cooking gas adoption and the impact of India's Ujjwala programme in rural Karnataka. *Nature Energy, 4.*

Karlan, D., Osei, R., Osei-Akoto, I., & Udry, C. (2014). Agricultural decisions after relaxing credit and risk constraints. *The Quarterly Journal of Economics, 2,* 597–652.

Kassem, D. (2020). *Does electrification cause industrial development? Grid expansion and firm turnover in Indonesia.*

Kochanova, A., Hasnain, Z., & Larson, B. (2018). Does e-government improve government capacity? Evidence from tax compliance costs, tax revenue, and public procurement competitiveness. *The World Bank Economic Review, 1,* 101–120.

Kojima, M., & Trimble, C. P. (2016). *Making power affordable for Africa and viable for its utilities.* World Bank, Washington DC.

Landrigan, P. J., Fuller, R., Acosta, N. J. R., Adeyi, O., Arnold, R., Basu, N., Balde, A. B., Bertollini, R., Bose-O'Reilly, S., Boufford, J. I., Breysse, P. N., Chiles, T., Mahidol, C., Coll-Seck, A. M., Cropper, M. L., Fobil, J., Fuster, V., & Greenstone, M. (2018). The lancet commission on pollution and health. *The Lancet, 391*(10119), 462–512.

Lee, K., Brewer, E., Christiano, C., Meyo, F., Miguel, E., Podolsky, M., Rosa, J., & Catherine, W. (2016). Electrification for "under grid" households in rural Kenya. *Development Engineering, 1,* 26–35.

Lee, K., Miguel, E., & Wolfram, C. (2020). Experimental evidence on the economics of rural electrification. *Journal of Political Economy, 128*(4), 1523–1565.

Lemay, N. V., Sullivan, T., Jumbe, B., & Perry, C. (2012). Reaching remote health workers in Malawi: Baseline assessment of a pilot M-health intervention. *Journal of Health Communication, 17*(Suppl 1), 105–117.

Levy, F., & Murnane, R. J. (2004). *The new division of labor: How computers are creating the next job market.* Princeton University Press.

Lewis-Faupel, S., Neggers, Y., Olken, B. A., & Pande, R. (2016). Can electronic procurement improve infrastructure provision? Evidence from public works in India and Indonesia. *American Economic Journal: Economic Policy, 3,* 258–283.

Linden, L. L. (2008). *Complement or substitute? The effect of technology on student achievement in India.* InfoDev working paper, 17.

Lipscomb, M., Mobarak, A. M., & Barham, T. (2013). Development effects of electrification: Evidence from the topographic placement of hydropower plants in Brazil. *American Economic Journal: Applied Economics, 5*(2), 200–231.

Lowder, S. K., Skoet, J., & Raney, T. (2016). The number, size, and distribution of farms, smallholder farms, and family farms worldwide. *World Development, 7,* 16–29.

Lucas, A. M. (2010). Malaria eradication and educational attainment: Evidence from Paraguay and Sri Lanka. *American Economic Journal: Applied Economics, 2*(2), 46–71.

Maloney, W. F. & Molina, C. A. (2016). *Are automation and trade polarizing developing country labor markets, too?* Policy Research Working Paper; No. 7922. World Bank, 2016.

Manacorda, M., & Tesei, A. (2020). Liberation technology: Mobile phones and political mobilization in Africa. *Econometrica, 88*(2), 533–567.

Martinez, B., Ixen, E., Hall-Clifford, R., Juarez, M., Miller, A., Francis, A., Valderrama, C., Stroux, L., Clifford, G., & Rohloff, P. (2018). M-health intervention to improve the continuum of maternal and perinatal care in rural Guatemala: A pragmatic, randomized controlled feasibility trial. *Reproductive Health.*

McCall, B. (2019). Sub-Saharan Africa leads the way in medical drones. *The Lancet, 3,* 17–18.

McKenzie, D. (2017). How effective are active labor market policies in developing countries? A critical review of recent evidence. *World Bank Research Observer, 32*(2), 127–154.

McRae, S. (2015). Infrastructure quality and the subsidy trap. *American Economic Review, 105*(1), 35–66.

Michler, J. D., et al. (2019). Conservation agriculture and climate resilience. *Journal of Environmental Economics and Management*, 93, 148–169.

Michaels, G., Natraj, A., & Van Reenen, J. (2014). Has ICT polarized skill demand? Evidence from eleven countries over twenty-five years. *The Review of Economics and Statistics, 96*(1), 60–77.

Miguel, E., & Kremer, M. (2004). Worms. *Econometrica*, 72(1), 159–217.

Mohammed, O. (2015). *Africa has the world's fastest-growing labor force but needs jobs growth to catch up*. Quartz Africa.

Moneke, N. (2019). *Can big push infrastructure unlock development? Evidence from Ethiopia.*

Morten, M. & Oliveira, J. D.. (2018) *The effects of roads on trade and migration: Evidence from a planned capital city.*

Muralidharan, K., Niehaus, P., & Sukhtankar, S. (2016). Building state capacity: Evidence from biometric smartcards in India. *American Economic Review, 06*(10), 2895–2929.

Muralidharan, K., Singh, A., & Ganimian, A. J. (2019). Disrupting education? Experimental evidence on technology-aided instruction in India. *American Economic Review, 109*(4), 1426–1460.

Nilekani, N. (2018). *Giving people control over their data can transform development*. World Bank Blogs.

O'Donovan, J., & Bersin, A. (2015). Controlling Ebola through M-health strategies. *The Lancet. Global Health, 3*, 22.

O'Neill, P. H., Ryan-Mosley, T., & Johnson, B. (2020). *A flood of coronavirus apps are tracking us. Now it's time to keep track of them*. MIT Technology Review.

Olken, B. A. (2006). Corruption and the costs of redistribution: Micro evidence from Indonesia. *Journal of Public Economics, 90*(4), 853–870.

Olken, B. A., & Pande, R. (2012). Corruption in developing countries. *Annual Review of Economics, 4*(1), 479–509.

Pangotra, P., & Shukla, P. R. (2012). *Infrastructure for low-carbon transport in India: A case study of the Delhi-Mumbai dedicated freight corridor*. UNEP.

Peyton, N. (2020). *Coronavirus seen as trigger for mobile money growth in West Africa*. Reuters.

Qin, B., Stromberg, D., & Wu, Y. (2017). Why does China allow freer social media? Protests versus surveillance and propaganda. *The Journal of Economic Perspectives, 31*(1), 117–140.

Redding, S. J., & Sturm, D. M. (2008). The costs of remoteness: Evidence from German division and reunification. *American Economic Review, 98*(5), 1766–1767.

Republic of Rwanda Ministry of Infrastructure (MININFRA). (2020) *Solar energy in Rwanda.*

Robotics and Automation News. (2019). *UPS foundation supports Ghana's vaccine drone delivery network.*

Rodrik, D. (2016). Premature deindustrialization. *Journal of Economic Growth, 1*(1), 1–33.

Romer, P. M. (1990). Endogenous technological change. *Journal of Political Economy, 98*(5), S71–S102.

Rosenzweig, M. R., & Udry, C. R. (2019). *Assessing the benefits of long-run weather forecasting for the rural poor: Farmer investments and worker migration in a dynamic equilibrium model*. Working Paper.

Rud, J. P. (2012). Electricity provision and industrial development: Evidence from India. *Journal of Development Economics, 97*(2), 352–367.

Schlogl, L. & Sumner, A. (2018). The rise of the robot reserve army: Automation and the future of economic development, work, and wages in developing countries. SSRN Electronic Journal.

Solow, R. M. (1957). Technical change and the aggregate production function. *The Review of Economics and Statistics, 39*(3), 312–320.

Stacey, K. (2018). *Pakistan's pivot to coal to boost energy gets critics fired up*. Financial Times.

Stevenson, J. R., Serraj, R., & Cassman, K. G. (2014). Evaluating conservation agriculture for small-scale farmers in sub-Saharan Africa and South Asia. *Agriculture, Ecosystems Environment, 187*, 1–10.

Storeygard, A. (2016). Farther on down the road: Transport costs, trade and urban growth in sub-Saharan Africa. *The Review of Economic Studies, 3*(3), 1263–1295.

Sundin, P., Callan, J., & Mehta, K. (2016). Why do entrepreneurial M-health ventures in the developing world fail to scale? *Journal of Medical Engineering & Technology, 40*(7–8), 444–457.

Suri, T. (2011). Selection and comparative advantage in technology adoption. *Econometrica, 79*(1), 159–209.

Suri, T. (2017). Mobile money. *Annual Review of Economics, 9*(1), 497–520.

Suri, T., Jack, W., & Stoker, T. (2012). Documenting the birth of a financial economy. *Proceedings of the National Academy of Sciences of the United States of America, 109*, 10257–10262.

Syverson, C. (2011). What determines productivity? *Journal of Economic Literature, 49*(2), 326–365.

Power Technology. (2020). The grand renaissance hydroelectric project.

Tsivanidis, N.. (2018). The aggregate and distributional effects of urban transit infrastructure: Evidence from Bogotá's Transmilenio.

United Nations (UN). (2019). *Report of the secretary-general: Progress towards the sustainable development goals.*

United Nations (UN). (2020). *SDGs: Sustainable development goals.* Sustainable Development Knowledge Platform, 2020.

Uwiringiyimana, C. (2020). *Rwanda uses drones to help catch lockdown transgressors.* Reuters.

Vodafone. (2020). *What is M-PESA?*

Webb, M. (2019). The impact of artificial intelligence on the labor market. *SSRN Electronic Journal.*

Wolfram, C., Shelef, O., & Gertler, P. (2012). How will energy demand develop in the developing world? *Journal of Economic Perspectives, 26*(1), 119–138.

World Bank. (2003). World development report 2004: Making services work for poor people.

World Bank. (2016). World development report 2016: Digital dividends.

World Bank. (2018a). Combating corruption.

World Bank. (2018b). The state of social safety nets.

World Bank. (2018c). Why modern, fair and open public procurement systems matter for the private sector in developing countries.

World Bank. (2018d). World bank group commits $1 billion for battery storage to ramp up renewable energy globally.

World Bank. (2019). World development report 2019: The changing nature of work.

World Health Organization (WHO). (2017). Tracking universal health coverage: 2017 global monitoring report.

World Health Organization (WHO). (2019). World malaria report 2019.

Zipline. (2020). Zipline's Covid-19 response.

Chapter 3
A Practical Framework for Research

Temina Madon ⓘ, Anustubh Agnihotri, and Ashok J. Gadgil ⓘ

1 Introduction

This chapter outlines a practical framework for designing scalable technology solutions that solve development challenges. We begin with an overview of the common constraints to sustainable development that often are encountered in the context of poverty. These constraints are based on a large body of research in development economics, political economy, psychology, and other social sciences; and they help to explain why engineering innovations so frequently fail to achieve outcomes when implemented in the real world. In the second part of this chapter, we provide a framework for implementing development engineering projects, consisting of four key activities: innovation, implementation, evaluation, and adaptation. Combining these activities in an iterative (and usually nonlinear) path allows the researcher to anticipate and design around the most common pitfalls associated with "technology for development."

T. Madon (✉)
Center for Effective Global Action, University of California, Berkeley, Berkeley, CA, USA
e-mail: temina@ocf.berkeley.edu

A. J. Gadgil
Department of Civil and Environmental Engineering, University of California, Berkeley, CA, USA

A. Agnihotri
University of California Berkeley, Berkeley, CA, USA
e-mail: anustubh@berkeley.edu

© The Author(s) 2023
T. Madon et al. (eds.), *Introduction to Development Engineering*,
https://doi.org/10.1007/978-3-030-86065-3_3

59

2   Innovation Under Constraints

To find solutions to thorny development challenges, researchers must begin with building a deep understanding of context and environment. To some extent, this can come from direct observation—from researchers embedding themselves within representative communities, observing the cadence of daily life, learning how it is to walk in the shoes of the potential users of a future innovation. This approach is central to the success of product design firms like IDEO (Kelley, 2005).

Until recently, direct observation (and other elements of human-centered design) has remained relatively uncommon in the technocentric approaches to "engineering for development" found in many elite universities. Yet development economists, political scientists, and others in the social sciences have invested decades in such work, and in the collection of descriptive data across populations and contexts. This has resulted in generalizable findings about the market systems, institutions, behaviors, and social norms governing life in many low-resource settings. Learning to systematically apply these insights to the design of a novel technology is an essential thrust of development engineering.

This section provides an overview of the common constraints encountered in many developing countries—and to some extent in low-resource communities throughout the world. Without judgement, these constraints are actually just alternatives to the "ideal" market systems and institutions imagined to exist in wealthy countries. In some cases, they have emerged as critical adaptations to local conditions (such as resource scarcity, conflict, colonization, and ethnic diversity). In the engineer's mindset, we can think of these conditions as design requirements, because they can affect the adoption, performance, impact, and scaling of a technological solution. We can also think of these constraints, themselves, as targets for intervention (Soss et al., 2011). For example, predatory policing, which may be observed as a constraint to economic development, could be directly targeted through the design of mobile applications and political reforms that empower citizens to monitor and report police activity.

In a sense, the most basic constraint faced by people living in poverty is income uncertainty. For survival, humans require continuous access to food, water, heating, cooling, and shelter. Yet poor households, by definition, experience scarcity—not just lack of income but also income that is lumpy across time (Collins et al., 2009). This makes it difficult to invest in basic needs, let alone new technologies or other assets. In urban settings, this lumpiness may take the form of irregular income from small family-owned enterprises. These businesses are often constrained by a lack of formal access to capital (in the form of savings or credit products). As a result they find it difficult to invest in the inventory, marketing, supply chain tools, and other inputs needed to build more reliable profits.

It is a more complex story for households reliant on farming for survival. Agriculture employs the majority of the world's poor, typically on small family-owned farms. Income from agriculture is seasonal by nature: profits are generated largely at harvest time. This cyclic pattern of production creates lumpiness in household consumption (Mobarak and Reimao, 2020). In addition, productivity is dependent on weather and climate conditions, which are highly unpredictable and can vary substantially from season to season or from year to year. This uncertainty makes it difficult for households to purchase goods or services on a regular basis, and it can also deter households making investments in new technologies (even when the longer-term economic benefits of a technology are well understood).

Beyond lumpy consumption, households face a lack of access to savings, credit, and insurance products, all of which are useful for managing risk and smoothing household consumption. This unmet need for financial services—combined with unpredictable "shocks" like climate change, illness, and death—means that many poor households are risk averse when it comes to spending on new products and services.

There are a host of other constraints encountered by low-income households in developing countries (and in many developed countries). In this chapter, we will outline three classes of constraints that the development engineer should consider: (1) market constraints; (2) institutional constraints; and (3) behavioral and social constraints. You may not encounter all of these constraints in a given project, and those encountered may not be binding (meaning that they may not be the bottleneck we need to target). However, they are useful as diagnostic and design tools (Hausmann et al., 2008), and they can help explain why technologies that have worked in "developed" settings may fail when transplanted to a new setting.

These constraints can be considered as design requirements, but they can also be direct targets for technological innovation. Where markets fail to meet the needs of poor households, there may be a technology—say, the capture of real-time information on prices—that can level the playing field for disadvantaged consumers. When institutions have been captured by elites (creating conditions for inequality), there may be innovations that enable the decentralization of assets or force transactions to be more transparent to citizens. While some of these problems require policy reform, others may be amenable to technological intervention.

2.1 Market Constraints

Markets are the mechanisms through which goods and services are produced, distributed, and consumed; and well-functioning markets can generate clear signals of supply and demand, partially transmitted in the form of prices. In reality, all markets operate imperfectly, and every country suffers from market distortions (or "failures") that result in the inefficient allocation of resources. Yet the developing world is particularly complex.

In most developing countries, the economy is dominated by the informal sector, which consists of market activity that is not organized, monitored, or regulated by government. This informality, combined with challenges like weak infrastructure and high transport costs, inhibits the development of modern, market-based economies. Informality also reduces government tax revenue and the state's ability to redistribute resources through public benefits programs. As a result, markets in developing countries often fail to efficiently allocate the supply of goods and services to those with greatest demand.

Informality may enhance resilience in some communities and contexts; however, it also intensifies the uncertainty that poor households already deal with. Understanding informal and imperfect markets—and anticipating their effect on the performance and sustainability of a technology—is key to designing a product or service that will achieve development impacts. A brief summary of common market constraints is outlined in Table 3.1.

2.2 Institutional Failures

Organizations, in particular government bureaucracies and nongovernmental organizations (NGOs), play a critical role in delivering basic services to people in developing countries—from water, sanitation, and education to pensions and social protection schemes. Many low-income households rely on these formal institutions, whose operations are guided by written rules and laws, for their welfare. At the same time, people in resource constrained settings also rely on informal arrangements[1], like social networks, based on kinship or caste for accessing services. For example, it is common for villagers in rural settings to finance loans or emergency support from family members or money lenders within the village.

Both formal and informal institutions can introduce inefficiencies and distortions in implementation of new technologies or policies (Helmke & Levitsky, 2006). For example, ethnic or provincial community leaders may hold socially important positions in communities, limiting the power of government appointees to manage local affairs. In the absence of effective community oversight, these local leaders can control the functioning of the state apparatus and capture public goods for private benefits (a process known as "elite capture"; see Bardhan & Mookherjee, 2000). Governments may find that their policies fail to achieve outcomes for disadvantaged communities, or that outcomes differ from a policy's stated objectives. For example, states in resource constrained settings tend to generate less tax revenue than targeted, due to weak collections and audit capacity as well as missing infrastructure (Acemoglu et al., 2011).

In all parts of the world where formal institutions are inefficient or weak, informal institutions remain relevant and effective at meeting the needs of citizens. Indeed informal institutions, like their more codified counterparts, can establish and enforce rules, negotiate disputes, distribute shared resources, and constrain social behavior. However, informality is also challenging for the scale-up of a technology: Informal institutions often follow tacit rules, known only to "insiders." By their very nature, informal norms and institutions (particularly those without written record) require context-specific understanding. Researchers who want to successfully implement and scale up new technologies need to invest time and resources in trying to understand how informal institutions behave. A few examples of commonly encountered constraints are given in Table 3.2.

---

[1] See Helmke and Levitsky (2006) for understanding the role of informal arrangement. They describe informal institutions as "created, communicated, and enforced outside of officially sanctioned channels."

Table 3.1 Examples of market constraints

Lack of insurance (risk transfer) markets: People living in poverty face a diversity of risks, made worse by a lack of formal insurance products. This naturally reduces the appetite for risk-taking. Even innovations that demonstrably improve welfare may seem too risky for upfront investment by households, which is why money-back guarantees, free trials, and warranties can be useful (Fuchs et al., 2020). Unmitigated risk is a particularly important issue for agricultural businesses, because actuarially-priced crop insurance is still too expensive for most small-scale farmers (Cole & Xiong, 2017). Without insurance, the investment in a yield-enhancing technology can be lost to unexpected floods or drought.

Capital constraints and weak credit markets: Firms and households in low-income communities often lack access to the upfront capital needed to invest in a new technology. They may lack savings, or they may face high interest rates for credit (Banerjee and Duflo 2007). In part this is because lenders incur fixed costs when servicing loans, and partly it is because of asymmetric information and high default rates: people who are unbanked (or underserved by formal financial services) lack conventional credit histories, which makes it difficult for lenders to assess creditworthiness. When developing a solution that requires upfront costs, researchers may need to design smart subsidies, cost-sharing arrangements, or innovative financing to be deployed alongside the solution.

Missing information: Buyers in low-income settings may lack access to information about the products available in markets, particularly if they have limited literacy or live in remote areas. This missing information reduces agency and can prevent households from adopting affordable technologies that could improve their welfare outcomes (Fuglie et al., 2019). Sellers in these settings may also lack access to information, including demand signals, as a result of missing market data. While digital receipts are the norm in developed countries, these enabling technologies have not yet penetrated the majority of small merchants in developing countries. As a result, sellers may not have the consumer insights needed to stock the right inventories. Finally, SMEs often have weak knowledge of management practices; where such practices are widely adopted, they can increase the efficiency of production and trade (McKenzie, 2020). An example is the adoption of improved agricultural practices by smallholder farmers; farmers may lack information about how to optimally weed or manage pests, and simply providing information provision can increase yields (Fabregas et al., 2019). Of course it matters how the information is presented and many have observed that conventional trainings for farmers or SMEs fail to improve outcomes (Bridle et al., 2020, McKenzie, 2020).

High transaction costs: For remote households, the travel to markets to buy a product can be prohibitively costly, even where subsidized public transportation is available. In addition to transport costs, households may face high opportunity costs when accessing certain products; losing a day's income on travel to a distant market can have serious implications for daily earners. In addition, there are the costs of searching for the right product to meet needs; it can be difficult to gather information about product quality and prices, since they are often opaque or negotiable in informal markets. Collectively, these transaction costs can put welfare-enhancing technologies outside the reach of low-income communities.

High transport costs and shallow markets: Lack of physical infrastructure (like warehousing and roads) increases the cost of transporting goods. This can make it prohibitively expensive to transport goods to market, particularly those requiring refrigeration. These high transport costs make it difficult for buyers and sellers to enter into transactions. As a result, small rural markets are often insulated from larger markets, leading to price variation and spikes in supply and demand. Middlemen with access to transport often exploit price variations as arbitrage opportunities; this undermines more inclusive development by pushing profits for agricultural production toward traders, rather than producers.

(continued)

Table 3.1 (continued)

Labor market failures: Inefficiencies in the labor markets of the developing world are driven in part by the high costs of job search for would-be workers, as well as asymmetric information between employers and job seekers. It is difficult for job candidates to signal their skills and experience, in part because of unregulated of unregulated vocational training firms and firms and counterfeit training certifications. There are also lumpy labor supply cycles in agricultural settings, due to harvest cycles: for parts of the year, there is too much labor, and at other times, there is too little. Seasonal migration from rural to urban centers can partially address this challenge, but there are high upfront costs for laborers looking to migrate. These frictions in labor markets make it difficult for businesses to grow, consolidate, and achieve economies of scale (Nilsson, 2019). For the developer of a new technology, labor market failures can also introduce difficulties in establishing the technical workforce needed to operate, support, and maintain a solution.

Input and output market inefficiencies: To maintain profitability, firms must sustain business operations (or other productive activities) over time. However, interruptions and inefficiencies in business operations are often introduced by upstream and downstream failures. For example, in agriculture, a lack of reliable access to fertilizer and other inputs can reduce crop yields, while a lack of access to output markets reduces farmers' bargaining power over the prices they receive. Traffic and transport delays can introduce uncertainty into the delivery times for key inputs and outputs, resulting in wastage and other inefficiencies. Also an issue is the seasonality (or temporal variation) of input and output markets, particularly in agriculture and related businesses. Seasonality introduces time sensitivity, by requiring that inputs or outputs be available at specific points in time.

Market-distorting policies: Even well-intentioned government policies can create distortions in how markets function. For example, governments often procure staple crops from farmers at a minimum support price. This can limit the development impact of technologies that improve farm-to-market supply chains, since higher input costs are not recovered from market prices (Jayne et al., 2013). Government subsidies, taxes, and mandated commodity pricing can all introduce inefficiencies in markets and need to be accounted for by researchers developing technologies that leverage market-based processes for successful implementation

Lack of quality grading: The lack of standards and certifications in informal and less developed markets—for example, the lack of technology to grade the quality of agricultural produce—can affect the price received for goods and services (Bernard et al., 2017). When quality information is not signaled, the market does not return the expected premium to producers of higher quality goods. As a result, the incentive to provide higher quality goods is eliminated, and producers are less likely to adopt quality-improving technologies.

Missing human capital: A common challenge in low-resource communities is underinvestment in people's skills, knowledge, and, which begins at primary school and carries through to higher education. For people's skills, knowledge, and education, this can mean a lack of access to trained workers to produce, distribute, or support a product. It may also mean that you need to invest more in onboarding users and building their confidence in using a new product. Ultimately, missing skills and expertise can limit the ability of new technologies to achieve impact and can affect the efficiency of firms, particularly in areas of management (McKenzie, 2020).

Table 3.2 Examples of institutional constraints

Elite capture: A pervasive challenge in developing countries is the capture of formal (and informal) institutions by elites. Elites capture resources for private benefit rather than for (intended) public use (Bardhan & Mookherjee, 2000). Elite capture is also linked to political patronage networks where resources are used to monitor and control how citizens and communities elect political representation (Stokes et al., 2013; Kitschelt & Wilkinson, 2007). Thus, officials may often be unable to adhere to rules and guidelines while implementing public policies.
Intermediaries: Intermediaries are pervasive lower-level actors that use private information or networks to impose "rents" on individuals seeking their rightful access to public services (Bertrand et al., 2007). This takes the form of bribes or speed-money paid by citizens to get their requests processed by an institution or bureaucracy. Intermediaries rely on social norms and strong social networks to help citizens get access to government services (Witsoe, 2012). The presence of intermediaries is also linked to the inability of citizens to directly hold the bureaucracy accountable. There is an analogous network of intermediaries in markets, for example agricultural middlemen who use private information about market prices to exploit smallholder farmers (Goyal, 2010).
Weak contracting environments: In resource-constrained contexts, the ability to enforce contracts is often weak or absent, due to limited resources invested in courts and other mechanisms of conflict resolution. Thus, the process of seeking judicial redress for lack of adherence to contracts can take years if not decades. Even contracts with governments may not be implemented properly due to lack of transparency, red tape, or corruption (Gupta, 2012). This introduces uncertainty into business transactions and can reduce trust in formal institutional processes
High transaction costs: There are often heavy user costs for "free" services provided by government agencies. Because formal institutions in developing countries are often inefficient, they often impose costs on those seeking access (in the form of high transport costs to reach a government agent, long delays in the administration of benefits, or expensive documentation required to qualify for benefits). Thus, many citizens avoid these high transaction costs by seeking the help of intermediaries to get government approvals.
Principal-agent problems within government: Principal agent problems arise when the goals of the "principal" (a person or group with authority) are misaligned with the incentives of the "agent" performing a service on the principal's behalf. Governments regularly face this challenge, for example, when programs formulated by political leaders need to be implemented by local officials. Ensuring that the local official implements the program appropriately (rather than shirking duties, skimming resources, or altering protocols) requires sophisticated monitoring systems. Without the ability to monitor the principal cannot ensure that agents are performing tasks that align with the goals. The lack of monitoring results in absenteeism among officials and increases the corruption and leakage from public programs resulting in poor performance in the government's delivery of public services (Pande, 2020).
Principal-agent problems among citizens: Voters are the principal (they are the ultimate source of authority). Yet they can only hold politicians (the agents) accountable at the ballot box, which gives just one opportunity for accountability every 4–5 years. In the interim, communities should have access to effective grievance redressal systems—channels through

(continued)

Table 3.2 (continued)

which citizens can voice their dissatisfaction with government. Yet, these are often unavailable to disadvantaged communities, who may also lack the information and incentives to engage in political action (Pande, 2020). The same is true of civil society. A well-funded civil society (including watchdog groups, advocacy organizations, universities, and religious institutions) can act to hold the government accountable. In wealthier countries, this is funded by a combination of philanthropists, government grants, and individual citizens—as in the case of news subscriptions supporting independent journalism. There is limited civil society capacity in many developing countries, and within most disadvantaged communities.
Asymmetric information (and disinformation): The lack of literacy and education limits the extent to which written information can be expected to spread across different segments of a community or society. While the advent of smartphones and greater Internet penetration is rapidly changing how information is disseminated. Individuals may lack the ability to seek out credible information sources, due to limited social networks or lack of connection to government decision-makers. This can result in rapid spread of misinformation (Badrinathan, 2021).
Collective action failures: Cooperation among individuals, households and firms may be required to capture economies of scale, for example through bulk purchasing, collective marketing, or industry-wide standardization. Coordination is also important to increase bargaining power (e.g., labor unions among workers, farmer cooperatives for smallholders), invest in public goods (e.g., infrastructure and education), and manage shared resources (e.g., water for irrigation). Disadvantaged or marginalized groups often find it difficult to bargain collectively for their rights; they may also struggle with cooperation due to cultural factors (Ostrom, 2010). Part of the challenge is due to limited social capital. Societies with limited social capital (i.e., reduced trust and sense of reciprocity among individuals) find it difficult to leverage collective action. Ethnic diversity also contributes towards the challenge of collective action, with more homogeneous societies being able to overcome collective action challenges (Habyarimana et al., 2007)
Weak social capital: In communities that have experienced colonialism, civil war, and other forms of violence, there is often a lack of trust among individuals. This loss of social capital leads to lack of trust in institutions. Individuals may not view institutions as fair, or impartial, and working in the common public interest. Limited social capital can reduce the ability to act collectively (Nunn and Wantchekon, 2011).

2.3 Social Norms and Behaviors That Constrain Development

Communities and individuals living in poverty face unique behavioral and cognitive constraints that affect their decision-making about technology. Further, social norms at the community and the household level also have the potential to shape decision-making around technology adoption in resource constrained settings. Some of these are outlined in Table 3.3.

Table 3.3 Examples of norms and behavioral constraints

Cognitive biases: While lack of information impacts how individuals in resource-constrained settings make economic decisions (about their jobs, investment in children's education, household expenses, savings, etc.), this alone does not predict a household's decisions. A large body of research from psychology and behavioral economics suggests that decision-making is influenced by cognitive biases that emerge under the pressure of resource constraints (Mullainathan, 2013; Banerjee & Duflo, 2011, Kremer et al., 2019). These biases are associated with cognitive scarcity or to reduced attention available to allocate to tasks (given high levels of uncertainty and routine income shocks that poor people must weather). Cognitive biases can prevent or slow adoption of promising technologies, and researchers implementing new technologies should take into account the biased nature of decision-making (Mani et al., 2013)
Intra-household bargaining: While products or technological innovations are targeted towards individuals, the decision to adopt a product is often taken at the level of the household. The nature of intra-household bargaining has a large and adverse impact on women, who often lack the economic and social agency to make the decision to adopt new technologies without the consent of other household members (elders like mother-in-law or husband). For example, while mobile phones have rapidly spread across developing countries access to them is often gendered with men being able to access them more easily (Joshi et al., 2020; Women, 2018). Intra-household bargaining and women's limited control over resources need to be accounted for when designing new technologies
Social norms: Individual decisions are heavily constrained by societal norms and rules, especially in communities where individuals have limited anonymity and interact with other members on a regular basis. In such settings the decision to adopt a new product can both be sanctioned or influenced by community norms. For example, in poor villages people with limited resources take loans to organize a lavish wedding in order to meet expectations from the community (Banerjee and Duflo, 2007). Similarly, in Brazilian communities social norms related to family size were induced by telenovelas, which depicted the lives of female characters with small families (La Ferrara et al., 2012).
Social learning: Learning and gathering information is often a social process. Women's self-help groups have often been used to disseminate products (e.g., microfinance interventions or sanitary products are often introduced through SHGs), since they facilitate social learning In addition, social learning is playing an increasing role in the adoption of agricultural technologies (BenYishay and Mobarak, 2019).
Aspirations: Aspirations are affected by societal expectations, peer pressure, and sanction within communities. These mental constructs can influence individuals' decisions, including about investments of effort, attention, and other limited resources (Genicot and Ray, 2020)
Mental models: There are assumptions about what researchers know and what power they have, derived from people's experiences with colonialism, inequality, and poverty. These experiences are internalized and can deprive individuals of agency and a sense of self-efficacy, instead endowing outsiders with authority (Wuepper and Lybbert, 2017).
Mood disorders (depression): Mounting evidence suggests that poverty is a driver of depression, which has disabling effects on people's well-being, productivity, and ability to provide care for others (Ridley et al., 2020). In some cases, it may be useful to design a new solution with the expectation that end-users are suffering from depression, which can affect decisions and behaviors

Box 3.1: Reviews of Market, Institutional, and Behavioral Constraints
Below is a collection of practical white papers outlining the constraints faced by households, institutions, and markets in developing economies. These are organized by sector, making them a useful resource for engineers and development practitioners. Additional reviews of the evidence from international development research can be found at the International Initiative for Impact Evaluation (http://3ieimpact.org).

Agriculture

Experimental Insights on the Constraints to Agricultural Technology Adoption (2019), accessed at https://escholarship.org/uc/item/79w3t4ds
Market inefficiencies and the adoption of agricultural technologies in developing countries (2013), accessed at https://escholarship.org/uc/item/6m25r19c

Governance

Governance Initiative Review Paper, J-PAL Governance Initiative (2019). J-PAL Working Paper, accessed at https://www.povertyactionlab.org/sites/default/files/review-paper/GI_review-paper_2019.pdf

Digital Identities

Digital Identification & Finance Initiative Africa: An Overview of Research Opportunities (2019). J-PAL Working Paper, accessed at https://www.povertyactionlab.org/sites/default/files/review-paper/DigiFI_framing-paper_june-2019.pdf

Post-Primary Education

Expanding access and increasing student learning in post-primary education in developing countries: A review of the evidence (2013). J-PAL Working Paper, accessed at https://www.povertyactionlab.org/sites/default/files/2020-03/PPE_review-paper_executive-summary_2013.05.07.pdf

Gender and Women

What Works to Enhance Women's Agency: Cross-Cutting Lessons from Experimental and Quasi-Experimental Studies (2020). J-PAL Working Paper, accessed at https://www.povertyactionlab.org/page/what-works-enhance-womens-agency

Labor Markets

Reducing search barriers for job seekers (2018). J-PAL Policy Insights, accessed at https://doi.org/10.31485/pi.2234.2018

(continued)

Box 3.1 (continued)
Urban Services
Improving Access to Urban Services for the Poor: Open Issues and a Framework for a Future Research Agenda (2012). J-PAL Working Paper, accessed at https://www.povertyactionlab.org/sites/default/files/2020-03/USI_review-paper.pdf

Youth and Employment

J-PAL Youth Initiative Review Paper (2013). Abdul Latif Jameel Poverty Action Lab Working Paper, accessed at https://www.povertyactionlab.org/sites/default/files/documents/YouthReviewPaper_March_2013_0.pdf

J-PAL Skills for Youth Program Review Paper (2017). Abdul Latif Jameel Poverty Action Lab Working Paper, accessed at https://www.povertyactionlab.org/sites/default/files/review-paper/SYP_review-paper_2017.pdf

3 Framework for Research

As the previous section outlines, the success of any innovation requires deep understanding of the constraints—market, institutional, and behavioral—that can prevent adoption of a development solution and its impact at scale. These constraints inform the design of technology, and they can also affect our ability to implement high quality research. Years of field work have resulted in a set of best practices that we present here, as a *practical framework* for advancing promising technologies from the lab to the field. The most important guiding principle behind this framework is its emphasis on feedback and iteration and the avoidance of a linear implementation process.

The practice of development engineering focuses on the entire arc of innovation—from problem discovery and technological invention, to prototyping and pilot testing, impact evaluation, and finally adaptation for scale-up. These stages are part of a continuum, and they are not necessarily carried out in sequence. In some cases, the real-world evaluation of an existing product will lead to the design of an entirely new technology, based on iterative feedback from users. The chlorine water dispensers developed by Miguel, Kremer, and colleagues is one example (Null et al., 2012). In other cases, a novel technology will enable measurement of development outcomes at higher frequency or resolution, leading to the discovery of new problems and opportunities (Blumenstock et al., 2016; also see Chap. 15 in this textbook by Wilson).

Each of the following chapters in this textbook will describe a unique research workflow, but they all fit within the framework of four activities: innovation, implementation, impact evaluation, and adaptation for scale (see Fig. 3.1).

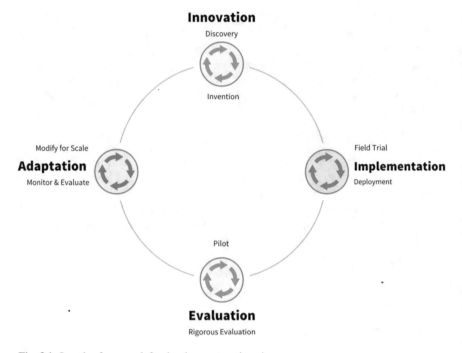

Fig. 3.1 Iterative framework for development engineering

Innovation

Innovation is at the heart of every development engineering intervention or solution. It is the process of discovering and characterizing a problem and then developing a generalizable technological solution—one that can address the challenge at scale. The innovation can lie in adapting technology to solve a new problem (e.g., by bundling an emerging or existing technology with a novel economic, political, or behavioral intervention); or it might lie in designing an entirely new technology around any of the constraints outlined earlier in this chapter. The innovation may even lie in creating new ways to measure development outcomes, either through instrument design (like a wireless cookstove sensor) or the design of new analytic techniques (like the use of remotely sensed imagery to predict household asset wealth).

However, the discovery of a suitable problem, and the development of the design requirements for a solution, is never linear; it requires a critical and evolving understanding of local context. As an example, we consider the design of a treatment system for removal of arsenic from drinking water. First, we investigate the experiences and environment of households affected by arsenic contamination (mostly people living in rural Bangladesh and Eastern India). In this context, shallow tube wells are the main source of drinking water, household asset wealth and consumption are low, and willingness to pay for Arsenic-safe water is lower than willingness to pay for piped water (Khan et al., 2014). Perhaps there are collective action failures in the maintenance of existing water infrastructure, and community trust of outside commercial providers is limited (Alfredo and O'Garra, 2020).

The engineer's goal is to solve the problem of access to clean water, using a combination of technological and socio-economic innovations to overcome these hypothesized constraints. The solution must address market failures, institutional challenges, and the preferences and behaviors of people facing arsenic poisoning—in addition to their public health needs. These constraints can be shaped into the solution's design space, in the form of performance parameters (e.g., failure tolerance, reliability, salience, desirability, cost, accessibility). It is also important to anticipate any negative externalities created by the innovation, such as wastewater production and environmental contamination.

Once the problem has been defined, the constraints characterized, and a solution posed and prototyped (often just on paper), the researcher can begin to articulate a theory of change: a set of hypotheses about how the proposed solution overcomes observed constraints. This is the key output of the innovation activity: prototype brainstorming and a set of hypotheses that are to be tested. But how do we get here?

To develop basic insights about a community's development challenges, researchers often use qualitative approaches like ethnographic observation and human-centered design (HCD).[2] Much has been written about these methods, and we refer readers to a few resources in Box 3.1. In addition to these methods, development engineers often consult existing data and survey research to understand their targeted communities. Well-designed surveys can offer a quantitative and representative view of users' perceptions and preferences. Examples include nationally representative datasets like the Living Standards Measurement Surveys (Grosh & Glewwe, 1995) and the Demographic and Health Surveys (Corsi et al., 2012), as well as survey research projects published in journals of development economics and political economy. In addition, national statistical offices in most countries publish census data at intervals, and they may release other de-identified administrative datasets. Large nongovernmental organizations (NGOs) often publish their own datasets, in areas like education (Banerji et al. 2013; Mugo et al., 2015), health (Murray et al., 2020), and politics (Afrobarometer Data). Resources like these are described and referenced in many of the chapters in this book.

Of course there are many challenges in collecting reliable survey data in low-resource settings (Iarossi, 2006), and these have been deeply documented across a range of domains—from health, nutrition, and gender to welfare and politics (Caeyers et al., 2012; Glennerster et al., 2018; Lupu & Michelitch, 2018). Issues

---

[2] Human-centered design (HCD) is a pervasive new approach that offers detailed toolkits for designing solutions that are participatory and attend to human values (Holeman and Kane, 2020). It often encourages practitioners to set aside knowledge of existing solutions, instead entering the design process with as few assumptions as possible. It focuses on listening and responding to users and building empathy for their lived experiences. However, users can provide incomplete, biased, or irrelevant information, and there is growing concern that these newer approaches lack evidence of effectiveness, despite their popularity (Robertson & Salehi, 2020; Sloane et al., 2020; Thomas et al., 2017).

with survey data include sampling errors, respondent biases, and small sample sizes (because budgets for survey research are always limited). There are other sources of unexplained variation in survey data, as well—artifacts that are introduced throughout the surveying process. These range from the selection of interviewers (West & Blom, 2017) and nuances in the wording or order of survey questions (Blair et al., 2020) to the length of a survey and how you compensate study participants (de Weerdt et al., 2020).

To improve the reliability and reproducibility of survey results, some researchers carry out intensive qualitative research with random samples of individuals surveyed in quantitative surveys, particularly in cases where survey questions address sensitive issues like corruption, crime, or other risky behaviors (Blattman et al., 2016). Some researchers now publish their survey protocols or use published questionnaires that have been validated against more reliable methods of data collection (Meyer et al., 2015). Still, there are few repositories that allow you to browse and search for questionnaires by geography, population, or topic. You will often need to sift through the supplementary materials published as part of academic journal articles (in fields as diverse as sociology, anthropology, economics, political science, and public health) to discover existing survey instruments.

To supplement survey datasets, researchers increasingly leverage digital technologies that capture complementary and (in some cases) less subjective information about context within their communities of interest. For example, financial transactions using mobile phones or debit cards can offer a view into consumer behavior including purchasing patterns (Bachas et al., 2017), loan repayments (Björkegren & Grissen, 2018), and social insurance mechanisms (Blumenstock et al., 2016). Remotely sensed data—like satellite imagery or drone video footage— allow us to directly observe agricultural yields and management practices from the sky (Lobell et al., 2020). Data extracted from social media platforms can expose relative poverty (Fatehkia et al., 2020) as well as popular sentiment and prevailing social norms, using natural language processing to automate analysis (Calderon et al., 2015). Technology companies like Facebook have begun leveraging internal, georeferenced user interaction logs to produce a range of datasets, from human population density to international firm surveys (Stevens et al., 2019; Schneider, 2020). Anonymized call detail records and geolocation data from mobile phones can also reveal household outcomes—from consumption patterns and asset wealth to migration decisions and response to violence (Aiken et al., 2020; Blumenstock et al., 2015; Chi et al., 2020; Blumenstock et al., 2018).

Insights obtained using these large-scale datasets can be useful to understand context and also to measure development outcomes. However, few of these methods have been extensively validated against "ground truth." In addition, these digital data sets carry their own biases, for example, based on who has access to mobile technology or on which communities have been surveyed enough to train a machine learning algorithm based on satellite imagery.

Ultimately, the data take us only so far. It takes decades to build deep knowledge of the development constraints facing any country or community. It requires knowledge of domestic and regional politics and economic history; it requires familiarity with local views about colonialism and its legacies. It also requires understanding a nation's struggles with ethnic and gender identity. Perhaps this creates a natural imperative to collaborate with researchers, policymakers, and civil society organizations based in the communities you wish to empower. Inclusive, respectful partnerships with local actors are key to many successful development engineering projects, and this success relies on the alignment of incentives for all participants.

Box 3.2: Resources for Defining Development Challenges

Qualitative and Design Methods

- Field work for development research (Scheyvens, 2014)
- Qualitative research methods (Marshall & Rossman, 2014)
- Participatory action research (Kindon et al., 2007)
- Human-centered design methods (Holeman & Kane, 2020)

Quantitative survey research

- Academic research surveys of specific populations, including in-person enumerated surveys or mobile surveys (Rossi et al., 2013)
- Nationally representative surveys and census data (Nikolov, 2009)
- International surveys like the demographic and health surveys (Corsi et al., 2012) and Living Standards Measurement Study (Grosh & Glewwe, 1995)

Administrative Data

- Transactions records generated in the delivery of services by health systems, schools, government agencies, cooperatives, and other public organizations (Meyer & Mittag, 2019)
- Customer transaction records generated by retailers and other private sector firms (Di Clemente et al., 2018)

"Big" Data

- Social media traffic scraped from public sources or accessed through agreements with technology firms (Calderon et al., 2015)
- Satellite imagery, including public and private assets (Jean et al., 2016)
- Anonymized call detail records accessed through agreements with mobile network operators (Blumenstock et al., 2015)
- Geolocation data captured through consumer smartphone applications (Williams et al., 2015)
- Networked sensors (e.g., personal activity monitors, grid electricity sensors, and precision agriculture devices) that capture large volumes of environmental or behavioral data (Ramanathan et al., 2017)

Implementation

Implementation is the process of piloting a new innovation, monitoring its technical performance in the field, and understanding the factors that influence "effective" implementation. This is in sharp contrast with the linear (and often idealized) model of the engineer moving directly from invention to impact. An iterative approach is required—a multistage progression from invention, to pilot, to full deployment—with feedback loops. We must frequently return to earlier stages of our research to update our hypotheses and theories of change. Each assumption is revisited in light of the data and insights gained from the previous iteration. This iteration—the process of advancing, updating, and retrialing—can start within the lab. But it is also part of the move from the lab to field trials or the expansion from one market to another.

In industrialized countries, such iterative loops are relatively commonplace in product development. Technology firms have specialized teams focused on product marketing, user interaction, design, product management, engineering, quality control, sales and growth, and financing. Because these activities are well-resourced and have been well-studied, they result in relatively reliable processes of iteration. In the context of development engineering, often the same team is blessed with the burden of taking a technological invention from the lab, all the way to product development, evaluation, and distribution. There are plenty of stumbles and failures along the way. This framework therefore focuses on training the practitioner to cultivate a learning mindset, treating each iteration as a valuable learning opportunity, and remaining ready to investigate and pivot when outcomes diverge from expectation.

The implementation stage does not focus on technological prototypes alone. It also involves the design, testing, and refinement of different business models (or delivery models) for a technological solution. At this stage, it is useful to test hypotheses about users' willingness to pay or about how they access information. This is a good time for using experimental methods that reveal the demand for a new product or service, including pricing experiments as well as behavioral games (like take-it-or-leave-it studies) that can reveal people's preferences (Dupas et al., 2013).

At this stage, the development of sound partnerships also becomes paramount. Innovating in a resource-constrained setting can be challenging because it often requires coordination across multiple partners, each with differing standards, norms, and incentives. One partner may support small-scale manufacturing, while another carries out field testing. Still other partners may be needed to collect user feedback or implement rigorous evaluations. The researcher is often reliant on local partners to understand the local context and effectively implement studies, deployments, and experiments. These partners may deprioritize the project or deviate from agreed plans, because of internal challenges or as a response to the external environment.

The weakness of government or community institutions can also constrain the ability to implement a project effectively. Piloting of new technologies may be regulated by governments, and research involving humans is always overseen by local review boards. Yet a lack of transparency can make it challenging to obtain the necessary permissions for experimentation. Researchers must learn local processes and find ways to overcome institutional challenges. Iterative implementation allows the researcher to discover the optimal implementation strategy over time, learning how the solution will ultimately perform in the targeted setting.

Evaluation

The evaluation component of the development engineering framework focuses on using scientific approaches (i.e., randomized controlled trials and quasi-experimental methods) to isolate the causal impacts of an innovation. We are interested in understanding the effects of a new solution on human and economic development; we also want to accumulate knowledge that can generalize to new contexts. Evaluation plays a central role in development engineering, in part because the field is in strong need of better evidence. If we fail to learn from our work, we will perpetuate the "valley of death" between tech-for-good innovations and their successful scale-up.

In this framework, we emphasize evaluations that test hypotheses and rigorously investigate how a technology affects health, economic, and other outcomes in the "real world." We are also interested in exposing barriers to technology adoption, and testing our theoretical models about the optimal delivery of a technology. In some cases, we may want to measure spillovers (or unintended consequences) of an innovation or to assess long-term effects. While a short-term evaluation might show strong initial take-up, later follow-ups often expose disuse, environmental costs, and other failures that dampen benefits.

We may use surveys to collect self-reported outcomes, or we may use sensors and other networked devices to automate the monitoring of outcomes. We may choose to instrument our solution, so that ongoing evaluation is incorporated into operations. Regardless, by designing your evaluation carefully, you can investigate whether the failure of a technological solution stems from technology design itself or from the delivery model. If the failure is due to a flawed business model, evaluation results can be used to modify pricing or design a new financing scheme. If the failure has to do with technology design—for example, if the volume of human waste collected from a community is too limited to support continuous urea extraction—the results of evaluation can be used to refine design parameters and develop a more appropriate solution. A good evaluation allows for design iteration—with researchers incorporating the feedback into redesign—and also yields generalizable knowledge that can be applied in new contexts.

While many researchers will evaluate their solution at pilot scale, with careful control over implementation, there can also be value in deploying and evaluating a technology at large scale (e.g., nationwide). This teaches us something about the effectiveness of a solution at scale, when it is implemented under less controlled conditions (see Chap. 20, on evaluation of *Aadhaar*, India's national digital authentication program). There is also value in evaluating a solution deployed across multiple contexts, in tandem, through portfolios of field experiments that test a shared hypothesis. This teaches us about the variations in effectiveness across conditions and can also reveal the sorts of adaptations that are required for an innovation to succeed at large scale (ref meta-keta, WSH eval).

Adaptation

Scaling up involves taking an innovation from the evaluation stage (with evidence of positive impact, albeit on a limited number of users) and adapting it to reach a larger number of users and to reach users in new geographies. The process of maintaining a technological solution at scale comes up with its own unique challenges. For example, for scale-ups that rely on market processes, the business model becomes a critical factor determining long-term sustainability. There will be challenges in managing deep supply chains, which requires strategies for mitigating risks from the market frictions commonly found in developing countries. Operations and mainte-nance, along with monitoring for quality assurance, will require critical attention. In addition, customer "success" may require not only technical support but also costly investments in user training or onboarding, particularly for communities that have not interacted extensively with your class of technologies. Some innovations will require intellectual property (IP) protections to ensure broader use and scale-up. However, IP regimes in resource poor countries may be poorly designed or weakly implemented, making IP protection a risky choice for researchers.

For scale-ups implemented in partnership with governments, it is key to navigate the political economy of institutions (and the incentives of those with vested interests) as well government regulation, legal challenges, and the role of civil society in oversight. Further, public institutions responsible for implementing or disseminating services may fail to adhere to the researcher's well-defined standards, which can compromise the fidelity of implementation. This is an issue in many under-resourced communities, where there are high rates of absenteeism among frontline workers who are also overburdened with administrative tasks and responsibilities (Finan et al., 2017). The prevalence of corruption, along with weak monitoring of government workers, can also make implementation of projects a challenge.

The success of scale-up efforts is closely linked to the concepts of evaluation and iterative implementation. Evaluations conducted as part of small-scale field pilots allow the researcher to understand the challenges of implementation and gather evidence of a solution's impact. Ideally these evaluations also reveal the mechanisms through which a product acts and expose any required or enabling conditions. This generalizable knowledge enables scale. As we move from the pilot context to a larger scale or from one country to another, we can then test whether the conditions for intervention success are found in new target environments and target scaling efforts where the innovation is most likely to achieve impact (Bates & Glennerster 2017).

Several successful examples of scale-up, from chlorine dispensers to provide clean water to deworming tablets, were first piloted at a smaller scale. There are also examples where attempted scale-up without evaluation led to failure. For example, the Embrace infant warmer developed to work in poor countries with

Fig. 3.2 The path of development engineering

limited healthcare facilities (Pg 71, Jugaad) proved effective in pilots but failed to find traction after initial adoption.

The case studies that follow are written to tie these processes—innovation, iterative implementation, evaluation, and adaptation—together. They demonstrate how feedback from one stage informs the next. The framework (and this textbook) will also undergo iteration, as new ideas are incorporated over time (Fig. 3.2).

4 Additional Resources

In addition to this textbook, there is an expanding pool of resources available to researchers in Dev Eng. These include the open access peer-reviewed journal *Development Engineering: the Journal of Engineering in Economic Development*. This journal publishes original research across multiple areas of Dev Eng, including:

- Engineering research and innovations that respond to the unique constraints imposed by poverty
- Assessment of pro-poor technology solutions, including field performance, consumer adoption, and end-user impacts
- Novel technologies or tools for measuring behavioral, economic, and social outcomes in low-resource settings
- Lessons from the field, especially null results from field trials and technical failure analyses
- Rigorous analysis of existing development "solutions" through an engineering or economic lens

References

Acemoglu, D., Ticchi, D., & Vindigni, A. (2011). Emergence and persistence of inefficient states. *Journal of the European economic association, 9*(2), 177–208.

Afrobarometer Data. Available at http://www.afrobarometer.org.

Aiken, E. L., Bedoya, G., Coville, A, & Blumenstock, J. E. (2020) *Targeting development aid with machine learning and mobile phone data: Evidence from an anti-poverty intervention in Afghanistan*. In Proceedings of the 3rd ACM SIGCAS Conference on Computing and Sustainable Societies (COMPASS '20) (pp. 310–311). New York: Association for Computing Machinery.

Alfredo, K. A., & O'Garra, T. (2020). Preferences for water treatment provision in rural India: Comparing communal, pay-per-use, and labour-for-water schemes. *Water International, 45*(2), 91–111.

Bachas, P., Gertler, P., Higgins, S., & Seira, E. (2017). *How debit cards enable the poor to save more* (No. w23252). National Bureau of Economic Research.

Badrinathan, S. (2021). Educative interventions to combat misinformation: Evidence from a field experiment in India. American Political Science Review, 115(4), 1325–1341.

Banerjee, A. V., & Duflo, E. (2007). The economic lives of the poor. *Journal of economic perspectives, 21*(1), 141–168.

Banerji, R., Bhattacharjea, S., & Wadhwa, W. (2013). The annual status of education report (ASER). *Research in Comparative and International Education, 8*(3), 387–396.

Banerjee, A. V., & Duflo, E. (2011). *Poor economics: Rethinking poverty & the ways to end it.* Random House India.

Bardhan, P. K., & Mookherjee, D. (2000). Capture and governance at local and national levels. *American Economic Review, 90*(2), 135–139.

Bates, M. A., & Glennerster, R. (2017). The generalizability puzzle. *Stanford Social Innovation Review, 2017*, 50–54.

Bertrand, M., et al. (2007). Obtaining a driver's license in India: An experimental approach to studying corruption. *The Quarterly Journal of Economics, 122*(4), 1639–1676.

BenYishay, A., & Mobarak, A. M. (2019). Social learning and incentives for experimentation and communication. The Review of Economic Studies, 86(3), 976-1009.

Bernard, T., De Janvry, A., Mbaye, S., & Sadoulet, E. (2017). Expected product market reforms and technology adoption by Senegalese onion producers. *American Journal of Agricultural Economics, 99*(4), 1096–1115.

Björkegren, D., & Grissen, D. (2018). *Behavior revealed in mobile phone usage predicts loan repayment*. Available at SSRN 2611775.

Blair, G., Coppock, A., & Moor, M. (2020). When to worry about sensitivity bias: A social reference theory and evidence from 30 years of list experiments. *American Political Science Review, 114*(4), 1297–1315.

Blattman, C., Jamison, J., Koroknay-Palicz, T., Rodrigues, K., & Sheridan, M. (2016). Measuring the measurement error: A method to qualitatively validate survey data. *Journal of Development Economics, 120*, 99–112.

Blumenstock, J., Cadamuro, G., & On, R. (2015). Predicting poverty and wealth from mobile phone metadata. *Science, 350*(6264), 1073–1076.

Blumenstock, J. E., Eagle, N., & Fafchamps, M. (2016). Airtime transfers and mobile communications: Evidence in the aftermath of natural disasters. *Journal of Development Economics, 120*, 157–181.

Blumenstock, J., Ghani, T., Herskowitz, S., Kapstein, E. B., Scherer, T., & Toomet, O. (2018). *Insecurity and industrial organization: Evidence from Afghanistan*. The World Bank.

Bridle, L., Magruder, J., McIntosh, C., & Suri, T. (2020). *Experimental insights on the constraints to agricultural technology adoption*. Working paper, agricultural technology adoption initiative. UC Berkeley: Center for Effective Global Action. Retrieved from https://escholarship.org/uc/item/79w3t4ds

Caeyers, B., Chalmers, N., & De Weerdt, J. (2012). Improving consumption measurement and other survey data through CAPI: Evidence from a randomized experiment. *Journal of Development Economics, 98*(1), 19–33.

Calderon, N. A., Fisher, B., Hemsley, J., Ceskavich, B., Jansen, G., Marciano, R., & Lemieux, V. L. (2015, October). *Mixed-initiative social media analytics at the World Bank: Observations of citizen sentiment in Twitter data to explore "trust" of political actors and state institutions and its relationship to social protest*. In 2015 IEEE International conference on big data (big data) (pp. 1678–1687). IEEE.

Chi, G., Lin, F., Chi, G., & Blumenstock, J. (2020). A general approach to detecting migration events in digital trace data. *PLoS ONE, 15*(10), e0239408.

Cole, S. A., & Xiong, W. (2017). Agricultural insurance and economic development. *Annual Review of Economics, 9*, 235–262.

Collins, D., Morduch, J., Rutherford, S., & Ruthven, O. (2009). *Portfolios of the poor*. Princeton University Press.

Corsi, D. J., Neuman, M., Finlay, J. E., & Subramanian, S. V. (2012). Demographic and health surveys: A profile. *International Journal of Epidemiology, 41*(6), 1602–1613.

De Weerdt, J., Gibson, J., & Beegle, K. (2020). What can we learn from experimenting with survey methods? *Annual Review of Resource Economics, 12*, 431–447.

Dupas, P., Hoffmann, V., Kremer, M., & Zwane, A. P. (2013). *Micro-Ordeals, Targeting and habit formation*. Unpublished manuscript.

Fabregas, R., Kremer, M., & Schilbach, F. (2019). Realizing the potential of digital development: The case of agricultural advice. *Science, 366*(6471).

Fatehkia, M., Coles, B., Ofli, F., & Weber, I. (2020). The relative value of Facebook advertising data for poverty mapping. *Proceedings of the International AAAI Conference on Web and Social Media, 14*(1), 934–938.

Finan, F., Olken, B. A., & Pande, R. (2017). The personnel economics of the developing state. In *Handbook of economic field experiments* (Vol. 2, pp. 467–514).

Fuchs, W., Green, B., & Levine, D. I. (2020). Optimal arrangements for distribution in developing markets: Theory and evidence. Available at SSRN 2957288.

Fuglie, K., Gautam, M., Goyal, A., & Maloney, W. F. (2019). *Harvesting prosperity: Technology and productivity growth in agriculture*. World Bank Publications.

Genicot, G., & Ray, D. (2020). Aspirations and economic behavior. Annual Review of Economics, 12, 715-746.

Glennerster, R., Walsh, C., & Diaz-Martin, L. (2018). *A practical guide to measuring women's and girls' empowerment in impact evaluations*. Gender Sector, Abdul Latif Jameel Poverty Action Lab.

Grosh, M. E., & Glewwe, P. (1995). *A guide to living standards measurement study surveys and their data sets* (Vol. 120). World Bank Publications.

Goyal, A. (2010). Information, direct access to farmers, and rural market performance in central India. *American Economic Journal: Applied Economics, 2*(3), 22–45.

Gupta, A. (2012). *Red tape: Bureaucracy, structural violence, and poverty in India*. Duke University Press.

Hausmann, R., Rodrik, D., & Velasco, A. (2008). *Growth diagnostics* (pp. 324–355). Towards a new global governance.

Habyarimana, J., Humphreys, M., Posner, D. N., & Weinstein, J. M. (2007). Why does ethnic diversity undermine public goods provision? *American Political Science Review, 101*(4), 709–725.

Helmke, G., & Levitsky, S. (Eds.). (2006). *Informal institutions and democracy: Lessons from Latin America*. jhu Press.

Holeman, I., & Kane, D. (2020). Human-centered design for global health equity. *Information Technology for Development, 26*(3), 477–505.

Iarossi, G. (2006). *The power of survey design: A user's guide for managing surveys, interpreting results, and influencing respondents*. The World Bank.

Jayne, T. S., Mather, D., Mason, N., & Ricker-Gilbert, J. (2013). How do fertilizer subsidy programs affect total fertilizer use in sub-Saharan Africa? Crowding out, diversion, and benefit/cost assessments. *Agricultural economics, 44*(6), 687–703.

Joshi, A., Malhotra, B., Amadi, C., Loomba, M., Misra, A., Sharma, S., . . . Amatya, J. (2020). Gender and the digital divide across urban slums of New Delhi, India: Cross-sectional study. *Journal of Medical Internet Research, 22*(6), e14714.

Kelley, T. (2005). *The ten faces of innovation: IDEO's strategies for beating the devil's advocate & driving creativity throughout your organization*. Crown Business.

Khan, N. I., Brouwer, R., & Yang, H. (2014). Household's willingness to pay for arsenic safe drinking water in Bangladesh. *Journal of Environmental Management, 143*, 151–161.

Kitschelt, H., & Wilkinson, S. I. (Eds.). (2007). *Patrons, clients and policies: Patterns of democratic accountability and political competition.* Cambridge University Press.

Kremer, M., Rao, G., & Schilbach, F. (2019). Behavioral development economics. In Handbook of Behavioral Economics: Applications and Foundations 1 (Vol. 2, pp. 345–458). North-Holland.

La Ferrara, E., Chong, A., & Duryea, S. (2012). Soap operas and fertility: Evidence from Brazil. American Economic Journal: Applied Economics, 4(4), 1-31.

Lobell, D. B., Azzari, G., Burke, M., Gourlay, S., Jin, Z., Kilic, T., & Murray, S. (2020). Eyes in the sky, boots on the ground: Assessing satellite-and ground-based approaches to crop yield measurement and analysis. *American Journal of Agricultural Economics, 102*(1), 202–219.

Lupu, N., & Michelitch, K. (2018). Advances in survey methods for the developing world. *Annual Review of Political Science, 21*, 195–214.

Lybbert, T. J. (2017). Perceived self-efficacy, poverty, and economic development. Annual Review of Resource Economics, 9, 383–404.

Mani, A., Mullainathan, S., Shafir, E., & Zhao, J. (2013). Poverty impedes cognitive function. *Science, 341*(6149), 976–980.

McKenzie, D. (2020). *Small business training to improve management practices in developing countries: Reassessing the evidence for "Training Doesn'T Work".* Policy Research Working Papers. The World Bank.

Meyer, B. D., Mok, W. K., & Sullivan, J. X. (2015). Household surveys in crisis. *Journal of Economic Perspectives, 29*(4), 199–226.

Mobarak, A. M., & Reimao, M. E. (2020). Seasonal poverty and seasonal migration in asia. *Asian Development Review, 37*(1), 1–42.

Mugo, J. K., Ruto, S. J., Nakabugo, M. G., & Mgalla, Z. (2015). A call to learning focus in East Africa: Uwezo's measurement of learning in Kenya, Tanzania and Uganda. *Africa Education Review, 12*(1), 48–66.

Mullainathan, S., & Shafir, E. (2013). *Scarcity: Why having too little means so much.* Macmillan.

Murray, C. J., Aravkin, A. Y., Zheng, P., Abbafati, C., Abbas, K. M., Abbasi-Kangevari, M., … Abegaz, K. H. (2020). Global burden of 87 risk factors in 204 countries and territories, 1990–2019: A systematic analysis for the Global Burden of Disease Study 2019. *The Lancet, 396*(10258), 1223–1249.

Nilsson, B. (2019). The school-to-work transition in developing countries. *The Journal of Development Studies, 55*(5), 745–764.

Null, C., Kremer, M., Miguel, E., Hombrados, J. G., Meeks, R., & Zwane, A. P. (2012). *Willingness to pay for cleaner water in less developed countries: Systematic review of experimental evidence.* The International Initiative for Impact Evaluation (3iE).

Nunn, N., & Wantchekon, L. (2011). The slave trade and the origins of mistrust in Africa. American Economic Review, 101(7), 3221–52.

Ostrom, E. (2010). Analyzing collective action. *Agricultural Economics, 41*, 155–166.

Pande, R. (2020). Can democracy work for the poor?. Science, 369(6508), 1188–1192.

Ridley, M., Rao, G., Schilbach, F., & Patel, V. (2020). Poverty, depression, and anxiety: Causal evidence and mechanisms. *Science, 370*(6522), eaay0214.

Robertson, S., & Salehi, N. (2020). *What if I don't like any of the choices? The limits of preference elicitation for participatory algorithm design.* arXiv preprint arXiv:2007.06718.

Schneider, J. W. (2020). *Future of business survey methodology note.* Facebook, November 2, 2020.

Sloane, M., Moss, E., Awomolo, O., & Forlano, L. (2020). *Participation is not a design fix for machine learning.* arXiv preprint arXiv:2007.02423.

Soss, J., Fording, R. C., & Schram, S. F. (2011). *Disciplining the poor: Neoliberal paternalism and the persistent power of race.* University of Chicago Press.

Stevens, F. R., Reed, F., Gaughan, A. E., Sinha, P., Sorichetta, A., Yetman, G., & Tatem, A. J. (2019, July). *How remotely sensed built areas and their realizations inform and constrain gridded population models*. In IGARSS 2019-2019 IEEE international geoscience and remote sensing symposium (pp. 6364–6367). IEEE.

Stokes, S. C., et al. (2013). *Brokers, voters, and clientelism: The puzzle of distributive politics.* Cambridge University Press.

Thomas, V., Remy, C., & Bates, O. (2017, June). *The limits of HCD: Reimagining the anthropocentricity of ISO 9241-210*. In Proceedings of the 2017 workshop on computing within limits (pp. 85–92).

West, B. T., & Blom, A. G. (2017). Explaining interviewer effects: A research synthesis. *Journal of Survey Statistics and Methodology, 5*(2), 175–211.

Witsoe, J. (2012). Everyday corruption and the political mediation of the Indian state: An ethnographic exploration of brokers in Bihar. *Economic and Political Weekly*, 47–54.

Women, G. C. (2018). *The mobile gender gap report 2018*.

Chapter 4
Asking the "Right" Questions

Temina Madon ⓘ and Kentaro Toyama

1 The Paradox of the Development Engineer

In any problem-solving endeavor, identifying the right problem and asking the right questions is at least half the challenge. A well-posed problem can suggest an obvious, effective solution, while a poorly chosen problem can lead to dead-end non-solutions that leave no one better off. In this chapter, we consider important questions that should be asked with respect to potential beneficiaries or collaborators, the larger context of a problem, the type of impact, approaches to scale, and ethical considerations.

To begin with, it is worth recognizing that very few people involved with development engineering are intentionally trying to do the wrong thing. If anything, it is the opposite – we are involved because we hope to improve the world in some way. But, international development and development engineering have plenty of critics, too (Sainath, 1996; Collier, 2007; Easterly, 2007; Dambisa, 2009; Morozov, 2011; Toyama, 2015). After three quarters of a century and over $2 trillion dollars of aid, few low-income countries have crossed the threshold into middle income (Collier, 2007; Dambisa, 2009). Despite decades of innovation with cook stoves, cold chains, and communication technology, many developing world communities are not healthier, more empowered, or more informed as a result (Easterly, 2007; Morozov, 2011; Toyama, 2015). Whether we agree with the critics or not (see, e.g., Kenny, 2012; Sachs, 2005), there can be value in engaging with their critique that technological innovation has not always delivered on its socioeconomic promises.

T. Madon (✉)
Center for Effective Global Action, University of California, Berkeley, Berkeley, CA, USA
e-mail: temina@ocf.berkeley.edu

K. Toyama
University of Michigan School of Information, Ann Arbor, MI, USA

If we accept this reality – if not as universal fact, then as possibility – then we can move on to more productive questions: Why is it that, despite our consciously positive intentions, development engineering often fails to meet its objectives? And, what can we do to increase our chances of success? Since later chapters are devoted to the second question, we focus in this chapter on the first.

If our conscious intentions are positive, but the outcomes fall short, there are only four possibilities:

(1) We have counterproductive *unconscious* intentions.
(2) Our approach is flawed.
(3) There was bad luck.
(4) The problem admits no solution, at least not one that we, specifically, are able to offer.

(Give it some thought – there really are no other possibilities.) Taken together, these possibilities should instill in us a deep humility: Either we are at fault (possibilities 1 and 2), or the circumstances are beyond us (3 and 4).

And, that brings us to the development engineer's paradox. All engineering, indeed, all problem-solving, requires a certain confidence, even arrogance; we must believe that we can improve on all prior attempts. Yet, development engineering also requires profound humility. Humility in the face of difficult odds. Humility because of our own deficiencies of skill, knowledge, or character. Humility in the presence of communities, contexts, and constraints that are often not our own.

How can the development engineer's paradox be mitigated? Confidence without humility leads to overconfidence and arrogance, which in turn closes off our senses to the full range of information that could be available in any creative endeavor: What do potential beneficiaries think? How do our collaborators feel? Are we testing a proposed solution in all of the ways it should be tested? Yet, humility without confidence leads to paralysis and disengagement. The only viable path is one in which we use our confidence as faith in our eventual ability to reach a solution, while we apply our humility to temper every step along the way. We have to believe a solution exists, while doubting – and therefore intensively seeking confirmation for – each element of any proposed solutions. (The converse is a disaster – we would doubt the possibility of a solution, while being blithely confident in whatever options occur to us.)

2 Beneficiaries and Aspirations

So, humility is essential throughout any development engineering project, but it is perhaps most important when it concerns our understanding of potential beneficiaries and what is good for them. In fact, the very phrases "beneficiaries" and "what is good for them" trigger unease among experienced development practitioners. Of course, anyone who engages in development is seeking to do work that is good for someone else, so we cannot fault the attempt. The problem lies, rather, with the presumption that we know "what is good for them," and even more fundamentally, with the presumption that we could *ever* actually know.

Indeed, the most common, most accurate criticism leveled at international development may be that of misguided paternalism: When someone who presumed to know what was good for another person, another community, or another nation, turned out to be wrong. The history of development is full of such examples, large and small. For decades, Western nations pushed "Washington Consensus" policies that encouraged low-income countries to lower trade barriers in the name of economic growth (Williamson, 2011); with hindsight, we see that countries that complied with the consensus opened themselves up to exploitation, while those that strategically defied it, like China, protected their homegrown businesses and thrived (Serra & Stiglitz, 2008).

Development engineering has its share of such failures, as well: The Play Pump (Costello, 2010), the Soccket (Kenny & Sandefeur, 2013), One Laptop per Child (Villanueva-Mansilla & Olivera, 2012; Beuermann et al., 2015), and many other technologies have failed to deliver on grand promises, even as rich-world proponents pushed them into low-income communities. There are also homegrown efforts, like the Computador Popular and the Simputer (Fonseca & Pal, 2006), which have disappointed.

The obvious way to avoid presumption is to engage deeply and continuously with potential beneficiaries – to understand what they want, what they are constrained by, what resources they have, what strengths they can build on, and what dreams they have for the future. Development engineers and, sometimes, their critics have therefore refined a host of approaches to the design of solutions informed by beneficiaries: participatory design (Schuler & Namioka, 1993), cooperative design (Bodker & King, 2018), co-design (David et al., 2013; Ramachandran et al., 2007), participatory action research (Kemmis, 2006; Kemmis & Wilkinson, 1998), community-based participatory design (Braa, 1996), ethnographic design (Blomberg et al., 2009), human-centered design (Putnam et al., 2016), user-centered design (Putnam et al., 2009), asset-based community development (Kretzmann & McKnight, 1996; Mathie & Cunningham, 2003), and so on. What underlies all such approaches is a respect for potential beneficiaries as people who deeply understand the problem context, who have their own creative talents, and whose buy-in is required for uptake, impact, and sustainability of the solution. Partly for this reason, many practitioners of these methodologies prefer to refer to beneficiaries as "partners" or "collaborators."

In earlier chapters, we briefly reviewed participatory and human-centered design approaches, but here we mention a mindset that we have found useful in our own work and which we find counters some of the worst pathologies of "traditional" engineering: Instead of focusing solely on solving material needs, we should look for ways to align solutions with beneficiary *aspirations*.

We define an aspiration as "a desire that is persistent and aiming for something higher" (Toyama, 2018). It is useful to contrast aspirations with *needs*, the latter a common focus of both engineering and international development, in which the goal is to understand and address human needs through "needs assessments." While needs are often defined in relation to negative experiences – such as pain, hunger, illness, or poverty – aspirations are optimistic and forward-looking. Needs are also highly volatile, intensely felt but tending to vanish upon being met; in contrast, aspirations sustain over the longer term. Thus, when projects connect to beneficiary aspirations, beneficiaries are more likely to engage productively and for the longer term.

Among other things, being guided by beneficiary aspirations helps avoid the presumptions of paternalism. For example, in some very low-income communities, parents prioritize sons over daughters for food and education; they are betting on their boy children for future income – unfortunate but sensible in patriarchal societies. Of course, the girls need nutrition and school as much as the boys, but few parents feel the tug of that need. Outside efforts to reprimand parents or to otherwise push schooling rarely stick (Herz et al., 2004). In one Indian context, however, it was found that the example of just a handful of women per village being recruited to high-paying white-collar jobs caused improved schooling and nutrition outcomes for *other* local girls (Jensen, 2012). In other words, when parents saw that girls could also meet family aspirations for income security, they invested more in their daughters.

Another example is encouraging the use of toilets. Some communities have a cultural aversion to using toilets even when outsiders bother to build them in their neighborhoods. The sanitary need is not felt. (Or perhaps, the modern ideal of cloistering in a tiny room with one's own waste is not as attractive compared with going in the big outdoors!) In Haryana, India, however, great improvements in toilet building were seen in response to the "no toilet, no bride" campaign, in which women were encouraged not to marry men who did not have an indoor toilet (Stopnitzky, 2017). Celebrities were recruited to endorse the idea, thereby aligning aspirations for marriage and middle-class lifestyles with sanitation.

Finally, there is the option of addressing needs and aspirations simultaneously through unconditional cash transfers, in which households are given cash to spend as they see fit. It turns out that for a range of contexts, families – free of externally imposed donor preferences or judgements – not only relieve short-term needs but apply the funds toward longer-term aspirations (Weidel, 2016).

3 Framing the "Problem"

So far we have discussed the bottom-up approach to discovering development challenges, by observing people's daily lives and understanding their aspirations. This clearly plays an important role in defining new questions and opportunities in development engineering. Yet to holistically characterize a development challenge, researchers must incorporate "top-down" insights as well, by exploring the market, institutional, and social failures that create and reinforce poverty. These are often complex, high-level challenges that communities deal with every day – from the high costs of moving rural goods to urban markets to the hassles of accessing healthcare from unreliable public clinics. Yet the forces underlying these failures are not always evident to observers on the ground.

Nevertheless, their dynamics are first-order in defining the problem to be solved and in finding a solution that scales. Fortunately, there are existing resources that can help us understand the behavior of markets and institutions in many low-income settings. These include national statistics and international surveys, as well as quantitative academic research published in social science journals. When we combine on-the-ground observations and anecdotes with "top-down" models and insights, we can more thoroughly characterize the complex problems facing communities.

Of course there is always the risk that we become too reductionist in our definition of the problem. In trying to simplify a complex challenge, we may over-rationalize, failing to notice the details that really matter (Scott, 2020). We may design a solution so artificial that it spectacularly fails, resulting in great calamity for the "beneficiaries" (Duflo, 2017). As it turns out, most development problems are complicated, multifaceted, and interconnected. They can be framed through multiple lenses. For example, in Chap. 16, Tarpeh and co-authors discuss the challenge of urban sanitation in East Africa. It can be framed as a public health problem, a food production puzzle, an issue of failed governance, or a matter of planetary boundaries (Rockström et al., 2009). How you decide to frame the problem is often a moral, political, or personal choice that cannot be reduced to simple cost-benefit analysis. You may think of urban sanitation as an opportunity to capture economic value from waste, or you may see it as part of a global strategy to reduce human reliance on synthetic fertilizers. It is equally valid to view it as a public health emergency, through the lens of health as a human right. Asking the "right" question often depends on your personal values and motivations.

This, in turn, requires us to acknowledge our own identities and the privileges we benefit from. Privilege is a sort of unearned power, and it often accrues to people with higher education, many of whom will be engaged in research at some point in their careers (Adhikari et al., 2018). If you are reading this book, you likely are endowed with a privilege that creates blind spots. Privilege can make it difficult for you to directly observe or experience a community's preferences, because there is a cultural distance that will always mark you as separate. You may get less direct feedback, your ideas may be challenged or vetted less thoroughly, and your casual requests may be prioritized over others' substantive needs.

There are productive ways to navigate and mitigate these risks. One of the most effective is to work with partners who have fewer blind spots than you. A useful example is the model used by Digital Green, an NGO that advises small-scale farmers on best practices using a combination of digital technology and grassroots partnerships. The core insight of Digital Green is that smallholder farmers are themselves great experimenters and teachers (Gandhi et al., 2016). The platform enables farmers to teach each other about agricultural practices that boost income, through local production and dissemination of video content. This model taps into the credibility that local farmers carry within their communities; it also helps to overcome the blind spots of the engineers working on Digital Green's backend infrastructure.

Other strategies include working for or alongside community organizations before engaging in research (which can improve the richness and relevance of your research questions) and proactively discussing and addressing power imbalances in research relationships. Researchers from elite, wealthy institutions often benefit from greater financial resources and access to government leaders; this comes with complex ethical obligations. Researchers in less resourced institutions may have unique knowledge, perspectives, and community relationships (Naritomi et al., 2020); at the same time, they often face difficult-to-navigate community expectations. Each of these contributions (and accompanying constraints) must be materially recognized, even when they are intangible or difficult to quantify.

If you are a researcher with privilege, it is useful to actively solicit feedback, invite questions about your own contributions to a project, and look for opportunities to understand and invest in your colleagues. There are also a range of professional resources emerging over the last few years: the African Academy of Sciences, Mwazo Institute, and EASST Collaborative are all developing models for cross-national research collaboration (Hoy, 2018; Naritomi et al., 2020), including tools to more equitably control decisions about funding, co-authorship, and study design.

In summary, to define a problem well it is valuable to spend time on the ground, connecting with the communities you seek to empower and participating in their lived experiences. It is enriching to read deeply about the history, politics, and markets of the countries and communities you work in (Adhikari et al., 2018) and to form mutually respectful partnerships with local organizations (Tindana et al., 2007). These experiences will shape your own framing of problems, even if unconsciously. They may help you embrace complexity, avoiding the constraints of the "rational."

4 Conducting Ethical Research

Once a problem has been identified, how does one proceed ethically and responsibly with research? What does "ethical research" mean for a development engineer? The reality is that there is no consensus definition of ethics in research. More than anything, it is a framework for examining trade-offs in the decisions we make. First, we must recognize that engineers are interventionists: Just like economists and clinicians, engineers intervene on people's lives. Development engineers, in particular, seek to solve human problems; this work necessarily involves interaction with people and communities.

Here we will not address whether it is ethical for outsiders to conduct research in low-resource countries; much has been written on this, particularly in the domain of global health (Tindana et al., 2007). We simply note that there are compelling opportunities to contribute in the sphere of economic and human development; and while we will face ethical challenges whenever we intervene, we can engage with humility.

This will require us to ask difficult questions, like:

- Do the benefits to participants in research outweigh potential harms? How are benefits and risks being defined: by outsiders or by the participating community itself?
- Will the benefits of research be distributed fairly to all parties, in accordance with their risks and contributions?
- Will beneficial downstream products of research (including generalizable knowledge) be made accessible to people after the study is completed? What promises are being made, either explicitly or implicitly, in this regard?
- Have human subjects provided locally meaningful and substantive "informed consent" for their participation in research?

Asking questions like these can help minimize the risks of exploitation (El Setouhy et al., 2004). But there are several other practical considerations when it comes to responsible research. When conducting research in less developed environments – where legal and regulatory protections may be limited, opaque, or altogether absent – the responsibility for oversight is often shifted to the researcher (Alper & Sloan, 2014). For example, in the absence of a strong environmental regulator, you may need to develop your own checklists and protocols for monitoring and mitigating the potential environmental impacts of your study. You may need to build the financial management practices of local partners, so that they can legally accept research funding from external donors. You may even need to help establish a local institutional review board (IRB). In all cases, it is essential to involve (and compensate) local researchers, to ensure that your practices and protocols are appropriate and adapted to local norms.

Research Involving People

Any research that involves human participants – for example in interviews, surveys, or usability testing – must obtain prior approval from an IRB or Research Ethics Committee (REC). In some cases, an expedited review may be possible, particularly if the intervention poses little risk of harm to humans. Indeed some IRBs are beginning to innovate, recognizing that the evaluation of social interventions generally poses less risk of harm than the testing of novel clinical treatments (Schopper et al., 2015). Still, the process of completing a submission to an IRB is a valuable one and pushes the researcher to systematically evaluate whether a project's benefits to the community will actually outweigh its risks.

Engineers operating in areas of medicine and public health are advised to follow practices developed and adopted by the global health community, which draws from a rich body of scholarly work in international bioethics (Pinto et al.). For development engineers operating outside health, an excellent resource on running responsible research in low-income countries is the *Oxford Handbook of Professional Economic Ethics* (2016). A chapter by Glennerster and Powers focuses on the ethics of randomized evaluations; another chapter by Alderman, Das and Rao takes a broader view of field work, including issues related to trust, transparency, and privacy.

Finally it is important to communicate ethically with partners and subjects about your research interests and motivations. Creating hope around a promising innovation – one that will not necessarily continue after the study ends – can be harmful to communities. In many developing communities, outsiders are seen as wealthy power brokers, and locals may be eager to comply and please (even something as simple as serving tea and milk to researchers, with milk they would otherwise give to their kids). Are we being sensitive to these power dynamics?

Taking a Broader View

Researchers often implement their studies in cooperation with relatively nimble, efficient nongovernmental organizations. What is the obligation, then, to connect with local government staff as you proceed? There may be little incentive to engage with bureaucracies. Yet the eventual scale-up of any "proven" innovation will likely require some government input – whether for the actual scaling of the intervention (e.g., in the case of a novel public service) or in the form of approvals for the authority to operate at scale.

In general, researchers also find it valuable to engage local government agencies and even informal community institutions when implementing a research project. These leaders can provide valuable insights and support that can promote the success of research projects. It can also build the government's capacity to regulate new classes of technology, if required. The recent expansion of machine learning algorithms (and their penetration into areas of consumer finance, retail, and even government services) has broadened most governments' understanding of public safety and consumer protection. There is a rich literature on the fairness, accountability, transparency, and ethics of machine learning in the "real world," and development engineers would benefit from exposure to these ideas (Abebe et al., 2020).

Finally, maintaining a safe research environment must consider the risks experienced by students, enumerators, and other research staff. Research can introduce risks to students, and the communities they engage with, and these should be anticipated and managed (Pinto & Upshur, 2009).

The Money

As the saying goes, "money talks," and in development engineering, whoever provides the funding has significant, sometimes subtle, influence. Whether the funding comes from a multilateral agency, a national government, a philanthropic foundation, an individual donor, or a private enterprise, money almost always has its own agenda, and that agenda may not fully align with the goal of altruistically serving beneficiaries. That tension can be the cause of ethical challenges. Just to provide some examples:

- Donors often want to take credit for large-scale impact. To this end, they may prefer that millions of people are "reached" or "touched" by an intervention, whether or not it has any positive impact.

- Governments are sometimes mired in politics, whereby projects underwritten by one administration are redirected, canceled, or sabotaged by another administration.
- For-profit companies need to break even to survive. They may cut corners, seek richer customers, or misrepresent impact in a bid to become viable.

These are just a few examples, but all of them have consequences for fund recipients. The ethical development engineer will be forced to consider and make difficult decisions. In the best of cases, a funder may be appeased with minor tweaks to the optimal path for beneficiaries; in the worst, practitioners may feel they have no choice but to end their involvement (only to see less ethical colleagues take up the project). Sometimes, even the option to withdraw from a project may cause additional ethical problems.

Like many ethical quandaries, there are no easy answers to these dilemmas, but there are heuristics to minimize or mitigate them. First, many ethical issues can be avoided through sufficient advance dialogue, planning, and transparency among key stakeholders. If it is known up front what a donor most hopes for with their funds, alternatives can be considered and expectations can be managed; and again, in the worst case, it is possible to walk away from the funds without having engaged a vulnerable community. If a donor agrees to a set of plans for a project, including the expected trajectory *after* the period of funding, then they have little basis to complain when the plan is carried out. (Of course, a thorough plan would also figure in a range of contingencies.)

Next, ongoing transparency and honesty can also help. When unexpected situations arise, communicating them to donors can avoid worse disappointments down the line. Most donors engaged in development work genuinely care about beneficiary communities – and even those who do not will still prefer to appear to care. If a project cannot go as planned because it turns out it would harm beneficiaries, a change of course can often be negotiated.

Finally, it is worth remembering that despite the inherent power imbalance between funder and funded, sponsors of development engineering need the engineer. When anything requiring an ethical breach is requested, the engineer has power to push back, even if based on the threat of ceasing the work. Exercising that power when necessary ensures that the development engineer is not complicit in problematic action.

5 Impact and Scale, Take 2

At this point, we revisit the issues of impact and scale that were discussed in Chap. 3 but incorporating some of the discussion in this chapter. One critical skill for the development engineer is the ability to make judicious trade-offs among ideals of practice. Ultimately, that judgment comes from experience and reflection (another reason for humility), but we offer some suggestions below.

Scope

Few problems are purely technology problems, because technology does not work itself – it depends on capable users as well as ongoing maintenance and upkeep, all of which in turn requires favorable social, cultural, economic, institutional, and political conditions. Human beings are creatures of habit, but any new technology necessitates that *someone* do something differently, whether it is for homemakers to adapt to a new type of cookstove, for healthcare workers to perform their rounds in a new way, or for bank regulators to devise new policies for mobile payments. All of these efforts are extra-technological and therefore require something beyond technical engineering. Any engineering project in which these other efforts are not addressed is bound to disappoint.

But, neither is it generally realistic to expect engineers of development to train users, supervise healthcare workers, or affect policy. That brings us to a critical question that we believe all development engineers should ask early in a project: "Is there a potential organizational partner with shared impact goals, that has the relationships, the capacity, and the commitment to address most extra-technological challenges?" If not, it is worth reconsidering the project. If so, working with such a partner is perhaps the most effective way to ensure meaningful impact. There are many benefits to working with a capable partner aligned with one's objectives: A good partner will often have good insight into beneficiary context and aspirations; they will serve as a sounding board; and they often have critical relationships necessary for larger-scale impact. Perhaps most of all, partners free the development engineer to focus on the technical, economic, and social innovations they are best suited to contribute, while ensuring that implementation is of high quality. The alternative is for the development engineer to establish such an organization themselves, but that requires a skillset and level of commitment well beyond development engineering.

External Validity and Scale

Do outcomes that hold for one community would hold for another? That is the question of external validity (Cartwright, 2011; Banerjee et al., 2017), an issue critical for development projects that seek impact beyond their pilot community. Strictly speaking, even the most rigorously run evaluations cannot claim validity beyond the group from which participants were sampled or the context in which the trial was conducted (Toyama, 2015). Of course, some external validity can be inferred based on hypothesized mechanisms (Pearl & Barenboim, 2014) – the impact of a new medicine will likely transfer from one human group to another, based on universally shared biology; behavioral interventions may transfer based on shared psychology. But, little can be taken for granted (Gauri et al., 2019).

And, because context often changes with scale, external validity is also a concern as projects scale up. To some extent, the computer software industry has internalized this lesson. Large multinational companies routinely involve "growth" experts at the earliest stages of engineering design, to ensure that new products incorporate enough flexibility to operate within varied market conditions, supply chains, and regulatory regimes. The problem to be solved by a product may be near-universal, like paying for electricity using a mobile phone; but the underlying technology and feature set will vary across geographies and cultures.

An entire profession – alternately called internationalization, localization, or globalization – has emerged around the alignment of software designs with the diverse social norms, business practices, laws, and technical constraints found in different countries (Aykin, 2004). This aspect of product design is complementary to user-centered design and equally important. For anyone who has worked in the field, it is not simply a matter of translating product manuals into new languages. It touches hardware design, database design, system architecture, algorithm development, service level agreements, and so much more (Jimenez-Crespo, 2013). The upfront investment in flexibility allows companies to build for sustainability and scale; it enables expansion to new markets and policy environments without a complete re-engineering of the product.

Whether the goal is to transplant a solution to another setting or to scale up a solution, the approach to external validity is theoretically simple, though often a challenge in practice: Repeat the same innovate-implement-evaluate cycle discussed in Chap. 3 for each new setting and as the intervention is scaled up. What works in one village may need adaptation to work in another or at the district level; what works at the district level will need tweaking for the state or province level; what works for a state will need modification for national impact; what works for one country may need adjustment for another country. And, the adaptations are likely to require both technological and socio-political adaptation. What can be handcrafted at one scale may need assembly line or factory production at higher scales. User engagement that is ad hoc at one scale may need to become systematized, institutionalized, and possibly legalized at higher scales. A useful rule of thumb for development engineering projects is that new research is required at every order of magnitude or two of scale.

Similarly, at each new context or level of scale, it is essential to ensure that desired impacts are continuing, through additional evaluations. Many programs succumb to sociologist Peter Rossi's "Iron Law of Evaluation: The expected value of any net impact assessment of any large scale social program is zero" (Rossi, 1987). Though perhaps overstated, the iron law points to the very real tendency for project impact to dissipate with displacement or scale, as indifferent bureaucrats and less-informed beneficiaries play a proportionally larger role in implementation.

Cost-Benefit Analysis

Assuming that evaluations demonstrate an intervention to be effective, another important question is whether the intervention is *cost-effective* especially compared with alternative solutions. Especially for well-focused goals such as increasing the number of vaccinated children or improving educational test scores, it should be possible to capture project effectiveness with a cost-benefit analysis. The underlying idea is simple – compute the financial cost per unit of impact and compare it against cost-benefit ratios of interventions with the same objective (Dhaliwal et al., 2013). In practice, this can be somewhat difficult as cost data must be carefully gathered, and an honest accounting of fixed costs (that could be distributed over a large program) and variable costs (that are incurred on a per-unit-of-impact basis) must be made; any comparative analysis would also need this information for alternative interventions (Levin & McEwan, 2000; Brown & Tanner, 2019).

What further complicates cost-benefit analyses is that an intervention's sum total benefits and side effects are rarely enumerable, not to mention measurable. Most cost-benefit analysis requires some estimation of intangible factors, some of which may outweigh tangible benefits in importance.

Even when imperfect, however, some rudimentary analysis is worthwhile to gauge cost-effectiveness. One point of reference is that of the aforementioned unconditional cash transfer, in which beneficiaries are given cash – typically about $1000 – with no strings attached (Weidel, 2016). Evidence is accumulating that such gifts have a range of long-term benefits for poor households, with little negative effect. Proponents have begun to call for such transfers to be considered the benchmark when evaluating development programs (Blattman & Niehaus, 2014) – if a program's cost-benefit profile does not at least match that of cash transfers, why not just give the costs of the program directly to beneficiaries?

Unanticipated Negative Effects

Critics of development often note that projects have negative *unintended consequences* (Merton, 1936). Unintended consequences can arise for many reasons. Sometimes, a technology or intervention may cause direct harm as a side effect, as was the case of DDT, a powerful pesticide that turned out to be toxic to many animal species (and humans, when used in large doses, see Carson, 1962). In some cases, a technology can enable harmful forms of mass misuse, as with the global spread of misinformation and extremism on social media (Singer & Brooking, 2018; Fernández-Luque & Bau, 2015). In other cases, an intervention effective at the small scale can backfire at larger scales, as occasionally happens with improvements in agricultural productivity that lead to regional gluts and a subsequent decline in prices (Burke et al., 2019).

Yet another class of unintended consequences occur when one person's benefit causes someone *else* harm or the perception of harm. Projects aimed at empowering women or minority groups, for example, can backfire, by incurring hostility and possibly violence from the oppressing group (Sultana et al., 2018). Similarly, poverty alleviation efforts can elicit envy and resentment from neighbors who might also be impoverished but unable to take part in an intervention. Especially with technological interventions, which tend to amplify underlying human forces, existing inequalities may be exacerbated (Toyama, 2015). The resentment that results from growing inequality can provoke conflict, especially when layered on existing divisions of caste, race, religion, or ethnicity.

These are just a few examples of unintended consequences. By definition, we cannot know all of a project's unanticipated effects in advance, so the best we can do is to "stay with the trouble" (Haraway, 2016), to continue to engage and address negative consequences. This, again, reinforces the need to engage with beneficiary communities throughout, and possibly even beyond, the project lifecycle. Knowing as we do now that there are always unintended consequences, not to do so is neglect and indifference.

References

Abebe, R., Barocas, S., Kleinberg, J., Levy, K., Raghavan, M., & Robinson, D. G. (2020, January). Roles for computing in social change. In *Proceedings of the 2020 conference on fairness, accountability, and transparency* (pp. 252–260).

Adhikari, S., Elorrieta, J. I., & Pomeranz, D. (2018) *Working in emerging markets: Opportunities and blind spots.*

Aker, J. C., Ghosh, I., & Burrell, J. (2016). The promise (and pitfalls) of ICT for agriculture initiatives. *Agricultural Economics, 47*(S1), 35–48.

Alderman, H., Das, J., & Rao, V. (2016). Conducting ethical economic research: complications from the field. In G. F. DeMartino & D. N. McCloskey (Eds.), *The Oxford handbook of professional economic ethics* (pp. 402–422). Oxford University Press.

Alper, J., & Sloan, S. S. (Eds.). (2014). *Culture matters: International research collaboration in a changing world: Summary of a workshop.* National Academies Press.

Aykin, N. (2004). *Usability and internationalization of information technology.* Lawrence Erlbaum Associates.

Banerjee, A. V., Duflo, E., & Gueron, J. M. (2017). *Handbook of field experiments.* North-Holland.

Beuermann, D. W., Cristia, J., Cueto, S., Malamud, O., & Cruz-Aguayo, Y. (2015). One laptop per child at home: Short-term impacts from a randomized experiment in Peru. *American Economic Journal: Applied Economics, 7*(2), 53–80.

Blattman, C., & P. Niehaus. (2014). Show them the money: Why giving cash helps alleviate poverty. *Foreign Affairs,* May/June 2004.

Blomberg, J., Burrell, M., & Guest, G. (2009). An ethnographic approach to design. *Human-Computer Interaction,* 71–94.

Bødker, S., & Kyng, M. (2018). Participatory design that matters – Facing the big issues. *ACM Transactions on Computer-Human Interaction (TOCHI), 25*(1), 1–31.

Braa, J. (1996, November). *Community-based participatory design in the Third World.* In Proceedings of the participatory design conference (Vol. 96).

Brown, E. D., & Tanner, J. C.. (2019, November). *Integrating value for money and impact evaluations: Issues, institutions, and opportunities.* World Bank Group, Policy research working paper #9041.

Burke, M., Bergquist, L. F., & Miguel, E. (2019). Sell low and buy high: Arbitrage and local price effects in Kenyan markets. *The Quarterly Journal of Economics, 134*(2), 785–842.

Carson, R. (1962). *Silent spring.* Houghton Mifflin Company.

Cartwright, N. (2011). Evidence, external validity and explanatory relevance. *Philosophy of Science Matters: The Philosophy of Peter Achinstein,* 15–28.

Collier, P. (2007). *The bottom billion: Why the poorest countries are failing and what can be done about it.* Oxford University Press.

Costello, A. (2010). Troubled water. *Frontline/World.* Available at: http://www.pbs.org/frontlineworld/stories/southernafrica904/video_index.html.

Dambisa, M. (2009). *Dead Aid: Why aid is not working and how there is a better way for Africa.* Farrar.

David, S., Sabiescu, A. G., & Cantoni, L. (2013, November). *Co-design with communities. A reflection on the literature.* In Proceedings of the 7th International development Informatics Association Conference (IDIA) (pp. 152–166).

Dhaliwal, I., Duflo, E., Glennerster, R., & Tulloch, C. (2013). Comparative cost-effectiveness analysis to inform policy in developing countries: A general framework with applications for education. In *Education policy in developing countries* (pp. 285–338).

Duflo, E. (2017). The economist as plumber. *American Economic Review, 107*(5), 1–26.

Easterly, W. (2007). *The white man's burden*. Oxford University Press.

El Setouhy, M., Agbenyega, T., Anto, F., Clerk, C. A., Koram, K. A., English, M., ... Mfutso-Bengu, J. (2004). Moral standards for research in developing countries from "reasonable availability" to "fair benefits". *The Hastings Center Report, 34*(3), 17–27.

Fernández-Luque, L., & Bau, T. (2015). Health and social media: perfect storm of information. *Healthcare Informatics Research, 21*(2), 67–73.

Fonseca, R., & Pal, J. (2006, May). Computing devices for all: Creating and selling the low-cost computer. In *2006 international conference on information and communication technologies and development* (pp. 11–20). IEEE.

Gauri, V., Jamison, J. C., Mazar, N., & Ozier, O. (2019). Motivating bureaucrats through social recognition: External validity – A tale of two states. *Organizational Behavior and Human Decision Processes*.

Glennerster, R., & Powers, S. (2016). Balancing risk and benefit: ethical tradeoffs in running randomized evaluations. In G. F. DeMartino & D. N. McCloskey (Eds.), *The Oxford handbook of professional economic ethics* (pp. 402–422). Oxford University Press.

Haraway, D. J. (2016). *Staying with the trouble: Making Kin in the Chthulucene*. Duke University Press.

Gandhi, et al. (2016). Digital green. In D. Hanks & R. Steiner (Eds.), *Harnessing the power of collective learning*. Taylor & Francis.

Heeks, R. (2002). *Information systems and developing countries: Failure, success, and local improvisations*.

Herz, B., Herz, B. K., & Sperling, G. B. (2004). *What works in girls' education: Evidence and policies from the developing world*. Council on Foreign Relations.

Hoy, A. Q. (2018). Africa cultivates innovation to boost global reach. *Science, 360*(6387), 391–392.

Jensen, R. (2012). Do labor market opportunities affect young women's work and family decisions? Experimental evidence from India. *The Quarterly Journal of Economics, 127*(2), 753–792.

Jimenez-Crespo, M. A. (2013). *Translation and web localization*. Taylor & Francis.

Kemmis, S. (2006). Participatory action research and the public sphere. *Educational Action Research, 14*(4), 459–476.

Kemmis, S., & Wilkinson, M. (1998). Participatory action research and the study of practice. In *Action research in practice: Partnerships for social justice in education* (Vol. 1, pp. 21–36).

Kenny, C. (2012). *Getting better: Why global development is succeeding – And how we can improve the world even more*. Basic Books.

Kenny, C., & Sandefur, J. (2013). Can Silicon Valley save the world? *Foreign Policy, 201*, 72.

Kretzmann, J., & McKnight, J. P. (1996). Assets-based community development. *National Civic Review, 85*(4), 23–29.

Levin, H. M., & McEwan, P. J. (2000). *Cost-effectiveness analysis: Methods and applications* (Vol. 4). Sage.

Mathie, A., & Cunningham, G. (2003). From clients to citizens: Asset-based community development as a strategy for community-driven development. *Development in Practice, 13*(5), 474–486.

Merton, R. K. (1936). The unanticipated consequences of purposive social action. *American Sociological Review, 1*(6), 894–904.

Morozov, E. (2011). *The net delusion: How not to liberate the world*. Perseus Books.

Moyo, D. (2009). *Dead aid: Why aid is not working and how there is a better way for Africa*. Macmillan.

Namioka, A., & Schuler, D. (1993). *Participatory design: Principles and practices*. Taylor & Francis.

Naritomi, J., Sequeira, S., Weigel, J., & Weinhold, D. (2020). RCTs as an opportunity to promote interdisciplinary, inclusive, and diverse quantitative development research. *World Development, 127*, 104832.

Pearl, J., & Bareinboim, E. (2014). External validity: From do-calculus to transportability across populations. *Statistical Science*, 579–595.

Pinto, A. D., & Upshur, R. E. (2009). Global health ethics for students. *Developing World Bioethics, 9*(1), 1–10.

Putnam, C., Rose, E., Johnson, E. J., & Kolko, B. (2009). Adapting user-centered design methods to design for diverse populations. *Information Technologies & International Development, 5*(4).

Putnam, C., Reiner, A., Ryou, E., Caputo, M., Cheng, J., Allen, M., & Singamaneni, R. (2016). Human-centered design in practice: Roles, definitions, and communication. *Journal of Technical Writing and Communication, 46*(4), 446–470.

Ramachandran, D., Kam, M., Chiu, J., Canny, J., & Frankel, J. F. (2007). *Social dynamics of early stage co-design in developing regions*. In Proceedings of the SIGCHI conference on human factors in computing systems (pp. 1087–1096).

Rockström, J., Steffen, W., Noone, K., Persson, Å., Chapin, F. S., III, Lambin, E., . . . Foley, J. (2009). Planetary boundaries: Exploring the safe operating space for humanity. *Ecology and Society, 14*(2).

Rossi, P. H. (1987). The Iron Law of evaluation and other metallic rules. In *Research in social problems and public policy* (Vol. 4, pp. 3–20). JAI Press.

Sachs, J. (2005). *The end of poverty: How we can make it happen in our lifetime*. Penguin.

Sainath, P. (1996). *Everybody loves a good drought: Stories from India's poorest districts*. Penguin Books.

Schopper, D., Dawson, A., Upshur, R., Ahmad, A., Jesani, A., Ravinetto, R., . . . Singh, J. (2015). Innovations in research ethics governance in humanitarian settings. *BMC Medical Ethics, 16*(1), 1–12.

Schuler, D., & Namioka, A. (Eds.). (1993). *Participatory design: Principles and practices*. CRC Press.

Scott, J. C. (2020). *Seeing like a state: How certain schemes to improve the human condition have failed*. Yale University Press.

Serra, N., & Stiglitz, J. (Eds.). (2008). *The Washington consensus reconsidered: Towards a new global governance*. Oxford University Press.

Singer, P. W., & Brooking, E. T. (2018). *LikeWar: The weaponization of social media*. Eamon Dolan Books.

Stopnitzky, Y. (2017). No toilet no bride? Intrahousehold bargaining in male-skewed marriage markets in India. *Journal of Development Economics, 127*, 269–282.

Sultana, S., Guimbretière, F., Sengers, P., & Dell, N. (2018, April). *Design within a patriarchal society: Opportunities and challenges in designing for rural women in bangladesh*. In Proceedings of the 2018 CHI conference on human factors in computing systems (pp. 1–13).

Tindana, P. O., Singh, J. A., Tracy, C. S., Upshur, R. E., Daar, A. S., Singer, P. A., . . . Lavery, J. V. (2007). Grand challenges in global health: community engagement in research in developing countries. *PLoS Medicine, 4*(9), e273.

Toyama, K. (2015). *Geek Heresy: Rescuing social change from the cult of technology*. PublicAffairs.

Toyama, K. (2018). From needs to aspirations in information technology for development. *Information Technology for Development, 24*(1), 15–36.

Villanueva-Mansilla, E., & Olivera, P. (2012). Institutional barriers to development innovation: Assessing the implementation of XO-1 computers in two peri-urban schools in Peru. *Information Technologies & International Development, 8*(4), 177.

Weidel, T. (2016). Philanthropy, cosmopolitanism, and the benefits of giving directly. *Journal of Global Ethics, 12*(2), 170–186.

Williamson, J. G. (2011). *Trade and poverty: When the Third World fell behind*. MIT Press.

Part I
Expanding Access to Affordable and Reliable Energy, While Minimizing the Environmental Impacts

Kenneth Lee ⓘ

It is impossible to imagine modern life without access to affordable and reliable electricity. Yet this is the reality for the roughly 800 million people living in energy poverty today, the majority of whom are located in South Asia and sub-Saharan Africa.[1] As the economies in these regions grow, it is inevitable that demand for energy will rise.

How will developing countries meet this growing demand for energy? If the answer lies in conventional, highly polluting fossil fuel technologies – as has largely been the case in the economic development of India and China – then the environmental consequences will be severe. At the local level, rapidly expanding the use of fossil fuels for energy production, industrial manufacturing, and transportation, among other activities, will result in higher levels of air pollution to the detriment of public health. And at the global level, increasing energy demand will produce a commensurate rise in carbon emissions, further accelerating climate change and the impacts of global warming which are already being felt in greater frequencies across the world. This is the fundamental dilemma. So many people still lack access to basic energy services or require greater levels of energy consumption to support a modern standard of living, yet the worst consequences of climate change are just around the corner.

Recent advancements in technologies, especially those in the areas of clean energy, remote sensing, and mobile payments, offer reasons for hope. Solar lanterns, for instance, have never been more affordable and accessible than they are today. This was made possible by a decline in the cost of photovoltaics, improvements in batteries, and financial technology ("FinTech") innovations that now allow poor

---

[1] See: IEA (International Energy Agency). 2019. World Energy Outlook 2019. Paris: IEA.

K. Lee
Chief Research and Evaluation Officer, The Pharo Foundation, Nairobi, Kenya
e-mail: klee@pharofoundation.org

consumers to acquire new energy devices through "pay-as-you-go" programs, thus alleviating the credit constraints that often limit the adoption of promising technologies in rural settings. In Kenya, the use of solar lanterns and solar home systems has become so widespread that kerosene lanterns – not long ago an essential item in rural communities across the developing world – are now on track to finally become obsolete.[2]

For development engineers, the energy and environment challenges are threefold. First, we need strategies to deploy existing and emerging technologies to expand access to electricity and then ensure that it remains reliable and affordable. Second, we need strategies to limit the environmental impacts, like air pollution and carbon emissions, that will be caused by this expanded access to energy. And third, as the planet continues to warm, we require a host of solutions to help people adapt to emerging and future climate-related threats, like extreme heat, flooding, and water scarcity.

This section includes three case studies on recent or ongoing research projects in Kenya, Ghana, and India. The first chapter, which covers the question of how best to expand access to electricity in Kenya, narrates the story behind an interdisciplinary project that began as a study on solar microgrids, but evolved into a study on the economics of last-mile grid electrification. The second chapter, which focuses on the frequent blackouts that have plagued the electricity grid in Ghana, describes a project to design, deploy, and operate a large network of remote sensors to measure power outages and quality at households and firms. Then, shifting to the problem of air pollution in India, third chapter describes how the installation of continuous emissions monitoring systems led to improvements in the government's ability to monitor and regulate the particulate matter emitted by local industries.

The chapters in this section share some common lessons and themes. First, the case studies highlight the exciting opportunity that new technologies have in developing countries to address market failures or imperfections (like information asymmetry and credit constraints) and solve other issues related to the performance of public institutions. Second, the case studies demonstrate how the initial beliefs we may hold regarding the viability of (or even the demand for) a particular technology may be drastically revised (or even discarded) following some critical piloting experience in the field. Third, from these examples, we learn that the success of a new technology is not tied solely to the ingenuity or technical quality of its design. Rather, the overall impact of a new technology depends on what happens when it is deployed at scale; how it interacts with the various economic, political, and behavioral forces in play; and the motivations or incentives of key decision makers and actors. Ultimately, what the case studies in this section reveal is that technology cannot be separated from the underlying context – which is perhaps a lesson that extends to the broader field of *Development Engineering*.

---

[2] In the 2009 Kenya Population and Housing Census (KPHC), 75.7% of Kenyan households stated that they were using kerosene or fuel wood as their main source of lighting. By the 2019 census, only 19.3% of households were using these same sources. Over the period, usage of solar increased from 1.6% to 19.3% of all households.

Chapter 5
Expanding Access to Electricity in Kenya

Kenneth Lee (ID)

1 The Development Challenge

In 2012, we began a study on solar microgrids in rural Kenya. Over time, it evolved into an experiment that randomized the expansion of the national electricity grid instead. In this chapter, I tell the story behind this project, focusing on the pivots and iterations that shaped the path of our research on the economics of electrification over nearly a decade.

When we started our project, access to electricity was widely seen as a major driver of economic development, just as it remains today. Then-United Nations Secretary General Ban Ki Moon famously referred to it as the "golden thread" connecting economic growth, social equity, and an environment where people could thrive. Supporting this outlook was the well-known, near-perfect correlation between electricity consumption and GDP per capita, which is shown in Fig. 5.1.

The academic literature proposed many plausible channels through which electricity access could improve lives. Electric lighting, for instance, could extend the workday, increasing labor supply and income. Since lighting would also give children more time to do their homework, educational attainment might even improve. Some pointed out the potential issues with rural electrification programs as well. In a review of the history of electrification in sub-Saharan Africa, Bernard (2012) emphasized the limited productive uses of electricity that had often been

For the full results of this study, see Lee et al. (2016a), and Lee, Miguel, and Wolfram (2016b, 2020a, b). I am grateful to Shipra Karan for excellent research assistance in preparing this chapter. All errors are my own.

K. Lee (✉)
Chief Research and Evaluation Officer, The Pharo Foundation, Nairobi, Kenya
e-mail: klee@pharofoundation.org

101

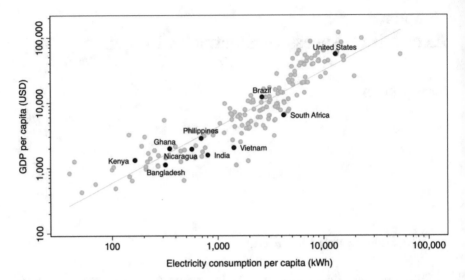

Fig. 5.1 The positive correlation between electricity consumption and GDP per capita
Notes: Both variables are presented on a logarithmic scale. 2014 data obtained from the World
Bank DataBank Reprinted from Lee et al. (2020b)

observed across rural Africa. Although people seemed to use electricity for lighting
and communications, they were less likely to use it for agriculture, handicrafts,
and other activities that could be profitable. Khandker et al. (2014) proposed an
additional issue that the gains from rural electrification could be much greater for
wealthier households, which could exacerbate economic inequalities.

At the time, over a billion people still lacked access to electricity. The question
of how governments could best expand access to power remained front and center.
Moreover, developing countries were expected to drive a considerable amount of
growth in global energy consumption (Wolfram et al., 2012). As a result, expanding
access in these countries using conventional fossil fuel technologies would certainly
accelerate global warming. The development challenge was clear: In countries
with high rates of energy poverty, how could electricity access be expanded while
mitigating the consequences on the global environment?

In the spring of 2012, UC Berkeley's *Development Impact Lab* brought together
a team of economists and engineers to work on this problem.[1] The basic goal of
the collaboration was to improve the design of the solar microgrid technologies

---

[1] The initial team included principal investigators Eric Brewer (Electrical Engineering and
Computer Science), Edward Miguel (Economics), and Catherine Wolfram (Business); researchers
Carson Christiano (Center for Effective Global Action), Matthew Podolsky (Technology and
Infrastructure for Emerging Regions), and Francis Meyo (Innovations for Poverty Action);
and graduate students Javier Rosa (Electrical Engineering and Computer Science) and myself
(Agricultural and Resource Economics).

that were being developed for poor countries. By bringing together engineers and economists, the iterative process of engineering design could be merged with microeconomic survey data and evidence obtained using the randomized control trial (RCT) approach that had become widespread in development economics. We believed the results of such a collaboration could inform the design of technologies and public policies in unique ways.

A couple aspects of the partnership generated a great deal of excitement. First, the engineers had been working with start-up companies to design solar-powered microgrids that featured novel, prepaid, smart metering technologies. These devices offered a trove of high-frequency, electricity usage data. We thought about using these data to predict the kinds of appliances people were using in their homes. Combined with data collected through household surveys, we could perhaps unlock the precise mechanisms through which electrification improved well-being. And in places where electricity theft was an issue, we thought about measuring the impacts of the smart metering technologies on monitoring and enforcement. There were many possibilities for data science.

One of the start-up companies working on this technology was based nearby, in Oakland. They had just built a pilot microgrid in Kenya and were looking to scale up. After a series of meetings, we agreed to explore measuring the impacts of their solar microgrids in an RCT. Kenya was a good fit, since it was a place where the economists had extensive prior field experience. From the company's perspective, an independent team documenting the beneficial impacts of their product could be useful in their public communications and marketing efforts, as well as in their search for venture capital funds to propel their rapid expansion across Africa. Of course, if they partnered with us, our research grant would pay for (or at least subsidize) the cost of some of their microgrid installations.

Second, from a research standpoint, we were excited about measuring the causal impacts of electrification in an experimental setting. Until then, the applied microeconomics literature covering this topic was limited. Nearly all of the existing work was nonexperimental and tended to rely on administrative or observational data, which made it a challenge to distinguish causal effects from correlations. In any situation, one could imagine a host of unobserved factors that could be correlated with someone's access to electricity (the cause) and the changes they might experience over time (the possible effects). An exception at the time was Dinkelman (2011), which showed that rural electrification substantially improved female labor supply in South Africa. The study had creatively, and quite convincingly, applied an econometric technique to isolate a causal effect of electrification, using only administrative data. Thus, in the early days of our study, there was growing scholarly interest in building upon the evidence base on the impacts of electrification using ever more rigorous approaches to estimation.

More broadly, it seemed just a matter of time before billions of dollars would be directed towards electrification programs across the world. In the face of climate change, solar microgrids had great potential. We thought about how an infrastructure experiment could yield new benchmarks on the causal effects of electrification. Perhaps these could serve as useful inputs in the large-scale, infrastructure investment decisions that would surely be made in the future.

This case study chronicles our research timeline and how our views on the development challenge evolved along the way. Initially, we set out to build solar microgrids in off-grid villages in Kenya. But instead, we partnered with the Kenyan government and connected hundreds of randomly selected rural households to the national grid for the first time. We encountered a number of implementation challenges, many of which played a role in shaping our eventual conclusions. As it turned out, it was relatively expensive to build electricity network lines to rural homes. And around 3 years later, we had found no evidence indicating that household access to the grid had meaningfully changed a set of pre-defined economic and noneconomic outcomes. The project spurred a number of follow-up projects, which are still ongoing.

The bulk of our project was carried out between 2012 and 2017. Over this period, major policies would be introduced to accelerate the rural electrification of Kenya. In certain parts below, I bring up interactions we had with donors and policymakers at various points in time. I share these stories for a few reasons. First, I think these meetings imparted on us a number of timely perspectives that helped guide and refine our research focus. Second, I think our experience in Kenya offers an example of how the slow and deliberate process of academic research can sometimes be outpaced by rapidly shifting policy priorities in developing countries. Finally, and on a more personal note, I think the linkages between our research project and the wider policy environment in Kenya encapsulate what makes this line of work so exciting, and I hope some of this comes across in this chapter.

The remainder of this case study is organized as follows. The next section discusses the technology choices available to policymakers at the start of our project. Then, in the following sections, I describe the important decisions we needed to make to set up an experiment; the things we learned that influenced our research questions and intervention design; and how we made sense of our findings given the evolving policy context. The final section offers a view on some of the important research questions for the future.

2 Innovations in the Technology Landscape

There are several ways to address the development challenge of expanding access to electricity. Traditionally, governments have addressed this challenge by investing in expansions of their national grids. All developed countries have reached universal rural electrification in this way. The issue moving forward, of course, is the extent to which the grid can supply electricity from nonfossil fuel sources of energy.

The 2000s introduced various improvements to an array of decentralized, renewable energy alternatives, including solar lanterns, solar home systems, and renewable energy microgrids. There was hope that these novel technologies could allow people living in the Global South to gain access to electricity, while minimizing the negative consequences on the environment. Across sub-Saharan Africa, the rapid adoption of mobile phones had made landline telecommunications infrastructure obsolete. By 2012, many entrepreneurs, donors, and observers were talking about how this improved set of decentralized, renewable energy solutions would allow off-grid households to similarly leapfrog the grid.

A couple trends seemed to be driving this growing level of enthusiasm. First, it was becoming much cheaper to manufacture these products. With each passing year, there were increases in appliance efficiencies, reductions in the cost of photovoltaics, and improvements in battery capabilities. By around 2012, off-grid solar began to be seen as a potential alternative to the grid. Second, in countries like Kenya, rapid growth in mobile phone usage had been accompanied with widespread adoption of mobile money platforms like M-PESA. Around 2010, new start-up companies like M-KOPA began integrating pay-as-you-go technologies directly into their solar lanterns and solar home systems. Poor, rural consumers could now buy these products on credit, unlocking each day of usage with a small payment sent over their mobile phones. This was seen as a gamechanger in rural settings, where credit constraints had often limited the take-up of promising, new technologies.[2]

2.1 Prepaid, Smart Metered Solar Microgrids

Microgrids, which connect small networks of users to a centralized and stand-alone source of power generation and storage, were also generating substantial interest. Microgrids could provide longer hours of service and higher capacities than solar lanterns and solar home systems, making it feasible to use power more productively. Furthermore, they could be powered with clean energy sources, like solar, wind, and hydro. Despite their potential, microgrids had not yet been deployed at scale in developing countries. In fact, a number of early microgrid pilot deployments had completely failed.

For example, in the early 2000s, dozens of microgrids had been set up in rural villages across Rajasthan, India. The microgrids were built to connect rural hamlets to a 10-kilowatt capacity solar panel system. After several years, many of the microgrids had fallen into disrepair. There were a number of reasons why this happened. For instance, the Rajasthani microgrids, which offered users just enough electricity to power lighting and small appliances, did not meter households individually. Instead, each grid would be switched on and off at certain hours of the day. In return, each user needed to pay a monthly fixed fee. But when certain users refused to pay, there was no way to terminate their service, leading to a downward spiral in revenue collection. Making matters worse, the battery banks powering the microgrids could not withstand the hot and humid environmental conditions

---

[2] See Lee, Miguel, and Wolfram (2020b) for a discussion of historical rural electrification initiatives from around the world, as well as the current technology landscape.

of Rajasthan. When the batteries failed, the lack of payment enforcement meant there was little to no cash reserve available to cover the cost of a replacement. The primary culprit for the failure of the Rajasthani microgrids was not some electrical engineering issue, but rather a misalignment of economic incentives. Simply put, the early microgrids were in need of a better business model.[3]

In Kenya, our microgrid partner had taken advantage of the technological trends to develop a next-generation, village-scale solar microgrid that allowed consumers to pay-as-they-go using their mobile phones. They marketed their technology as one that could empower consumers to make real-time decisions about their energy consumption, while alleviating credit constraints. Importantly, each user had their own smart meter that would send information about power consumption and credit balances over text messages. Depending on how each system was sized, they promised power that would be more reliable than the national grid.

In 2012, official estimates of the national household electrification rate in Kenya ranged from 18 to 26%.[4] We were intrigued by the potential market for this microgrid. And as our discussions with our partner progressed, it became easy for us to imagine the thousands of off-grid villages across Kenya where this technology would thrive. We took it for granted that the people living in these off-grid villages would be receptive to this novel technology. Soon, we would discover that we were wrong.

3   Iterative Learning: A Major Pivot

In the summer of 2012, we traveled to Western Kenya to scope out a potential research project. The experiment we envisioned was straightforward. First, we would identify a hundred or so off-grid villages, randomly assigning half of them into a treatment group. In these villages, our microgrid partner, with contributions from the engineers on our team, would set up their prepaid, smart metered microgrids. As an additional experimental feature, we thought about varying the price of each connection in order to estimate a demand curve for electricity access.[5] Later on, a team of enumerators would administer detailed, household-level microeconomic surveys. In theory, comparing survey data between households in

---

[3] This account is based on my August 2012 interviews with former microgrid operators and individuals living in several rural communities in Rajasthan, India, where these early microgrids had been built.

[4] In 2012, the World Bank Databank reported that Kenya's electrification rate was 18%. That year, Government of Kenya reported that the electrification rate was 26%. The information on the World Bank Databank has since been updated. Currently, the 2012 and 2013 reported electrification rates are much higher, at 40.8 and 43.0%, respectively.

[5] In economics, the demand curve depicts the relationship between price and quantity. When compared against a supply curve, policymakers can assess the welfare implications of different policy decisions.

the treatment communities and their counterparts in the control communities would yield unbiased, causal estimates of the impacts of electrification.

Our microgrid partner suggested that we find communities with a couple important features. One, we needed villages with a high density of potential users. The microgrids would be sized to supply power to roughly 50 customers. If customers were clustered close together, the line losses on the microgrid's low-voltage network would be minimized. Two, we needed villages with many unelectrified businesses since these were likely to use more electricity, thus increasing revenue to our partner. From our standpoint, we also wanted villages that were far away from existing national grid infrastructure. The last thing we wanted was to invest our resources and time in villages that would soon receive grid electricity from Kenya Power, the national electricity distribution utility.

This was no easy task. After visiting a local Kenya Power office in the Western county of Busia, we learned that Kenya Power had yet to geotag the locations of its infrastructure, meaning there was limited administrative data that could help us locate a sample of off-grid villages. In lieu of actual data, we were given permission to photograph the aging infrastructure maps that were displayed on the walls. In addition, we were provided with an assortment of tips on where we could find the distant yet densely populated communities that would meet our criteria.

3.1 On-grid, Off-Grid, or Under-Grid?

As we drove across Western Kenya searching for rural, off-grid villages, we noticed something peculiar. Although the vast majority of rural homesteads lacked access to electricity, nearly every off-grid village we visited seemed to have a power line running nearby. Rather than being "off-grid," much of what we observed appeared to be *underneath* the grid.

Why were so many rural households left unconnected to these electricity lines? We learned that a major barrier was the high cost of connection. In fact, during the decade leading up to the start of our study, any household in Kenya within 600 m of a low-voltage distribution transformer could apply for an electricity connection at a fixed price of 35,000 Kenya shillings (KES), which was worth roughly $398 USD at the time. This seemed far too expensive in Kenya, where annual per capita income was below $1,000 for most rural households. At the same time, the cost to the utility of supplying a single connection in an area with grid coverage was estimated to be several multiples higher.

Our trip to Kenya that summer was not much of a success. Instead of finding a hundred villages, we found just a handful. However, the experience triggered a shift in the way we viewed the development challenge. Until then, we had been thinking about electricity access as a binary variable. Households were either on-grid or off-grid. What naturally followed was an assumption that off-grid households were too far away to connect to a national electricity network and therefore required novel solutions. We wondered whether this assumption had played a role in the growing enthusiasm among engineers, entrepreneurs, and donors for the new generation of off-grid, distributed energy solutions, most of which were being designed and manufactured outside of the Global South. The solar microgrids that were being developed in Oakland, for instance, had essentially been designed with remote users and communities in Africa in mind.

Suddenly, it seemed plausible that a substantial share of the 600 million people lacking access to electricity were not off-grid but were instead "under-grid," which we defined as being close enough to connect to a low-voltage line at a relatively low cost. This distinction seemed important because the policy implications for off-grid and under-grid communities were quite different. In under-grid communities, it might be preferable to design policies that could leverage existing infrastructure, as opposed to promoting an independent solution like a microgrid.

The argument against grid power seemed to hinge on the extent to which the grid delivered dirty, fossil fuel power. But across sub-Saharan Africa, installed generating capacities were still relatively low, and substantial capacity additions were slated for the future. Importantly, a large share of these additions was expected to feature nonfossil fuel technologies. In Kenya, where fossil fuels represented about a third of installed capacity at the time, several major geothermal and wind projects were already under development. Given the trends, why not focus on expanding electricity access through a grid that might soon be channeling a higher share of clean energy?[6]

3.2   *Private Versus Public Infrastructure*

Upon our return to Berkeley, we ran into problems agreeing on an acceptable research design with our microgrid partner. From a research standpoint, we needed to randomly select the villages where our partner would build their microgrids. In addition, we required access to all of the data generated by their smart meters; we needed to publish our findings, regardless of how favorable the conclusion; and of course, we planned to make all of the data and analysis involved in our work freely available to public.

All of this is highly undesirable for a start-up company. Our microgrid partner needed to prove that it had a viable and scalable business model. There was upside to having the benefits of their technology rigorously and independently measured and published. But there were obvious downsides to giving up control over consequential business decisions, like where they could build their microgrids.

We had other differences in opinion as well. For example, some of the communities that met our microgrid partner's various criteria (e.g., density of structures,

---

[6] See Lee, Miguel, and Wolfram (2016b) for a discussion of whether increases in energy access should be driven by investments in electric grid infrastructure or small-scale "home solar" systems (e.g., solar lanterns and solar home systems).

number of small businesses, etc.) looked to be—at least from our perspective—under grid. And while there may have been good business reasons for building a private microgrid directly underneath the public grid, this did not make sense to us. Moreover, to achieve a certain degree of statistical power in our experiment, we needed our microgrid partner to rapidly scale up its operations. New systems needed to be built in scores of villages, as soon as possible. But the difficulties we had faced in locating suitable sites seemed to portend inevitable delays and slow progress.[7]

We began to consider whether it made more sense to study the economics of expanding national grid access. After our summer travels, we had no doubt that Kenya's future would revolve around its grid. Although the cost of a Kenya Power connection was exorbitantly high, it was roughly in line with the per household cost of our microgrid partner's technology. If we shifted our focus to the grid, we could design a research study with a similar sample size, without requiring an increase to our budget.

In early 2013, our collaboration with our microgrid partner began to taper off. Fortunately, we had begun a promising round of discussions with Kenya's Rural Electrification Authority (REA). Created in 2007, REA was a government agency that had been established to accelerate the pace of rural electrification. Using funds from the central government, international development agencies, and a small tax on every Kenya Power electricity bill, REA had been responsible for rapidly electrifying the majority of the country's rural secondary schools, markets, and health clinics. It was the single reason why so many of the rural households we had observed that previous summer looked to be under-grid.

We proposed an experiment to REA to randomly connect households to the grid and then experimentally measure the impacts of electricity access on a variety of social and economic outcomes. They were interested in the idea. As a public infrastructure agency, REA could adopt a longer-term investment horizon, meaning they did not face the same pressures as a start-up company. Moreover, our project fit with their basic mandate to achieve universal rural electrification. In other words, there was a clear path towards justifying the incremental costs that would be incurred in order to participate in our research project. By the summer of 2013, the CEO of REA had agreed in principle to explore the viability of a randomized experiment.

4 Randomizing Access to Grid Electricity

Infrastructure investments tend to involve high fixed costs, relatively low marginal costs, and long investment horizons. As a result, they tend to be owned and regulated

---

[7] This turned out to be somewhat accurate. By 2017, roughly five years after we first began our partnership discussions, our microgrid partner was operating just four systems, providing power to 300 households in total.

by the government. As mentioned earlier, it is difficult to measure the causal effects of electrification. There are a number of factors that tend to be correlated with the presence of electricity infrastructure (like the placement of roads), and many of these factors could also contribute to changes in key economic outcomes, like employment. In addition, it is only natural that governments would target expensive infrastructure investments towards regions predicted to enjoy the greatest rates of economic growth, or areas that were preferable to the ruling government party and were thus in line to reap a number of other rewards. In these situations, there is a problem of omitted variable bias (sometimes called confounding or selection bias). Whatever we may believe to be the effects of electrification may actually be caused by unobserved variables. If these are not addressed econometrically, the effects of electrification can be overemphasized.

At the outset of our project, rigorous microeconomic evidence in this area was limited. Two papers that caught our attention included the Dinkelman (2011) paper mentioned above, as well as Lipscomb et al. (2013), which estimates the long-term impacts of electrification in Brazil. Both studies relied on administrative data and addressed the issue of omitted variable bias using a similar econometric approach.[8] From a research standpoint, we thought about how a field experiment could remove concerns about omitted variable bias entirely, yielding a new set of causal estimates that developing country policymakers could use as they weighed the relative costs and benefits of investments in health, education, energy, and other areas.

How could we randomly assign access to electricity, without forcing some people to connect to the grid and others to remain in the dark? In a methodological note that was greatly influential for our project, Bernard and Torero (2011) offered a solution: By providing treatment households with a subsidized electricity connection, we could randomly encourage them to connect to the grid. If many people responded to these offers, there would be enough variation in electricity access to measure impacts. On top of that, by offering different subsidy amounts to different households, we could randomly assign the effective price of a connection, allowing us to trace out a demand curve.[9]

4.1 Conducting a Census Across "Transformer Communities"

Now that we had an outline of our experimental intervention in place, several additional decisions needed to be made. The first set of choices involved defining a sample and a unit of randomization. We needed to consider the potential economic

---

[8] See Lee, Miguel, and Wolfram (2020b) for a discussion of the different econometric approaches that have been used to estimate the impacts of electrification

[9] In our final experiment, we randomly assigned effective prices 0, 15,000, and 25,000 KES to the treatment households, while the control households had the option to connect at the prevailing price of 35,000 KES.

spillovers of our treatment. For example, a family living in an electrified household could easily share the benefits of their connection with some of their neighbors. In this case, a direct comparison of living standards between the treated households and neighboring control households would underestimate the size of impacts. To address this issue, we decided to assign our treatment at the community level. This choice would have the additional benefit of avoiding fairness concerns among neighbors, given the high monetary value of the subsidized connection offers.

We also needed to avoid situations in which households applied for power but were rejected for some technical reason, like being too far away from a transformer. So instead of defining communities along the lines of village boundaries—which is what most field experiments randomizing at a community level tended to do—we defined them as "transformer communities," encompassing the universe of structures that were within 600 meters of a central, low-voltage distribution transformer. This, of course, had the added bonus of being consistent with Kenya Power's existing fixed price connection policy.

Working with our partners at REA, we randomly drew a sample of 150 transformers, located at markets, secondary schools, and health clinics, spread across in the counties of Busia and Siaya in Western Kenya. Some of the transformers had been installed 4 to 5 years earlier, so they were not necessarily new.

The next step was to establish a sampling frame, or census, of all of the unconnected and "under-grid" households that could potentially be enrolled into our study. So, over the fall of 2013, a team of enumerators scouted the territory surrounding each transformer on foot, documenting the GPS location, some basic observable features (like the quality of roofs and walls), and the electrification status of every structure they were able to find. A number of iterations were required to establish a survey protocol that could produce maximum coverage of the area inside each transformer community. Examples of various transformer communities are provided in Fig. 5.2.

By December 2013, our field staff had geotagged over 20,000 structures, including households, enterprises, public facilities, transportation hubs, and other types of buildings. The data showed that electrification rates were extremely low, averaging just 5% for households and 22% for businesses. In addition, half of the unconnected households we observed were estimated to be within 200 m of a low-voltage power line.[10]

At that point, REA estimated it had connected 90% of the country's public facilities to the grid. In the highly populated regions of Central and Western Kenya, it was believed that the vast majority of households were within walking distance of multiple public facilities. If true, then the under-grid pattern we documented in Western Kenya could very well extend across vast swathes of the country. For policymakers, connecting people to a network that was already in place seemed like low-hanging fruit on the path towards universal electrification.

---

[10] See Lee et al. (2016a) for a discussion of "under-grid" households and the implications for electrification policies.

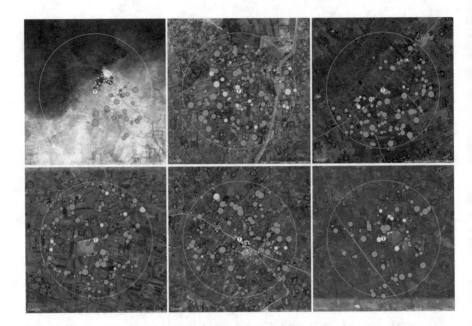

Fig. 5.2 Examples of transformer communities in Western Kenya
Notes: The white circle labeled "T" in the center of each transformer community identifies the location of the REA transformer. The larger white outline demarcates the 600-m radius boundary. Green circles represent unconnected households; purple squares represent unconnected businesses; and blue triangles represent unconnected public facilities. Yellow circles, squares, and triangles indicate households, businesses, and public facilities with visible electricity connections, respectively. Household markers are scaled by household size, with the largest indicating households with more than 10 members and the smallest indicating single-member households. In each community, roughly 15 households were randomly sampled and enrolled into the study. The average density of a transformer community is 84.7 households per community, and the average minimum distance between buildings (i.e., households, businesses, or public facilities) is 52.8 m. Reprinted from Lee et al. (2016a, b)

We reached out to a number of organizations that were involved in rural electrification efforts across Africa. It was an exciting moment in time. Earlier that year, President Obama had launched *Power Africa*, a bold new initiative that targeted, as one of its initial goals, 20 million new electricity connections in six African countries including Kenya. We presented our findings to the coordinator of Power Africa, who was based in Nairobi at the time. Perhaps our under-grid narrative could be useful in shaping their overall strategy. They did not appear that interested in our work. Later, we learned that Power Africa's strategy to meeting its short-term connection targets would be to support the distribution of decentralized solar technologies (mainly solar lanterns and solar home systems) through its *Beyond the Grid* initiative, as shown in Fig. 5.3.

We also presented our findings to the energy leads at the World Bank. From these interactions, we learned that the World Bank was in the process of planning a major

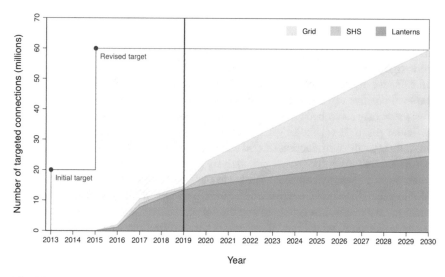

Fig. 5.3 How USAID Power Africa will meet its target of 60 million new connections
Notes: Based on USAID Power Africa Annual Reports

commitment to modernize Kenya's electricity system. Over the following months, we continued meeting with this team to share updates. As we later learned, this commitment from the World Bank would enable the Kenyan government to launch its transformative *Last Mile Connectivity Project* (LMCP), just a couple years later. The initiative, which is briefly mentioned later in the chapter, has already gone on to connect millions of people to the grid.

4.2   Binding Our Hands with a Pre-analysis Plan

In all of our discussions with potential donors, policymakers, and partners, there was always a great deal of interest in discussing the impacts of rural electrification. Everyone seemed to hold their own unique perspectives and predictions on what would happen when rural households accessed electricity for the first time. Existing academic studies tended to focus on the same major outcomes, like employment and educational attainment. This was in part due to their reliance on administrative records, as well as general household surveys, neither of which placed a large emphasis on energy outcomes, like the types of electrical appliances that people owned or wished to own. In our experiment, we would be free to draft our own survey instruments, meaning we could measure anything we wanted. In sum, this was a situation in which the conclusions of our experiment could seem arbitrary since they depended on the outcomes we chose to emphasize. With so many possibilities, how could we best assess the overall impacts of our electrification program?

There was also another issue. With REA, we were collaborating with individuals who had dedicated their entire working lives to expanding rural access to power. Like our former microgrid partner, there was upside to having the benefits of their work rigorously and independently evaluated. Yet in our interactions with REA officials, we sensed a high degree of confidence that our study would point to massive, positive effects. Furthermore, as a government agency, we could imagine how there may have been underlying political pressures or incentives to report a particular result. It was, after all, hard for any of us to picture rural life remaining the same after the introduction of grid electricity. But what if we were wrong?

Around that time, the use of pre-analysis plans in development economics was becoming increasingly common. The basic steps we followed were straightforward. Prior to accessing data, we wrote down our hypotheses; the econometric regression equations we planned to estimate; and the set of key outcomes we would consider. In our initial plan, which was filed before we analyzed the first round of follow-up survey data, we identified 77 outcomes of interest overall, across 10 broad families including energy consumption, productivity, education, and others. We then narrowed this list down to the 10 primary outcomes, shown in Table 5.1, that would guide our conclusions. We discussed these outcomes with our government partners, which allowed us to manage their expectations. Finally, the plan was registered online in order to be made accessible to future readers.[11] The pre-analysis plan disciplined our interpretations of impacts, limiting the scope for data mining or a biased presentation of results. As we would later find, our pre-analysis plan increased the space we had to draw our conclusions and made it easier to defend our findings, given the lack of impacts we eventually found.

5 Iterative Learning: Unexpected Field Challenges

In December 2013, we signed a Memorandum of Understanding (MOU) with REA's Chief Executive Officer. REA agreed to honor the subsidized connection prices in our experiment, giving us the green light to proceed with the main component of our project. The MOU, which took 6 months to sign, was the result of countless in-person visits to the REA headquarters in Nairobi. At each visit, we would deliver a presentation to the CEO and other key managers at REA, updating them on what we were learning from our activities in Western Kenya and slowly building up our relationships with key individuals. At times, we would need to re-pitch the basic objectives of our experiment. At others, REA would provide helpful feedback on the information that would be most useful to them in our research.

In many of these meetings, a sticking point was the cost of the connection. We agreed to use a substantial amount of our research funds to cover the cost

---

[11] The pre-analysis plans for this project can be accessed at https://www.socialscienceregistry.org/trials/350

Table 5.1 Defining ten primary outcomes of interest in a pre-analysis plan

ID	Outcome	Unit	Type	Description	Ref.
P.1	Grid connected	HH	Indicator	Indicator for main household connection	1.1
P.2	Grid electricity spending	HH	Total	Estimated prepaid top-up last month or amount of last postpaid bill	1.7
P.3	Employed or own business – Household	HH	Proportion	Proportion of household members (18 and over) currently employed or running their own business	4.5
P.4	Total hours worked	Resp.	Total	Total hours worked in agriculture, self-employment, employment, and household chores in last 7 days	4.11
P.5	Total asset value	HH	Estimated value	Estimated value of savings, livestock, electrical appliances, and other assets	5.6
P.6	Annual consumption	HH	Value	Estimated value of annual consumption of 23 goods	6.2
P.7	Recent symptoms index	Resp.	Index	Index of symptoms experienced by the respondent over the past 4 weeks	7.3
P.8	Life satisfaction	Resp.	Scale	Life satisfaction based on a scale of 1 to 10	7.8
P.9	Average test score	Child	Z-score	Average of English reading test result and Math test result	8.3
P.10	Political and social awareness index	Resp.	Index	Index capturing the extent to which the respondent correctly answered a series of questions about current events	9.4

Notes: Following Casey et al. (2012), we registered pre-analysis plans for the experiment which are available at http://www.socialscienceregistry.org/trials/350. This table summarizes the ten primary outcomes of interest for analysis using the first two rounds of follow-up survey data

of subsidies, which guaranteed that REA would receive no less than 35,000 KES per connection. In turn, REA would need to cover the difference between the cost of construction and the 35,000 KES raised from each accepted offer. The uncertainty came from the wide variance in the cost of supplying each connection, since this depended on factors like the location, density, and terrain conditions of each community. Although there were economies of scale in connecting multiple households at the same time, it was difficult to predict how many households in each community would accept the offers. Given the many fixed commitments in REA's organizational budget, there were understandable concerns about the total cost of what many at REA viewed as just an academic exercise (which was not inaccurate).

Over time, we established a stronger relationship with our government partners. Crucially, agreed on a compromise. Instead of offering a subsidized price to any unconnected household within the boundaries of a transformer community, we agreed to limit our sample to the households that were no more than 400 m away from a low-voltage line, which worked out to 84.9% of all of the households recorded in our census. According to REA's estimates, this would substantially reduce the construction costs, improving the financial viability of the project.

Our experiment would require contributions from a number of individuals working out of REA's offices in Nairobi and Kisumu at various levels of the organizational hierarchy, each facing a different set of incentives. Given the innate challenges of working in Western Kenya (e.g., travel distances, road conditions, etc.), there were many potential bottlenecks that could delay our progress. Although our MOU was not legally binding, it proved to be surprisingly effective as a document that we could point to in the face of an unanticipated challenge. To our luck, many of our interactions with REA would be facilitated by an ambitious, data-driven, and public-minded bureaucrat with whom we would work closely. This person not only took a keen interest in seeing the academic results of our study through but also proved instrumental in clearing several administrative roadblocks we faced. His involvement ensured that much of our work would be executed smoothly. Still, we encountered a number of unexpected problems, some of which required us to revise our intervention protocols and others that simply stalled our progress. Throughout the process, we carefully documented each of these issues, which helped us make sense of our eventual results.

5.1 Connecting Households in Areas with No Electricians

In the spring of 2014, our team of enumerators made rapid progress enrolling households into our study and collecting baseline social and economic data. Meanwhile, we worked out the details of our randomized pricing intervention, which was scheduled to commence that summer. The process was as follows. After completing the baseline survey, treatment households would receive a letter from REA describing a time-limited opportunity to connect to the grid at a subsidized price, which was randomly assigned at the transformer community level. At the end of the offer period, our staff would verify payments and provide REA with the list of households that needed to be connected. REA would then dispatch designers to each community to sketch out the low-voltage network that needed to be built. Based on these drawings, such as the one shown in Fig. 5.4, REA would earmark the appropriate amount of materials, including wooden poles, low-voltage feeder lines, and service drop lines. Next, a construction team would arrive on site to connect households. Each household would then register an account with Kenya Power in order to receive a prepaid meter. Once complete, households could use power.

There was one unanticipated problem. Although REA could attach a service line to a building, households would still need outlets for plugging in their appliances, sockets for light bulbs, fuses, and other internal wiring. But in rural areas with low electrification rates, there were few electricians.

Panel A

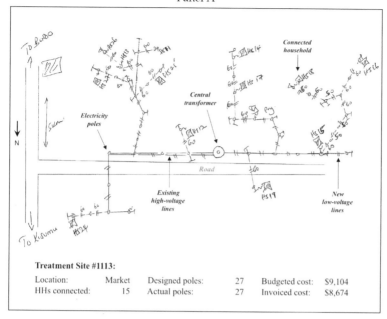

Treatment Site #1113:

Location:	Market	Designed poles:	27	Budgeted cost:	$9,104
HHs connected:	15	Actual poles:	27	Invoiced cost:	$8,674

Panel B

Fig. 5.4 Example of a REA design drawing and the electrification of a treatment household
Notes: After receiving payment, REA designers visited each treatment community to design the local low-voltage network. The designs were then used to estimate the required materials and determine a budgeted estimates of the total construction cost. Materials (e.g., poles, electricity line, service cables) represented 65.9% of total installation costs. Reprinted from Lee et al. (2020a)

To solve this problem, we located a manufacturer in Nairobi that could produce a customized "ready-board," an all-in-one household wiring solution. Each ready-board, which is shown in Fig. 5.5, featured a single light bulb socket, two power outlets, and two miniature circuit breakers. The ready-board was designed to be installed on the indoor side of an exterior wall, so that an outdoor service line could pass through a hole in the wall and connect directly into the back of the ready-board. The ready-boards were designed to be modular as well, thus providing households with the option of installing additional boards as their consumption needs grew. As it turned out, the wiring issue was a relatively easy challenge to address. However, it illustrated how the success of a particular technology in a developing country could be hindered by the lack of a key supporting service.

5.2 Major Supply-Side Issues: Blackouts and Construction Delays

Blackouts were another issue. Most of the time, the blackouts in Western Kenya would last just minutes or a few hours. But sometimes, a blackout would cut power from an entire community for months. In 2014, as REA carried out its connection work, we began noting the frequency, duration, and primary reason for all of the long-term blackouts experienced in our sample. In total, 19% of the transformers in our sample experienced at least one long-term blackout, which lasted 4 months, on average. The transformers seemed prone to burnouts, and other technical failures caused by severe weather conditions. In some cases, Kenya Power would temporarily relocate a transformer to another community, unannounced. There were also reports of vandalism, as well as theft due to the perceived value of the copper and oil components inside the transformers.

Overall, it took an extraordinarily long time to connect households in our experiment. The first household was metered in September 2014, just a couple months after it had completed payment. The last household was metered over a year later in October 2015. The average connection time was seven months.

Major delays arose at each stage of the process, as shown in Fig. 5.6. The longest average delays occurred during the design phase (57 days) and the metering phase (68 days). The design delays were caused in part by a sudden government announcement in 2015 to provide free laptops to all Grade 1 students, nationwide. A presidential election was on the horizon, and the incumbent would once again be running for office. Since only half of Kenya's primary schools were electrified at the time, REA suddenly found itself under pressure to connect more primary schools. As a result, there were less REA designers available to work on less-prioritized projects, like ours.

The metering delays occurred because of unexpected issues at Kenya Power. There were lost meter applications, shortages in prepaid meters, competing priorities for Kenya Power staff, and in some cases, expectations that bribes would be paid.

Panel A

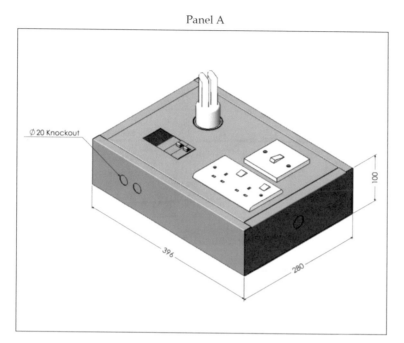

Panel B

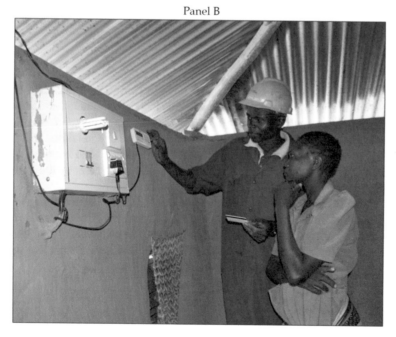

Fig. 5.5 *Umeme Rahisi* ready-board solution designed by Power Technics
Notes: Treatment households received an opportunity to install a certified household wiring solution in their homes at no additional cost. Each ready-board, valued at roughly $34 per unit, featured a single light bulb socket, two power outlets, and two miniature circuit breakers. The unit is first mounted onto a wall and the electricity service line is directly connected to the back. The hardware was designed and produced by Power Technics, an electronic supplies manufacturer in Nairobi. Reprinted from Lee et al. (2020a)

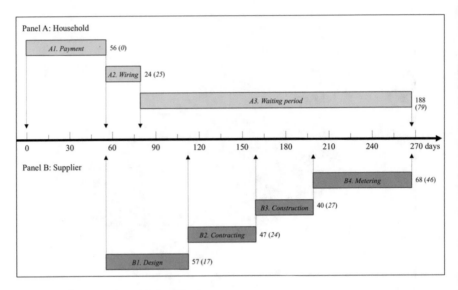

Fig. 5.6 Timeline of the rural electrification process
Notes: Panel A summarizes the rural electrification process from the standpoint of the household, divided into three key phases. Panel B summarizes the process from the standpoint of the supplier, divided into four key phases. The numbers to the right of each bar report the average number of days required to complete each phase (standard deviations in parantheses). Households were first given 56 days (8 weeks) to complete their payments. Afterwards, it took on average 212 days (7 months) for households to be metered and electricity to flow to the household. Reprinted from Lee et al. (2020a)

Other delays were caused by a general shortage in construction materials at REA storehouses. Heavy rains made roads impassable in some communities. Difficulties obtaining wayleaves, which permit electricity lines to pass through private property, required drawings to be reworked, additional trips to the storehouse, and further negotiations with contractors. In some cases, households that had initially declined the ready-board changed their minds; in an unfortunate case lightning struck, damaging a household's electrical equipment; and so on.

Of course, these kinds of problems are not unusual in developing country settings. As researchers, there was little we could do to address what were primarily supply-side issues. So, we instructed our project staff to send weekly and persistent reminders to REA and Kenya Power, and we took notes. It's possible that without these reminders, the delays could have been much worse.

5.3   Discovering Cost Data and Investigating Potential Leakage Issues

As we strove to better understand the root cause of each delay, we became deeply familiar with REA's internal administrative procedures. For instance, we learned that once the design drawings were complete, the total cost of required materials was tallied up in an official budgeted cost slip. Similarly, following the construction work, a final invoiced cost slip was generated and stored in REA's database. With REA's permission, we were able to access and integrate these data into our analysis, which added an unexpected cost angle to our study.

Essentially, our intervention consisted of bundling electricity connection applications together. As a result, the pricing variation we had introduced at the community level—in order to estimate a demand curve—had the bonus effect of creating local construction projects at various scales. With REA's cost data, we could now trace out the average and marginal costs at different community coverage levels, allowing us to study economies of scale. By combining the demand and cost data, we could assess the social surplus effects of electrification, using the textbook framework of electricity distribution as a natural monopoly.[12]

The cost data allowed us to think about another issue. Midway through our intervention, we began hearing stories that some of the poles had gone missing. There were rumors that a contractor had appropriated some of the assigned materials and sold them back to REA's suppliers. In the cost data, however, the budgeted and invoiced cost slips reported nearly identical figures. So, to investigate, we asked our field staff to return to each community to count the number of electricity poles in the ground and compare these numbers with the budgeted and invoiced cost slips. As it turned out, more than 20% of the poles were missing in the field.

In economic theory, it can be debated whether this type of leakage is economically harmful. REA's loss, after all, might be offset by the contractor's private gain. However, we learned that using less poles could cause lines to sag, and this not only lowered service quality but also increased the risk that poles fell over. Thus, it seemed that leakage could be an important issue affecting the long-run reliability of the grid and the overall economic returns to grid infrastructure.

5.4   The Gap Between Demand and Costs

By the end of 2014, we had collected enough data to trace out the demand curve and plot a few points of the cost curve. In our pre-analysis plan, we had recorded our

---

[12] In microeconomics, an industry is a natural monopoly if the production of a particular good or service by a single firm minimizes cost. The classic microeconomic textbooks often mention water, fixed line telecommunications, or electric power, all of which require physical distribution networks, as real-world examples of natural monopolies (see Mankiw, 2011 for an example).

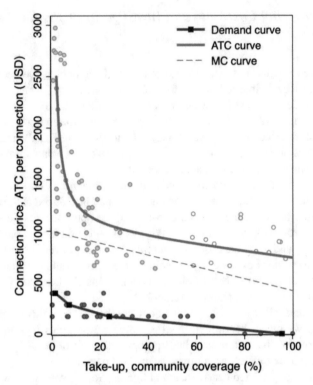

Fig. 5.7 Experimental evidence on the demand for and costs of rural electrification
Notes: The experimental demand curve is combined with the population-weighted average total cost per connection (ATC) curve corresponding to the predicted cost of connecting various population shares, based on the nonlinear estimation of $ATC = b_0/M + b_1 + b_2M$. Each point represents the community-level, budgeted estimate of ATC at a specific level of coverage. Reprinted from Lee et al. (2020a)

own prior estimates of demand at various prices. In addition, REA had shared an internal memo with us that included the government's own predictions of demand. By comparing our experimental estimates with these two sets of priors, we could conclude that demand was much lower than expected. Our cost curve, which was incomplete at the time, appeared to show substantial economies of scale from connecting numerous households at the same time. The average cost of a connection in our experiment was around $1,200. However, we did not yet have the data to know that the economies of scale would quickly level off, and that at full community coverage, the average connection cost would still be two to three times higher than the status quo connection price of 35,000 KES. The final demand and costs curves are shown in Fig. 5.7.[13]

---

[13] See Lee, Miguel, and Wolfram (2020a) for a full description of the experimental results.

From our perspective, the gap between demand and cost, which we referred to as social surplus, was important. In economics, the area under our demand curve could be interpreted as the present value of all future benefits accruing from grid access over the population. If the demand curve was much lower than the supply curve, then this would suggest that the benefits of rural household grid access were not high enough to offset the immediate cost of supply.

We needed time to process and debate these early results. But in March 2015, we were encouraged to present our findings to a number of audiences in Nairobi, including the recently assembled *National Electrification Strategy* committee. The committee featured many familiar faces from REA and Kenya Power, as well as representatives from various government departments. The purpose of the committee was to determine the specific details of the government's rural electrification plan, which included the work that would be supported by the World Bank. The response to our presentation was positive. Soon afterwards, a committee member shared with us a draft version of the national electrification strategy. The document had been written around that time, with key contributions from our collaborators at REA. It estimated the economies of scale from a mass connection program would yield cost savings of 30%, in line with the early results we had shared with our partners. It also recommended a reduction in the connection price from 35,000 to 15,000 KES, one of the experimental price points in our study. A couple months later, President Kenyatta announced the *Last Mile Connectivity Project* (LMCP), a $364 million program funded by the World Bank and the African Development Bank. The connection price would soon be reduced for everyone.

In July 2015, we were invited to present our findings at a workshop organized to launch the World Bank's $458 million commitment to modernizing Kenya's electricity system, which would provide support for the LMCP, among other investments. We pointed out the large and potentially problematic gap between demand and cost, as well as some of the field challenges described above. The response was again positive. Then, later in the year, additional details of the LMCP were released to the public. It was made clear that unlike in the past, clusters of potential customers would now be connected to the grid at the same time. In addition, all customers requiring internal wiring would be provided with a free ready-board.

In retrospect, 2015 introduced momentous changes to Kenya's rural electrification outlook. And given the role of REA in drafting the national electrification strategy—as well as our frequent presentations to Kenya Power, the World Bank, and others—it is possible that our research influenced some of the assumptions and decisions that were being made at the time. We will never know for sure, but the timing of our discussions and the later policy decisions seems consistent with our research exerting at least some level of influence during the key moments.

That said, in 2015, we could not answer any questions about the economic impacts of our intervention. Due to the connection delays, which had been ongoing, follow-up survey data remained months away. Moreover, we had not yet agreed on a satisfying way to interpret the gap between demand and supply—an explanation that factored in the potential budget and credit constraints at the consumer level, as well as the organizational performance issues we were observing in the field. And with each passing month, it began to feel as though our research progress was falling more and more behind. We had been there in Kenya at the start of all of these consequential policy choices. But by the time we had our survey evidence on impacts a couple years later, the national electrification plan was set and the LMCP was well under way.

6   No Meaningful and Statistically Significant Impacts

By November 2016, we had completed our first round of follow-up household surveys, which took place 16 months after connection on average. With a pre-analysis plan in place, it was fairly easy to calculate the impacts of our intervention on our primary set of outcomes. As expected, energy consumption had increased for treated households, but only by a miniscule amount. People had not really acquired any new appliances either. In fact, there were no detectable effects on assets, consumption, health, student test scores, and a host of other outcomes. After nearly a year and a half, we had no evidence of any meaningful and statistically significant impacts of electricity access, a result that differed from much of the earlier literature on the topic. Although this was perhaps fascinating from a research standpoint, it was a result that also felt depressing and demoralizing on a human level.

In the spring of 2017, we met with the Principal Secretary of the Ministry of Energy and Petroleum in Nairobi, as well as the heads of Kenya Power, REA, and the other parastatals comprising Kenya's electricity sector. We discussed tariff reform, and in particular, the need to eliminate a monthly fixed charge that had made the prepaid meters difficult for consumers to understand and use.

We wanted to know how the government planned to do to encourage productive use, as it approached its goal of universal access. We presented our preliminary impact results, which the secretary found convincing. He noted, quite memorably, that the question of universal electricity access was not an economic one, but a political one. It was an election year, and the LMCP was a daily topic in the news. Our conversation highlighted the political economy considerations that often determine which and when certain groups benefit from government programs.

By the end of the year, we completed a second round of household surveys, which took place roughly 32 months post connection. The data revealed a similar pattern of no meaningful and statistically significant impacts. The evidence was consistent with the large gap between demand and cost that we had estimated in our experiment, several years earlier.

We were challenged to consider what might happen to our demand and cost analysis if the surrounding institutional and economic context was more favorable. That is, how much higher would the demand curve be if we could eliminate the credit constraints, blackouts and delays that may have suppressed demand? And what would happen to our estimate of social surplus if we eliminated leakage and incorporated a moderate level of income growth?

To answer this question, we referred heavily to the notes we had kept during the construction process, which provided us with the average duration of connection delays, the average amount of time the transformers had blacked out, and other statistics. Using this information, we projected the incremental effect on social surplus of fixing each issue. For example, we estimated that reducing the waiting period from 188 to 0 days would increase the consumer surplus by about 30%. In our final calculation, we predicted that if a number of improvements were simultaneously introduced, the area under the demand curve would finally exceed total costs, reversing our conclusion. This was, of course, an ideal scenario. In the real world, the LMCP was being rolled out across rural Kenya at one of our subsidized price levels. Surely, the plethora of issues that had affected demand and costs in Western Kenya existed elsewhere across rural Kenya.

The overall interpretation of our experimental results was that providing poor households with electricity access alone was not enough to improve economic and noneconomic outcomes. This stood in stark contrast to the previous, non-experimental studies that had documented large and beneficial gains from electrification. Perhaps in those settings, there were other factors, either correlated with or visibly part of the electrification efforts, that had influenced the direction of the results.

7 Looking Ahead

Since 2000, there has been tremendous progress across the world in reducing the number of people living without access to electricity. But nearly all of the global gains have been achieved in India. By 2030, roughly 500 million people in sub-Saharan Africa will still be without power (IEA, 2019). It is likely that Kenya's model of mass electrification will serve as a blueprint for a number of African countries in the years to come.

With our experiment largely complete, we began thinking about new areas that seemed ripe for further research and that would have relevance to other settings in sub-Saharan Africa. For example, there is the question of whether the impacts of electrification are concentrated in certain types of individuals, like people who have the means to make the most out of a new connection. There was some evidence of this in our data. The impacts appeared larger for households that had a high willingness to pay for a connection, which was positively correlated with income and education at baseline. But due to limitations in our sample size, our results were suggestive at best.

There are also questions about the interaction between a technology like grid electrification and surrounding contextual factors. Would the impacts of our experiment have been greater if electricity access were paired with some kind of complementary input? What if the beneficiaries of Kenya's LMCP also received subsidies to purchase different kinds of electrical appliances? Would this encourage people to experiment with new activities, thus paving the way towards greater levels of consumption and impacts?

Moreover, there is the question of how to make it easier for new users to consume and pay for electricity consumption. This is perhaps the most pressing challenge facing Kenyan energy policymakers today. Consider for a moment that between 2012 and 2019, half of Kenya's population gained access to grid electricity for the first time. (More than three quarters of Kenyans are now connected to power.) This is a historic achievement. Yet the meteoric rise in electricity connections (shown in Panel A, Fig. 5.8) has been matched by a plunge in average electricity consumption per connection (Panel B) as a greater share of poor, rural households are connected to the grid (as is perhaps predicted in our research).

The problem is that each new connection requires a small expansion in the national low-voltage network. But if the marginal customer is generating little to no revenue, there is less money, on average, to maintain high-quality connections across the expanded network. Panel C illustrates this point by plotting the recent, sharp decline in gross profit per customer and gross profit per kilometer of low-voltage network lines.[14] Based on the reliability and leakage issues in our study, it is easy to imagine this problem spiraling into the future, until the grid is completely overrun with service quality issues. And as some have argued, poor-quality electricity service can lower incentives to pay for consumption, leading to a self-reinforcing, vicious circle of blackouts, and bankruptcies.[15] How will Kenya encourage greater rates of paid consumption in order to minimize the consequences of electrifying too fast?

Finally, in the face of climate change, the development challenge looms ever as large. How can developing countries expand access to energy while minimizing the costs to the environment? As this case study describes, we initially set out to build solar microgrids but ended up focusing on the grid instead. Thinking back, I still believe this was the right move. But in a way, our conclusions on the economics of rural electrification are less than satisfying. Not only was the grid a costly intervention, but our experiment produced no evidence of meaningful and statistically significant impacts, at least in the medium run. Meanwhile, new innovations will continue pushing forward the technological frontier of decentralized, renewable energy alternatives. Did we turn away from solar microgrids too soon? Or is the future calling for a more coordinated deployment of private and public infrastructure? These are just a few of the questions that deserve further consideration.

---

[14] I use gross profit, which is defined as revenue minus energy power purchase costs, as a crude proxy for the maximum amount of cashflow available for Kenya Power to pay annual maintenance expenses. We do not see the same decline in gross profit per kilometer of 11 kVA lines, since the LMCP targeted Kenya's under-grid population and thus did not require new transformer installations.

[15] See Burgess et al. (2020) for an example of a vicious circle in the context of rural electrification.

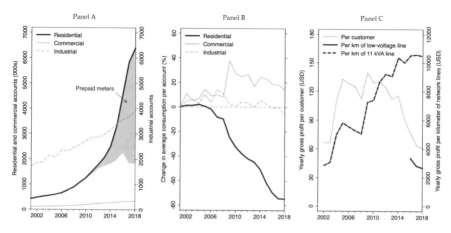

Fig. 5.8 Electricity access and consumption trends in Kenya

Notes: Based on Kenya Power annual reports and my own calculations. Panel A illustrates the recent expansion of the Kenyan electricity grid, particularly across residential consumers. Panel B shows how average consumption per customer is rapidly falling (since growth in the customer base is coming from poor, rural households). Panel C plots gross profit—defined as revenue minus power purchase costs—per customer, per kilometer of low-voltage network lines (data only available from 2016 and beyond), and per kilometer of 11 kVA network lines. Panel C suggests that the ability of Kenya Power to pay for line maintenance using operating revenue is declining

8 Discussion Questions

1. The researchers find no evidence of meaningful and statistically significant impacts of household grid electrification in the short and medium run in rural Kenya. What do you think are the implications of this result on the broader United Nations Sustainable Development Goal (SDG7) of universal access to affordable, reliable, sustainable, and modern energy for all by 2030?

2. The author hypothesizes that the impacts of grid electrification may have been greater if electricity access had been paired with complementary inputs. What complementary inputs could have the highest potential for impact? How would you design a research study to measure the impacts of supplying these inputs? Assuming that the provision of these inputs is effective, how would you revise the design of a mass electrification policy, such as the LMCP, in Kenya?

3. The author suggests that making it easier, for new users to consume and pay for electricity consumption may be the most pressing challenge for rapidly electrifying countries such as Kenya. Which interventions, policies, and technologies do you think hold the greatest potential to address this challenge?

4. A recurring issue in this project was striking a balance between meeting the goals of the infrastructure provider, while meeting the standards of academic research in applied economics. What were the key decisions in designing and implementing this study? And what changes do you think could have improved

the general study outcome, both in terms of the study results as well as the policy impact?
5. Did the researchers pivot away from solar microgrids too soon?

References

Bernard, T. (2012). Impact analysis of rural electrification projects in Sub-Saharan Africa. *World Bank Research Observer, 27*(1), 33–51.
Bernard, T., & Torero, M. (2011). Randomizing the "Last Mile": A methodological note on using a voucher-based approach to assess the impact of infrastructure projects. *International Food Policy Research Institute (IFPRI) discussion papers, 1078.*
Burgess, R., Greenstone, M., Ryan, N., & Sudarshan, A. (2020). The consequences of treating electricity as a right. *Journal of Economic Perspectives, 34*(1), 145–169.
Casey, K., Glennerster, R., & Miguel, E. (2012). Reshaping institutions: Evidence on aid impacts using a preanalysis plan. *Quarterly Journal of Economics, 127*(4), 1755–1812.
Dinkelman, T. (2011). The effects of rural electrification on employment: New evidence from South Africa. *American Economic Review, 101*(7), 3078–3108.
IEA (International Energy Agency). (2019). *World Energy Outlook.*
Khandker, S. R., Samad, H. A., Ali, R., & Barnes, D. F. (2014). Who benefits most from rural electrification? Evidence in India. *Energy Journal, 35*(2), 75–96.
Lee, K., Brewer, E., Christiano, C., Meyo, F., Miguel, E., Podolsky, M., Rosa, J., & Wolfram, C. (2016a). Electrification for 'under grid' households in Rural Kenya. *Development Engineering, 1,* 26–35.
Lee, K., Miguel, E., & Wolfram, C. (2016b). Appliance ownership and aspirations among electric grid and home solar households in Rural Kenya. *American Economic Review: Papers and Proceedings, 106*(5), 89–94.
Lee, K., Miguel, E., & Wolfram, C. (2020a). Experimental evidence on the economics of rural electrification. *Journal of Political Economy, 128*(4), 1523–1565.
Lee, K., Miguel, E., & Wolfram, C. (2020b). Does household electrification supercharge economic development? *Journal of Economic Perspectives, 34*(1), 122–144.
Lipscomb, M., Mushfiq Mobarak, A., & Barham, T. (2013). Development effects of electrification: Evidence from the topographic placement of hydropower plants in Brazil. *American Economic Journal: Applied Economics, 5*(2), 200–231.
Mankiw, N. G. (2011). *Principles of economics* (5th ed.). Cengage Learning.
Wolfram, C., Shelef, O., & Gertler, P. (2012). How will energy demand develop in the developing world? *Journal of Economic Perspectives, 26*(1), 119–138.

Chapter 6
Measuring Grid Reliability in Ghana

Noah Klugman, Joshua Adkins, Susanna Berkouwer, Kwame Abrokwah, Matthew Podolsky, Pat Pannuto, Catherine Wolfram, Jay Taneja, and Prabal Dutta

1 The Development Challenge

The power grid is arguably the most complicated machine humanity has built, and the payoffs from this marvel have been transformative. No country has achieved economic industrialization without significant increases in energy use. Hospitals, schools, factories, and homes across the world depend on electricity for their daily

N. Klugman (✉) · J. Adkins · P. Dutta
Electrical Engineering and Computer Science, University of California, CA, USA

nLine, Inc, Berkeley, CA, USA
e-mail: nklugman@berkeley.edu; adkins@berkeley.edu; prabal@berkeley.edu;
https://www.nline.io

S. Berkouwer
The Wharton School, University of Pennsylvania, Philadelphia, PA, USA
e-mail: sberkou@wharton.upenn.edu

K. Abrokwah
Brixels Company Limited, Accra, Ghana

M. Podolsky
Electrical Engineering and Computer Science, University of California, CA, USA

P. Pannuto
Computer Science and Engineering, University of California, San Diego, CA, USA
e-mail: ppannuto@ucsd.edu

C. Wolfram
Haas School of Business, University of California, Berkeley, CA, USA
e-mail: cwolfram@berkeley.edu

J. Taneja
Electrical and Computer Engineering, University of Massachusetts, Amherst, MA, USA
e-mail: jtaneja@umass.edu

© The Author(s) 2023
T. Madon et al. (eds.), *Introduction to Development Engineering*,
https://doi.org/10.1007/978-3-030-86065-3_6

operations. As such, the developing world has seen tremendous investments in the electricity grid in recent years.

Investments in the electricity sector in the developing world have often focused on increasing access to electricity by expanding the grid. There has been less focus on increasing the quality of electricity provided by the existing grid. However, poor reliability is often associated with a reduction in the demand, utilization, and social benefit of electricity (Gertler et al., 2017). Reassuringly, the importance of grid reliability is increasingly being recognized: the UN Sustainable Goals now specify that access to electricity must also be reliable (McCollum et al., 2017).

Improvements in electricity reliability can be harder to achieve than improvements in access. Improving a grid's reliability requires fine-grained information about how different attributes of the grid perform. However, these data are historically expensive to collect and not prioritized (The GridWise Alliance, 2013; Taneja, 2017). Many electrical utilities, investors, and energy regulators are under-resourced for the enormously complex and expensive task of planning, extending, and operating the grid at consistent reliability of service, especially in the presence of rising consumption and increasingly adverse weather patterns. Thus, the development challenge is to reduce the economic and technical barriers to monitoring the quality of existing electricity networks in the developing world, so that this data can inform actions taken to improve grid reliability.

Developing a novel technology in an academic laboratory setting is not sufficient for addressing critical information gaps: The widespread deployment of these technologies brings a unique and unpredictable set of challenges. This chapter describes a project that transferred research out of the academic laboratory and into a field setting. Our interdisciplinary team of engineers and economists designed, deployed, and continues to operate a large sensor network in Accra, Ghana, that measures power outages and quality at households and firms. The deployment began in June 2018 and consisted of 457 custom sensors, nearly 3500 mobile app installations, and nearly 4000 participant surveys, as well as user-incentive and deployment management meta-systems developed by our team, which will continue collecting reliability data until at least September 2021.[1] Grid reliability data collected by our deployment in Accra will enable multiple impact evaluations of investments in the grid that, in part, explore the effects of improvements in reliability on socioeconomic well-being.

Entering the field, we anticipated we would have to pivot our technology and deployment methodology, both initially designed in the lab, to respond to unanticipated cultural, technological, and organizational requirements. This indeed happened, as we describe throughout this case study. However, we also held the incorrect assumption that the non-technology dimensions of this work would stabilize over time and wouldn't meaningfully contribute as a barrier to scale as compared to overcoming engineering hurdles. However, in practice, we found that

---

[1] This research has Institutional Review Board approval from the University of California at Berkeley (CPHS 2017-12-10599).

pain points experienced at each deployment scale were fundamentally interdisciplinary, often manifesting as new or more stringent cultural and organizational requirements, each of which required innovation to overcome. This case study therefore focuses mainly on the consequences of changing deployment context, in this case by changing scale, and distilling lessons we wish we had learned before forecasting budgets, human resource requirements, and project timelines. These lessons remain critical for our team as we continue to scale our deployment beyond the work described here. We hope that learning about our case study encourages other researchers and helps them hit the ground running as they pursue highly granular measurements of the world's critical systems.

2 Context

We briefly introduce the context in our deployment site of Accra, Ghana, touching on the local energy context and introducing our implementing partners. We then introduce the research questions that data from our deployment was designed to explore, providing further context for the design decisions described in the rest of this case study.

2.1 Energy Environment in Ghana

Ghana's electric grid has roughly 4.74 million connections and experiences a peak load of 2881 MW, a supply capacity of 4695 MW, and an estimated 24.7% distribution loss rate (Energy Commission Ghana, 2019). The distribution utility in the capital city of Accra is the Electricity Company of Ghana (ECG).

Reliable electricity has the potential to provide huge social and economic benefits (Bowers, 1988; McCollum et al., 2017; Hamidu & Sarbah, 2016) In Ghana, however, the grid at times falls short of enabling these, causing frustration that has even boiled over to civil unrest (Aidoo & Briggs, 2019; Ackah, 2015). From 2013 through 2015, the country experienced drastic electricity undersupply, culminating in outages between 6 and 24 h during 159 days in 2015. This period is known as "Dumsor," the Twi word for "off-on." While Dumsor has largely been remedied with the introduction of new generation capacity (Clerici et al., 2016; Millennium Challenge Corporation, 2008), Ghana still reports electricity reliability metrics that underperform relative to countries with similar GDPs (Millennium Development Authority, 2018).

2.2 Partners

Mirroring the increasingly global focus on energy reliability, current investments in the Ghanaian electricity sector are primarily aimed at improving the *reliability* of electricity distributed on the existing grid. Partially in response to the Dumsor crisis, the Millennium Challenge Corporation (MCC) and the Government of Ghana signed the Ghana Power Compact in 2014, a USD 308 million[2] investment designed to improve the grid generation, transmission, and distribution systems in Ghana, to be implemented by the newly created Millennium Development Authority (MiDA) (Millennium Development Authority, 2018). This investment has multiple goals, including cutting operational costs, reducing transmission and distribution losses, increasing affordable access to grid connections, and improving reliability. The country's current work to improve grid reliability motivated our selection of Ghana as our deployment site (Millennium Challenge Corporation, 2008). One of the goals of our deployment is to work with both MCC and MiDA to provide data to support their monitoring and evaluation goals.

2.3 High-Level Goals

2.3.1 Improving Energy-Reliability Data Quality

$$SAIDI = \frac{\text{Total duration of sustained interruptions in unit time}}{\text{Total number of consumers}} \qquad (6.1)$$

$$SAIFI = \frac{\text{Total number of sustained interruptions in unit time}}{\text{Total number of consumer}} \qquad (6.2)$$

To improve reliability, it is important to measure it (McCollum et al., 2017; Sustainable Energy for All, 2018). Two widely used key performance indicators that capture the overall reliability of the grid are the system average interruption duration index (SAIDI) and the system average interruption frequency index (SAIFI) (IEEE Std 1366, 2001). The calculation of SAIDI and SAIFI is shown as Eqs. (6.1) and (6.2). The monitoring and evaluation teams at MCC and MiDA consider changes in SAIDI and SAIFI to be indicative of the impact of the Ghana Power Compact (Millennium Development Authority, 2018).

---

[2] When the Compact was first signed in 2014, it totalled USD 498 million – this amount was lowered to USD 308 million in 2019 when a 20 year concession of the utility was cancelled.

In Accra, the highest spatial and temporal resolution measurements of grid interruptions are collected by the supervisory control and data acquisition (SCADA) system operated by ECG. This SCADA system has limited reach, covering only high-voltage transmission lines and some portion of the medium-voltage distribution network (Nunoo & Ofei, 2010). To improve monitoring, ECG has recently deployed some smart meters, but economic and social challenges raise barriers to achieving broad smart meter coverage in the short term (Acakpovi et al., 2019; Millennium Development Authority, 2014; Banuenumah et al., 2017). ECG recently completed a much larger effort to upgrade their prepaid meters; however, many of the prepaid meters do not regularly communicate power quality measurements back to the utility (Quayson-Dadzie, 2012; Electricity Company of Ghana., 2019).

Measurements of low-voltage outages come primarily from customer calls, yet data collected from the national call center suggest that this data stream is imperfect. Frequent outages may reduce willingness to report, as calling the utility to report an outage can require significant effort without a guarantee that it will shorten the duration of an outage: 90% of respondents to a survey we conducted had not reported a single outage to ECG in the preceding 3 months, despite also acknowledging in the same survey that they had experienced outages during that time. There are few call center reports during the night while people are asleep, and dips in reporting during the day while people are at work. While some of these patterns may reflect reality (a grid will fail more often when it is hot and being used at capacity, conditions more likely in the middle of the day (He et al., 2007)), these data are likely not truly representative.

Therefore, our deployment aims to generate more accurate estimates of SAIDI and SAIFI by placing sensors in the field that extend beyond the SCADA system to automatically report the location and duration of power outages in the low-voltage network, as well as further upstream. We also capture voltage fluctuations and frequency instabilities at the low-voltage level, as these can have significant impacts on the value of appliances and machinery.

2.3.2   Providing Utility-Independent Measurements

To understand how well infrastructure improvements impact reliability, it is important to have utility-independent measurements (Millennium Development Authority, 2018). Many widely used tools, including SCADA and smart meter technologies, are dependent on utility participation, in part because they directly interface with utility property. From an academic perspective, independence is important as it allows for unbiased research output. Independence is often desired by regulators as well, who may want to verify measurements provided by the utility, as the utility has incentives to report favorable reliability metrics. Currently, no high-resolution source of independent data about grid reliability in Accra exists. Our deployment was designed to evaluate the feasibility and efficacy of a novel sensing methodology for monitoring the reliability of the electricity grid while working independently of the utility. Our physical sensor is designed to be installed at outlets in households and business, and our sensing app personal smartphones, allowing us to choose deployment sites and deploy sensors without utility involvement. The data returned from our deployment is truly independent.

2.3.3 Exploring Impacts of Reliability

The causal relationship between electricity reliability and socioeconomic well-being is not well understood. Anecdotally, frequent outages constrain economic well-being by reducing the benefits from welfare-improving appliances, such as fans and refrigerators, or income-generating assets, such as welding, sewing, or other productive machinery. Our deployment was designed in part to generate both reliability and socioeconomic data to support an ongoing economic study that aims to estimate the causal impact of power quality and reliability on socioeconomic outcomes, such as productivity and health, for residents and firms in Accra.

3   Innovate, Implement, Evaluate, Adapt

Starting in May 2018, we conducted three deployments at three different scales: a small-scale pilot, a medium-scale deployment, and a large-scale deployment.[3] To date, 3400 individuals in Accra downloaded our mobile app, called DumsorWatch; 457 people installed our plug-in sensor, called PowerWatch; and over 4000 surveys were performed to directly measure socioeconomic outcomes. In December 2019, we surveyed an additional 462 participants to understand their experiences and to collect updated measures of time-varying socioeconomic outcomes.

We present the technology deployed, the design of our deployment, and where our planning and assumptions failed or caused unexpected problems. While doing so, we attempt to categorize our key challenges and describe the steps we have taken to overcome each of these challenges, emphasizing issues that arose that were more complex and/or costly than originally forecasted. We find that our experiences differed depending on the scale of the deployment, each scale uncovering its own complexities.

3.1   Innovation: Data Collection Instruments

We developed two types of data collection instruments – sensors and surveys – to achieve the goals described in Sect. 2.3. These instruments collect the required data to estimate SAIDI and SAIFI independently of utility participation.

---

[3] As of November 2020, we continue to expand with additional deployments.

3.1.1 Sensors

We developed two different sensors that detect the presence and absence of grid power: an app called DumsorWatch that is installed on a participant's mobile phone and a sensor called PowerWatch that is plugged into a power outlet at a household or business.

DumsorWatch is an Android app installed on the everyday-use smartphone of a participant who lives and/or works in Accra. "Dumsor," the local word for power outages, was used for branding and association with power outages in the Ghanaian context. DumsorWatch automatically senses power outages and power restorations through a combination of on-phone sensors and cloud services (Klugman et al., 2014).

PowerWatch, our plug-in sensing technology, integrates power reliability sensors with a GSM radio to send measurements in near-real time to a cloud database (Fig. 6.3). By designing PowerWatch to plug into a participant's home or business, as opposed to connecting directly to the electric grid, we avoid the need for prior approval or cooperation to deploy the sensors and therefore maintain independence from the utility, a primary goal of our deployment. PowerWatch senses power outages and power restorations timestamped to the millisecond, as well as GPS-based location, voltage, and grid frequency.[4] PowerWatch contains a battery to allow for continuous reporting throughout a power outage and will queue data if there are GSM-network-connectivity problems, to be uploaded once GSM connectivity is restored.

How Do PowerWatch and DumsorWatch Work?
PowerWatch consists of an outage detection sensor that plugs into an outlet and is installed in homes and businesses and reports the state of the grid over a cellular backhaul to the cloud. Every minute, the sensor takes a reading of power state, grid voltage, grid frequency, GPS, and cellular quality. It also records the number of nearby Wi-Fi signals as secondary validation, as wireless hotspots may be grid powered. In addition, upon changes in power state, the device records the timestamp (from an on-board real-time clock) and current acceleration. All these measurements are stored locally on an SD card and transmitted to the cloud when a cellular connection is available. Acceleration signals that a participant is interacting with the device, making it likely that any charge-state change at that time is a false positive and allowing us to more easily reject the data point. The sensor contains a 2000 mAh battery, which can run the sensor for several days, longer than most outages

(continued)

---

[4] PowerWatch internally records voltage measurements 150 times per AC cycle; then every 2 min, it reports the RMS voltage and grid frequency averaged over 30 AC cycles.

in Accra. When the sensor is on battery power, it still reports data at the same frequency to our servers, a feature necessary for calculating outage duration.

The primary sensors used by the DumsorWatch app are the phone's location sensors (GPS), the phone's charging state which monitors if the phone is connected to a power source or not (and notifies DumsorWatch when that state changes), and the system clock to give the time of any observed events. Secondary sensors in the app help refine the likelihood that a change in charge state corresponds to a power outage or restoration. For example, the accelerometer can measure if the phone was moving when its charge state changed (as it would if a charging cable was inserted or removed from the phone side), and the phone's Wi-Fi radio can report the presence or absence of wireless hotspots. On-phone processing and analysis of microphone recordings may be able to detect the presence of a 50 Hz "hum" of grid mains. Additionally, the users of the app can manually report a power outage and power restoration by pressing a button in the app.

For both technologies, a cloud-based analytic system searches for outage reports from multiple devices to ensure the validity of an outage. To perform this search, we cluster outage reports into density-based clusters in both space and time. This lets us reject noise from a single sensor (i.e., a single participant unplugging a device or a pre-paid meter running out of credit) and ensures that only true outages are reported by the system.

3.1.2 Surveys

A 60-min socioeconomic survey accompanied the deployment of each PowerWatch device, and all participants who received a PowerWatch device also downloaded the DumsorWatch app. A shorter survey was administered to participants who solely downloaded the DumsorWatch app. All surveys were completed using SurveyCTO, and all participants received an airtime transfer as a thank you for participation (SurveyCTO, 2019). We conducted high-frequency checks to address any obvious data quality issues. Example data collected includes:

1. Demographics: name, age, education, income.
2. Electricity attributes: appliance and surge protector ownership, usage of electricity and generators.
3. Recall of power quality in the past 2, 7, and 30 days.
4. Social media usage and perceptions of the energy crisis.

Along with providing data for the economic study, the survey was used to support the development and deployment of the technology itself. For example, the survey recorded a unique code for the PowerWatch device and DumsorWatch app and the participant's phone number and GPS location. To inform DumsorWatch debugging, we asked how they used their mobile phones, how many phones and SIM-cards they use, and how frequently they upgrade their phones. To inform the PowerWatch deployment, we asked whether the participant turns off their electricity mains at night and whether they had any safety concerns about PowerWatch.

3.2 Innovation: Deployment Methodology

To support our deployment as scale increased, we designed and implemented a novel set of deployment management tools. While our methodology evolved to support each deployment scale, its general structure remained fairly consistent. First, we developed criteria for site selection that allows us to answer specific socioeconomic questions. Next, we devised a sampling procedure that gave sufficient coverage of each chosen site, as well as sufficient redundancy to enable cross-validation of the new measurement technology. Finally, we worked with a team of field officers to deploy in the chosen sites, employing our deployment management tools to maintain and monitor the system. The rest of this section considers each of these components in detail.

3.2.1 Site Selection

We selected a subset of the sites where infrastructure upgrades are planned ("treatment sites") and then quasi-randomly selected a set of sites that are comparable in observable characteristics ("control sites"). For each site, we defined a geographic surveying area that is the intersection of a 200-meter radius from the site centroid and a 25-meter region extending from the low-voltage network being measured. We wanted the area to be relatively small, so that we could have a high degree of confidence that customers within the area were all connected to the same infrastructure, but it needed to be large enough to have a sufficient number of residents or firm owners for us to enroll into the study. We performed this analysis using GIS tools operating on a newly created map of Accra's grid constructed by an independent firm that had been contracted by MiDA as part of the improvements funded by the Ghana Power Compact. Using these GIS maps, we produce a series of maps marking the geographic area bounding each site. Field officers used these maps, along with the GPS coordinates for the sites, to identify the surveying area and deploy sensors accordingly.

3.2.2 Sampling Strategy

We deployed our sensors at the home or place of work (or both, if these are co-located) of Accra residents, targeting a 50/50 split between households and firms. Installing PowerWatch at consumer plugs and DumsorWatch on consumer phones

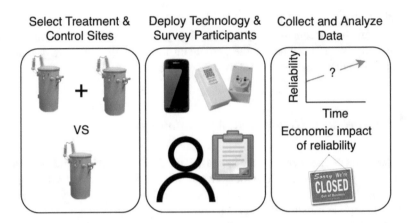

Select Treatment & Control Sites Deploy Technology & Survey Participants Collect and Analyze Data

VS

Reliability

?

Time

Economic impact of reliability

CLOSED

Fig. 6.1 Deployment overview
Notes: To support the goals of the deployment, our team selects sites that are being improved by the Ghana Power Compact and control sites. The technology is deployed in both sites along with surveys at the beginning and end of the deployment. This lets us meet our goals of evaluating the impact of grid improvements to power reliability and the socioeconomic impact of that reliability on consumers

allows us to not depend on direct access to utility infrastructure such as transformers or lines and to measure power quality at the point where it is least understood: the customer (Fig. 6.1).

We planned a deployment of three PowerWatch devices and 20 DumsorWatch app downloads at each site. Our strategy is built around redundant sampling such that multiple sensors are placed under a single transformer. When multiple sensors in this group report an outage at the same time, we can be confident it was due to an issue affecting the transformer rather than a single customer. Further, when we observe sensors below multiple transformers reporting outages simultaneously, we can infer the outage occurred at a higher level of the grid. This sampling strategy is shown in Fig. 6.2.

3.2.3 Deployment and Surveying Team

We hired local staff who supported our continuously operating deployment. One team member works full time as a field manager to oversee initial roll-out and ongoing maintenance of the system and an auditor to follow up with participants who report problems or whose sensors are no longer functioning.

To implement our medium- and large-scale deployments, we temporarily employed a team of ten field officers and three team leads. Prior to the start of the deployment, the field officers were trained extensively to ensure the correct protocols were used to obtain consent, conduct surveys, plug in the power strip and PowerWatch device at the respondent's home, download the app onto the respondent's phone, and conduct any necessary troubleshooting related to the

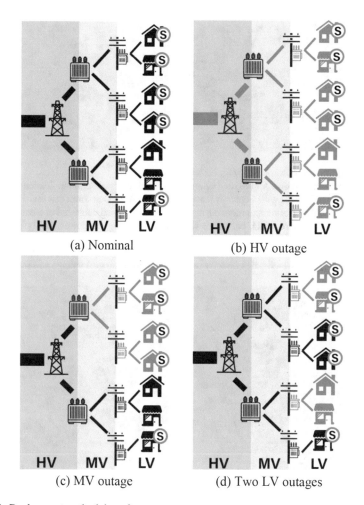

(a) Nominal   (b) HV outage

(c) MV outage   (d) Two LV outages

Fig. 6.2 Deployment methodology for sensors
Notes: Sensors, either the PowerWatch sensor or DumsorWatch app (both labeled S), monitor
power at homes and firms. By randomly sampling outlets at households and businesses, they
detect a significant portion of low-voltage (b), medium-voltage, (c) and high-voltage (d) outages.
In the aggregate this data can be used to estimate the average frequency and duration of outages,
including both single-phase and multi-phase outages, by looking at reports from sensors that are
close together in space and time. Additionally, this spatiotemporal analysis allows identification of
what voltage level experienced the fault. Undersampling can lead to missed outages when sensors
are not present in any of the affected units, shown in the bottom outage of (d)

technologies. Field officers find potential participants, get informed consent, and
screen their eligibility. They then conduct the survey, install the sensors, and
answer any participant questions. We conducted multiple training exercises where
each team member learned about the technologies being deployed and practiced
completing the survey and deploying the technologies.

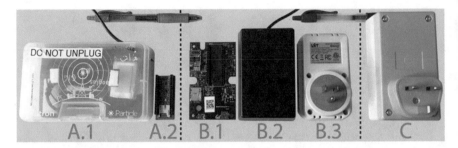

Fig. 6.3 Evolution of PowerWatch with each deployment

Notes: PowerWatch revision A consisted of an off-the-shelf compute/communication module and enclosure (A.1) and paired with a custom sensor front-end (A.2). Data from this revision informed the need for a better enclosure and more casing in revision B, which consisted of a custom sensing and communication board (B.1), enclosure with externally plugged power supply (B.2), and a separate grid voltage and frequency sensor (B.3). While the separate grid voltage and frequency sensor allowed for easier assembly, its complications led us to build revision C, a completely encased custom sensor which plugs directly into the wall, to sense grid voltage and frequency

Fig. 6.4 Field officers in uniform

Notes: Providing consistent branding built trust in the community as field officers visited potential participants. During the medium-scale deployment, choosing a color scheme inspired by our university accidentally resulted in a color scheme similar to that of the local power utility, causing some confusion. While we were easily able to choose new colors for the large-scale deployment, we highlight that it is important to consult with local experts before making branding decisions

Field officers visited sites in groups of two to alleviate safety concerns. We provided team uniforms, shown in Fig. 6.4, to make clear they are part of an official project. We also provided backpacks to carry supplies, tablets to conduct the survey, Wi-Fi hotspots to upload the survey and download the DumsorWatch app, flashlights for safety, and feature phones to verify the phone numbers of participants to ensure we know where to send the participation incentives.

3.2.4 Dependence on Participants

The placement of PowerWatch sensors directly in homes and firms – where participants may unplug them, run generators, or experience power shutoffs due to nonpayment – increases the noise of our data relative to a deployment on utility-owned equipment such as transformers. Similarly, the DumsorWatch app may be uninstalled from respondents' phones, reducing coverage and leading to a potentially under-sampled signal. A key challenge was thus ensuring that we only enrolled participants who had the ability and desire to participate for the full study duration and then to minimize any cause for participants to choose to withdraw consent from participation.

In a preemptive attempt to decrease the statistical noise caused by human factors, we screen participants for specific criteria including owning a phone with Android version 4.1–8.1 and being an active customer on the grid. In order to minimize attrition, we explain the goals, risks, and benefits of the project, as part of the consent process. Finally, we provide a phone number to call if participants have any questions or concerns.

To further encourage continued participation, we compensate participants monthly with airtime credits on their mobile phone. All participants whom we recruited to download the DumsorWatch app received 5 Ghana Cedi (0.93 USD) of airtime for initial recruitment and 4 Ghana Cedi (0.75 USD) monthly for keeping DumsorWatch installed. Participants who also installed a PowerWatch device received an additional 10 Ghana Cedi (1.86 USD) for installing the sensor and 5 Ghana Cedi (0.93 USD) monthly for keeping PowerWatch installed. Additionally, participants who have a PowerWatch sensor placed at an outlet in their home receive a power strip so that the sensor does not take up a needed outlet.

3.2.5 Deployment Management Tools

We developed three software subsystems to support the deployment: (1) an automated incentive system to transfer the airtime incentives; (2) a deployment management system to (a) track sensor and participant status and (b) display deployment health to the field management team; and (3) a data visualization and analysis system. We discuss these systems, and the experiences that led us to develop them as the deployment scaled, in Sect. 3.5.

3.3 Evaluation: Overview

For each of the small-, medium-, and large- scale deployments, we report problems that occurred, techniques used to mitigate their impacts, and the effectiveness of the mitigation. To emphasize the parallels between challenges exposed at different scales, we organize this discussion around four categories of challenges: organizational, cultural, technical, and operational. Organizational challenges relate to procurement, hiring, and finances; cultural challenges relate to how cultural considerations impacted the deployment and operation of the technology; technical challenges relate to the development, manufacturing, and functioning of the technology; and operational challenges relate to the successful deployment and operation of the technology.

3.4   Evaluation: Small-Scale Pilot

The first activity we performed was a deployment of 15 PowerWatch sensors and 5 DumsorWatch app downloads. The goal of this deployment was to validate that the technology can reliably sense power outages and transmit this information over many weeks in the field. We performed no survey work and no site selection work for the small-scale pilot: devices were not deployed with participants enrolled from the public but in the private homes of our research partners. The primary challenges were related to producing the technology, connecting the PowerWatch sensors to the cellular network, and building enough local capacity to deploy PowerWatch and DumsorWatch.

In addition to testing the technology, we worked to build relationships to support future scaling. We reached out to local stakeholders for feedback on the assumptions driving our sensor design, speaking with engineers and managers at ECG, MiDA, and several independent contractors involved in the Ghana Power Compact. We also received data from ECG that helped validate our hypothesis that their existing estimates of SAIDI and SAIFI could benefit from higher-resolution measurements.

Even at a small scale, we experienced unanticipated technical challenges. To connect the PowerWatch devices to the cellular network, we initially used SIM cards sold by Particle, the US-based manufacturer of the cellular technology used in PowerWatch, in part because these SIM cards had been advertised to work in any country. But in practice, their ability to maintain a stable network connection was worse than that of a local SIM. We therefore decided to use SIM cards from the largest local carrier (MTN), but we encountered a three-SIM-card-per-person limit upon purchase. Although we were able to circumvent this by visiting different stores, purchasing SIM cards in stores was not an option for future scale.

Another challenge was keeping SIM cards functional over the full study period. Prepaid SIM cards require data plans, which are purchased using an unstructured supplementary service data (USSD) application that can only be run from within Ghana; there is no web-based account management or top-up available. We initially solved this problem by purchasing a 90-day data plan, the longest available. This was sufficient for our small-scale pilot but would not be viable for future deployments.

Table 6.1 Pain points of different scales

Type	Small scale	Medium scale	Large scale
Organizational	Local SIM procurement	Hiring local staff	
		Contracting local companies	
		Paying outside free tier	
Technical	Global SIM operation	Custom hardware	Assembly
		Firmware development	Site selection
		App development	
		SIM operation	
Operational	SIM top-up	Transportation	Deployment management
		Field officers	
		Incentivizing participants	
		Data sharing	
Cultural	Learning local context	Local leader approval	Unexpected phone usages
			Survey design

Notes: At each deployment scale, we encountered pain points – complexities that were more difficult than we expected from a simple increase in deployment size. Many occurred at the transition to medium scale, when local capacity had to be built, expenses to operate the technology increased, lack of technical reliability became more apparent, and systems that could once be human-operated had to be automated. Large scale brought new problems, most notably the need for automated deployment management tools to track deployment state

3.5 Adaptation from Small-Scale Pilot Experience and Evaluation of Medium-Scale Deployment

In our medium-scale deployment, 1981 individuals downloaded the DumsorWatch app, and 165 individuals installed PowerWatch sensors. After an initial 1-week training, field officers first deployed the PowerWatch sensors, which included also administering a detailed socioeconomic survey and installing the DumsorWatch apps on the phones of PowerWatch recipients. Once the deployment of the PowerWatch sensors was complete, field officers spent 3 weeks deploying the DumsorWatch apps among an additional set of participants, which also included a shorter socioeconomic survey. Once initial deployment was complete, the monitoring activity continued for 7 months.

Unlike the small-scale deployment, this scale required implementing our full deployment design, including hiring a full local implementing team, recruiting and incentivizing participants, choosing deployment sites, extracting value from the data streams, and implementing the survey instruments. We enumerate the changes experienced as we increased from small- to medium-scale in Table 6.1, paying particular attention to the challenges extracted.

3.5.1 Organizational

The medium-scale deployment was large enough that the financial responsibilities were significant. We had to start managing multiple monthly payments for cloud services and payments to local companies for cell network connectivity and incentive transfers. Most of this increase in complexity was ultimately handled by staff at the University of California, Berkeley, but establishing payment schedules took a large effort from the research team. Bureaucratic requirements at the university also caused frequent delays in payments, especially when payment was needed in a short time frame (1–2 weeks). Increased flexibility, for example, at an independent organization or private sector company, might better suit technological deployments that have some complexity in administrative streams like finances and hiring.

Because prepaid SIM cards were not available at the quantities we now needed, we had to enter into a contract with the cellular provider, MTN. To alleviate concerns about whether our application was legitimate, we visited the MTN main office in our university shirts, gave a technical demo, and answered questions about our backgrounds and affiliations.

At medium scale, many of the cloud-based software services our systems were built on were no longer eligible for free-tier usage. For one service, this meant that we would be unable to continue without signing a multiyear contract that extended beyond the length of the deployment. We found a workaround for this deployment by applying to a special program within the company, but in future deployments we would more carefully consider pricing models for ancillary services.

3.5.2 Cultural

Visiting households and firms requires permission from the relevant local district assemblies. We wrote letters of introduction and visited these assemblies to receive permission. Receiving this permission also increased participant trust.

We also worked with the field officers to refine our survey design. During training activities, the field officers had the opportunity to react to questions and provide suggestions for improvement. We used this feedback to make the survey as culturally appropriate and in line with our research objectives as possible. As field officers entered the field, we received continuous feedback on ways to improve our survey and deployment procedures.

Finally, we learned that a uniform would be valuable for building participant trust. We provided DumsorWatch-branded shirts and backpacks for the field officers, so they would look official when approaching participants. These are shown in Fig. 6.4. Field officers also carried identification cards that they could show participants in case of any questions.

3.5.3 Technical

At medium scale, frequently visiting sensors for debugging was no longer feasible, so we prioritized sensor stability and remote failure detection and mitigation. This included developing a full custom embedded system for PowerWatch (shown in Fig. 6.3 B.1) with built-in mechanisms to reset the device on failure. Additionally, we spent considerable time implementing and testing more reliable firmware, incorporating error collection libraries, and building dashboards displaying the health of both PowerWatch and DumsorWatch. We assembled this version of PowerWatch over 3 days with the help of fellow graduate students.

Another technical challenge concerned mobile phone heterogeneity. We had little insight into what types of mobile phones and versions of Android were most common in Accra. Thus, we implemented DumsorWatch to be backwards compatible to 4.0.0, a version of Android no longer supported by Google (Google, 2018). Backward compatibility took considerable engineering effort and had side effects such as making DumsorWatch incompatible with many modern Google cloud services, including Google's bug tracking tools, making app failures much harder to correct. Further, we chose to support older versions of Android at the expense of supporting the newest Android version at the time, Android 8.1, a decision that made those with the newest phones no longer eligible for participation. While this did not reject large numbers of participants during this deployment, as more people moved to newer devices, it would have more impact on recruitment, necessitating future engineering costs.

Finally, we experienced two challenges related to SIM card operations. First, we could not identify a way to test PowerWatch sensors in the United States using MTN postpaid SIM cards, which were not configured to allow PowerWatch to connect to cellular networks outside of Ghana. We therefore built a US-based testbed for sensor development that used US-based SIM Cards. However, to perform final assembly and quality assurance, steps that require PowerWatch to be connected to the MTN network, we needed to wait until the sensors were in Ghana for a deployment, compressing the timeline of these tasks and increasing risk if quality assurance failed. Second, MTN's process for provisioning the SIM cards required additional oversight and took much longer than expected, which delayed deployment and made clear that a different partner would be required to manage large fleets of SIM cards assigned to a single customer.

These problems led us to continue exploring global SIM card options, and we tested a small number of Twilio SIM cards during this deployment. We found they had similar problems to the Particle SIMs previously evaluated. We contacted Twilio support and found their documented list of Ghanaian network operators was out of date, making unlisted providers unavailable on the Twilio network and leading to a drop in service quality. This theme of global solutions lacking local service quality is explored further in Sect. 4.2.

3.5.4 Operational

The operational challenges started with transporting our equipment to Ghana. We carried the PowerWatch sensors, power strips (handed out to participants as incentives), and equipment for field officers into Ghana in suitcases over multiple trips from the United States. PowerWatch sensors were carried on the plane whenever possible to minimize their chance of being lost. This method of transportation worked but led to multiple questions from airport security in the United States and customs in Ghana. We were able to overcome these hurdles by creating documentation about our project and providing this along with letters of invitation from MiDA, but even still this transportation method depended on our team being persistent and prepared with documentation, unwrapping all equipment from its packaging to make it look less likely to be resold, labeling all equipment with tags indicating it was property of the university and not for resale and only traveling with a few suitcases at a time. More generally, ensuring safe and timely transport of nonconsumer technology across borders will likely require additional measures depending on the local context.

Implementation of our site selection methodology required GIS maps of the electric grid. We worked with stakeholders to determine where the best maps of the grid were maintained, a task made more complicated by maps being held by multiple subcontractors of the Ghana Power Compact. With MiDA's support we were given access to maps that, while not perfect, included enough detail for our site selection procedures. At this medium scale, which was also relatively concentrated geographically, it was feasible for a member of the research team to study the GIS maps visually, identify the proposed treatment sites, manually identify control sites quasi-randomly to match treatment sites based on observable characteristics such as grid characteristics and satellite imagery, and produce site maps that field officers could use to identify potential respondents at each site. This process met the requirements for this scale deployment but would prove exceedingly complicated for a larger deployment.

At medium scale we felt it was not feasible to transfer recurring incentives to participants by hand. We had anticipated this problem and designed an incentive-management system to support this goal. The system was designed to capture user behavior (e.g., whether a participant completed a survey, installed DumsorWatch, kept DumsorWatch installed, etc.) and to transfer airtime automatically. The actual transfer of airtime took place through a third-party API. We developed and tested the incentive transfer system alongside our deployment activities (Klugman et al., 2019).

Finally, at medium scale, the data collected were significant enough that they became valuable to stakeholders in the region. Because many of these stakeholders would be responsible for helping the project achieve further scale, we made an effort to develop and share anonymized visualizations and summary statistics.

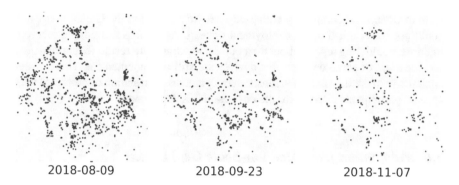

2018-08-09   2018-09-23   2018-11-07

Fig. 6.5 Relative locations and number of Android app events over time
Notes: Starting in August 2018 (left), we were receiving events from 989 phones in our deployment
area; however, the number of participants fell to 573 by September (middle) and 310 by November
(right). Because of these deployment challenges, we were unable to fully longitudinally test the
app technology

3.5.5 Failures and Missteps

One class of failures experienced at medium scale is attributable to simple technical
immaturity. For example, we found (and are still finding today) bugs both in our
automated incentive-transfer system and in the third-party payment API used to
incentivize participants. This API is provided by a small company, but we believed
it to be the best option for transferring airtime in Ghana. Both technologies should
have been more aggressively tested prior to launch. There is a clear need for a
fleet of testing phones in Ghana for continuous integration and automated testing
of incentive transfers. However, as with most hardware-based testing systems, this
is difficult to implement in practice. As a result, most participants experienced late
payments, which we hypothesize caused the significant number of DumsorWatch
uninstalls shown in Fig. 6.5.

More fundamental were issues with effectively recording, connecting, and
correcting critical deployment metadata, such as the location and status of each
device and payment, which we collectively refer to as the *state* of the deployment.
We had not anticipated the complexity of managing data about participants, devices,
and app installs, each of which was collected by a different system and some of
which informed each other.

This led to an ad hoc sharing of information through our encrypted shared drive.
The field team uploaded surveys containing participant- and deployment-placement
information on a daily basis. The research team downloaded and cleaned these
periodically and provided the resulting CSV files to the individual engineer handling
either sensor management or the payment system. Errors in the surveys (common
due to typos in long unique IDs) were communicated back to the field team via
phone calls and emails, and the resultant corrections in the field were not always
communicated back to the research team. This process was ineffective while we

were in Ghana and completely collapsed after we returned to the United States and could not focus full time on deployment upkeep. As devices moved, we received multiple, conflicting reports about their current location. As a result, we permanently lost the *state* of some devices; five devices are still unaccounted for. These issues continue to make data analysis, sensor debugging, and correlation of problems with specific participants difficult for the devices in this deployment.

3.6   Adaptation from Medium-Scale Deployment and Evaluation of Large-Scale Deployment

Beginning in February 2019, we built upon our medium-scale deployment and added 292 new PowerWatch devices and 1419 new app downloads in 3 districts of Accra, resulting in a combined large-scale deployment of 457 PowerWatch devices and 3400 DumsorWatch apps.

3.6.1   Organizational and Cultural

The organizational and cultural challenges did not change from the medium-scale deployment. Existing service contracts were sufficient or easily renegotiated, and the field team scaled linearly with the size of deployment.

3.6.2   Technical

The increased number and technical complexity of the new PowerWatch sensors constructed for the large-scale deployment precluded relying on other graduate students to help assemble devices as we did with the medium-scale deployment; however, the scale was still too small to be cost- or time-effective for contracted assembly. Our solution was to build our own assembly line and hire 10 undergraduates to assemble devices. This required us to develop discrete assembly steps, a training guide, and quality assurance techniques. The PowerWatch assembly line can be seen in Fig. 6.6. This assembly line produced the 295 PowerWatch sensors over 4 weeks and 110 person-hours of total work, with a 2.4% error rate, which was far below what we were anticipating. Although this activity was successful, difficulties in recruiting and paying students hourly, and challenges with the academic schedule, mean this model would not scale much beyond 400 units.

The larger number of sites meant site selection was no longer easy to do manually. This led us to develop a GIS-based site selection system, which generates sites based on our site selection rules, labels these sites, and creates site location images for the field officers. This system requires cleaning the GIS maps of the grid collected from the utility and was designed and maintained by a dedicated graduate student.

Fig. 6.6 PowerWatch assembly line
Notes: Over 4 weeks, 10 undergraduates worked 110 person-hours to assemble 295 PowerWatch sensors. They were responsible for assembling the plug; screwing together the enclosure; attaching the circuit board; connecting the battery, antenna, SIM card, and SD card; and provisioning the device with base firmware. They worked from team-created assembly manuals and training materials

We continued exploring global SIM card options, using Aeris SIM cards for a subset of this deployment. We found that due to Aeris' focus on global IoT connectivity and the number of customers they have in sub-Saharan Africa, their SIM cards work significantly better in Ghana than Particle or Twilio SIMs.

3.6.3 Operational

The largest operational change was addressing the issues described in Sect. 3.5 with our custom deployment management software, described further in Sect. 4.1.

3.7 Adaptation from Large-Scale Deployment to Sustainable Large-Scale Deployment

After the completion of our large-scale deployment, our team was asked by MiDA to scale the deployment again, this time to over 1200 PowerWatch sensors, with the continued goal of estimating SAIDI and SAIFI and providing this data to multiple impact evaluations of the Ghana Power Compact.

Before scaling up further, a subset of our team's engineers created a company. This was not an easy decision. The academic research lab context afforded space, materials, and access to other researchers. Building a new organization has overhead, potentially taking resources away from solving deeper problems. Working across disciplines lets our team address an important data gap for economists with a new technology provided by engineers, and developing the technology as researchers allowed us to be slow, make mistakes, iterate, and get to know the stakeholders without the pressures placed on a subcontractor delivering a product. Most importantly, freedoms enjoyed within the academy around transparency, knowledge transfer, and independence contribute greatly to our personal drive to perform this work, and the financial obligations that come with forming a company often deprioritize these goals.

We recognized, however, that this project was no longer a good fit for the goals of an academic research lab. We anticipated fewer innovations related to the sensor and deployment methodology and thus less research suitable for academic publication, while also anticipating the need to spend more time supporting the expanding deployments. Further, we recognized the need to free the project from institutional dependencies that had been too slow for the rapid pace of field work, a pace that would only increase with scale.

Since starting the company, the overhead of establishing an organization – including hiring employees, establishing accounting systems, navigating conflicts of interest, and trying to package our measurements as a product – has been significant. However, we have been more nimble and responsive when faced with challenges, have successfully scaled our deployment again, and, most importantly, have a structure in place to allow this work to exist for longer than just the length of a PhD thesis.

Pivot: Mobile Phone Sensing to Plug-in Technologies?

One original research question tested during this project was whether we could repurpose daily-use smartphones, already deployed in the hands of billions of people, as low-resolution grid reliability sensors. Leveraging the widespread ownership of cellphone devices in developing countries would allow cash-constrained utilities to improve the accuracy of reliability measurements at lower cost than widespread smart meter deployment. In fact, a primary objective of the deployment of PowerWatch, our plug-in sensor, was to provide ground-truth outage data against which the DumsorWatch app's measurements could be compared! However, as we scaled up both technologies, we saw PowerWatch succeed, and DumsorWatch underperform in two unexpected ways, one due to unanticipated participant behavior and the second due to a changing technical context. The combination of these limitations led us to prioritize PowerWatch and put DumsorWatch on the back-burner just as it was starting to return some positive results,

The first unanticipated result we observed is that DumsorWatch was uninstalled from participant phones at a higher rate than originally anticipated, even when participants were financially incentivized to keep the app installed. Most frequently, this happened automatically – for example, because the respondent reset, replaced, or lost their phone – and the respondent did not

(continued)

reinstall the app. Many participants also uninstalled their apps intentionally, for example, to save space on their phones or because they had privacy concerns from having an unknown app installed on their private device. As a result, just 3 months after the original deployment, DumsorWatch was only detecting around 10% of the outages that it was at the start. On the flipside, participants also expressed PowerWatch was less invasive than an app on the phone, potentially explaining why it regularly remained plugged in over long periods of time.

The second unanticipated action that impacted the performance of DumsorWatch was that the Android operating system changed to limit long-running background services. DumsorWatch depended on a background service that would wake the app up whenever the phone was plugged in or unplugged. This function was eliminated mid-deployment by Google in an effort to improve the user experience, since some long running apps have an outsized impact on battery life or quietly collect large amounts of data on users, burning data allocations and leaking privacy. We tried to get around this by limiting the versions of Android we recruited participants for to those that still supported background services. However, the new OS limitations ensured the sunset of DumsorWatch as an application layer technology, which led to us to reprioritize our engineering efforts to improve the PowerWatch system.

It is worth noting that our implementation of DumsorWatch successfully detected power outages! For as long as it was operational, participant phones running DumsorWatch demonstrated that uncoordinated smartphones experiencing charge state changes at the same time correlate with ground truth grid reliability measurements as provided by PowerWatch. However, based on both factors explained above, we believe the best path forward for the ideas captured in DumsorWatch is for these types of measurements to be taken as an OS level service in Android and to be aggregated as a primary crowd-sourced measurement (similar to how Google Maps captures and exposes traffic data (Google, 2009)). This pivot was made easier by the presence of PowerWatch, which could answer many of our remaining questions. While it remained a hard choice, it lets us better prioritize our team's and our funders' resources.

4   Lessons Learned

Each level of scale brought unique complexities for both engineering and operational tasks. While some of these complexities were one-time costs, many can be attributed to either the continuous nature of operating and managing a sensor deployment at scale or a context in the local culture we were not anticipating. This combination of scale and continuity stretched the administrative ability of the university system; exploded the continuous data stream that must be maintained to manage the sensors requiring automation; amplified errors in data collection; and ultimately required us to develop automated tools to facilitate tasks that we and our field team could not handle at scale. We hope these lessons will inform future efforts to deploy continuous monitoring and evaluation systems in developing regions.

4.1 Continuous Monitoring Requires Continuous Upkeep

Continuous operation of a sensor network and phone application requires significant metadata and upkeep not required for a large survey deployment. Sensor deployment times and locations must be recorded and correlated with participant information. Unique app identifiers need to be collected to ensure app installation compliance. Participant phone numbers need to be stored so participants can be appropriately incentivized. All of this information needs to be effectively communicated to the field officers for debugging on an ongoing basis and updated over time because participants and their devices are in constant flux. As we describe in Sect. 3.5, maintaining a high quality of sensor deployment and implementation of our experimental design requires a systematic approach to tracking these data at scale.

At a fundamental level, the introduction of continuous monitoring systems into a deployment introduces feedback loops that are not present in a large surveying effort. These feedback loops, shown in Fig. 6.7, have two major implications for a deployment:

1. Errors introduced into the feedback loop by incorrect metadata from a survey are important and often amplified if not addressed.
2. The deployment's *state* (e.g., the location and status of each payment and device) is kept across multiple systems and is likely to become inconsistent if feedback is not automated.

For our large-scale deployment, we addressed these problems and have seen major improvements in our deployment results. The first correction was to prevent surveying errors on critical metadata. We implemented barcodes to record the unique IDs of sensors and phone applications, and we equipped the field officers with feature phones so they could text the participant to verify the participant's phone number and take a picture of the sent text message.

The second correction was to develop custom software responsible for automatically (1) keeping *state* consistent across all databases, (2) communicating errors to the field team, and (3) implementing corrections to survey data when updates are submitted by the field team. The field team completes a set of deployment, retrieval, and debugging surveys in SurveyCTO, and the deployment management software automatically consumes these surveys using the SurveyCTO API. The data from the surveys is then verified and the information distributed to the appropriate databases. Information about surveys with errors, along with a list of non-operational devices, is available to the field team through a web interface, and field team error corrections

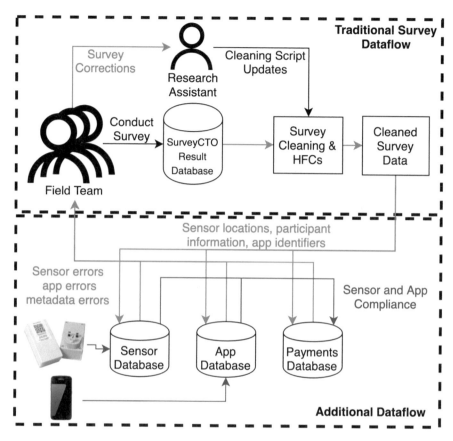

Fig. 6.7 Deployment dataflow
Notes: While traditional surveying methods have a linear data flow where data is exported for later analysis, the integration of continuous sensing in the deployment generates feedback loops. This in turn creates more separate places where state is stored, creates more necessity to communicate this state, and amplifies the problem of errors during surveying. We implemented a deployment management system to alleviate these problems. Blue arrows represent data flows that we automated from the beginning because we anticipated their complexity before the medium-scale deployment. Red arrows represent data flows that we originally attempted to perform manually and that we now automate or facilitate with a deployment management tool

are communicated back to the software through a final correction survey. The deployment management software is represented by the red arrows in Fig. 6.7.

This architecture makes the deployment significantly easier to manage. Systems like these are necessary for both deploying and maintaining a continuously running sensor network, especially one in which the *state* of the deployment is constantly changing due to direct interaction with participants.

4.2   Global Solutions May Miss Local Context

Several times in our deployment we were forced to consider tradeoffs between using technology and services developed and operated locally and similar solutions developed by larger companies targeting global scale. Specifically, we made this decision in both our choice of the cellular network provider and the service used to send airtime incentives to participants. Unsurprisingly, we found local service providers were more likely to provide high-quality service in Ghana compared to US-based companies, which had only nominal ability to operate globally (and little experience or market in doing so). Even our largest scale was not large enough to get dedicated support contracts with these US-based companies.

At the same time, we found local providers did not handle our medium- or large-scale deployments flawlessly. Our airtime top-up provider was not technically ready for the scale of our medium and large deployments, and neither the airtime provider nor MTN was prepared to bill and support our enterprise accounts. Therefore, to continue to scale, we went back to evaluating mobile virtual network operators (MVNO) and global airtime top-up providers, aiming to find companies with demonstrated experience in Ghana and similar geographies. After evaluating several MVNOs and airtime top-up firms, we found a set of global companies that provide a good mix of technical maturity, experience in handling enterprise customers, and reliable service in Ghana and other countries.

4.3   University Lacks Financial Agility

One of our primary organizational problems was the inability to pay for the various services necessary to perform our deployment. This was not for the lack of available funding but a mismatch with the administrative capacity in academia.

While our university policy dictates a single-day turnaround on wire transfers, in practice this often took over 15 days. Setting up subcontracts with companies, especially companies with which the university had never contracted before, often took months, and our deployment required numerous contracts because our technology relies on external service providers. As a result, changes to our deployment plan – even weeks in advance – would often cause major issues. Even if we thought there was enough time for payment prior to a deployment, members of the research team still frequently resorted to using private financial resources, maxing out personal ATM limits in Ghana to support deployment activities.

Additionally, the university does not have good mechanisms for supporting recurring but inconsistent costs (such as a pay-per-use cloud service) because every change in cost requires approval. We found it significantly easier and more reliable to front payments for these critical services via personal credit card.

If we were to plan for this deployment again, we would build in significantly more time for delays and send more money than necessary to our stakeholders in

(a)

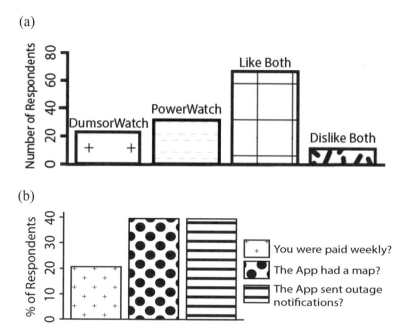

Fig. 6.8 Participant perception of sensors (**a**) Was DumsorWatch or PowerWatch preferred? (**b**) Are you likely to download DumsorWatch when . . . ?

Ghana early in the deployment so they could better handle later delays in payment from the university. Even still, it would be difficult to imagine the deployment running at its described pace without the research team extending personal credit.

4.4   Technology Usage Patterns Impact Design

Our system depends on participants to download apps and install sensors in their homes or businesses. To validate our methodology in the local context, we completed an endline survey with 462 participants from the medium-scale deployment before launching the large-scale deployment. The results of this survey proved surprising and critical for the design of the next level of scale.

We asked participants what they thought of the sensors. Figure 6.8a shows that participants liked both PowerWatch and DumsorWatch, with a slight preference for PowerWatch, challenging our assumption that the mobile app would seem less invasive than a physical device. Better understanding this inversion remains future work, but one hypothesis is that mobile phone resources are scarce and highly valued.

We then explored how incentives influence participation. Figure 6.8b shows that real-time information about local outages was valued highest, indicating a strong

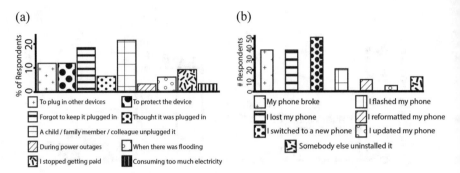

Fig. 6.9 Participant-reported reasons why sensors were uninstalled (**a**) PowerWatch (**b**) Dumsor-Watch

local desire for energy-reliability data and suggesting data alone could be effective in incentivizing participation.

Even so, many participants either uninstalled DumsorWatch from their phone or unplugged PowerWatch from the wall. We asked participants about the root cause of these behaviors. Figure 6.9a shows that people unplugged PowerWatch for many different reasons, some of which could likely be addressed through better information sharing ("to protect the device," "during power outages," "consuming too much electricity") or through more careful user interface design ("thought it was plugged in"). These lessons will be incorporated in field officer training for future deployments.

More challenging are the results from Fig. 6.9b, which indicate a high degree of fluidity in mobile phone usage. In particular, formatting and "flashing" (resetting) phones were significant user interactions that our team was not familiar with. Also, large numbers of phones broke. Our methodology never asked a participant to reinstall the app because we assumed it would stay installed, and this assumption did not map to the local context.

The results of this survey were important for improving system performance and would have been even more effective if we had run it earlier. The successful deployment of any technology hinges on its suitability to the local context. Developing an early understanding of the potential risks, for example, through qualitative surveys or focus groups, can help inform the design of a technology while there is still ample scope to do so.

5 Conclusion

When first approached with the opportunity to run a deployment at scale in Accra, our team was naively confident. We were able to decompose the larger task of a deployment into subsystems, each of which we could effectively engineer. However, well-designed subsystems are not enough. Critically, we overlooked the human links between these systems, leading to problems not due to sensors malfunctioning but instead from the complexities of sensor placement and upkeep. This meta-task of deployment management was not forgotten but neglected for the more traditional engineering tasks, like pushing for a more fully featured firmware in PowerWatch or a better-tested implementation of DumsorWatch.

Despite these hurdles, we conducted a largely successful deployment that met all of our design goals. This was only achieved through effort from a large and creative team, a resource that many research groups cannot easily obtain.[5] In reaction to specific pain points at larger scales, we developed meta-tools, not to replace the human links but to assist them.

We hope that by identifying and describing these meta-tools, as well as our broader collection of lessons learned, we lower the barrier to entry for conducting similar-scale deployments. This is a goal worth working towards. Insights gathered from direct deployments of sensors around the world will be critical for achieving deeper understanding how critical life sustaining systems operate and, eventually, how they can be improved. To realize this, the development engineering community will have to work to reduce the costs and complexities of performing deployments. Steps towards this may include continually sharing lessons learned, making software meta-tools available and open-source and setting up incentive systems within academia to help engineering researchers value innovation in methodology, reliability, and scale-ability as highly as innovation in any other more traditional high-order subsystems.

Discussion Questions

1. When planning an interdisciplinary project, how should you deal with differences in timelines, goals, or expectations between stakeholders from different disciplines or with different incentives?
2. When planning a real-world deployment, how do you balance and prioritize your expected contributions to science (direct professional benefit) and your expected contributions to the population under study (direct societal benefit)?
3. What do you do when a project unexpectedly requires a skill you have not received training in?

---

[5] This work was supported in part by the Millennium Challenge Corporation through the Development Impact Lab (USAID Cooperative Agreement AID-OAA-A-13-00002), part of the USAID Higher Education Solutions Network, in part by funding from the Department for International Development Energy and Economic Growth initiative (DFID EEG Project A0534A), and in part by the CONIX Research Center, one of six centers in JUMP, a Semiconductor Research Corporation (SRC) program sponsored by DARPA. We additionally thank CEO Martin Eson-Benjamin, CFO John Boakye, Director Dr. Kofi Marfo, Sharon Abena Parku, and the entire team at the Millenium Development Authority (MiDA); Jeff Garnett, Hana Scheetz Freymiller, and the rest of our partners at the Millennium Challenge Corporation (MCC); Karen Notsund; Nana Bonku Eshun; Kobina Aidoo; Ivan Bobashev; Aldo Suseno; Revati Thatte; Lane Powell; and staff from the Electricity Company of Ghana, SMEC Ghana, and ESB International. Finally, we thank the Field Supervisors and Field Officers in Accra who put in many hours of hard work to implement, improve, and maintain our deployment.

4. How do you assess the efficacy of a service provider when you will be the first to empirically test their capacity to scale in a given market?
5. After your research study is complete, what stakeholders will manage your deployment? How will you ensure the infrastructure is useful and usable for them? What is your handoff plan, both for short-term management and long-term maintenance?
6. Where is information stored in your system, what are the sources of this information, and what other systems depend on this information? Are systems in place to ensure that information is correct and consistent for different data consumers?

References

Acakpovi, A., Abubakar, R., Asabere, N. Y., & Majeed, I. B. (2019). Barriers and Prospects of Smart Grid Adoption in Ghana. Procedia Manufacturing. In *2nd International Conference on Sustainable Materials Processing and Manufacturing*.

Ackah, C. (2015). Electricity insecurity and its impact on micro and small businesses in Ghana. In *Study presented at Institute of Statistical, social and economic research (ISSER) workshop*. University of Ghana, Legon.

Aidoo, K., & Briggs, R. C. (2019). Underpowered: Rolling blackouts in Africa disproportionately hurt the poor. *African Studies Review, 62*, 3.

Banuenumah, W., Sekyere, F., & Dotche, K. A. (2017). *Field survey of smart metering implementation using a simple random method: A case study of new Juaben municipality in Ghana*. IEEE PES PowerAfrica.

Bowers, B. (1988). Social benefits of electricity. *IEE Proceedings A (Physical Science, Measurement and Instrumentation, Management and Education, Reviews), 135*, 5.

Clerici, C., Taylor, M. S., & Taylor, K. (2016). *Dumsor: The electricity outages leaving Ghana in the dark*. Al Jazeera.

Electricity Company of Ghana. (2019). *Prepayment Metering*.

Energy Commission Ghana. (2019). *National Energy Statistics 2000–2019*.

Gertler, P. J., Lee, K., & Mobarak, A. M. (2017). *Electricity reliability and economic development in cities: A microeconomic perspective*.

Google. (2009). *The bright side of sitting in traffic: Crowdsourcing road congestion data*. Official Google Blog.

Google. (2018). Google Play services discontinuing updates for API levels 14 and 15. (Dec 2018). https://android-developers.googleblog.com/2018/12/google-play-servicesdiscontinuing.html

Hamidu, M., & Sarbah, A. (2016). An assessment of the electrical energy needs of beauty saloon industry in Ghana. *American Journal of Management Science and Engineering, 1*(2).

He, J., Cheng, L., & Sun, Y. (2007). Transformer real-time reliability model based on operating conditions. *Journal of Zhejiang University-Science A, 8*(3).

IEEE Guide for Electric Power Distribution Reliability Indices. (2001). IEEE Std 1366, 2001 Edition (2001), i–. https://doi.org/10.1109/IEEESTD.2001.94438

Klugman, N., Rosa, J., Pannuto, P., Podolsky, M., Huang, W., & Dutta, P. (2014). Grid watch: Mapping blackouts with smart phones. *ACM HotMobile, 14*.

Klugman, N., Correa, S., Pannuto, P., Podolsky, M., Taneja, J., & Dutta, P. (2019). The open incentive kit (OINK): standardizing the generation, comparison, and deployment of incentive systems. *ACM ICTD '17*.

McCollum, D., Gomez Echeverri, L., Riahi, K., & Parkinson, S. (2017). *SDG7: Ensure access to affordable, reliable, sustainable and modern energy for all* (pp. 127–173). International Council for Science. https://doi.org/10.24948/2017.01

Millennium Challenge Corporation. (2008). *Guidance on Common Indicators.*

Millennium Development Authority. (2014). *Power distribution feasibility studies*, Ghana Phase II Feasibility Assessment.

Millennium Development Authority. (2018). *Ghana monitoring and evaluation plan.*

Nunoo, S., & Ofei, A. K. (2010). *Distribution automation (DA) using supervisory control and data acquisition (SCADA) with advanced metering infrastructure (AMI). IEEE Conference on Innovative Technologies for an Efficient and Reliable Electricity Supply.*

Quayson-Dadzie, J. (2012). *Customer perception and acceptability on the use of prepaid metering system in Accra west region of electricity company of Ghana.* Ph.D. Dissertation.

SurveyCTO. (2019). Because your data is worth it. https://www.surveycto.com/. Accessed on 03/15/2019.

Sustainable Energy for All. (2018). *Sustainable energy for all: Monitoring, evaluation, and learning framework.*

Taneja, J. (2017). *Measuring electricity reliability in Kenya.* Technical report. Working paper. Available at http://blogs.umass.edu/jtaneja/files/2017/05/outages.pdf

The GridWise Alliance. (2013). *Improving Grid Reliability and Resilience: Workshop Summary and Key Recommendations.* https://www.energy.gov/sites/prod/files/2015/03/f20/

Chapter 7
Monitoring Industrial Pollution in India

Anant Sudarshan

1 The Development Challenge

As India has developed, so has her demand for energy. This demand has largely been met with abundant, inexpensive, and highly polluting fossil fuels. These choices have had fundamental environmental consequences, imposing costs that significantly threaten the country's economic prospects. Perhaps the most significant of these is the dangerously rapid deterioration in the quality of India's air, with satellite data suggesting an increase of over 70% in the concentration of particulate matter 2.5 (PM 2.5) between 1998 and 2016.[1]

Air pollution now poses one of the most severe public health challenges to the country (Balakrishnan et al., 2019). Study after study has pointed to the general health risks of air pollution, in terms of life expectancy for example, and additional new research suggests that the effects of poor air quality may even extend to reduced crop yields, lower labor productivity, and decreased cognitive skills (Chang et al., 2019; Bharadwaj et al., 2017; Burney & Ramanathan, 2014). In other words, the apparent trade-off between environmental protection and economic growth is something of a Hobson's choice—the Indian growth story cannot continue without cleaning up the air.

Notwithstanding the poor quality of its environment, India has a fairly strong and wide-ranging set of environmental laws. Nevertheless, these have not been sufficient to effectively control industrial air pollution. Empirical evidence from the

---

[1] Global time-series data of particulate air pollution concentrations is available as part of the Air Quality Life Index, available at: https://aqli.epic.uchicago.edu/the-index.

A. Sudarshan (✉)
Energy Policy Institute at The University of Chicago, University of Chicago, Chicago, IL, USA
e-mail: anants@uchicago.edu

© The Author(s) 2023
T. Madon et al. (eds.), *Introduction to Development Engineering*,
https://doi.org/10.1007/978-3-030-86065-3_7

highly industrialized Indian states of Maharashtra and Gujarat highlight the degree to which factories have been found to be violating pollution norms. In the cities of Surat and Ahmedabad in Gujarat, for instance, Duflo et al. (2013) collected data from hundreds of industrial plants and found that about 35% were polluting above the legally-prescribed limits. In Maharashtra, Greenstone et al. (2018) digitized over 13,200 regulatory pollution tests, spanning the period from September 2012 to February 2018, and found over half to reveal exceedances in the regulatory limits.

One reason why Indian manufacturing emissions pose a particularly thorny challenge to understaffed environmental regulators is that it is difficult to monitor a large number of highly polluting but relatively small factories.[2] In the state of Maharashtra, for instance, data collected between 2017 and 2020 showed that even when considering only the largest plants, the frequency of pollution testing remained well below once per year, on average.

The development challenge therefore has two parts. First, how can we reduce industrial air pollution in an expansive and rapidly developing country like India? And second, how can we improve the quality of air pollution data available to regulators, and can these types of improvements help reduce air pollution?

One promising solution to this challenge is the use of Continuous Emissions Monitoring Systems (CEMS). These instruments are installed in the smokestacks of factories where they continuously measure the concentration (or mass) of the air pollution that is being emitted. The data collected are then transmitted in real time to a remotely located computer server, vastly improving the quality of information available to the environmental regulators.[3]

This case study discusses a project that aimed to evaluate the impacts of CEMS in a large-scale randomized control trial (RCT) conducted in partnership with the Gujarat Pollution Control Board (GPCB), the environmental regulator in the state of Gujarat in India. The goal of the project was to see whether the "big data" generated through CEMS could spur improved regulatory actions, thus lowering air pollution. While CEMS is not a new technology per se, its deployment in the Indian setting was novel. As we learned from our experience, technology acts through the interaction of human beings and hardware. As such, its effectiveness cannot be divorced from the incentives and capabilities of the people using it. Consequently, a central theme running through this chapter is the importance of evaluating a technology within the specific institutional and economic context in which it will be used.

The remainder of this chapter is structured as follows. In Sect. 2, we provide background information on the setting for the project, including how it was conceptualized and implemented, as well as findings from related research. In Sect. 3, we describe the CEMS technology in more detail and how the pilot implementation

---

[2] In India, 33% of manufacturing output comes from small-scale plants, located in about 6,000 clusters scattered across the country (Shah et al., 2015).

[3] An analogous technology to CEMS is the smart-meter, which measures electricity consumption, although the gap between CEMS and the status-quo of periodically manually sampling air pollution emissions is greater than the gap between smart-meters and the older electromagnetic metering technology.

was designed to enable rigorous measurement of its benefits. In Sect. 4, we provide some novel results on the impacts of CEMS on industrial air pollution in India. In addition, we discuss what it means for a technology intervention to "work" or to be "successful" in a developing country context. Section 5 concludes.

2   Context

There is no single cause for the problem of excessive industrial air pollution in India. As is mentioned above, the performance of India's existing regulatory framework has been far from perfect. Low institutional capacity and expertise, high transaction costs in taking legal action, corruption, high compliance costs for industry, and poor data have all contributed to this problem in varying degrees.

The starting point for this project is the ground-breaking RCT by Duflo et al. (2013, 2018), which studied the role of information and monitoring in regulating industrial air pollution in Gujarat. When Duflo et al. began this work in 2010, air pollution levels were rising across industrial cities in Gujarat. When third-party laboratories were used to audit the emissions of these factories, however, there appeared to be widespread compliance with environmental standards. This was a puzzling result. How could these test results be reconciled with the common sight of black air rising out of hundreds of small chimneys in and around Gujarati cities?

To answer this question, Duflo et al. ran a 2-year RCT in partnership with the Gujarat Pollution Control Board, the state environmental regulator. First, the study shed light on how the regulator was crippled by a persistent culture of data falsification. It documented collusion between the industrial plants and the auditors that were supposed to report on their performance. Next, the study evaluated an intervention designed to resolve this problem of collusion by severing the conflicts of interest, namely, by shifting the auditor hiring and payment decisions away from the individual plants and towards the government regulator.

Although the researchers were able to provide a partial solution to the monitoring problem (by making the audit reports more accurate), manual testing remained an infrequent and expensive method of gathering data. The regulators would still be unaware of how much the industries were emitting on a regular basis. In addition, the status quo emissions audits could not say anything about what the factories were doing when they were not being actively tested. In consequence, they were perhaps best viewed as simply an assessment of how industrial plants performed when on their best behavior.[4] Overall, Duflo et al. (2013, 2018) suggest that high-quality,

---

[4] Differences between in-test and in-operation pollution levels have been found in other settings. For instance, in 2015 the United States Environment Protection Agency uncovered systematic and deliberate differences between on-road and in-test pollution from diesel Volkswagen cars, resulting in worldwide fines and vehicle recalls.

continuous information might be necessary to know which plants were polluting the most and that these monitoring improvements could potentially reduce emissions.

2.1 How Was the Project Location Chosen?

Large field experiments involving partnerships between researchers and the government are often opportunistic, and a function of the initial interests of individual politicians and bureaucrats. In this case, India's Minister of Environment, Forest, and Climate Change felt that CEMS might not only solve the information problem but could also be a first step towards implementing market-based methods of environmental regulation. The project described here was thus the first attempt in India to carry out a systematic field evaluation of the effectiveness of CEMS. It also became the precursor to India's first cap-and-trade market for industrial particulate pollution, which was launched in Surat in 2019.

The location of the CEMS pilot itself shines some light on the myriad forces that determine how and where a policy innovation is first implemented. The project initially involved the state governments of Gujarat, Maharashtra, and Tamil Nadu, three of the largest industrial states in India. However, requiring industrial plants to install CEMS proved to be politically and administratively difficult for a number of reasons. For example, the technology was untested, and several implementation challenges could be foreseen. For risk-averse regulators, there was a potential downside to being the "first mover" on a project that might not end well. In addition, the installation of CEMS devices would impose a significant cost on polluters—in the order of several thousand dollars—and this would surely result in pushback and reluctance on the part of the industrial factories.

Furthermore, up to that point, there was no precedent to using CEMS data as the basis for prosecuting industrial polluters. Indian environmental law is built upon criminal penalties. The basis upon which plants can be found to be in violation of pollution limits is enshrined in the laws. New methods of monitoring cannot be easily used as the legal basis for regulatory actions without evidence of their reliability and accuracy. In sum, requiring plants to install equipment that had not been granted legal status posed challenges that would understandably create industry resistance.

Eventually, the government of Gujarat was able to prevail upon the industry association of the city of Surat to support (or at least not oppose) the use of CEMS devices. Surat is a hub for the textile industry and is the site of a dense manufacturing cluster of hundreds of relatively small-scale plants that typically burn solid fuels.

One potential reason for their agreement is that the government informed them that CEMS would later be used to introduce emissions markets. This acted as a powerful incentive for the manufacturing plants. It changed the impression surrounding the equipment installation mandate from a pure negative (e.g., regulation would not change, while monitoring would increase and expensive equipment would have to be purchased) to a potential opportunity. Market-based regulation had been repeatedly recommended in India as a potential solution to the inflexibility, uncertainty, and the high costs associated with status-quo regulation.[5]

2.2   Identifying the Experimental Sample

Having identified Surat as a project site, a decision needed to be made on how to select the initial sample of plants that would be transitioned to the new technology. This filtering process was driven by the potential of a plant to harm people through air pollution, as well as by the limitations of the technology itself. As a starting point, the sample of eligible plants was first restricted to factories located within 30 km of the city center, in order to focus attention to the plants that were located in the most densely populated areas. For a sense of scale, the metropolitan population of Surat was about 6.5 million, with a density of over 4000 people per square kilometer.

Conditional on being located within this radius, a factory needed to be burning solid or liquid fossil fuels and needed to have a stack (i.e., chimney) large enough to accommodate a CEMS device. This led to a sample of 373 plants. Most of these plants were operating in the textile sector (over 94%) and were burning coal (37%) or lignite (27%). Over the course of the multiyear project, 42 plants closed down due to an economic downturn in Gujarat, which correspondingly reduced the pilot sample size.

In the case of the other two states, Maharashtra and Tamil Nadu, there was no suitable set of plants that could be transitioned to CEMS, despite multiple years of efforts on our part. Thus, the geographic scope of our overall CEMS research effort shrank from three states to one.

The point of recounting this history is to underscore that the location of technology pilots, as well as the population involved, emerge through multiple layers of deliberate or accidental selection. This means that it is worth paying attention to the external validity concerns for any given technology deployment. In this case, the impacts of rolling out CEMS in Gujarat may not apply in Maharashtra or Tamil Nadu. For instance, if the plants that are successfully able to resist CEMS also happen to be politically well-connected and more difficult to regulate, then what we learn about the impacts of CEMS on our pilot population would not apply to other populations.

---

[5] For example, in the previous year, India's Ministry of Environment, Forests, and Climate Change had constituted a *High Level Expert Committee* to review environmental legislation and regulation in the country. The committee had recommended a paradigm shift in the tenet of pollution control from the existing "command and control" approach to a "market-based" approach that would include innovations like "cap-and-trade" programs.

2.3 How Does CEMS Work?

Continuous Emissions Monitoring Systems (CEMS) for particulate emissions consist of a network of hardware devices and software programs that link monitored industrial plants to the environmental regulator in a manner that allows emissions data to be securely transmitted at regular intervals. The CEMS hardware components required at each industry site consist of the following:

1. Particulate matter (PM) CEMS analyzer and flow meters to measure the mass of pollutants emitted.
2. Data logger unit for saving records on-site, in case of Internet failure.
3. Data acquisition system (DAS), normally consisting of an on-site computer and a server at the site of the regulator.
4. Software to visualize and analyze CEMS emissions data.

The CEMS analyzer for particulates is a device that relates the physical properties of emissions from a factory chimney to the concentration (or mass flowrate) of suspended particles in the air. For example, optical devices measure the attenuation in the intensity of a laser beam sent through smoke. An alternative technology exploits the so called "triboelectric" effect and relies on measuring the electric charge induced by the movement of particles near a probe.

These approaches to measurement are "indirect" since they measure a property of the gas that is used as a proxy for the presence of solid particles. As a consequence, these devices must first be calibrated against manual readings that directly measure the weight of particles in a specified volume of exhaust. In other words, an electrical signal generated by the analyzer must be mapped to a value of particulate emissions. Typically, this mapping is obtained by fitting a linear model relating a set of manual measurements taken at different levels of boiler loads (typically nine readings) to the corresponding measures of current produced by the analyzer. This delivers an equation $y = a + bx$ where x is a current reading in amperes, y is an estimate of the concentration of pollutants, and a and b are parameters estimated during an initial calibration process.

3 Innovate, Evaluate, Adapt

There is a well-known management cliche that, "if you can't measure it, you can't manage it." There is some truth to this when it comes to regulating industrial emissions. When the government is unaware of how plants are behaving most of the time, they have little ability to target regulatory actions, and may have no legal basis to penalize otherwise polluting factories who manage to pass the occasionally scheduled tests. Furthermore, from the point of view of human health—the ultimate motivation for environmental regulation in the first place—what matters is not how much pollution a plant emits at any given point in time but the cumulative mass released into the ambient air over a period of time. A highly polluting plant operating for only a few days in the year might fare the worst on a one-off manual test. But it may in fact be a lot less harmful than a "cleaner" plant that is operating 24 h a day, 365 days a year.

All of this suggests that if regulators are to do their job properly, they need information that identifies which plants are polluting. Unfortunately, the standard method for measuring factory air pollution, which is commonplace around the world including in India, involves a painstakingly manual process. To measure the concentration of suspended particulate emissions, a team of engineers must extract a sample of air from the smokestack of a factory while carefully following a prescribed protocol. This air sample must then be transferred to a laboratory where solid particles are dried to remove moisture and then weighed. The mass of these collected particles is then converted into a concentration measure after "normalizing" the volume of air extracted by standardizing it to a particular temperature and pressure. The whole procedure is labor- and time-intensive. The CEMS technology improves matters by transmitting emissions data in real time, directly from the chimney of a plant to the regulator. In comparison to the manual, point-in-time, method of measurement, the CEMS technology offers a potentially dramatic improvement.

There are a number of possible benefits from CEMS. Improving the information available to environmental regulators could allow existing rules to work better, for example, by allowing the government to target inspections and use calibrated penalties on the worst offending plants. CEMS could also open up the possibility of new forms of regulation that are made possible by higher quality and higher frequency data. For instance, emissions trading schemes or pollution taxes can be introduced only when there is an accurate count of the quantity being traded or taxed. These policy tools may have important advantages over existing "command and control" approaches to regulation but are feasible only if there is high-quality of data. Thus, on the surface at least, the CEMS technology would seem to provide everything needed to resolve the challenge of monitoring plants and, in so doing, to reduce the pollutants they release.

3.1 Setting Up an Evaluation

The goal of our pilot was to understand the impact of mandating CEMS on plant and regulator behavior. On the side of the plants, the primary outcomes of interest were pollution and reporting quality. On the side of the regulator, we were interested in whether the government interacted differently with these plants, enforcing more penalties, for instance. We were able to do this by integrating an RCT into the implementation of the pilot project.

The evaluation initially began with a small group of installations serving as a technical dry run. The purpose was to field-test the CEMS installation and calibration protocol stipulated in the Government of India's Central Pollution

Control Board (CPCB) specifications. We selected 11 industrial plants (which we refer to collectively as "Phase I") from a cluster of factories in Surat. Next, four different vendors supplied these eleven plants with CEMS devices, consisting of nine DC triboelectric-based devices (measuring particulate matter (PM) mass flow), and two electrodynamic-based devices (measuring PM mass concentration). The 11 industries were chosen to represent the diversity of the full sample and to cover a range of boiler sizes and types of installed air pollution control devices (e.g., numbers of cyclons, bag filters, etc.).

The field tests were important because they showed that the error associated with a CEMS measurement taken at any point in time disappears as multiple observations are added together. These sorts of tests provided confirmation that even if a device noisily measured instantaneous levels of emissions, it could still provide valuable information to determine long-run aggregate or average levels of emissions.

For the remainder of the sample, we used an RCT design to rigorously quantify the impact of CEMS. The simplest possible design would have involved dividing plants into a treatment group and a control group and then mandating that factories in the former group install CEMS, while factories in the latter group are regulated and monitored as usual. But this design would have required an enforced pause in the rollout of CEMS, after the treatment group plants had installed their devices and begun sending the regulator real-time data.

In our context, an enforced pause was not feasible for several reasons. First, installing CEMS devices was a time-consuming affair because the devices were not readily available in a still nascent market. Saturating the treatment group of factories alone took several months due to delays in purchase orders being fulfilled, delays in calibrating devices, and data connectivity problems that appeared at scale (and were not encountered during implementation for the limited number of Phase I industries.) Losing many more months of time in order to enforce a pause was therefore unacceptable to the government. Also, it would have probably resulted in the small number of CEMS vendors operating in the market to exit due to lack of business, or declining to deploy the necessary staff to the field for required after-sales services.

Second, as we have mentioned, it was not easy to ensure that the industries in Surat would comply with the new requirements to install hardware. It was believed that the likelihood of smoothly saturating the entire population of industrial plants would be significantly reduced if there was a long period of time during which half of the plants in the sample were held off from placing new equipment orders.

Finally, as our work progressed, it became increasingly likely that the GPCB would want to launch an emissions trading scheme in Surat. Thus, the effort to set up these devices had the dual goals of understanding the impact of improved data on environmental outcomes, and eventually measuring the impact of a market-based approach, like cap-and-trade, to reducing industrial air pollution. All of these considerations led both the government and the research team to want to avoid long delays in implementation.

To meet these constraints and allow for a rigorous evaluation design, the remaining plants were randomly assigned into three groups (Phases II, III, and IV). Each group was mandated to install a CEMS and begin sending data to the regulator

Fig. 7.1 Timeline of CEMS rollout and data collection
Notes: Each survey was 3 months long

in a sequential manner. In this type of staggered, phased-in research design, the plants that were mandated to install CEMS in the later phases could serve as a comparison group for the plants that had installed CEMS earlier. For example, once Phase II installations had begun sending data, and while Phase III plants were installing and calibrating their devices, the Phase IV plants would still be continuing business as usual. Thus, the outcomes for Phase II and Phase IV plants could be compared against one another in order to estimate the causal effects of CEMS. This was possible due to random assignment to each group.

3.2 Implementation Challenges in the Field

On paper, the experimental design was fairly straightforward. In practice, implementation was anything but simple. Figure 7.1 provides a timeline for the pilot as a whole. The pilot began in early 2014 and ran until late 2018. From an experimental research standpoint, the period between the second baseline survey (or midline survey, carried out in 2016) and the endline survey (carried out in 2018) constitutes the effective duration of the treatment. This is just a fraction of the overall duration of the project. The 2-year period between the first baseline survey and the second baseline survey represents a period of time in which a mature technology—that had been readily deployed in large plants in developed countries—was effectively unusable in Surat.

Why was this the case? The primary challenge was the interaction between the technology and the institutions and systems around it. The most notable unexpected challenges appeared in the process of installation and calibration of the CEMS devices, which I now detail below.

3.2.1 Problems Encountered While Installing CEMS

The first step in setting up a functioning CEMS network is for plants to equip themselves with the necessary hardware. This proved to be a rocky experience, in part because no well-functioning market in CEMS devices was in place when the pilot started. This was not surprising, given that there was no demand for CEMS before the equipment mandate. This posed several challenges to pilot implementation.

The most important of these was the near complete absence of after-sales service and trained personnel. Initially, the goal of the equipment vendors was to make a one-time sale, under the assumption that the pilot was likely to be an isolated, one-off experiment. Investing in trained staff to be based in Surat likely made little business sense for many of these companies. High-quality service contracts were likewise unavailable. The consequence was a host of equipment problems that could not be properly addressed and industry dissatisfaction with the technology, creating a fragile ecosystem prone to breakdowns. Only after several months of persuasion and pressure by the government on the manufacturers and vendors of CEMS devices did the situation improve.

A second area which proved to have several teething problems was the implementation of a robust data acquisition system to transfer emissions information from industrial sites to the state regulator. Two aspects of the prevailing ecosystem made this hard. First, different CEMS vendors used varying, proprietary data storage formats, making it difficult to create a common platform that might be able to read and transfer data. Second, industries in Surat frequently had poor Internet connections, old computers, no dedicated IT staff, and faced occasional power outages that would force a complete reset of the system. These issues are commonplace in developing countries but are not necessarily the focus of manufacturers of CEMS technology. Creating IT and data storage systems that were robust to these weaknesses proved to be challenging and time-consuming.

Finally, since the equipment mandate merely required industrial polluters to install a CEMS device, many industries chose the lowest price and lowest quality option. This type of equipment choice enabled plants to initially meet the regulatory requirement without incurring a large upfront payment but heightened the costs associated with the lack of after-sales service. Thus, several months after initial installation, some of these plants found their devices to be dysfunctional and new purchases became necessary, which set back the overall timeline for the experiment.

3.2.2 Problems Encountered While Calibrating CEMS

CEMS devices also need to be calibrated before being used. If the initial calibration parameters are misreported, the subsequent measurement of emissions will be biased. The calibration step requires plants to work with equipment vendors and environmental auditors to make manual measurements and then correlate these with the electrical output from the CEMS analyzers. The calibration requirement is crucial because it represents an important mechanism through which data can be falsified, in a similar way to the corruption noted in Duflo et al. (2013).

Calibration will not cause problems in a laboratory environment. However, it is exactly the sort of weakness that can render a technology unreliable in the

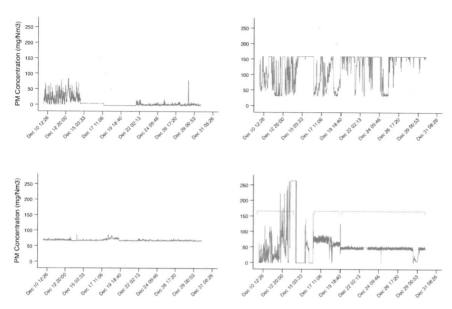

Fig. 7.2 Examples of potentially invalid CEMS data
Notes: Suspicious data transmitted by four CEMS devices during the month of December 2019. Clockwise from top left. (1) Zero values indicate device disconnection in a Type-1 device; (2) flatline values can result either from voltage top-censoring or a probe which has not been cleaned; (3) abnormally low variance suggests wrongly calibrated device or voltage amplification settings; (4) sudden decreases in overall emission levels are suspicious, and sudden changes in variance suggest tampering of voltage amplification settings

field. Importantly, the calibration process requires information from outside the CEMS itself, since CEMS must be used in conjunction with data from a different method of measurement. This means that the issues that can affect the manual data collection process can also affect CEMS. In sum, we cannot evaluate CEMS without considering the underlying economic incentives of the calibration agents and the institutions that influence them.

Early on in the project, there were several instances of CEMS transmitting unreliable data, whether deliberately falsified or not. Figure 7.2 provides examples of data that can be identified as unreliable through visual inspection or a series of automated data checks. Mitigating these problems in the field was not easy and required a combination of measures including: regular data validation; statistical tests on the emissions time series; double-blind calibration, where the lab involved in testing manual samples is separated from those collecting samples; encryption of transmitted data; and regular spot-checks following calibration. All of these measures needed to be developed from the ground up, significantly reducing the degree to which these devices could be treated as a "plug-and-play" solution to the information problem.

Table 7.1 Regulatory actions tied to emission measurements with CEMS

Action	Criteria
Send SMS warning	Among the five highest polluters in the last week
Send formal warning letter	Among the five highest polluters in the last 2 weeks
In-person meeting with plant management	Among the five highest polluters in the last 3 weeks
Regulator site visit and pollution test	Among the five highest polluters in the last 4 weeks
Legal show-cause notice	Industry fails on-site pollution test

3.2.3   Linking Data to Regulatory Practice

The installation of a CEMS analyzer in a factory chimney is unlikely to be useful in isolation. It is not uncommon for governments to accumulate large quantities of data without using them to guide policy. In order for the CEMS technology to achieve its potential, the environmental regulator needs to determine how the resulting data will be analyzed and used. Defining data use protocols in advance would seem to be an essential factor contributing to the success of this technology.

The GPCB chose to institute—but did not perfectly implement—a detailed *Action Matrix* linking the data they received to regulatory action. Plants were penalized if they did not transmit data regularly and if their pollution levels exceeded the specified thresholds. Table 7.1 summarizes the key features of this protocol.

4   Did CEMS Work?

In a narrow sense, we might regard a technology as "working" if it delivers in the field a performance that is very close to the design specifications determined in a lab or via controlled engineering trials. This definition is not very useful from the perspective of a policymaker. Thus, a slightly broader definition is to define CEMS as "working" if it delivers *useful* data. The extent to which CEMS data is useful depends on both its information content and its potential value to regulators.

Yet even this broader definition may also be inadequate. Even if CEMS delivered valuable information, it might not result in lower levels of pollution. Changes in the behavior of industrial plants are likely to occur only if the regulator uses the new information to change the incentives for plants to pollute. Institutional weaknesses, rent-seeking, political patronage, and other implementation failures could lead to regulators ignoring to act upon these improved sources of information. In this section, I report on some of the descriptive and experimental results from our deployment, in order to assess the technical performance, usefulness, and impact of CEMS in Gujarat.

4.1 Assessing the Technical Performance of CEMS

Suppose we define a working CEMS device as an instrument that reliably delivers a signal that accurately represents plant emissions. An important insight from our experiment was that the CEMS devices did not automatically work until they were coupled with regulatory measures that appropriately aligned the incentives of plants and vendors with those of the regulator. As discussed, the mandate to install CEMS devices was followed by a host of implementation challenges. Some of these reflected difficulties that are widespread in developing countries, including poor infrastructure or the absence of trained staff in factories. Others, such as the paucity of after-sales service, arose out of market weaknesses that were a consequence of mandating purchases in a very nascent market. Still others arose from the opportunity for data falsification created by the calibration requirement. Although improper calibration does not reflect an engineering failure, the ease with which regulatory devices can be gamed is quite fundamental to an assessment of the technology's effectiveness.

However, although the new monitoring system was not technically perfect at the outset, it improved steadily and significantly until it reached a point where a meaningful experimental evaluation could be conducted. Some of this improvement occurred after the environmental regulator introduced the *Action Matrix*, shown in Table 7.1. This is because this protocol created the first real set of consequences for plants that did not maintain their devices, failed calibration audits, or did not transmit data regularly. Statistical checks were later developed to automatically identify plants whose data suggested the possibility of miscalibration or tampering, further improving matters. Figure 7.3 shows that after the final calibration, a strong relationship existed between CEMS readings and concurrently measured manual measurements.

Figure 7.4 shows the history of CEMS uptime rates for 66 industrial plants, with the regulatory measures taken by the Gujarat Pollution Control Board marked with vertical lines. The performance of CEMS appears to improve with the introduction of each measure. Eventually, a high level of data availability is achieved.[6] This evidence has important implications for how a new technology is introduced and used. It seems intuitive that new regulation should be introduced only after the underlying technology is working very well. However, our experience suggests that the presence or absence of regulation cannot be disconnected from the performance of the enabling technology. In other words, in order to achieve an equilibrium involving a well-functioning CEMS network, it may be necessary to introduce regulation that commits to using the data generated by CEMS, even before all of the technical problems are resolved.

---

[6] Data availability is only one aspect of a CEMS device working well, the other being the accuracy of data which in turn requires high quality calibration.

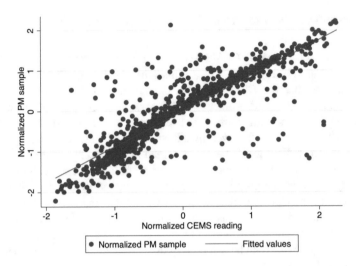

Fig. 7.3 Accuracy of CEMS data
Notes: Correlation between CEMS reading and PM sample. The figure shows data the data from calibration of CEMS devices. The figure shows the normalized values of the PM samples and CEMS readings per industry, and the red line shows the fitted values

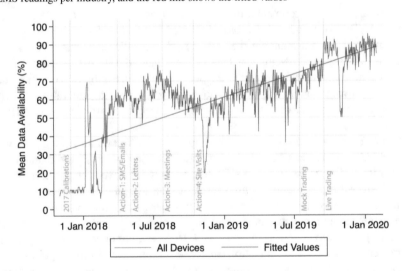

Fig. 7.4 Average CEMS data availability at 66 industrial plants
Notes: Average percentage data availability from CEMS devices over time where periods of no transmission from a plant are set as zero. The introduction of the action matrix in February 2018 was followed by a significant improvement in performance. The introduction of an emissions trading scheme in July 2019 produced similar results

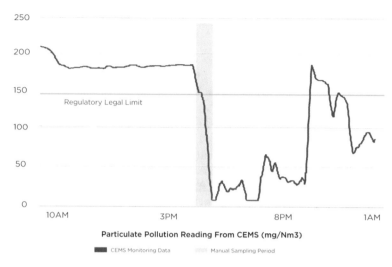

Fig. 7.5 CEMS data transmitted before, during, and after an in-person emission inspection
Notes: This plant appears to be strategically reducing emissions while being tested

4.2 Assessing the Usefulness of CEMS

CEMS are intended to augment an existing testing protocol in which the main problem is that information is only collected at a single point in time. The case for CEMS is based in part on the notion that continuous data is more useful as a measure of environmental performance than a single snapshot.

To investigate this, we examined periods of time in which we could compare readings from CEMS devices to manual samples collected by the GPCB. Recall that these manual samples formed the basis of regulation in the status quo. There are two potential sources of bias. First, there is the falsification of data identified in Duflo et al. (2013). Second, there is possible bias stemming from the fact that plants may behave differently when they are being tested to when they are operating normally.

Figure 7.5 shows a comparison between manual and continuous data for a plant in Surat. For this plant, it is striking how the period of time in which the manual sampling takes place is entirely unrepresentative of pollution at all other times. This suggests that relying on manual tests alone could result in significantly underestimating how much a plant is contributing to air pollution. Furthermore, the variation in emissions levels over time suggests that even if there were no systematic bias associated with pollution performance during testing periods, it is unlikely that extrapolations based on a single observation could reasonably approximate the true levels of pollution released by the plant.

4.3   Estimating the Impact of CEMS on Pollution Emissions

The third measure of success relates to whether or not CEMS reduced pollution emissions, which is the ultimate goal of the policy intervention. According to this measure, we found limited evidence that the technology worked.

The direct effect of requiring plants to install CEMS on subsequent pollution levels can be estimated due to the RCT design we implemented. To rigorously quantify the effects of CEMS, we need to not only measure what happened to factories after installing these devices but also identify a counterfactual describing what outcomes *would* have occurred absent CEMS. The phased rollout of the pilot, combined with random assignment to different groups, meant that at the start of the experiment, the plants that were asked to install CEMS early on (Phase 2) were on average identical to those that were asked to install CEMS later on (Phase 4). For this reason, any *differences* in pollution emissions that are observed between the two groups after CEMS data is sent to the regulator can be attributed to the installation of CEMS. All other factors unrelated to the technology mandate would affect both groups equally, and thus would not bias any comparisons between the two groups.

Table 7.2 presents a comparison of baseline characteristics of treatment and control plants in the study sample. The randomization check shows that plant characteristics are balanced across treatment groups. Out of all of the baseline measures reported, there are just two variables that show a statistically significant difference between treatment groups, at the 10% level (these are the cost of recent modifications and the number of inspections performed by the GPCB in 2014). In Panel A we examine various costs associated with purchasing or using air pollution control equipment. The variable costs from operations and maintenance associated with using this equipment are significant, roughly 40% of capital costs on average. This becomes important because the *presence* of equipment is easily observed and enforceable when granting permissions for plants to begin manufacturing. Accordingly, all these factories in our sample do possess pollution control devices (Panel B), with cyclones being the most common. The high levels of pollution that we observe are nevertheless influenced by the costs of *operating* equipment. Panel C reports statistics on inputs and revenue from these plants and boiler and thermopack capacities (which influence fuel use and pollution). Panel D shows that pollution levels at baseline are remarkably high with mean levels of pollution concentration above 300 mg/Nm^3. We also measure levels of regulatory engagement in terms of the number of inspections conducted and legal actions required in the year of the survey.

Table 7.3 reports the treatment effects of CEMS at endline, based on estimating the following regression equation:

$$y_{i1} = \alpha + \beta y_{i0} + \gamma T_i + s_{it}$$

Table 7.2 Differences in industrial plant characteristics at baseline

	Treatment mean	Obs	Control mean	Obs	Difference
A					
Operating cost	7.24	130	7.20	128	−0.056
Maintenance cost	2.15	130	2.25	128	0.200
Capital cost	25.12	130	25.85	128	1.337
Cost of modifications	0.40	130	0.07	128	−0.336*
B					
Number of cyclones	2.12	130	2.14	128	0.053
Number of bag filters	0.99	130	0.86	128	−0.128
Number of electrostatic precipitators	0.02	130	0.09	128	0.063
Number of scrubbers	1.01	130	1.12	128	0.114
C					
Asset value, Excl. Land	626.69	107	802.81	105	187.222
Yearly revenue	2933.72	123	4182.33	119	944.664
Number of employees	235.21	126	261.63	121	26.863
Boiler capacity (TPH)	6.17	130	11.48	128	4.692
Thermopack capacity (mKcal)	32.81	130	17.64	128	−16.432
D					
Ringelmann mean score	1.72	130	1.62	128	−0.091
PM concentration (mg/Nm3)	368.10	123	335.10	128	−37.536
Inspections	2.39	130	2.70	128	0.332*
Legal	0.68	130	0.67	128	0.012

Notes: The column *Difference* is calculated from the difference of the means of both groups controlling by cluster fix effects. Initial randomization assigned 141 industries to the treatment group and 139 industries in the control group. The baseline includes 130 industries in the treatment group and 128 in the control, because several plants closed between sample selection and surveying. The variable *In- spection* reports the number of inspections done by GPCB the year of baseline (2014), and the variable *Legal* reports the number of legal actions an industry has to comply with during the year of baseline, 2014. All cost figures relate to the use (or purchase) of air pollution control devices. All monetary variables are in 100,000 s INR. The Ringelmann score is on a 1–5 scale of smoke opacity, where 1 is the least opaque (polluted). Asterisks indicate statistical significance level (2-tailed) of difference between means: * $p < 0.10$, ** $p < 0.05$, *** $p < 0.01$

Here y_{i1} is the outcome variable for plant i measured during the endline survey, y_{i0} is the baseline value, and T is a dummy indicating membership of the treatment group. We use this model to examine changes in three outcomes: PM readings from a manual sample taken at the plant; the logged value of the plant test result; and a binary measure of compliance, indicating whether the test resulted in a PM concentration reading below 150 mg/Nm^3. This value is the fixed regulatory limit for the concentration of particulate emissions from an industry smokestack. Note that since the treatment itself was the installation of CEMS, there was no continuous data available for control plants.

In Column 4 of Table 7.3, we report changes in a fourth outcome variable, the results of Ringelmann tests that we carried out at periodic intervals throughout the

Table 7.3 Impact of CEMS on pollution and compliance

	(1) PM	(2) Log PM	(3) Compliance	(4) Ringelmann
CEMS treatment	−18.87	0.06	−0.05	0.04
	(28.65)	(0.10)	(0.06)	(0.06)
Baseline value	−0.04	−0.04	−0.01	0.47***
	(0.03)	(0.06)	(0.04)	(0.07)
Weather controls	No	No	No	Yes
Ringelmann survey FE	No	No	No	Yes
No. of Obs.	241.00	241.00	241.00	1476.00
R-squared	0.05	0.06	0.06	0.18
Mean outcome	184.53	4.92	0.67	1.58

Notes: In column (1) the dependent variable is the endline PM sample. In column (2) the dependent variable is the log of the PM samples at endline. In column (3) the dependent variable is a dummy that takes the value of one if the firm is complying to the norm of PM of less than 150 mg/Nm^3. In column (4) the dependent variable is the Ringelmann score for rounds after 2018. Standard errors clustered at the plant level in parenthesis and all models include fixed effects for the industrial region within Surat where the plant is located. *Baseline value* refers to the outcome variable for each column as measured in the baseline survey. Model 4 uses panel data from three Ringelmann surveys conducted in the year of the endline survey (2018). Models 1–3 use data measured during the endline. Asterisks indicate coefficient statistical significance level (2 tailed): * $p < 0.10$, ** $p < 0.05$, *** $p < 0.01$

experiment. A Ringelmann test is a simple visual measure of the color of smoke in the chimney stack, which is a crude proxy for the level of particulates in emissions. These tests were repeated between the baseline and endline, providing a panel dataset. To estimate the impacts of the treatment on this outcome, we estimate a similar regression model to the one above, but with additional weather controls and with multiple observations for each plant over time, rather than values captured only at baseline and endline.

Across all specifications, we find no strong evidence that plants in the treatment group polluted less than their counterparts in the control group. Between baseline and endline, there was an overall reduction in both the level and variance of the manual spot measurements of pollution. The average PM concentration at endline was 185 mg/Nm^3 with a standard deviation of 211. At baseline, the average concentration was 338 mg/Nm^3 with a standard deviation of 374. However, this reduction occurred in both factories with and without CEMS making it difficult to attribute the change to the new technology. This finding is a cautionary tale underscoring why the use of a treatment and control group is so important. Had we only looked at pollution changes in factories with CEMS, we might have noticed they became cleaner and then credited this to the technological intervention. As it happens, unless plants in later phases pre-emptively cut pollution in anticipation of a future mandate (possible, but unverifiable), it may be the case that better monitoring alone is not sufficient to solve the development challenge at hand.

Why was factory pollution unresponsive to the installation of CEMS? One answer might lie in how the data was actually used, or not used, by the regulator.

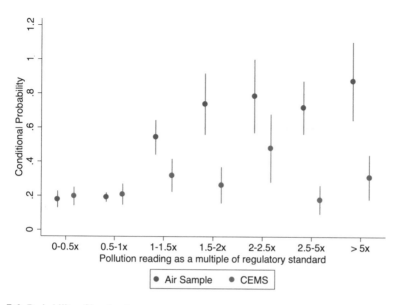

Fig. 7.6 Probability of legal action
Notes: Conditional probability of legal action by the Gujarat Pollution Control Board as a function of pollution measured by CEMS instruments and manual samples expressed as multiples of regulatory standard. Dots represent mean values in bins and bars are 95% confidence intervals

Recall Table 7.1, which describes the protocol that the GPCB stated would determine how it acted upon the continuous pollution data it received. The protocol envisages actions of steadily increasing severity, rising to a legal notice. However, upon examining the history of regulatory interactions at the end of the pilot, we found that the regulators' bark was worse than its bite. Although the first few actions in the schedule (e.g., SMS messages, in-person meetings, and letters) were executed, there was no significant difference between the treatment and control groups in the probability that a plant received a legal notice or even an in-person pollution test. In other words, when it came to taking actions that involved more than "cheap talk," the regulator blinked.

Interestingly, this lack of follow-through action in response to high CEMS readings was not perfectly mirrored in the other data source available to the regulator, namely, the results of traditional manual tests. This difference is clearly visible in Fig. 7.6. The conditional probability of plants receiving a legal notice significantly increases when their last manual test reveals very high readings, but no such pattern exists when examining CEMS data. In other words, although the continuous monitoring systems generate useful, even superior, information, the regulator chose not to use it for targeting enforcement. If the plants understood that this would be the case, it is unlikely that CEMS would have had an observable effect on behavior.

5   Conclusion

The experiences recounted in this case study underscore the complexity involved in translating technology from a developed country to a developing country context. The Surat experience with CEMS demonstrates how technical solutions to development problems cannot be divorced from the institutional context in which they are used. It is often impossible to identify the effect of these interactions either in the lab or through iterative design processes in the field, because they are only apparent in applications at scale.

In the present example, the technical pilots that preceded the RCT provided an example of small-scale field trials based on which several technical tweaks were made. Although this step was necessary for implementation, it was far from sufficient. Other challenges such as calibration problems or staff capacity constraints became apparent only when several hundred devices were deployed at once.

Development engineering requires a blend of the social sciences and engineering. The Surat CEMS deployment was evaluated using an RCT design, drawing upon the expertise of a team of economists. However, without continuous technical innovations designed in situ, there would have been no intervention to test. The eventual results were mixed, placing the spotlight on additional questions, some of which have to do with the design of regulatory institutions and the legal framework underlying environmental regulation in India.

In 2019, the Gujarat Pollution Control Board launched its Emissions Trading Scheme (ETS) across a cluster of industrial plants in Surat. The Surat ETS is the first cap-and-trade market in particulate emissions across in the world. The data generated by the Continuous Emissions Monitoring Systems that were set up as a part of the project we have described here went on to form the data foundation for the ETS.

In the past, India's environmental regulations have been criticized for being blunt and inflexible, proving both costly for industry and difficult for the government to implement and enforce. Market-based instruments, such as emissions trading schemes, provide an alternative that could meet the dual challenge of economic growth and environmental safety. Thus, the Surat ETS represents a paradigm shift in Indian regulation, stemming in part from the recognition that no matter how well these instruments worked, if they were used within an archaic regulatory framework, pollution was unlikely to be reduced. In a sense, the most important impact of these devices may have been to open up the option of using modern market-based regulation in India, something that would have been completely impossible without a transformation in how pollution from industry was monitored.

6 Discussion Questions

1. As a development engineer, what criteria would you apply before declaring a technological solution a success? What types of disciplinary skills would a research team need to possess to give them the best chance of meeting these targets? What are the differences between the criteria for success that an R&D lab, a venture capitalist in a startup, or a policymaker might apply when evaluating a technology solution to a development problem?
2. The adoption of technology designed to solve development problems often requires government support through tax-incentives, regulatory mandates, or subsidies. On the one hand, if such policy assistance is to be entertained, then rigorous evidence on effectiveness seems essential. On the other hand, it may be difficult for any technology to improve without learning-by-doing and the opportunity to iterate and improve given results in the field. Choose a technology other than CEMS (perhaps from another chapter in the book) and discuss how you would approach this balancing act? As policymakers, what strategies would move us most quickly towards effective, cheap, reliable technology solutions, and what do you think a phrase like "data-driven policy-making" should mean in this context?
3. Can you think of ways to make it hard for plants to falsify data coming from a CEMS device, holding the technology constant? Is there a new rule the government might impose or additional data that could be used to cross-check the validity of data? How would you judge the cost-effectiveness and feasibility of these solutions in a developing country context?

References

Balakrishnan, K., et al. (2019). The impact of air pollution on deaths, disease burden, and life expectancy across the states of India: The Global Burden of Disease Study 2017. *Lancet Planet Health, 3*, e26–e39. https://doi.org/10.1016/S2542-5196(18)30261-4

Bharadwaj, P., Gibson, M., Zivin, J. G., & Neilson, C. (2017). Gray matters: Fetal pollution exposure and human capital formation. *Journal of the Association of Environmental and Resource Economists, 4*(2), 505–542. https://doi.org/10.1086/691591

Burney, J., & Ramanathan, V. (2014). Recent climate and air pollution impacts on Indian agriculture. *Proceedings of the National Academy of Sciences of the United States of America, 111*(46), 16319–16324. https://doi.org/10.1073/pnas.1317275111

Chang, T. Y., Zivin, J. G., Gross, T., & Neidell, M. (2019). The effect of pollution on worker productivity: Evidence from call Center Workers in China. *American Economic Journal: Applied Economics, 11*(1), 151–172. https://doi.org/10.1257/app.20160436

Duflo, E., Greenstone, M., Pande, R., & Ryan, N. (2013). Truth-telling by third-party auditors and the response of polluting firms: Experimental evidence from India. *The Quarterly Journal of Economics, 128*(4), 1499–1545. https://doi.org/10.1093/qje/qjt024

Duflo, E., Greenstone, M., Pande, R., & Ryan, N. (2018). The value of regulatory discretion: Estimates from environmental inspections in India. *Econometrica, 86*(6), 2123–2160. https://doi.org/10.3982/ECTA12876

Greenstone, M., Harish, S., Pande, R., & Sudarshan, A. (2018). The solvable challenge of air pollution in India. *Proceedings of the India Policy Forum, 14*, 1–40.

Shah, R., Gao, Z., & Mittal, H. (2015). *Innovation, entrepreneurship, and the economy in the US, China, and India: Historical perspectives and future trends.* Elsevier. https://doi.org/10.1016/C2014-0-01381-0

Part II
Market Performance: Technologies to Improve Agricultural Market Performance

Lorenzo Casaburi (iD)

Introduction

This section focuses on applications of Development Engineering in agriculture, with a geographic focus on sub-Saharan Africa. Agriculture employs a majority of the population in most African countries. According to the recent World Bank Development Indicators, in spite of the rapid urbanization, approximately 60% of people in sub-Saharan Africa live in rural areas, and the agriculture employment share is about 55%. Smallholders take a major part of the agricultural sector. Family farms operate 75% of the world's agricultural land. In poor countries, 60% of the landholdings are below 2 ha, and average plot size has been decreasing over the last few decades (Lowder et al., 2016).

Due to its centrality for economic development and poverty reduction, agriculture has rapidly become an important focus area for Development Engineering. Innovating, implementing, and adapting in the context of smallholder agriculture present very different challenges from the ones arising when working with larger farms. A long tradition in development economics has emphasized the importance of market constraints small farmers encounter in their production and marketing choices (Jack, 2013). In addition, at least since the seminal work by Bates (1981), scholars have highlighted political economy factors and institutional constraints – e.g., inadequate political representation of farmers' interests, elite capture, and weak contracting environments – as important obstacles to agricultural productivity growth. Finally, some features of agricultural production, like marked seasonality and high exposure to risk, imply that behavioral poverty traps may be particularly relevant for small farmers (Mani et al., 2013). The taxonomy of constraints proposed in Chap. 2 of this book thus provides a useful starting point to understand the domains for current and future Development Engineering contributions.

L. Casaburi
Department of Economics, University of Zurich, Zurich, Switzerland
e-mail: lorenzo.casaburi@econ.uzh.ch

The case studies in this section focus on some of the most pressing constraints farmers face. In the first chapter, Fabregas, Kremer, Harigaya, and Ramrattan discuss recent advances in digital agricultural extension in Kenya and Ethiopia. Digital extension helps overcome not only informational barriers (e.g., on which and how much fertilizer to apply) but also institutional constraints arising from the high ratio of farmers to government extension workers and to the difficulty in monitoring these workers. In the second chapter, Neza, Nyarko, and Orozco-Saenz focus on access to crop markets in Ghana. They review several interventions that aim at improving the prices farmers receive for their crops. After discussing an early evaluation of including price information systems, they review the promise of electronic trading platforms and commodity exchanges. In the third chapter, Aker reviews the development of rural fintech in West Africa, including credit, saving, and cash transfers. She then presents a case study from Niger on the potential of mobile money to improve the distribution of cash transfers.

These case studies share a number of common themes. First, in each of them, the innovations emerge from a strong collaboration between the academic researchers and local stakeholders, including governments (in the chapter by Fabregas et al.), private sector enterprises (in the chapter by Nyarko et al.), and NGOs (in the chapter by Aker). Second, the case studies provide clear examples of iterative implementation. Results from early-stage pilots informed the design of the next stages, either through an improved design of the original intervention (e.g., Aker describes how the results of a 2010 mobile cash transfer pilot shaped the design of the 2018 program) or through the development of more sophisticated interventions (e.g., Neza et al. describe the path from price information to trading platforms to commodity exchanges in Ghana; Fabregas et al. discuss the evolution from simple SMS-based digital extension to more complex formats). Third, the three chapters build on randomized controlled trials that evaluate rigorously the impact of the interventions on the outcomes of interest. They thus clearly describe on which dimensions the reviewed interventions worked (and on which ones they did not).

Naturally, these case studies discuss only a small portion of the growing body of Development Engineering research on agricultural markets. It is worthwhile to mention three areas of ongoing development (among many others): product traceability, land registration, and agricultural insurance for smallholders. Lack of product traceability along the supply chain prevents many farmers from accessing lucrative domestic markets and, especially, export markets. Foreign buyers often require exporters to report the detailed origin of their product and to prove compliance with sustainability standards. In highly fragmented value chains, this can be very hard. A number of recent products aim at digitizing the early steps of the value chain and thus reducing the cost of record-keeping and product-tracking. Several pilots are also using blockchains to improve traceability. Some of these products have reached a large number of farmers. However, more research is needed to understand how these innovations affect farmers' productivity and returns.

Another area of collaboration between government, private sector, and academics concerns land registration and property rights. A large body of evidence suggests that weak property rights on agricultural land hamper productive investments (e.g.,

Goldstein & Udry, 2008; Besley & Ghatak, 2010). Technological advancements are playing an important role on these issues. Recent experiences suggest that satellite imagery can improve the quality of land maps or facilitate the creation of new registries where they did not exist (see Ali et al., 2018, for an example from Rwanda). Several African countries (e.g., Kenya and Uganda) are in the process of digitizing their land registries, expecting that this will improve transparency and reduce frauds. However, to ensure fair access to these new records and avoid elite capture, it is important to complement these technological advancements with improvements in management capacity of local ministries and in regulations on data security and privacy protection (Deininger, 2018).

High transaction and claim verification costs represent important obstacles to the diffusion of insurance markets for smallholders. Over the last decade, governments, insurance companies, and academics have used remote sensing data to design insurance products based on normalized difference vegetation indexes (NDVI). These indexes predict quite well adverse events at the plot level. They thus provide a valuable alternative measure to determine insurance payouts, potentially substituting for expensive plot-level claim verification and for area-level crop-cutting exercises (see, e.g., Chantarat et al. (2013) for an early example of index-based livestock insurance in Northern Kenya). Ongoing trials are also evaluating "picture-based insurance"; in these products, companies use pictures of the plots, which farmers take with their phones, to verify claims (Ceballos et al. (2019) report results from a pilot in India). Researchers have also made remarkable progress in the use of remote sensing and machine learning algorithms to predict crop yields, even on smallholder plots (Burke & Lobell, 2017; Lobell et al., 2018). These developments are going to have important implications for poverty measurement, targeting of relief interventions, and impact evaluations, among many other areas.

References

Ali, D. A., Deininger, K., & Wild, M. (2018). *Using satellite imagery to revolutionize creation of tax maps and local revenue collection*. The World Bank.

Bates, R. H. (1981). *Markets and states in tropical Africa: The political basis of agricultural policies*. University of California Press.

Besley, T., & Ghatak, M. (2010). Property rights and economic development. In *Handbook of development economics* (Vol. 5, pp. 4525–4595). Elsevier.

Burke, M., & Lobell, D. B. (2017). Satellite-based assessment of yield variation and its determinants in smallholder African systems. *Proceedings of the National Academy of Sciences, 114*(9), 2189–2194.

Ceballos, F., Kramer, B., & Robles, M. (2019). The feasibility of picture-based insurance (PBI): Smartphone pictures for affordable crop insurance. *Development Engineering, 4*, 100042.

Chantarat, S., Mude, A. G., Barrett, C. B., & Carter, M. R. (2013). Designing index-based livestock insurance for managing asset risk in northern Kenya. *Journal of Risk and Insurance, 80*(1), 205–237.

Deininger, K. (2018). For billions without formal land rights, the tech revolution offers new grounds for hope. https://blogs.worldbank.org/developmenttalk/billions-without-formal-land-rights-tech-revolution-offers-new-grounds-hope

Goldstein, M., & Udry, C. (2008). The profits of power: Land rights and agricultural investment in Ghana. *Journal of political Economy, 116*(6), 981–1022.

Jack, B. K. (2013). *Market inefficiencies and the adoption of agricultural technologies in developing countries.* eScholarship, University of California.

Lobell, D. B., Azzari, G., Burke, M., Gourlay, S., Jin, Z., Kilic, T., & Murray, S. (2018). *Eyes in the sky, boots on the ground: Assessing satellite-and ground-based approaches to crop yield measurement and analysis in Uganda.* The World Bank.

Lowder, S. K., Skoet, J., & Raney, T. (2016). The number, size, and distribution of farms, smallholder farms, and family farms worldwide. *World Development, 87*, 16–29.

Mani, A., Mullainathan, S., Shafir, E., & Zhao, J. (2013). Poverty impedes cognitive function. *Science, 341*(6149), 976–980.

Chapter 8
Digital Agricultural Extension for Development

Raissa Fabregas, Tomoko Harigaya, Michael Kremer, and Ravindra Ramrattan

1 The Challenge

More than two billion people globally live in smallholder farming households, comprising a large proportion of the world's poor. Smallholder farm yields are estimated to reach only 25–50% of the potential in many parts of the world (Koo, 2014; Das, 2012). Increasing agricultural production is a necessity for meeting the growing demand for food in decades to come (World Bank, 2007), but the scope for expanding cropland is limited as ecological risks from deforestation and loss of biodiversity loom large (Lambin et al., 2013; Zabel et al., 2019). Raising agricultural productivity is, therefore, a critical component of improving the economic well-being of millions of people in developing countries. Simple technologies can significantly increase yields, and an increasing body of evidence suggests that access to agricultural advice can help farmers to improve their productivity. Yet, a vast

---

Ravindra Ramrattan lost his life in the 2013 attack on the Wastegate mall in Nairobi. During his professional career he contributed to several projects related to digital agriculture and digital finance for development. We remember him not only for his contributions in this area, but also for his curiosity, humor and kindness, which are sorely missed.

R. Fabregas (✉)
School of Public Affairs, The University of Texas at Austin, Austin, TX, USA
e-mail: rfabregas@utexas.edu

T. Harigaya
Senior Researcher, Precision Agriculture for Development (PAD), Boston, MA, USA

M. Kremer
University Professor in Economics and the College and the Harris School of Public Policy (Recipient of 2019 Nobel Prize for Economics), University of Chicago, Chicago, IL, USA

R. Ramrattan (*deceased*)
Ravindra Ramrattan wrote this book while at Kenya

© The Author(s) 2023
T. Madon et al. (eds.), *Introduction to Development Engineering*,
https://doi.org/10.1007/978-3-030-86065-3_8

majority of smallholder farmers do not have access to science-based agricultural information (Fabregas et al., 2019).

The rapid spread of mobile phones and other digital technologies present new opportunities to make quality agricultural information accessible at scale to farmers in developing countries (see Fig. 8.1). This chapter discusses why market failures

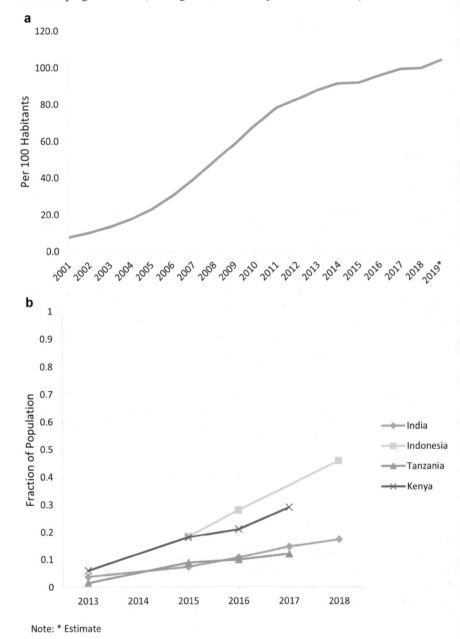

Note: * Estimate

Fig. 8.1 Penetration of mobile phones in developing countries (**a**) Shows mobile cellular subscriptions in the developing world per 100 inhabitants (2001–2019). (**b**) Shows the share of population who reports owning a smartphone. (Source: Financial Inclusion Insights, finclusion.org)

Box 8.1: From Research to Practice at Scale: Precision Agriculture for Development (https://precisionag.org)
Precision Development (PxD), a global nonprofit that provides actionable information to enable smallholder farmers to improve their wellbeing, was conceptualized and formed based on a series of research projects in India and Kenya conducted by academic researchers. PxD continues to learn, innovate, and scale its services, using technology, behavioral sciences, human-centered design, and experimentation (Illustration 8.1).

A Timeline of Precision Development (PxD)

2011

August
Researchers commence a study of an IVR-based advisory service, called Avaaj Otalo (AO), for cotton farmers in Gujarat, India (PxD co-founder Shawn Cole & Fernando, 2020)

September
Researchers commence a study of SMS advice to sugarcane farmers in Kenya (Casaburi, PxD co-founder Michael Kremer, Mullainathan, and Ramrattan, 2019)

2013

Researchers commence studies exploring farmers' valuation of local agricultural advice in Western Kenya (Fabregas, Kremer, Schilbach, 2020)

2015

Researchers commence the first of the six trials on SMS-based information services to promote agricultural lime in Kenya (Fabregas, Kremer, Lowes, On, and Zane, 2020)

December
Precision Agriculture for Development is legally incorporated by co-founders Heiner Baumann, Dan Bjorkegren, Shawn Cole and Michael Kremer

2016

April
 Launch of Krishi Tarang service (a rebranded version of AO) in Gujarat, India

Year End 27k
27k farmers reached

PxD initiatives active in two countries: Kenya and India. PxD staff hosted by Innovations for Poverty Action (IPA) and J-PAL India, respectively.

August
Research partnership is formed with One Acre Fund (OAF) in Kenya to analyse data from past experiments and design new trials to shed light on farmer behavior change challenges

2017

June
Launch of pilot service targeting paddy (rice) farmers in Odisha, India, testing PxD's model in a new geography and with a new crop

July
Partnership formed with Ethiopia's Agriculture Transformation Agency (ATA) to assist ATA to improve the 8028 Farmers Hotline through analysis of user experience, optimization of service delivery, and content development

September
Krishi Tarang service reaches 50,000 farmers

Year End
PxD initiatives active in five countries: India, Kenya, **Pakistan**, **Rwanda**, and **Ethiopia**. PxD Pakistan staff hosted by the Centre for Economic Research in Pakistan (CERP).

345k farmers reached   345k

2018

March
An experimental study commences in Gujarat, India to test whether customized fertilizer recommendations based on plot-level soil testing affect farmer behavior and outcomes

April
Partnership formed with the Government of Odisha's Department of Agriculture and Farmers' Empowerment to develop an agricultural advisory platform for the state's five million paddy farmers

Year End 858k
858k farmers reached

PxD initiatives active in six countries: India, Kenya, Pakistan, Rwanda, Ethiopia, and **Bangladesh**

May
Partnership formed with the Coffee Board of India to pilot a two-way digital advisory service (dubbed the Coffee Krishi Taranga service) with ~ 15,000 farmers in two districts in Karnataka

July
Kenya MoA-INFO service launched in partnership with the Kenyan Ministry of Agriculture, Livestock, and Fisheries (MoALF), international NGO, CABI, and telecoms company Safaricom

2019

April
Pakistan advisory service launched in partnership with the Government of Punjab to provide cotton and oilseed farmers with information about the state government's input subsidies and advice on recommended inputs

August
Partnership formed with CABI and the Zambian Ministry of Agriculture to provide farmers with pest management practices

2021

March
Colombia initiative launched in partnership with Rare, and with support from UKPact.

April
Precision Agriculture for Development (PAD) rebranded and re-registered as Precision Development (PxD)

June
PxD wins Brazilian government tender to deliver a digital agricultural extension service to 100,000 farmers in the North East region.

May
Partnership formed with the West Bengal Accelerated Development of Minor Irrigation Project (WBADMIP), a project supported by the World Bank, to provide advisory services to farmers enrolled in water user associations

3.53m

Year End
3.53 million farmers reached

PxD initiatives active in eight countries: India, Kenya, Pakistan, Rwanda, Ethiopia, Bangladesh, **Uganda**, and **Zambia**

2020

August
Partnership formed with the International Fund for Agricultural Development (IFAD) to deliver digital services to 1.7 million farmers and address the impact of COVID-19 in Kenya, Pakistan, and **Nigeria**. Partnership formed with the Brazilian Ministry of Agriculture and the Inter-American Institute for Cooperation on Agriculture to deliver digital services to 100k farmers in **Brazil's** Northeast Region

September
India services reach one million farmers

December
Nigerian service launched in partnership with IFAD and the Nigerian Federal Ministry of Agriculture and Rural Development (FMARD)

Year End 3.8m
3.8 million users reached

Services built by PxD active in eight countries: India, Kenya, Pakistan, Rwanda, Ethiopia, Bangladesh, Uganda and Zambia.

3rd Quarter 2021 5.2m
5.2 million users reached

Services built by PxD active in ten countries: India, Kenya, Pakistan, Rwanda, Ethiopia, Bangladesh, Uganda, Zambia, **Nigeria** and **Colombia**.

PxD
PRECISION
DEVELOPMENT

Illustration 8.1

and service delivery shortfalls undermine the creation and optimal flow of agricultural information and reviews ways in which digital innovations can overcome these barriers. We draw insights from existing evidence and use illustrative examples across geographies to highlight practical considerations and lessons learned through the iteration of these services. Many of these lessons come from our experience with Precision Development (PxD), a global nonprofit providing actionable information to smallholder farmers via mobile phones (see Box 8.1).

1.1 What Limits Access to Agricultural Information?

Many farmers have vast amounts of agricultural knowledge that has been transmitted throughout generations. Yet, the existence of new agricultural technologies and changes in farm conditions – from soil degradation, to market conditions – require ongoing learning and experimentation to optimize productivity. Farmers in developed countries have access to a wide range of technological developments – from high-density soil testing and moisture sensors to satellite and drone imagining – capable of gathering precise information about their farms and which enable them to apply inputs more efficiently. However, an overwhelming majority of smallholder farmers in developing countries do not have access to these technologies, and many would find it unprofitable to use them. Even relatively simple technologies, such as soil chemistry analyses, are well beyond the means of many smallholders. Additionally, experimenting with inputs in isolation is complicated because individual results are noisy and farmers may not know which dimensions to prioritize for experimentation (Hanna et al., 2014).

Several factors explain why the private sector often lacks sufficient incentives to create and offer appropriate information to farmers at the optimal scale. Information is generally a non-rival and non-excludable good. Once information is created, it can be easily shared with others at a very low cost. For example, a buyer of agricultural advice could share this information with many other farmers. If sellers of information cannot recover their investment costs, they will be unlikely to create it. Moreover, uncertainty on the part of buyers regarding the value of a particular piece of information may limit their demand.

Since many aspects of agricultural extension can be considered a public good, public provision is common in many countries (Anderson & Feder, 2004). Governments in developing countries spend millions of dollars every year creating agricultural knowledge and delivering extension services. Typical delivery approaches involve in-person visits and community events such as "farming workshops" and "field days" where technologies and input use are demonstrated (BenYishay & Mobarak, 2019; Emerick & Dar, 2020; Fabregas et al., 2017; Kondylis et al., 2017; Mueller & Zhu, 2021). However, scaling these services effectively poses a number of challenges. Foremost, in-person extension is often expensive, severely limiting its reach. Developing country governments maintain networks of over one million extension agents (Anderson & Feder, 2007), but ratios of farmers to extension agents generally remain high, leaving a majority of smallholders without adequate access to their services and information. Public extension services may also be affected by bureaucratic problems that limit the accountability of frontline information providers. Moreover, there are concerns that the most disadvantaged, including the poorest and women farmers, are often neglected (Saito et al., 1994; Cunguara & Moder, 2011). Other oft-used technologies, like radio, can reach farmers at scale, but these media make it difficult to tailor recommendations to local conditions.

In addition to service delivery challenges, generating agricultural content tailored to local conditions and the specific needs of individual farmers can be costly to produce (see Box 8.2). In many instances this leads to extension services providing blanket recommendations across large geographic areas. Cost-related considerations may also hamper the regular updating of dynamic conditions: in such cases advisory content quickly becomes obsolete. Finally, many extension systems focus on agronomic recommendations developed to maximize crop yields. However, farmers may optimize adoption decisions to maximize returns to investment (ROI), rather than crop yields, under real-world constraints, such as liquidity, transportation costs and expected market conditions, In addition, farmer's ROI calculation is influenced by individual characteristics such as risk and taste preferences. These considerations are rarely accounted for by traditional extension services.

Box 8.2: Why Do Smallholders Lack Weather Information?
Contributed by Hannah Timmis, Precision Development

Smallholders face substantial risk from fluctuations in the weather, and climate change only increases this risk. Weather unpredictability affects agricultural incomes directly, by varying the amount and quality of outputs produced from a given bundle of inputs and, indirectly, by compelling farmers to adopt costly risk mitigation strategies such as intercropping (Cole & Xiong, 2017). Accurate weather forecasts reduce this risk by enabling farmers to optimize their production based on future meteorological conditions. In India, for example, smallholders that live in areas with better seasonal forecasts calibrate their planting-stage investments to predicted rainfall and have higher profits on average (Rosenzweig & Udry, 2019). Despite these benefits, accurate forecasts are frequently unavailable in developing countries due to a combination of market and institutional failures. Producing a high-quality forecast involves large fixed costs. Global weather prediction models, which are operated by specialized meteorological centers, run many times a day and generate vast quantities of data. Forecast providers must access, assimilate, analyze, and disseminate this information, and some also deploy limited area models. The process is capital-intensive: high-performance computing, rapid data transmission systems, and highly skilled staff are all required. Yet, once the forecast is created, anyone can access and use the information at near-zero marginal cost. These production characteristics mean that basic weather forecasts are under-supplied by the market.

(continued)

Box 8.2 (continued)

Hence, by international agreement, national governments are responsible for weather forecasting. Members of the UN World Meteorological Organisation, which comprise 193 countries and territories, maintain dedicated public providers called National Meteorological and Hydrological Services (NMHS). The problem is that the capability of NMHS varies enormously. Many public agencies in developing countries frequently suffer from underfunding, failing infrastructure, outdated equipment, and inadequate expertise (Webster, 2013). The result is that valuable weather information does not reach the right people at the right time.

A case study from Pakistan illustrates the issue. In 2010, northern Pakistan suffered devastating floods, which killed 2000 people and destroyed $US500M in agricultural output. Researchers at Georgia Tech subsequently showed that the floods were predictable 8–10 days in advance if available data had been analyzed at the time (Webster et al., 2011). Yet, Pakistan's NMHS issued no warning.

1.2   The Potential of Digital Agriculture

The widespread adoption of mobile phones, combined with the advances in agricultural measurement and computational technologies presents new opportunities to address the barriers to making relevant information available for smallholder farmers. Information and communication technologies (ICT) and mobile phones, in particular, allow for low-cost, timely, and customized information delivery at scale. This medium can be particularly useful for the delivery of dynamic information which requires continuous updates, for example, weather information and market prices. Digital technologies also present comparative advantages for delivering information to farmers in remote areas with poor infrastructure, conflict-affected areas beyond the reach of in-person extension services, and in contexts affected by natural disasters in which the delivery of time-sensitive information can be life-saving.

Two features in particular make digital extension a promising area of innovation. First, ICT and other digital technologies allow for two-way communication with farmers. This can be leveraged to collect information about local conditions, farmers' backgrounds, and experiences with inputs. For most smallholder farmers, deploying hardware/sensor-based precision agriculture technologies would be prohibitively expensive. Mobile phone communication can facilitate information transmission in which farmers can ask specific questions and request information valuable to them. Even with recent technological efforts to reduce the cost of delivering precision agriculture in developing countries (Jain et al., 2019), there are many potential gains from collecting information directly from farmers. For instance, if data is collected at sufficient scale, it would enable extension systems to aggregate information more effectively, which in turn would allow for the generation of better recommendations to be made available to everyone in the system. Once aggregated, the information about individual's farm conditions could also be used to solve other informational frictions in supply markets. For example, it could be used to identify pest-prone areas or improve understanding of demand for inputs.

Second, digital extension can exercise large economies of scale to generate analytical insights and improve customization. In turn, the iteration of these insights and improvements can progressively increase impacts over time. Digital platforms generate large volumes of user data which can be utilized for constant experimentation and adjustments at low cost. The addition of more users and generation of more data allow for faster experimentation and advanced analytics – for example, through the use of machine learning – leading to faster improvements in the quality of customization and the magnitude of potential impacts. The progressive increase in returns to scale implicit in digital systems suggests that systems operating at scale, and leveraging data for constant learning, will likely derive the largest impacts.

Despite the potential of digital extension services many implementation challenges remain. For instance, many existing agricultural mobile-based systems are based on one-way, "push-only" approaches that focus on broadcasting one specific type of information (e.g., prices, specific recommendations for a crop, etc.). Not all information is useful, actionable, or accessible. The value of information depends on context. For example, advice on basic agronomic techniques is likely to be irrelevant for experienced farmers. Similarly, farmers who confront markets with constraints in the supply of labor may not adopt labor-intensive technologies. Operationalizing active two-way communication with farmers in a way that allows systems to learn about their needs could vastly improve the usefulness of the recommendations farmers receive. Moreover, relevant information uploaded to a digital platform may still not reach many smallholder populations in the absence of a user-centered design that facilitates access and comprehension for farmers across linguistic groups and takes into account low levels of literacy.

Similarly, cheap information delivery tools do not solve constraints in creating local and dynamic agricultural information. Few systems leverage experimentation or information created by farmers themselves or have mechanisms capable of facilitating local information creation. Finally, the current landscape of mobile-based agricultural information platforms is diverse, fragmented, and uneven in quality. Most ICT-based services only reach a few thousand farmers, and there has been little coordination to avoid duplication of information creation or to maximize gains from sharing. There are high fixed costs in setting up these systems – particularly with regard to information generation, software creation, establishing trust with farmers – which suggests that small-scale approaches are likely to be suboptimal.

Fully realizing the potential of digital agricultural extension will require addressing these issues through concerted efforts to develop and test a range of approaches. Successfully addressing these challenges will require interdisciplinary collaboration that incorporates lessons and insights from behavioral and data sciences, agriculture, economics, and engineering.

2 Implementing Mobile Phone-Based Agricultural Extension Services

Mobile-phone based agricultural extension services vary in the complexity of the design and the types of technologies used. In this section, we briefly discuss existing evidence on the impacts of current ICT-based extension approaches (see existing reviews for a more detailed scan of the literature, e.g., Aker, 2011; Nakasone et al., 2014; Aker et al., 2016; Fabregas et al., 2019). We then discuss selected aspects of implementation that are key to successful deployment.

2.1 Current Approaches via Mobile Devices

2.1.1 Services That Rely on Text Messages

Basic mobile devices with call and texting capabilities only are still the most common type of phone handset used in developing countries, and text messaging is still the cheapest way to reach people in many parts of the world. Text messages, or short message services (SMS), allow for written messages of 160 characters, which can be sent in bulk and broadcast in near real time to hundreds of thousands of people. This ability to reach farmers in resource-poor areas at very low cost makes text messaging an attractive option for closing information gaps.

This simple text messaging technology, however, has a number of limitations. First, there is limited scope for communicating complex information sending too many messages can annoy farmers (IDinsight, 2019) or overload cognitive capacity, potentially leading farmers to pay little attention to message content. Even in instances where farmers are eager to receive messages, illiteracy may limit the effectiveness of written information. Second, poorer farmers with low mobile literacy may find a significant barrier to direct two-way communication, especially in contexts where users pay for outgoing text messages. In particular, collecting accurate location data from farmers via text messaging, and tailoring information accordingly, poses a big challenge (see Box 8.3).

Existing evidence suggests that text messages can have modest, but positive effects on the likelihood of a farmer adopting recommended agricultural technologies. A text-message extension program offered to sugar cane farmers in Kenya found positive yield impacts in one trial but no effects in a second trial (Casaburi et al., 2019b). In Ecuador, text messages to potato farmers increased knowledge and self-reported adoption of integrated soil management practices (Larochelle et al., 2019). Similarly, delivering price information via SMS resulted in better farmer outcomes in Peru and India (Nakasone, 2013; Courtois & Subervie, 2015), but did not affect average crop prices obtained by farmers in Colombia and India (Camacho & Conover, 2011; Fafchamps & Minten, 2012). The existing evidence base offers limited insights on the heterogeneity of treatment effects. Are effects sensitive to the local conditions and the specific design features, or varying findings across studies are merely driven by sampling variation and imprecise impact estimates due to small samples? A meta-analysis of six experimental evaluations of text message services encouraging farmers to adopt an input to reduce soil acidity, implemented by three different organizations with thousands of smallholder maize farmers in Kenya and Rwanda, found that farmers who received texts were 19% more likely to follow the agricultural advice (Fabregas et al., 2021). While some of the individual experiments had statistically significant impacts and others did not, one cannot reject the hypothesis that the effects were the same across contexts. These results also suggest that one needs to be cautious when interpreting sources of impact heterogeneity across different studies.

Another key consideration for policy is how impacts compare to the costs of the programs. Since the marginal costs of sending an SMS is extremely low, even small effects can be cost-effective. While SMS is a well-known and simple technology widely used for digital extension, there are a number of opportunities that deserve further exploration. First, it is important to understand the extent to which more effective message design can increase impacts. Here, insights from marketing and behavioral economics could be useful. Second, when and how frequently messages should be targeted. Third, how to develop systems to collect farmer information through text message. All of these approaches require more experimentation.

Box 8.3: Customizing Advice to Farmers' Location via Text Message
A group of researchers working in Kenya, including several of the authors of this case study, partnered with a public entity, a social enterprise and an NGO to evaluate SMS-based agricultural services that recommended a specific type of input for acidic soils (Fabregas et al., 2021).

The government-run program recommended farmers first to test their soils to learn about the soil acidity of their plot before following the advice. However, individual soil testing can be prohibitively expensive for most smallholder farmers (approximately $20USD at the time of the study). Even though local recommendations could be generated using the available data on local soil tests in the region and shared directly with farmers through their phones, such customization required information about a farmer's location.

Establishing user locations faced a number of challenges. First, most smallholder farmers in rural Kenya do not have GPS-enabled smartphones. Therefore, potential users had to be people for whom information on location already existed (e.g., because they had already participated in other programs,

(continued)

Box 8.3 (continued)

and this had already been collected), or information about their location had to be collected through text message questionnaires. Second, many developing countries lack precisely defined physical addresses (Union, 2012). Collecting precise location information about specific farmers seemed almost impossible. In this particular case, the smallest unit that farmers could report uniformly was their village. However, village name spellings were not consistent and user text entry was often error-prone. Moreover, no GIS maps existed at the village level. At the end, data could only be reliably collected at a higher level of aggregation like the ward or sublocation.

The research team experimented with a number of solutions. In one project, farmers were recruited through partner organizations that already had information about farmers' locations. In a second project, participants were recruited through agricultural shops. Clients of these shops were invited to enroll into the agricultural extension program, and shopkeepers provided support in filling out a text-message questionnaire that asked farmers about their physical location.

These models have trade-offs. Recruiting through a partner organization lowers the cost of user acquisition, but the scale and the target farmer population are determined by the partner's reach. In contrast, coordinating a recruitment process through local agents can be costly and limited in scalability. A more scalable approach may be to partner with mobile network operators and use the data on cell tower locations. However, as discussed in the next subsection, there is often a large fixed cost in negotiating these agreements. As the adoption of GPS-enabled devices increases, more opportunities to gather farmers' locations at low cost and at scale will likely emerge.

2.1.2 Services That Rely on Interactive Voice Responses (IVR)

A second technology operating in basic phones which we highlight is interactive voice responses (IVR). This technology allows computers to interact with humans through voice. Several developing countries, including India, Madagascar, and Ethiopia already operate IVR phone-based government extension systems. These systems usually allow farmers to listen to prerecorded information and to record new questions (Fig. 8.2). This approach is likely to be more inclusive of users with low levels of literacy, though it requires users to listen to audio-recorded messages. It can also be more expensive to operate than text-based systems. Cole and Fernando (2021) evaluate an IVR mobile advisory system that provides agricultural advice to cotton and cumin farmers in India. The intervention increased self-reported adoption of recommended seeds, though it had no impact on the adoption of other inputs like pesticides or fertilizers.

2.1.3   More Advanced Technologies: Smartphones and Tablets

Smartphones and tablets offer new possibilities for sharing information and learning. For example, farmers could watch videos demonstrating new agricultural techniques or take pictures of pests affecting their crops and either request automatic identification and recommendations or raise questions with agronomists (Olson, 2018). Farmers could play with apps to better understand the risks associated with certain crops (Tjernström et al., 2019). However, access to smartphones and tablet devices is still limited in some parts of the world, and it may require innovative delivery approaches to reach scale, including engaging agricultural extension officers, agrodealers, and other local agents with familiarity with smartphones.

To date, a number of video-based interventions for farmers have also been found to have positive impacts at changing knowledge and self-reported farmer practices (Gandhi et al., 2007; Fu & Akter, 2016; Van Campenhout et al., 2018). Measured impacts on crop yields have been mixed, with some projects documenting null effects (Udry, 2019; Van Campenhout et al., 2019) and others documenting positive impacts (Van Campenhout et al., 2018; Arouna et al., 2019). However, a recent meta-analysis combining the effects of these existing projects suggests that, on average, yields increased by 4% as a result of these types of programs (Fabregas et al., 2019).

2.2   Implementation: Technological Considerations

Deploying mobile phone-based solutions requires coordination with a number of stakeholders including government, agricultural agencies, communications regulatory bodies, and local telco companies. In this section, we discuss key technological considerations for setting up and scaling a digital agricultural extension service.

2.2.1   Technology Infrastructure

Several options for technology infrastructure are available in most countries. These options include working directly with mobile network operators (MNOs), working through a mobile aggregator, or working with an existing mobile solutions provider. Negotiating individually with each MNO can be time-consuming and difficult. In contrast, many aggregators have an infrastructure and partnerships that allow them to send messages to subscribers across MNOs within a country, but regulations on mobile communications and the nascent market environment for aggregators vary across countries. Finally, mobile solutions providers may offer additional

IVR System Call Flow

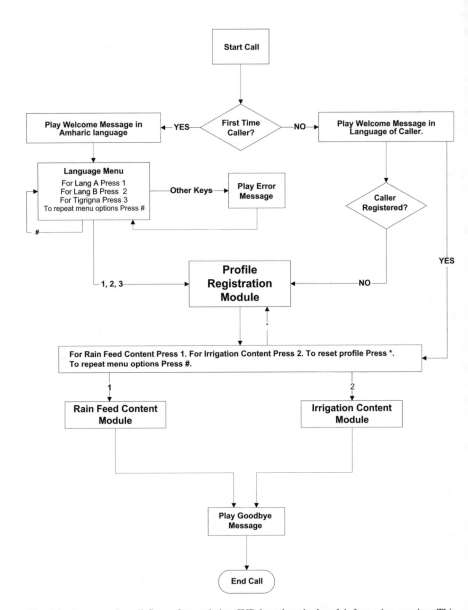

Fig. 8.2 An example call flow of an existing IVR-based agricultural information service. This figure illustrates the IVR menu structure for an existing IVR-based agricultural information service. A caller is first asked to select a language and answer a few profiling questions before being taken to the menu of agricultural content organized by irrigation access. (Source: PxD)

services such as dashboards for easy monitoring and a team of engineers to build customized service features. Below are some key considerations when considering these different options.

User Protection   Mobile phone subscribers receive many spam messages and calls. In some countries, unsolicited messaging in the absence of a user opting in to a service is prohibited or regulated. Noncompliance can result in the service being shut down or penalized. Even in the absence of such regulations, the opt-out right needs be considered as part of the user protection measure. A simple process that allows users to choose whether and when to opt out can reduce annoyance. Telecommunications authorities can also decide whether certain emergencies – such as severe pest outbreaks – warrant sending unsolicited messages.

Data Security and Privacy   Two-way digital agricultural extension systems can accumulate a large volume of private data about smallholder farmers. This often includes phone numbers, as well as more detailed information about a farmer's location, crop selection, and input usage, and may include sensitive information such as exact plot locations, agricultural sales, and credit history. Privacy protection for service users needs to be handled carefully. Some countries have data protection and security regulations that restrict how and where data about individual citizens can be stored and accessed. One must also consider obtaining consent from farmers to use their data for analysis or to share their information with third parties. It may be practical in some cases to obtain consent from users as they register into the service. However, with low mobile literacy among smallholder farmers, providing sufficient information and obtaining informed consent via digital messages can be difficult.

Access to Mobile Data   Mobile communications data owned by MNOs contain rich information about farmers, which can be used to improve the quality of a digital advisory service. For example, cell tower data provides user location information; call and message logs can be used to identify social networks or predict mobile literacy (Björkegren, 2019); and user profile and phone settings reveal user preference. Aggregators often do not have access to this kind of granular user-level data. More importantly, when a digital agricultural extension service is launched, usage data provides real-time feedback about demand for service and information. Whether working with MNOs or through aggregators, access to the basic usage data (e.g., whether the message was received, why it was not received, etc.) can contribute to building a robust monitoring and evaluation system as well as capacity to engage in rapid testing and iterative development.

2.2.2   Implementation Model: Product Considerations

A host of operational decisions need to be considered in setting up a digital agricultural extension service. How does one optimally recruit farmers? What types of implementing organizations make good partners? What are adequate revenue

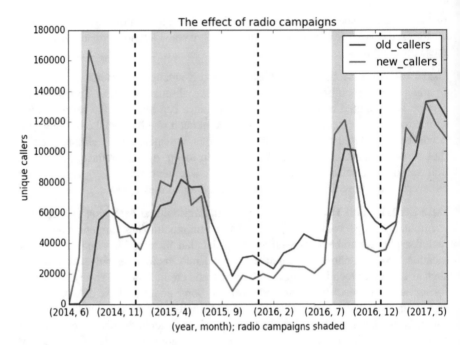

Fig. 8.3 Trends in access to a national agricultural hotline service. This graph illustrates how a mass radio campaign generated spikes in calls to a large public agricultural hotline service, serving as an effective tool to recruit farmers to digital agricultural advisory services. (Source: PxD)

models and financial paths to scale? Not all of these questions need to be answered right away, and implementation models may evolve as questions are answered through piloting and incremental scaling.

Customer Acquisition Farmer recruitment and partner selection are key design elements in the implementation of a digital agricultural advisory service. These decisions will affect the scale, density, and target farmer population as well as the scope of feasible customization and the cost of service delivery per farmer. Databases of farmer phone numbers may be available through – among others – government agencies, NGOs, and farmer associations. Some implementers already have valuable and detailed information about farmers (e.g., location, gender, primary crops, etc.), although they might work with a limited number of individuals. Alternatively, mass campaigns via radio advertisements and posters in village centers may be feasible. For example, a large public agricultural hotline service saw spikes in call volumes immediately after hotline numbers were broadcast via radio programs (Fig. 8.3). The cost per user acquisition through radio campaigns in this setting was $0.29/user in 2019.

Agronomic Content A vast body of agronomic sciences research and crop modeling informs evolving insights on new productivity-enhancing technologies and practices. However, the rigor of evidence supporting the use and impact of new technologies varies widely. Institutional willingness to take risks with recommendations and relative openness to incorporating new technologies into extension services may vary widely by implementing agency. For instance, government agencies might provide agricultural information that is agronomically correct but too technical to be adequately used by target farmers (Cole & Sharma, 2017; Fabregas et al., 2021). In vetting agricultural content, we emphasize the importance of considering farmer's profitability calculations in real-farm settings, and the potential risks associated with any technology. For example, customized advice generated by complex crop models may benefit from an empirical validation to corroborate whether customization, in fact, improves profitability among targeted farmers.

Costs and Financing of Digital Agricultural Systems   Mobile phones offer a low-cost means to reach smallholder farmers with information at scale. Moreover, the low marginal costs of distribution suggest that there can be high returns to scale. For instance, PxD's average cost of service has dramatically decreased from $5.20 per farmer per year in 2017 (serving approximately 345 K farmers) to $1.55 in 2019 (serving 3.53 M farmers).[1] Despite the very low marginal cost of scaling the service, however, setting up a digital agricultural system in a new setting might require nontrivial upfront capital.

There are several potential financing models for digital agricultural services. Many for-profit service providers in digital agriculture charge farmers a subscription fee. However, economic theory suggests that markets for information will often perform poorly and, therefore, financing models that solely rely on charging fees directly to farmers are likely to exclude a large proportion of smallholders who might still find these services valuable (Fabregas et al., 2019). Market failures arise because information differs from most other goods: it is non-rival (e.g., once created many people can benefit from it at minimal marginal distribution costs), it is non-excludable or partly excludable (e.g., once an individual has access they can share with many others), and there often exists asymmetric information in the market (e.g., buyers do not know the value of information sold to them). These features might suppress farmers' willingness to pay for information services and also make it difficult for service providers to recoup their costs, limiting their incentives to invest in generating informational products.

Limited empirical data suggests that farmer's willingness to pay varies widely across settings, but farmers are sensitive to prices. The IVR service in India studied by Cole and Fernando (2021) found that, despite a high rate of engagement, the average price which farmers were willing to pay was $2 when the cost of provision for a 9-month subscription for that particular service was $7. The percentage of farmers who took an offer at a randomly selected price varied from 6.7% at $4.13 to 85% at $0.68. In a study in rural Ghana, most farmers were willing to pay a low

---

[1] These estimates are based on PAD's total operating cost and the total number of farmers served across all initiatives, and the initiative-specific cost per farmer varies widely by the design of the service, partnership arrangement, and the scale at which it is operating.

price for digital information service ($0.10/month), but they were highly sensitive to price increases (Hidrobo et al., 2020). In a study of willingness to pay for local soil information in western Kenya, farmers were not willing to pay the full cost of local soil tests, but the aggregate valuation of all farmers for a given soil test in an area exceeded the cost of soil testing and distributing this information. This potentially makes investment in this information worthwhile from a social standpoint (Fabregas et al., 2017).

Other commercial models might help address some of these issues. A freemium model in which users receive free access to a basic service and pay for advanced features may increase access while still generating revenues. Alternatively, revenues may be raised from value-chain players with an aligned incentive to increase farmer productivity (e.g., contract farming companies, large agricultural corporations with corporate social responsibility). Overall, the above market failures provide a rationale for some public sector involvement in financially supporting these services.

Other Costs to Farmers Even in the absence of direct fees to use a digital agricultural service, farmers may face indirect (pecuniary and nonpecuniary) costs of accessing these services. For instance, phone signals can be weak and unreliable in remote areas; farmers may incur transportation costs to access electricity for charging phones; and some MNOs require a minimum phone credit to receive calls. In addition, challenges with mobile phone access and digital literacy may present greater barriers for marginalized populations, such as women and the poorest. Expanding digital agricultural services among those who may face high costs and barriers to access them will require a deliberate effort for the developer to address their specific barriers and meet their specific needs.

3   Iterative Development

Developing a user-centered service requires an iterative process guided by frequent user feedback. A variety of methods and approaches can be used to test, evaluate, and iterate the design of digital agricultural extension services to improve service delivery and impact (see Box 8.4).

3.1   Approaches

Human-Centered Design Approach Agricultural recommendations developed by scientists and experts are often technical and difficult to understand. Given the limited volume of content which a typical digital message can deliver at one time, the comprehensibility and actionability of a message are likely to be a critical driver of impact. The exact content of messages can be tested and iterated with target farmers through in-person or telephonic focus group discussions, interviews, and observations, before launching rigorous testing. For instance, one could share the messages with a small number of farmers and ask them to explain the recommendation to ensure that agricultural words used in the message are locally appropriate; farmers may be asked to call into an IVR service and interact with the system so that the developer team can observe the pain points in the interface.

Monitoring Implementation quality can be assessed by monitoring key performance metrics, such as user engagement, user satisfaction, or perceptions. Monitoring these outcomes can provide valuable information about program's aspects that may or may not be working well. For example, low user engagement suggests that the system might be failing to deliver agricultural information to a large number of farmers. Similarly, low user satisfaction can indicate that farmers are unlikely to utilize the service and the agricultural information it provides. Obtaining feedback from farmers and iterating on design may help identify and reduce any potential barriers. However, merely monitoring engagement and satisfaction does not tell us whether the service impacts farmer behavior and outcomes.

Box 8.4: Iterative Development of Customized Fertilizer Advice, PxD India

The Government of India invests a large amount of resources in soil testing of farmers' fields and distributing Soil Health Cards (SHCs). SHCs are physical soil report cards which provide detailed soil nutrient information and customized fertilizer recommendations. However, information presented in SHCs is highly technical and difficult for farmers to understand. For instance, previous research in Bihar shows that nearly 70% of farmers with sufficiently nutrient soils wrongly believed that SHCs recommended relevant fertilizer application (Fishman et al., 2016).

To address this challenge, in 2017 a research team set out to develop a digital support tool for SHCs. The team conducted a series of focus group discussions in Gujarat, India, followed by a "lab-in-the-field" experiment to develop and test supplemental materials including an audio aid for SHCs. Approximately 600 farmers across 12 villages were randomly assigned to be presented with (i) a (hypothetical) SHC only, (ii) a SHC with an audio aid, (iii) a SHC with a video clip, or (iv) a SHC with an agronomist on hand to explain the SHC. The SHC was presented as something for a farmer's friend. Farmers in groups (ii)–(iv) were also given a simplified SHC with fertilizer recommendations converted into a familiar local unit. The field team visited farmers door to door and administered short surveys at the beginning of the visit, after the SHC was presented, and after the supplementary materials were presented. The team found that all of the supplementary materials dramatically increased the proportion of farmers who understood the SHC

(continued)

Box 8.4 (continued)
content from 8 to over 40% and the level of trust in SHC recommendations by 5–7 percentage points (Cole & Sharma, 2017).

In the following season, researchers evaluated the impact of customized fertilizer recommendations on fertilizer application and yields among 1585 cotton farmers in Gujarat. Half of the farmers received a basic digital advisory service with weekly push calls on topics including planting, weeding, and pesticides, while the other half of treated farmers also received customized fertilizer recommendations via visual aids and weekly push calls. At the end of the first season, treated farmers reported more than two- to five-fold increases in the likelihood of using profitability-enhancing fertilizers, when compared to farmers in the control group (Cole et al., 2020). On aggregate, this intervention narrowed the gap between recommended and actual fertilizer use by 0.08 standard deviations.

The positive effects on fertilizer application, where previous efforts had failed, attest to the importance of an iterative approach. Despite the large impact on fertilizer adoption, the study observed no impact on self-reported cotton yields or satellite-based yields. Unfortunately, in the year of the trial, Gujarat had a historically low rainfall, potentially suppressing returns to fertilizers. This confounding factor highlights challenges associated with rigorously measuring agricultural impacts.

Continuous Experimentation Continuous A/B tests can be designed to answer a range of operational and product design questions. For example, one can use insights from behavioral sciences to experiment with different ways of framing a particular message or message contents to test influences on farmer behavior or the timing and frequency of messages to optimize user engagement (Fabregas et al., 2021). These rapid experiments on systems and tweaks in message design often focus on intermediate outcomes that are easy to measure: administrative data on system usage or self-reported outcomes on adoption, knowledge, comprehension, information sharing, and trust in the system. These outcomes can be used to optimize user experience. Experimenting with large sample sizes is necessary for detecting small effects and for harnessing the benefits of economies of scale in learning. For example, one could compare several experimental arms at once or use big data analysis to draw insights on heterogeneity or uncover other patterns in the data to inform service design, impact, and scope for improvements. We note that these design improvements may not add up linearly: user experience may drastically improve and lead to better outcomes when a number of tweaks remove major pain points at once. Therefore, testing the aggregate effect of many design improvements together, rather than individually, can be an effective way to approach product improvement. The effectiveness and targeting of messages might be significantly improved over time, when feedback loops and iterative learning tools are integrated into operations at scale (see Box 8.5). Indeed, evaluating too early may underestimate long-run impacts.

Impact Evaluations   Because impact can vary significantly by context and product design, building local evidence on the impact of a service can be important. However, implementing experimental evaluations to measure impact on downstream outcomes, such as yields and profits, can be complex and costly. In many instances, the effect sizes that would make these types of programs cost-effective are small, and detecting these effects in a study might require large sample sizes. Evaluations that have low statistical power are unlikely to detect effects that would still be considered cost-effective. Hence, large-scale evaluations may be suited to settings in which access to behavior and yield outcomes for a large number of farmers is accessible at low cost. For instance, in Kenya and Mozambique researchers worked with agribusinesses that regularly buy crops from farmers. Therefore, they could use a large sample of size of administrative data on yields (crop sales collected by these companies) to determine impacts on productivity (Axmann et al., 2018; Casaburi et al., 2019b).

Localization vs. Generalizability   Experiments designed to understand the impact mechanism of an intervention – why the intervention works – can often generate more generalizable insights than experiments that only assess whether or not an intervention works. Understanding *why* helps us formulate a broader conceptual model about conditions and constraints under which a particular intervention is likely to be effective. We iterate and refine the conceptual model and our understanding about farmer behaviors as we gather observations from similar experiments across multiple contexts with varying constraints. These broader lessons constitute global public goods that can inform policy and practice. In this sense, there is likely a large social value in experimenting and making results widely available.

3.2   Data and Measurement Issues

An effective feedback and iteration system leverages administrative outcome data while supplementing it with additional data collection. In addition, available advanced technologies, such as remote sensing, might be used to improve the cost-efficiency of outcome measurements over time.

Administrative Data on Usage   System usage data can provide reliable information about user engagement: however, not all systems offer this option. Pickup and listening rates for a push call service can offer insights on the amount of information each user accessed. However, it is often difficult for a service provider to obtain data on whether SMS advisory messages were opened by recipients.

Measurement on Input Adoption and Agricultural Practices Phone surveys, delivered via voice calls or text messages, allow for high-frequency data collection at a much lower cost than traditional in-person methods. However, low response rates and selection in attrition are common. For example, farmers who are more satisfied with the service might be more likely to provide feedback or more likely to respond to a phone survey. These biases in outcome data would make it difficult to draw appropriate inferences in A/B tests.

Simply measuring increases in farmers' knowledge or self-reported adoption of inputs or practices has a number of limitations. For instance, knowledge may not necessarily translate into any behavior change. Self-reports on whether farmers followed recommendations could lead to a biased estimation of effects. For example, farmers who received the service might overreport using suggested practices because of experimenter demand effects, or farmers' might fail to report using inputs if they believe that it might make them more likely to receive a program. A comparison of self-reported and administrative data use for four studies in Kenya found that the measured impact of mobile phone messages using self-reported data exceeded the impacts measured through administrative data (Fabregas et al., 2021).

To address concerns around experimenter demand effects, administrative data on purchases from input sellers could be used to measure farmer behavior. For instance, the text message program that encouraged farmers in East Africa to use locally appropriate inputs used both administrative data from input sellers and data from redemption of electronic discount coupons to understand whether farmers were more likely to purchase recommended inputs (Fabregas et al., 2021).

Yield Measurement Researchers might be most interested in estimating effects on farm profits or yields. However, the measurement of profits requires detailed data and assumptions about input and labor use and costs. Impacts on yields are often imprecisely estimated since it is often difficult for farmers to report yields precisely, and yields are dependent on a number of other environmental factors, such as seasonal rainfall. Moreover, small impacts on farmer behavior are likely to translate to modest improvements in yields, which might be difficult to detect.

A potentially promising approach for obtaining multiple seasons of yield data at limited cost (outside of contract farming settings) would be to obtain GPS location information for farmers' plots and then assess yields over multiple years using satellite data. Recent studies demonstrate a strong correlation between satellite yield measurements, crop cut data, and full plot harvests (Burke & Lobell, 2017; Lambert et al., 2018). An ongoing evaluation in India suggests that satellite yield measurements can reduce standard errors in estimates of treatment effects by over 50% when compared to farmer-reported data (Cole et al., 2020). This can substantially improve statistical power to detect impact (Fig. 8.4).

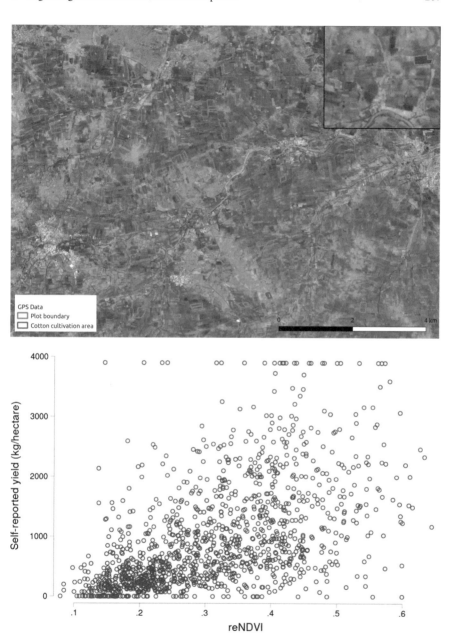

Fig. 8.4 Satellite-based yield measurement (**a**) Plot boundary data collected via Garmin; (**b**) Satellite-based (reNDVI) vs. self-reported yields. Cole et al. (2020) use satellite-based yield measures to evaluate the effect of customized fertilizer recommendations among cotton farmers in Gujarat, India. They collected plot boundary data using a Garmin GPS hand device (Panel a). Panel b illustrates the positive correlation between vegetation index and self-reported yield measurements. They calculate vegetation index values "*by taking the median value of each VI pixel contained in each sample plot for 5 Sentinel-2 images from 2018 . . . [and taking] the maximum value across the 5 satellite images*"

Box 8.5: Learning Through Feedback and Experimentation in a Government Agency

The use of mobile technologies in the public sector is usually discussed in the context of addressing traditional government failures, such as limited accountability and incentives for public-sector workers (Dal Bó et al., 2021; Callen et al., 2020; Muralidharan et al., 2020). To date, most mobile phone-based agricultural extension services rarely rely on public sector workers for delivery. Instead, the potential of these services lies in the large volume of user feedback data for rapid assessment, experimentation, and iterations. However, the government may face a new type of capacity constraints in leveraging the available data to monitor and improve the design of technology-based services. The example described in this box demonstrates that these constraints exist, but that they can be overcome.

An (anonymous) government agency has been operating an IVR-based agricultural information service for several years. When a farmer calls into the system through a toll-free number, the automated hotline service starts with a language selection and questions on farmer location and other characteristics for the first-time users. Only after this is completed, it takes farmers to the menu selection with a variety of agricultural topics for more than 20 crops. The system had been accessed by nearly three million farmers by mid-2017, but only a small fraction of farmers called back after the first try.

A research team conducted a diagnostic assessment of the system in 2017, in which they analyzed the existing administrative data on system usage to understand usage patterns and identify potential issues with the system (PxD, 2018). This exercise was followed by a series of in-person design sessions with farmers, where researchers observed farmers calling navigating the system in real to identify pain points. Additionally, a phone survey of users was conducted to assess the accuracy of farmer profiling data collected by the IVR service. This assessment revealed that the service was losing a nontrivial number of farmers in every required menu selection. The insights from the diagnostic assessment led to a number of ideas for system design improvements. The research team and the government agency started regular meetings to brainstorm ideas and agreed to implement A/B tests to experiment with different solutions. Over the following 2 years, 13 A/B tests were implemented, 6 of which have shown to significantly increase the likelihood of farmers successfully accessing agricultural content.

Selected insights from the diagnostic assessment and system tweaks for A/B testing

(continued)

Box 8.5 (continued)

Observation	Recommended system tweaks
Majority of first-time callers do not complete the registration and drop from the call	Remove registration questions for the first-time callers and postpone them to a later call
Majority of users select the menu option by pressing #1 on the keypad	Rotate the menu option seasonally to keep the most relevant topic as the first option
Many users do not press any key after a question; the system hangs up if no response	Add menu replay twice if no option is selected, before the application hangs up Add pause in between options for language menu Slow down the speed of recording
Most farmers don't access agricultural content	Add push call explaining how to use IVR service

4   Innovations to Improve Impacts

In this section, we discuss selected areas of innovation that offer significant potential for improving impacts for farmers.

4.1   Customization

Agricultural information can be customized across a range of dimensions. First, optimal agricultural practices vary widely in line with local conditions. For example, several field experiments have demonstrated a large spatial variation in yields and yield responses to inputs (Zingore et al., 2007; Seo et al., 2009; Suri, 2011; Tittonell & Giller, 2013), and agronomic research provides strong evidence that the nutrient composition in a particular soil informs which set of fertilizers, and quantities, thereof, will be optimal for maximizing yields (Sapkota et al., 2014). Second, the benefits farmers derive from advice are, in part, based on the applicability of the advice relative to real-time changes in the local environment, such as weather and pest outbreaks. Studies suggest that weather forecasts affect farmer investment decisions (Pandey, 1998; Chisadza et al., 2020) and that accurate forecasts increase farm profitability (Rosenzweig & Udry, 2019). Third, both the appropriateness of agricultural content and optimal message design may vary by farmer characteristics, such as land size and access to storage, as well as gender, age, and other individual characteristics.

The key challenge, across different types of customization, is to generate and gather relevant local data at scale (see Box 8.6 for some resources). In some cases, there is insufficient coordination and poor incentives among stakeholders to direct resources toward data generation. In other cases, the cost of collecting and aggregating local data and/or packaging it in a way that is accessible and easy to understand may act as barriers to scaling up.

Box 8.6: Examples of Publicly Available Remote Sensing Data Sources
Global Precipitation Measurement (GPM) Data – https://gpm.nasa.gov/data

The global precipitation measurement (GPM) mission, initiated by NASA and the Japan Aerospace Exploration Agency (JAXA), is an international network of satellites that provide global data on rain and snow fall. In the agricultural context, this data can be applied to analyze and forecast changes in water resources and, thereby, food security.

Sentinel-2 – https://sentinel.esa.int/web/sentinel/missions/sentinel-2

Sentinel-2 is an imaging mission dedicated to Europe's Copernicus program. The mission aims at monitoring variability in land surface conditions, including vegetation and soil and water cover, while also observing inland waterways and coastal areas. Publicly available data includes coverage of all continental land surfaces (including inland waters) between latitudes 84°N and 56°S, all coastal waters up to at least 20 km from the shore, all islands greater than 100 km^2, all EU islands, the Mediterranean Sea, and all closed seas (e.g., the Caspian Sea).

Google Earth Engine – https://developers.google.com/earth-engine/datasets/

Google Earth Engine combines a massive catalog of more than 40 years of historical imagery and scientific datasets with APIs and other analysis tools. The data catalog is organized into three categories, each with their own subcategories: climate and weather (surface temperature, climate, atmospheric, weather); imagery (Landsat, sentinel, MODIS, high-resolution imagery); and geophysical (terrain, land cover, cropland, other geophysical data).

Landsat – https://landsat.gsfc.nasa.gov/

The Landsat series of earth observation satellites is a joint NASA/US geological survey program that has continuously acquired images of the Earth's land surface since 1972. Landsat 8 is the latest mission with moderate-resolution (15–100 m, depending on spectral frequency) measurements of the Earth's terrestrial and polar regions in the visible, near-infrared, short wave infrared, and thermal infrared going back to 2013.

MODIS – https://modis.gsfc.nasa.gov/

(continued)

Box 8.6 (continued)

MODIS (Moderate Resolution Imaging Spectroradiometer) is an instrument that has been launched on the Terra (1999) and Aqua (2002) NASA satellites. It has a good temporal resolution, imaging the whole earth every 1 or 2 days. This makes it suitable to track large scale trends over time. For agriculture in particular, NDVI (normalized difference vegetation index) products can be useful, providing insight into vegetation changes over time. One MODIS-based NDVI product is MOD13A1.

The European Centre for Medium-Range Weather Forecasts (ECMWF) – https://www.ecmwf.int/en/forecasts/datasets

ECMWF is a European intergovernmental organization which generates medium, extended, and long-range forecasts using its own comprehensive earth system model and conducts research to improve forecasting skills.

Radiant MLHub – https://www.mlhub.earth/#home

Radiant MLHub is an open library for geospatial training data to advance machine learning applications on earth observations. It aims to be a repository of data and trained models for development. Currently, it has smallholder crop classification data but plans to add global land cover in the future.

Consultative Group on International Agricultural Research (CGIAR) Data Resources – https://bigdata.cgiar.org/wp-content/uploads/2020/05/Webinar-Slides-_-Secondary-Data-for-Crop-Modeling-2020-_-Presented.pdf

These slides for the CGIAR webinar: *Secondary data for crop modeling: Filling data gaps under lockdowns* include links to various data resources. Dataset topics include weather, soil properties, cropping calendar, management practices, evaluation data, phone surveys, household surveys, and satellite remote sensing.

Crowdsourcing The two-way nature of digital communication presents opportunities for aggregating relevant, real-time information through crowdsourcing. For example, a pest hotline can be used to identify pest outbreaks at an early stage, allowing faster detection of local outbreaks and alerting farmers in at-risk areas about pest prevention and management recommendations. Moreover, a "Yelp"-like system of customer service ratings could reduce information asymmetry in input markets (Hasanain et al., 2019). A small but growing literature suggests that crowdsourcing can be used successfully to reduce information scarcity in a variety of settings (Bailard & Livingston, 2014; Jame et al., 2016). However, evidence of its use and utility among smallholder farmer populations is scarce.

The relative advantage of using crowdsourcing to collect information is dependent in large part on whether a sufficiently large number of farmers contribute information with sufficient accuracy. There is ample room for research in this space to advance our understanding of technology design and (financial and nonfinancial) incentives for farmers to contribute high-quality information. A risk linked to crowdsourcing that requires attention in the design process is a potential data gap for less technology-adept farmers. If the needs and preferences of less technologically proficient or literate farmers are different from those who contribute information to the system, the resulting customization may result in making the information provided *less* relevant.

Data-Driven Customization A key advantage of digital agriculture is its ability to improve the quality of customized advice by using the data it generates. For example, in settings in which extensive data on farmer characteristics is available, analysis of large platform data may reveal differential patterns in system usage by farmer characteristics. These patterns could then be tested in A/B tests to inform systems and service iteration and improvements (see Box 8.5). Furthermore, when agricultural outcome data is available at large scale, customized recommendations based on agronomic trials or a crop model can be empirically validated in real farm settings and improved through subsequent experiments.

4.2   Using Digital Technology to Facilitate Social Learning

A large volume of literature suggests that social learning – learning from the experience of other farmers – is key to facilitating optimal technology adoption among farmers (Munshi, 2004; Bandiera & Rasul, 2006; Conley & Udry, 2010). Existing evidence suggests that mobile phone-based agricultural information services can generate information spillovers. For instance, in India, farmers who had not received the piloted services in the trial increased interactions with, and learned from others, who had (Cole & Fernando, 2021). Furthermore, directing the flow of information via mobiles phones can also affect existing dynamics of information networks (Fernando, 2021).

Beyond the diffusion of agricultural information through existing mechanisms, advanced communications technologies offer scope for increasing and directing the flow of information among farmers to facilitate more efficient learning. First, digital messages can be designed to spur conversations about agricultural practices and inputs within existing networks. Moreover, a farmer's beliefs about a particular input or practice may be influenced by the experience and beliefs of others. With or without novel information, increasing conversations about a particular input can potentially accelerate learning among farmers. Second, the two-way character of digital communications technology allows farmers to exchange information and learn from experts and other farmers beyond their networks of friends and neighbors. It is common in a radio program to solicit questions from listeners and broadcast responses as a way of facilitating learning. With digital technology, this type of learning can happen much more locally and in real time.

An important consideration when using digital tools to accelerate learning among farmers is the potential presence of behavioral factors in learning. Existing literature suggests that individuals confront a range of barriers when communicating factual information, experience, and perceptions and in interpreting information shared by others (e.g., Benjamin et al., 2016; Breza et al., 2018; Eyster et al., 2018). For example, if farmers who have had a bad experience with a new input talk more about their experience than those who had successful experiences with the same input, there may be convergence on an inaccurate belief that the input is ineffective. Given the nascent nature of these innovations, rigorous experimentation, assessment, and iteration – as discussed in the earlier section – will be critical for advancing the development of digital social learning tools capable of amplifying the impact of digital agricultural extension.

4.3 Digital Support for Existing Extension Systems

Digital agricultural extension can complement traditional extension systems. While many governments in developing countries maintain a network of extension workers, the evidence base on their impact on farmer outcomes is limited. (Anderson & Feder, 2004). Agricultural extension workers are difficult to monitor and incentivize: many tasks involve working independently, often in remote communities, with limited supervision. In addition, a lack of resources and poor institutional capacity limits the availability of training and technical support to extension workers. There are three broad mechanisms through which digital technology can potentially improve in-person extension services.

Extension for Extensionists Extension workers could be supported with better resources made available through digital devices. For example, extension agents could access detailed localized information through smartphones and could receive reminders to use appropriate messages for farmers based on the stage of the local agricultural season or to communicate important developments such as pest outbreaks, adverse weather conditions, or market disruptions. In a recent meta-analysis which estimated a positive impact of digital agricultural extension on farmer yields (Fabregas et al., 2019), four of the seven impact estimates were derived from an "indirect" model in which digital advice was delivered to farmers via extension agents or field officers.

Communication Between Farmers and Extension Workers Digital technology can facilitate communication between farmers and extension workers. An IVR system could aggregate local information to service commonly asked questions and equip extension workers with relevant information and recommendations. This could help extension workers determine which content is relevant for farmers in their area. Extension workers could also notify farmers about activities such as farmer field days or demonstration plots. Many extension workers already use digital communication channels, such as WhatsApp, to exchange information among themselves. However, these are nascent developments, and a dearth of rigorous evidence makes this a fertile area for future research.

Performance Management There is growing evidence on the use of mobile phones to help improve motivation and accountability of public-sector workers in developing countries, and a number of new initiatives have been successful at scaling up (e.g., Dimagi's CommCare, a data collection platform for frontline health workers). A study in Paraguay showed that increased monitoring of agricultural extension workers through the use of GPS-equipped mobile phones resulted in a 22% increase in the likelihood of visiting a given farmer over 7 days (Dal Bó et al., 2021). Calling beneficiaries to verify the delivery of cash transfers to farmers in Telangana, India, reduced nondelivery of the transfer by 8% (Muralidharan et al., 2020). In Pakistan, a smart-phone app to track activities of health facility inspectors increased the likelihood of rural health clinic inspection by 74% (Callen et al., 2020). In addition to increasing the effectiveness of monitoring, a mobile phone-based app that allows self-tracking has been shown to harness intrinsic motivation, and was associated with a 24% increase in performance (Lee, 2018).

5   Lessons Learned

In this chapter we discuss a number of issues that practitioners and researchers would need to consider when working with digital agricultural extension technologies. We provided insights from our work with several initiatives, implemented in different countries by a variety of organizations. While existing evidence suggests that these approaches can have positive impacts, delivering on the full promise of digital agriculture will require sustained iteration and testing. Moreover, as more sophisticated mobile technologies improve and are adopted over time, several more opportunities will open up.

While we identified a number of promising areas for future study throughout the chapter, we failed to discuss other important topics in digital agricultural extension. First, digital technologies can also help improve supply chains more widely. For instance, a hotline offered by a sugar company that contracted with sugarcane farmers led to an improvement in the delivery of inputs because farmers could report problems (Casaburi et al., 2019a). A system for agricultural supply dealers could be used to give better recommendations to farmers and gather data on which items to stock while facilitating price comparisons for farmers. Second, digital approaches might be particularly important during emergencies. Information could quickly get out (e.g., pests or weather shocks), but they could also help governments and other agencies gather information directly from farmers about critical needs. Third, crowdsourced information can be useful for a variety of purposes beyond agriculture. There might be complementarities with other sectors, or new ways of generating impacts, where data are responsibly shared for a variety of purposes.

We conclude by encouraging readers to actively engage with user needs and the constraints people face on the ground but also by having clear conceptual models or theories of change that can help guide the development and implementation of these technologies.

Discussion Questions

1. What are the trade-offs between improving customization and reaching scale in digital agricultural extension? What drives the trade-offs?
2. How could farmers who do not own a mobile device or do not own smartphones benefit from digital extension approaches?
3. Many development interventions follow the three-stage process – pilot, evaluate, and scale – but digital interventions may benefit from scaling quickly. What are the potential benefits and costs of this strategy? How do you ensure that the service delivers impacts to farmers?
4. Should digital agricultural extension services focus on solutions for basic phones because they would generate large benefits for the majority of poor smallholder farmers now or leverage the power of smartphones to create solutions that would generate large benefits in the future?
5. What are the potential distributional implications of digital agricultural extension?
6. Oftentimes agricultural information requires sending information about probabilities (e.g., the likelihood of rainfall) or potential risks. What are strategies to convey this information in an intuitive way to populations with low levels of education?

Acknowledgements We thank all of our partner organizations for making this work possible and the PAD team, especially Jonathan Faull, for valuable comments and support. Prankur Gupta and Diana McLeod provided superb research assistance.

References

Aker, J. C. (2011). Dial "A" for agriculture: A review of information and communication technologies for agricultural extension in developing countries. *Agricultural Economics, 42*(6), 631–647.

Aker, J. C., Ghosh, I., & Burrell, J. (2016). The promise (and pitfalls) of ICT for agriculture initiatives. *Agricultural Economics*. Wiley Online Library, *47*(S1), 35–48.

Anderson, J., & Feder, G. (2007). Agricultural extension. In R. E. Evenson (Ed.), *Agricultural development: Farmers, farm production and farm markets. Handbook of agricultural economics* (pp. 2343–2378). Elsevier.

Anderson, J. R., & Feder, G. (2004). Agricultural extension: Good intentions and hard realities. *The World Bank Research Observer*. Oxford University Press, *19*(1), 41–60.

Arouna, A., et al. (2019). *One size does not fit all: Experimental evidence on the digital delivery of personalized extension advice in Nigeria*. Available at: https://www.researchgate.net/publication/336927609_One_Size_Does_Not_Fit_All_Experimental_Evidence_on_the_Digital_Delivery_of_Personalized_Extension_Advice_in_Nigeria

Axmann, N., et al. (2018). *Agricultural extension, mobile phones, and outgrowing: Is it a win-win for farmers and companies?* Available at: http://ageconsearch.umn.edu/record/274234/files/Abstracts_18_05_23_08_25_40_08__172_16_20_153_0.pdf

Bailard, C. S., & Livingston, S. (2014). Crowdsourcing accountability in a Nigerian election. *Journal of Information Technology & Politics*. Routledge, *11*(4), 349–367.

Bandiera, O., & Rasul, I. (2006). Social networks and technology adoption in northern Mozambique. *The Economic Journal*. Oxford Academic, *116*(514), 869–902.

Benjamin, D. J., Rabin, M., & Raymond, C. (2016). A model of nonbelief in the law of large numbers. *Journal of the European Economic Association*. Wiley/Blackwell, *14*(2), 515–544. https://doi.org/10.1111/jeea.12139

BenYishay, A., & Mobarak, A. M. (2019). Social learning and incentives for experimentation and communication. *The Review of Economic Studies*. Oxford Academic, *86*(3), 976–1009.

Björkegren, D. (2019). The adoption of network goods: Evidence from the spread of mobile phones in Rwanda. *The Review of Economic Studies*. Oxford Academic, *86*(3), 1033–1060.

Breza, E., et al. (2018) *Seeing the forest for the trees? An investigation of network knowledge.* Available at: https://stanford.edu/~arungc/BCT.pdf

Burke, M., & Lobell, D. B. (2017). Satellite-based assessment of yield variation and its determinants in smallholder African systems. *Proceedings of the National Academy of Sciences of the United States of America, 114*(9), 2189–2194.

Callen, M., et al. (2020). Data and policy decisions: Experimental evidence from Pakistan. *Journal of Development Economics, 146*, 102523.

Camacho, A., & Conover, E. (2011). *The impact of receiving price and climate information in the agricultural sector*. IDB working paper series. Available at: https://publications.iadb.org/publications/english/document/The-Impact-of-Receiving-Price-and-Climate-Information-in-the-Agricultural-Sector.pdf

Casaburi, L., Kremer, M., & Ramrattan, R. (2019a) *Crony capitalism, collective action, and ICT: Evidence from Kenyan contract farming*. Available at: http://econ.uzh.ch/dam/jcr:e2ffc4e5-ab32-4405-bfa4-70b0e962aa81/hotline_paper_20191015_MERGED.pdf

Casaburi, L., et al. (2019b). *Harnessing ICT to increase agricultural production: Evidence from Kenya*, Working paper. Available at: https://www.econ.uzh.ch/dam/jcr:873845ce-de4d-4366-ba9a-d60accda577d/SMS_paper_with_tables_20190923_merged.pdf

Chisadza, B., et al. (2020). Opportunities and challenges for seasonal climate forecasts to more effectively assist smallholder farming decisions. *South African Journal of Science*. Academy of Science of South Africa, *116*(1–2), 1–5.

Cole, S., & Sharma, G. (2017). *The promise and challenges in implementing ICT for agriculture*. Working paper. Available at: https://precisionag.org/uploads/cole-sharma-july1-2017.pdf

Cole, S. et al. (2020). "Using satellites and phones to evaluate and promote agricultural technology adoption: Evidence from smallholder farms in India". Working Paper, September. https://2uy7xawu7lg2zqdax41x9oc1-wpengine.netdna-ssl.com/wp-content/uploads/2020/09/ATAI_Paper.sep18.2020.pdf

Cole, S. A., & Nilesh Fernando, A. (2021). Mobile'izing agricultural advice technology adoption diffusion and sustainability. *The Economic Journal, 131*(633), 192–219.

Cole, S. A., & Xiong, W. (2017). Agricultural insurance and economic development. *Annual Review of Economics*. Annual Reviews, *9*(1), 235–262.

Conley, T. G., & Udry, C. R. (2010). Learning about a new technology: Pineapple in Ghana. *The American Economic Review, 100*(1), 35–69.

Courtois, P., & Subervie, J. (2015). Farmer bargaining power and market information services. *American Journal of Agricultural Economics*. Oxford University Press, *97*(3), 953–977.

Cunguara, B., & Moder, K. (2011). Is agricultural extension helping the poor? Evidence from rural Mozambique. *Journal of African Economies*. Narnia, *20*(4), 562–595.

Dal Bó, E., Finan, F., Li, N. Y., & Schechter, L. (2021). Information technology and government decentralization: Experimental evidence from paraguay. *Econometrica, 89*(2), 677–701.

Das, S. R. (2012). *Rice in Odisha*. Available at: https://ageconsearch.umn.edu/record/287645/files/TB16_content.pdf

Udry. (2019). *Draft final report on disseminating innovative resource-based technologies for small scale farmers in Ghana*. Draft report. Available at: https://www.atai-research.org/

informationmarket-access-and-risk-addressing-constraints-to-agricultural-transformation-in-northernghana/

Emerick, K., & Dar, M. H. (2020). Farmer field days and demonstrator selection for increasing technology adoption. *Review of Economics and Statistics*, 1–41. https://doi.org/10.1162/rest_a_00917

Eyster, E., Rabin, M., & Weizsäcker, G. (2018). *An experiment on social mislearning*. Available at: http://ssrn.com/abstract=2704746

Fabregas, R., Kremer, M., Lowes, M., et al. (2021). "Digital information provision and behavior change: Lessons from six experiments in East Africa", Working Paper. Available at: https://drive.google.com/file/d/1cDqxx-0RNf6GA2QIPq3pIdm5loAw6GP8/view

Fabregas, R., Kremer, M., Robinson, J., & Schilbach, F. (2017). *Evaluating agricultural dissemination in Western Kenya, 3ie impact evaluation report 67*. International Initiative for Impact Evaluation (3ie).

Fabregas, R., Kremer, M., & Schilbach, F. (2019). Realizing the potential of digital development: The case of agricultural advice. *Science, 366*(6471). https://doi.org/10.1126/science.aay3038

Fafchamps, M., & Minten, B. (2012). Impact of SMS-based agricultural information on Indian farmers. *The World Bank Economic Review*. Oxford University Press, *26*(3), 383–414.

Fernando, N. A. (2021). Seeking the treated: The impact of mobile extension on farmer information exchange in India. *Journal of Development Economics*, 153, 102713, ISSN 0304–3878. https://doi.org/10.1016/j.jdeveco.2021.102713

Fishman, R., et al. (2016). *Can information help reduce imbalanced application of fertilizers in India?: Experimental evidence from Bihar*. International Food Policy Research Institute.

Fu, X., & Akter, S. (2016). The impact of mobile phone technology on agricultural extension services delivery: Evidence from India. *The Journal of Development Studies, 52*(11), 1561–1576.

Gandhi, R., et al. (2007). *Digital green: Participatory video for agricultural extension*. 2007 international conference on information and communication technologies and development. https://doi.org/10.1109/ictd.2007.4937388.

Hanna, R., Mullainathan, S., & Schwartzstein, J. (2014). Learning through noticing: Theory and evidence from a field experiment. *The Quarterly Journal of Economics*. MIT Press, *129*(3), 1311–1353.

Hasanain, A., Khan, M. Y., & Rezaee, A. (2019). *No bulls: Experimental evidence on the impact of veterinarian ratings in Pakistan*. Available at: https://armanrezaee.github.io/pdfs/livestock_3sept2019.pdf

Hidrobo, M., et al. (2020). Paying for digital information: Assessing farmers' willingness to pay for a digital agriculture and nutrition service in Ghana. https://doi.org/10.2499/p15738coll2.133591

IDinsight. (2019). *IVR and text message interventions to provide fertilizer information to farmers — Experiments from India*. IDinsight Blog. Available at: https://medium.com/idinsight-blog/ivr-and-text-message-interventions-to-provide-fertilizer-information-to-farmers-16e651402be6. Accessed: 26 May 2020.

Jain, A., et al. (2019). Low-cost aerial imaging for small holder farmers. In *Proceedings of the 2nd ACM SIGCAS conference on computing and sustainable societies*. New York, NY, USA: Association for Computing Machinery (COMPASS '19), pp. 41–51.

Jame, R., Johnston, R., & Markov, S. (2016). The value of crowdsourced earnings forecasts. *Journal of Accounting Research*. Wiley Online Library, *54*(4), 1077–1110.

Kondylis, F., Mueller, V., & Zhu, J. (2017). Seeing is believing? Evidence from an extension network experiment. *Journal of Development Economics*. Elsevier, *125*, 1–20.

Koo, J. (2014). Maize yield potential. In *Atlas of African agriculture research and development* (pp. 58–59). International Food Policy Research Institute.

Lambert, M.-J., et al. (2018). Estimating smallholder crops production at village level from Sentinel-2 time series in Mali's cotton belt. *Remote Sensing of Environment, 216*, 647–657.

Lambin, E. F., et al. (2013). Estimating the world's potentially available cropland using a bottom-up approach. *Global Environmental Change: Human and Policy Dimensions, 23*(5), 892–901.

Larochelle, C., et al. (2019). Did you really get the message? Using text reminders to stimulate adoption of agricultural technologies. *The Journal of Development Studies*, 548–564. https://doi.org/10.1080/00220388.2017.1393522

Lee, S. (2018). Intrinsic incentives: A field experiment on leveraging intrinsic motivation in public service delivery. *SSRN Electronic Journal*. https://doi.org/10.2139/ssrn.3537336

Munshi, K. (2004). Social learning in a heterogeneous population: Technology diffusion in the Indian green revolution. *Journal of Development Economics, 73*(1), 185–213.

Muralidharan, K., et al. (2020). Improving last-mile service delivery using phone-based monitoring. https://doi.org/10.3386/w25298

Nakasone, E. (2013). The role of price information in agricultural markets: Experimental evidence from rural Peru. *Agricultural and Applied Economics Association*. https://doi.org/10.22004/ag.econ.150418

Nakasone, E., Torero, M., & Minten, B. (2014). The power of information: The ICT revolution in agricultural development. *Annual Review of Resource Economics. Annual Reviews, 6*(1), 533–550.

Olson, P. (2018, 15 October). This startup built a treasure trove of crop data by putting a.I. in the hands of Indian farmers. *Forbes Magazine*. Available at: https://www.forbes.com/sites/parmyolson/2018/10/15/this-startupbuilt-a-treasure-trove-of-crop-data-by-putting-ai-in-the-hands-of-indian-farmers/. Accessed: 27 May 2020.

PxD. (2018). *Farmers listen to more advice when asked for call preference*. Available at: https://precisionag.org/wp-content/uploads/2019/12/Call-frequency_Listening-rates_Gujarat_2.2019.pdf

Pandey, S. (1998). Risk and the value of rainfall forecast for rainfed rice in the Philippines. *Philippine Journal of Crop Science (Philippines)*. Available at: http://agris.fao.org/agris-search/search.do?recordID=PH2001100781

Rosenzweig, M. R., & Udry, C. R. (2019). *Assessing the benefits of long-run weather forecasting for the rural poor: Farmer investments and worker migration in a dynamic equilibrium model*. National Bureau of Economic Research (Working Paper Series). https://doi.org/10.3386/w25894.

Saito, K., Mekonnen, H., & Spurling, D. (1994). *Raising the productivity of women farmers in sub-Saharan Africa*. World Bank.

Sapkota, T. B., et al. (2014). Precision nutrient management in conservation agriculture based wheat production of Northwest India: Profitability, nutrient use efficiency and environmental footprint. *Field Crops Research, 155*, 233–244.

Seo, S. N., et al. (2009). A Ricardian analysis of the distribution of climate change impacts On agriculture across agro-ecological zones in Africa. *Environmental and Resource Economics, 43*(3), 313–332.

Suri, T. (2011). Selection and comparative advantage in technology adoption. *Econometrica: Journal of the Econometric Society, 79*(1), 159–209.

Tittonell, P., & Giller, K. E. (2013). When yield gaps are poverty traps: The paradigm of ecological intensification in African smallholder agriculture. *Field Crops Research, 143*, 76–90.

Tjernström, E., et al. (2019). *Learning by (virtually) doing: Experimentation and belief updating in smallholder agriculture*. Available at: https://www.lafollette.wisc.edu/sites/tjernstrom/Learning_by_virtually_doing.pdf

Union, U. P. (2012). *Addressing the world – An address for everyone*. White paper. Available at: http://www.upu.int/fileadmin/documentsFiles/activities/addressingAssistance/whitePaperAddressingTheWorldEn.pdf

Van Campenhout, B., Spielman, D. J., & Lecoutere, E. (2018). *Information and communication technologies (ICTs) to provide agricultural advice to smallholder farmers experimental evidence from Uganda*. IFPRI Discussion Paper.

Van Campenhout, B., et al. (2019). The role of information in agricultural technology adoption: Experimental evidence from rice farmers in Uganda. *Economic Development And Cultural Change*. core.ac.uk. Available at: https://core.ac.uk/download/pdf/153429888.pdf

Webster, P. J. (2013). Improve the weather forecast for the developing world. *Nature, 493*, 17–19.

Webster, P. J., Toma, V. E., & Kim, H.-M. (2011). Were the 2010 Pakistan floods predictable? *Geophysical Research Letters, 38*(4). https://doi.org/10.1029/2010GL046346

World Bank. (2007). *World development report 2008: Agriculture for development.* World Bank. Available at: https://openknowledge.worldbank.org/handle/10986/5990

Zabel, F., et al. (2019). Global impacts of future cropland expansion and intensification on agricultural markets and biodiversity. *Nature Communications, 10*(1), 2844.

Zingore, S., et al. (2007). Soil type, management history and current resource allocation: Three dimensions regulating variability in crop productivity on African smallholder farms. *Field Crops Research, 101*(3), 296–305.

Chapter 9
Digital Trading and Market Platforms: Ghana Case Study

Keren Neza, Yaw Nyarko, and Angela Orozco

1 Development Challenge

In many poor nations and areas, the lack of markets is a major constraint to economic development. We will focus in this paper on smallholder agriculture, primarily in sub-Saharan Africa. In these areas, farmers would want to increase their outputs but worry that they will not find buyers for their crops at a good price. Buyers and traders similarly often have needs for agricultural goods and often cannot find farmers to supply those goods at the right quality and consistency over time. Potential agricultural food processing industrialists would want to set up their factories but also fear that they will not be able to reliably and consistently obtain the inputs for their goods.

In other words, in many poor nations, the lack of adequate markets is a big constraint on economic progress and growth. In economics, we normally refer to a market failure as a problem within economies which prevents Adam Smith's invisible hand to guide a nation or society to the optimum. Market failures are often defined as situations where there are gains from trade among different economic agents but where the markets are either nonexistent or have problems which prevent those gains from trade from being realized. These are situations in which if the market failure could be addressed there could be gains or benefits to the different market participants collectively.

This paper makes the case for the role of technology in addressing issues related to the lack of markets or the poor functioning of markets. We study how technology

K. Neza · A. Orozco
Center for Technology and Economic Development, New York University, New York, NY, USA

Y. Nyarko (✉)
Division of Social Science, New York University Abu Dhabi, New York University, New York, NY, USA
e-mail: yaw.nyarko@nyu.edu

could help engineer improvements in markets or the creation of markets where none existed before. As we mentioned earlier, we focus on sub-Saharan agricultural markets primarily, and that will be where we draw most of our examples. We believe, however, that the work presented here also applies to many other smallholder-dominated farming areas in developing countries.

So, in what precise forms does the inadequacy of markets take in our context of small holder farmers? In what ways is the "market failure" manifested? We will list a few of these now. The technological innovations we discuss in this chapter will address many of these failures.

This chapter is of course not the place to provide a lengthy description of agricultural markets in rural sub-Saharan Africa and similar emerging countries. For what we study though, there are two main economic agents we will identify. The first is the smallholder farmer. This farmer typically uses little machinery outside of his or her cutlass and produces on small tracks of land typically one or two acres. The other market participant is the trader or buyer. Most traders are small and travel to a limited number of villages to look for farmers with crops to sell and negotiate a price with them. The trader then takes the crops to bigger markets to sell. Sometimes, there are slightly larger buyers working on behalf of agro-processors or larger poultry companies who purchase crops from farmers in a manner similar to the traders but with slightly higher volumes. Due to poor road infrastructure, transportation by traders to farmers' farms or villages is relatively expensive and time-consuming for the trader.

With this brief picture of the context of the agricultural market structure we study, we now list some of the precise development challenges inherent in this system.

1.1 Matching Supply and Demand

In many rural areas, farmers wait on their farms (technically their "farm gates") with their crops waiting for traders to pass by and negotiate terms for a sale (see, e.g., Aker & Ksoll, 2016; Svensson & Yanagizawa, 2009; Drott & Svensson, 2010). Alternatively, they may send their goods to very small nearby villages or small-town markets again waiting for traders to show up. The traders in turn may live in bigger towns or cities. There are a large number of different villages each trader could visit to purchase crops. They may make the trek to one village, incurring the transport cost, only to find that there is very little good quality crop to be picked up there. There may be another village that the trader could have gone to with good crops at a good price; however, the trader did not have that information and so did not travel to that village. There is therefore a missing market or trades that should have taken place but did not (see Fafchamps & Minten, 2012; Demise et al., 2017; Ssekibuule et al., 2013). The farmer with the good crops and the trader with the need for those crops could not meet and trade because they did not know each other existed on that particular trading day. Some aspects of this interaction have been modeled as a search process (see, e.g., Nyarko & Pellegrina, 2020) – farmers are searching for traders and traders are searching for farmers, and many times each may fail to find the other. The lack of information on the existence or whereabouts of the farmer and trader is a major source of the market failure in this case. That is, when the farmer and trader do not succeed in finding each other, a "market" cannot be formed for them to trade. The farmer would then be stuck with his/her unsold produce ("postharvest losses"), and the trader may have incurred travel costs to a village and will have to return empty-handed. This is a "market failure."

In the literature, there is debate about the source of the market failure and even the existence of the market failure (i.e., where the market fails to function properly and enable people who want to trade to be able to find each other and be able to trade). For instance, Dillon and Dambro (2017) suggest that in these agricultural settings, there is no market failure and that the evidence has focused on measuring market integration (prices being set correctly) rather than market failure per se. They argue that there may be hidden risks or costs borne by traders which may cause markets to appear inefficient with less than optimal trading but that they are efficient if these risks and costs are factored into the analysis. For example, traders may have to factor into their calculations the risk of losing their goods when trucks break down on bad roads or if their goods are stolen. However, there is vast evidence that shows otherwise – smallholder farmers face constraints on both the supply side (inputs needed by farmers like fertilizers, seeds, and tractor services are in limited supply) and the demand side (buyers of their farm produce do not show up or provide good prices). On the supply side, some of the well-studied constraints faced by farmers include credit access (Banerjee, 2013; Harrison & Rodríguez-Clare, 2009) and lack of quality inputs (Bai, 2018). On the demand side, some of the constraining factors studied in the literature include access to high-income and high-price markets (Atkin et al., 2017; Verhoogen, 2008).

1.2   Price Information

There is a second aspect of lack of good markets which applies to smallholder farming. Typically, one side of the market does not have full information to make the appropriate economic decisions (see, e.g., Goyal, 2010; Allen, 2014; Startz, 2016). In our case of smallholder agriculture, it is usually the farmer who is stuck on his or her farm in a small village and has less information than the trader. The trader in contrast is the one who is often in a bigger town or else travels to many markets and so is up to date on the general trading conditions. When a trader goes to the farmer's farm to negotiate a trade, the trader usually has better information than the farmer. This often means that the trader will be able to offer the farmer unreasonably low prices because the farmer does not know that there are other markets where prices are higher. The price of a crop could be commanding high prices in a city. The trader who comes to visit the farmer may offer the farmer low prices for the farmer's goods. The farmer, not knowing that prices have recently risen in the cities, would accept the lower price offered by the trader. If the farmer had known of the better prices,

Question: Do you feel that you are well informed about market prices?

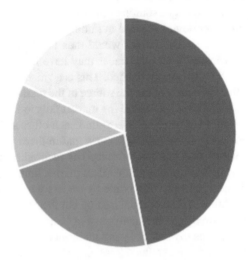

Fig. 9.1 Results of a survey of farmers on the question of knowledge of prices

the farmer would have bargained for a better price. Figure 9.1 shows the result of a survey of farmers on the question of knowledge of prices from Hildebrandt et al. (2020).

One could argue that, since both the farmer and the trader are poor, this is not a major problem because all this means is that traders can get a better price relative to farmers. In other words, what is the relative distribution of the gains from trade among these two market participants? Of course, to us as researchers, there are two responses to this. The first is that if the farmers could be guaranteed higher and more consistent prices, then the farmers would respond by increasing their output and perhaps investing more in their farms. That is, the lack of information on prices could be introducing risk and uncertainty to farmers making them unwilling to expand the scale of their operations. The second response is that we do often place more weight on the welfare of farmers than on traders. The farmers are usually poorer and have fewer alternative options for work in comparison with the traders and buyers of produce. So, both government policy and researchers often seek ways of improving the lot of farmers relative to the traders and buyers.

An important role of markets is what is often called "price discovery." It is meant to convey the belief that the markets communicate the "true" price of the commodity – that which will clear the market so farmers sell all they want and buyers buy all they want at the prevailing prices. In poorer communities and in particular in many smallholder agricultural communities, the trading processes do not result in the appropriate price discovery. The markets, to the extent that they exist, do not perform adequately their price discovery function. Markets do not inform the farmers of the potential true value of their crops, so they can make correct economic decisions. There could be high demand for a farmer's goods in a city (high prices), but nobody shows up to the farmer's farm in the village to ask for his/her goods. Alternatively, there could be abundant surpluses of a crop in one area or village (low prices), which is needed in a town, but the trader does not know of this so does not go to that area. The prices in the different areas are not transparent to (or known by) the farmers and traders. Or, technically, there is little "price discovery."

1.3 Information on the Quality of Crops

In many of the rural economies we study, there are typically complaints on both sides of the market exchange, farmers and buyers, about quality issues. On the one hand, buyers and traders complain that they do not get from farmers the quality of goods that they would desire. They say that the farmers are always trying to cheat them with inferior-quality goods. The prices traders pay to farmers, therefore, have to take into account the possible low quality of the goods they receive. On the other hand, many farmers do not believe that traders are honest in the assessment of the quality of their produce. Many farmers, because of this, do not believe that they receive the full benefits from improving the quality of their goods (Bagwell, 2007; Bai, 2018). Consider the example of smallholder transactions in maize (corn). Traders or buyers would want dry and clean maize free of pests and diseases like aflatoxins (a mold-like disease). Visual inspection of the maize is not always sufficient to check for disease and pests, especially in large bags. Farmers in turn do not believe that they will get better prices if they go through the work of properly drying the maize and fully clearing it of dirt and pests. Even if they believe there is some reward to this activity, they are not fully conversant with the price gradient – how much additional money they receive for the additional increase in the quality of their grains. Again, this is a failure of the market to adequately provide the price signals, in this case, the price gradient for quality (see Saenger et al., 2014; Bernard et al., 2017).

This market failure has important effects on the rural economy. Farmers often complain that they are not getting enough for their crops. They often say that they would put in more effort in their farming if they could be assured of a return on that investment. If the markets could create the price gradient in quality, the farmers would "climb" that gradient by producing better-quality goods, thereby increasing the return from their efforts and their crops. Traders and buyers too would benefit from the higher-quality goods. For example, many agro-processing industries cannot function without the reliable supply of consistently high-quality grains. In short, the economic development and transformation of the rural economies may be stymied by the market failure in pricing for quality of the crops.

1.4 Storage and Credit Market Failures

In rural economies, there is also often the failure of credit markets (World Food Program, 2010; Svensson & Yanagizawa, 2009). One way in which this happens and where there is the clearest manifestation of the market failure is in postharvest credit (Kaminski & Christiaensen, 2014). This is the situation where a farmer has successfully harvested the crop and has the crop bagged and ready to be sold. The crop, however, is being harvested at a time when most of the other farmers are storing their goods. If the crop is sold right after harvest, the farmer receives a low price for their crop. Indeed, a lot of the crop may go unsold if traders, inundated with many farmers all trying to sell their maize at the same time, do not come to their farm gates to purchase the maize. Agro-industries may similarly not need the produce of the farmers when there is a glut of crops in the market. Instead, they would most probably prefer the smoothing of the availability of crops across the year and seasons.

Both the farmers and the buyers would, therefore, wish for there to be a mechanism for sale of the crops at a future date. For this to work though, credit may be essential for the farmer. The farmer may need cash upfront to pay for unavoidable bills. The farmer will have household expenses and school fees for children, and they will need money to start planting for the next season. If the sale of the newly harvested grain is to be postponed, the farmer will have a demand for credit. Banks would want to supply that credit and to offer loans to such farmers. Banks, however, need to get collateral from the farmer without which the farmer may decide to default. There is the produce of the farmer which could be used as collateral. However, there is no way of easily and cheaply verifying the quality and hence value of the corn and also to verify that the corn will still be with the farmer when it is time to repay the loan. The farmer could always decide not to repay the loan, that is, to default. This is a classic credit market failure as both sides would want to trade if there could be credible certification and collateralization of the farmer's produce. In a well-functioning market, the farmer would want to take a loan from the bank, and the bank would want to offer the loan. However, the market will fail to be formed.

A consequence of lack of storage and credit facilities is that when there is a harvest, there may be a glut of food crops upon harvest which may not be sold and go to waste. This is a part of the postharvest losses which plague these markets. FAO (2019, page 32) estimates that in sub-Saharan Africa, 14% of food is lost between postharvest and retail distribution along the supply chain.

2 The Ideation

One may ask how we, the researchers, noticed these issues faced by farmers and came up with the idea of using technology to fix these problems. The answer of course is straightforward. The farmers in the communities told us their problems and explained their concerns to us. Many years ago, one of the authors of this chapter and his PhD students visited farmers in Ghana. Many of the farmers complained about how they were being cheated by the traders who gave them low prices for their goods. They spoke about not knowing what the prices in the big cities for their goods were. They mentioned that they engage in one-on-one bargaining with a trader who comes to their farm gate after they have completed harvesting or just before. The farmers said that they are in a weak negotiating position at that time as they have a perishable good, in addition to being the weaker informational position. Similar evidence was found by Eggleston et al. (2002), Aker and Fafchamps (2015), and Nakasone (2013).

All of that got our research team thinking about what is the best way of solving these development challenges faced by these poor and rural smallholder farmers. The team then started conversations with an African agricultural food services company with a strong technology focus, Esoko. This chapter discusses a lot of the initial work with Esoko on mobile phone-based price alerts. The chapter also discusses our work on commodity exchanges which was inspired by the initial work on price alerts. This chapter will not delve into the technological details per se. Instead, we will describe the impacts of the technology on the smallholder farmers and the lessons learned from various interventions, by the authors and many others.

3   Implementation Context

The research reported in this chapter took place primarily in Ghana although a lot of the early work and insights came from Ethiopia (see Minten et al., 2014). As mentioned earlier, the knowledge of the development challenges and the appreciation of their importance came from the farmers and traders in these countries when we undertook research visits to those nations over a number of years.

The first formal research conducted was with the company Esoko which provided price alerts to farmers in Ghana (see Hildebrandt et al., 2020). Our team was very bullish on the importance of the mobile phone in overcoming development challenges in our research communities. The mobile phone is ubiquitous in rural areas. Even in very small villages, we find farmers with mobile phones. When their own villages have no electricity or cellular network signals, the farmers go to the next small town near theirs on a regular basis to charge their phone or even make the calls. The national governments are committed to increasing mobile phone signal reception, so over time, access to mobile phones was expected to increase. By the end of 2018, sub-Saharan Africa had a mobile subscriber penetration rate of 44%, and 23% of the population used mobile Internet on a regular basis (GSMA Mobile, 2018). That explained our initial focus on mobile phones as a vehicle for addressing some of the development challenges.

In both Ghana and Ethiopia, the government and policy leaders were all keenly interested in improving the lot of the farmers in their countries. It was therefore easy to get the attention and the support at the highest levels for our research activities.

Our early research on mobile phones taught us the potential of technology could be high in our communities. As we moved to working on commodity exchanges, the personnel at the exchanges were enormously helpful to us. We had the support of the leader of the Ethiopian Commodity Exchange (ECX), first with the inaugural Chief Executive Officer (CEO) Eleni Gabre-Madhin through subsequent CEOs Anteneh Assefa and Ermias Eshetu and their teams. In Ghana, we had the support of the Ghana government (Ministry of Finance and Ministry of Trade and Industry), as well as the initial project staff and current leadership of the Ghana Commodity Exchange's (GCX), CEO Dr. Kadri Alfah, Chief Operating Officer Robert Owoo, and their teams.

4 Innovation

We divide our discussion on innovations into three sections: Sect. 4.1 Price Alerts Services, Sect. 4.2 Mobile Phone-Based Trading Platforms, and Sect. 4.3 Commodity Exchanges. Here, we will discuss the innovations themselves. In subsequent sections, we will discuss the implementation and evaluation of these interventions. An even later section will describe the results and lessons learned.

4.1 Price Alerts Services

One of the earliest innovations we were engaged in and an early area of interest in the academic literature is in mobile phone-based price alerts (see Hildebrandt et al., 2020; and the literature mentioned there). We mentioned in the development challenge in Sect. 1.2 that in many situations, smallholder farmers are at a price disadvantage when it comes to knowledge of prices of their commodities. Traders and buyers have more immediate knowledge of prices of foodstuff and crops across a nation, while the farmer, holed up on his farm, does not. Many researchers were therefore of the belief that mobile phone-based price alerts could either solve or lessen the impact of this problem.

Esoko is an agricultural services company in Africa. They provide farmers the prices of their commodities in different district, regional, and national markets (see Fig. 9.2). The company was started in 2009 based on the belief that there are gaps in the information flow for farmers, in particular on farmers getting access to price information. It was a startup with individuals recognizing this market failure, who formed a company to provide the information services. The Esoko business model, at least at the time of this research, was as follows. Market surveyors in the various district, regional, and national markets would collect on a weekly basis (or more frequently as needed) the market prices of all of the major food crops. Farmers who subscribe to the Esoko service would receive the prices of the crops they were interested in from the markets they are interested in. For example, a farmer would

Fig. 9.2 Left: Esoko text message weather alert. Right: Esoko text message price alert (Reproduced from Esoko, 2019)

tell Esoko that they are interested in the price of maize in both the regional market and in the national capital. The farmer would want that information so as to be in a better bargaining position when the traders show up on their farm gates or at the local markets to purchase their crops.

After the farmer has subscribed to the service, Esoko would then text the farmer the prices of the requested crops in the requested market on a monthly basis. They are able to see these prices and presumably use this information when bargaining with traders for their crops.

In the section on evaluation, we indicate how we evaluated this Esoko intervention. In Sect. 6, we review the results of the intervention and also indicate the results obtained from other researchers on other related price alert interventions for similar populations.

4.2 Mobile Phone-Based Trading Platforms

The next step in complexity but still using the mobile phone as the base is the use of the mobile phone in trading. We mentioned in our development challenge in Sect. 1.1 that matching supply and demand is a big problem in smallholder agricultural communities. Farmers often cannot find traders, and traders often cannot find farmers with the right good of the right quality and quantity. Another innovation in this space is the use of the mobile phone as a trading platform.

These platforms connect buyers and sellers in rather geographically fragmented markets and usually focus on a major cash crop of the selected region. In contrast to the traditional markets, these structured platforms allow farmers, traders, processors, and financial institutions to enter legally formalized trading and financial arrangements (Ochieng et al., 2020). They operate by digitizing and automating these price discovery mechanisms, which include detecting and declaring winners in auctions, disseminating price information, and garnering farmers to access alternative electronic markets. Consequently, these interventions and platforms have the ability to inform farmers about prevailing market prices, increase market competition, and enable transparency of the price search process. By rapidly connecting disparate market agents, they reduce information costs at various stages of the agricultural chain. This allows farmers to take advantage of previously untapped trade opportunities and to learn about previously unknown innovative practices. In consequence, the cost reductions yield welfare and income gains (Nakasone et al., 2014).

An example of this is Kudu, an electronic market platform for agricultural trade in Uganda (Newman et al., 2018). In this model, the farmers place their requests to either buy or sell goods in a centralized national database; then, the app processes to identify profitable traders, and the two sides are informed about it. Rather than allowing the buyers and sellers to browse through a list of potential trading partners, Kudu's matching algorithm connects bids based on maximizing the gains from trade that the platform can offer.

There are several features that make Kudu attractive to farmers. First, users not only place the desired price and quantity of their bid but also are able to narrow the options of their desired trade: this includes features of the grains such as shelled versus unshelled grains and wet versus cleaned maize. Second, the platform allows users to trade with anyone across the country, and the app's algorithm will take travel costs into account when proposing matches. The Kudu platform provides users with in-village support service and a call center, which enhance its reliability. Furthermore, Kudu does not charge for transactions, and while the creators have considered monetization options such as charging commissions for each transaction, they recognize that such changes are nontrivial. Finally, users can trade through the Kudu platform in four different ways: through short message service (SMS), using an Unstructured Supplementary Service Data (USSD) application, through a website, or speaking through the call center. All of these options provide its customers with the same features: buy, sell, and quote/request price information (see Fig. 9.3) (see Ssekibuule et al., 2013; Reda et al., 2010; for other mobile-based trading platforms in the context of development).

There are other mobile apps across sub-Saharan Africa that apply the same principle: provide a virtual market where farmers and buyers can trade their agricultural goods. Among those are M-Farm; DrumNet; and, more recently, Twiga Foods in Kenya (Baumüller, 2013; Mire, 2019) and Agro-Hub in Cameroon (Balashova & Sharipova, 2018). A variety of studies have confirmed the benefits, as well as the limitations of mobile trading apps. A survey of M-Farm users confirmed that receiving price information can help them plan for production; however, the

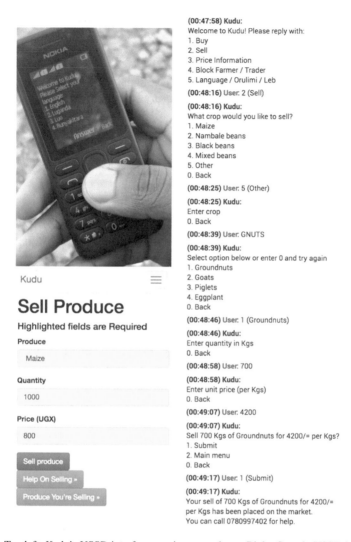

Fig. 9.3 Top left: Kudu's USSD interface running on a phone. Right: Sample USSD interaction for selling groundnuts. Bottom left: A user placing an ask on Kudu's web interface (Reproduced from Newman et al. 2018)

survey also revealed that there was a limited impact on expanding market linkages. This was driven by the fact that M-Farm could only provide single bilateral contracts between a farmer and a buyer, rather than establishing a full network that allowed for multiple connections (see Baumüller, 2013), where Agro-Hub is an agency that connects smallholder farmers with sustainable markets. More than 700 farmers reported an increase in productivity and income (Balashova & Sharipova, 2018). Twiga Foods is a mobile-based food distribution platform for small- and medium-size fruit and vegetable farmers. After both farmers and vendors sign up through the online and phone platform, the platform acts as a bridge between sellers (i.e., farmers) and buyers. It guarantees farmers a consistent market with higher prices, and similarly, through its food safety standards, it ensures a reliable and high-quality supply to vendors. It currently operates on a national scale and is the largest seller of bananas in Kenya. As of 2020, the company sources 245 tons of bananas each week from over 3000 farmers, which are distributed to over 14,000 vendors.

4.3   Commodity Exchanges

Ramping up the complexity of technological innovation, we move from the price alerts in Sect. 4.1, through the mobile phone-based platforms in Sect. 4.2, to arrive at the commodity exchange platforms. The commodity exchanges are usually national government–run institutions. They serve as national centralized markets for crops.

Commodity exchanges have existed for many years in many countries. The commodity exchanges in Africa and many areas with smallholder farmers are very recent, and many countries still do not have them. The improved technology in existence today has made modern commodity exchanges much easier to establish and is now within the reach of many poor nations. While big telephone and mainframe computer infrastructure would have been necessary in the past, today, relatively lower footprint technologies and cloud-based systems are able to run the exchanges in poorer nations at a fraction of the former cost.

The commodity exchanges work as follows. Farmers upon harvesting their crops send them to a warehouse close to their farms. The warehouse then inspects the grains and assigns a grade to them. The farmer then receives a receipt for the crops (called a Grain Receipt Note and/or, in the more advanced version to be described later, a warehouse receipt). The warehouses are usually owned or operated by the commodity exchange. Each farmer chooses a broker or is assigned one. The farmer gives instructions to the broker on when to sell their crops and at what price.

The buyers of grains also use brokers to purchase grains. The buyers and sellers of grains make offers on the commodity market. In some exchanges, this is at a set time during the day, while in others, it can occur at any time. Brokers look at the offers, and at some point, a buy-side and a sell-side broker will decide to trade. The price at which the broker trades will be entered onto the GCX platform for all to see. The buyer then picks up the grain at the warehouse that the buyer's broker has just purchased. On the other side of the market, the seller will receive cash through the seller's broker who has just sold the grains to the buyer's broker.

The commodity exchange is really nothing other than technology plus rules of trading, plus government-backed standards and warehouses. Software engineers write code that accepts bids from different market participants on their platform. When two sides of the market agree to trade, then the trading engine software matches them. Prior to the match, farmers would have deposited their produce at the warehouse, and this would be recorded by the software. Farmers would have sell-side brokers who seek to sell their commodities on their behalf. On the other side of the market are buyers working through buy-side brokers. All brokers are registered on the software and are screened and licensed.

Just as with the mobile phone-based trading platforms, the commodity exchange allows buyers and sellers to find each other. It therefore addresses the development challenge set out earlier, by enhancing the matching of buyers and sellers. In many of the commodity exchanges, rural farmers are able to access the commodity exchange through their mobile phones or through calls on the mobile phone to their brokers.

Since prices traded on the commodity exchange are made public, there is price information, so in principle, the problem described in Sect. 1.2 is solved, especially when the farmer has access to the prices either through price alerts issued by the exchange or through a direct phone call the farmers make to their brokers. Furthermore, the commodity exchange is a way of matching buyers and sellers. Any farmer, for example, looking for buyers simply needs to be in touch with his or her broker. The broker in turn has access to all the buyers on the exchange platform. Similarly, a buyer looking for farmers with grains simply needs to contact their broker who has access to all the farmers on the platform through the brokers representing the farmers.

In addressing the development challenges, the commodity exchange is like the price alerts and the mobile phone platforms. Where the commodity exchange has a unique advantage is in addressing the development challenges explained in Sects. 1.3 and 1.4.

The commodity exchange grades the commodity when the farmer brings the commodity into the exchange warehouse. The grading process ensures the grain is free of diseases and pests. It ensures the moisture level of the grain meets a minimum threshold, and as is, for example, maize, the drier grains receive a higher grade (the maximum allowed moisture content is 14%, and 12% moisture content can help get a grade I as opposed to a grade II or III or IV designation). The commodity exchange sales, or contracts as they are called, are all based on the grade. For example, there will be a price and trading day for grade I white maize contracts and a different price and trading day for the grade II white maize contract. In this manner, a market for different grades of the crop is created.

There will be a price gradient for quality, and farmers will know that when they invest their effort in producing better higher-grade maize, they will be rewarded for their effort. Traders will similarly know what the quality of their grain is and will know that they can pay more for better-quality grain. They will appreciate having the price gradient for different grades, so they can choose what grade to purchase rather than paying one price and not knowing the grade and therefore having the price according to a perceived average of many possible grades.

Finally, the commodity market is unique in its ability to address the market failure issues in Sect. 1.4 – storage and in particular credit market failures. The storage is straightforward – being a larger and national institution, the exchange is able to establish warehouses at strategic locations across the country. More importantly, the warehouse receipts the farmers receive upon depositing their grain at the warehouse can be used as collateral at a bank to obtain a loan. This will address the potential credit crunch that farmers face when they harvest their crops and prices fall with the glut of food in the market. The farmers are able to store their crops at the exchange warehouses to await a time when the prices improve. They are able to do that because they are able to take a loan from banks, collateralized by the warehouse receipts. Banks are willing to offer the loans because the warehouse receipt states the quality of the goods, as certified by the exchange, and there is a ready market for that collateral at the exchange. Historical prices give some indication of the value of the collateral. Höllinger et al. (2009) provide a description of the mechanics and structures required for a warehouse receipt system (WRS) with an emphasis on emerging and transition economies of Eastern Europe. Other studies on WRS include Miranda et al. (2019), Adjognon et al. (2019), and Katunze et al. (2017).

5 Evaluation

Most of the technological innovations mentioned in this chapter are evaluated using randomized control trial methodologies (Duflo et al., 2007). The basic idea can be illustrated in the intervention described by Hildebrandt et al. (2020) which we use here as our principal example. A number of communities were chosen, approximately 100. Through randomization, one-half of them, in this case 50, would be treatment communities with the other half (50) being the control communities. Farmers will be chosen as subjects in each of the 100 communities. The farmers in the randomly chosen treatment communities will be given the technology to be evaluated. In the work of Hildebrandt et al. (2020), it is the price alerts. In the current research of the authors of this chapter with other co-authors, it will be access to the services of a commodity exchange. Outcomes of interest of the subjects in all communities will be measured. These would be things like prices obtained for farmers' goods, production levels, and sales. Statistical techniques will then be used to determine whether there are observable differences in the treatment communities relative to the control communities. If there is, then we would have evaluated the technology and will record an impact.

Of course, what is described above is the very simple skeletal structure for the evaluation. In many of the experiments, there are two principal problems that need to be addressed. The first is spillover effects. For example, if there is a mobile phone price alert that is being evaluated, there need to be safeguards against one subject in a treatment community showing the price information to a subject in a control community. This problem of spillovers is typically addressed by clustering communities so that, for example, those which are geographically close to each other, other those with many farmers who are friends across communities, are pooled together and considered as one larger community. Again, one can see work of Hildebrandt et al. (2020) for ways of creating a connectivity measure between communities which was used to cluster those communities that were too close to each other by this metric.

The second big problem that comes up is one of balance. Since the communities are chosen randomly, it is possible that most of the treatment communities by chance happen to be in, say, the more prosperous part of the wider study area. In this example, the effects of the technology may be hard to disentangle from the effects of being in a more prosperous region. This problem of balance is solved using stratification. The wider area is divided into strata where issues like wealth levels and other characteristics are held fixed, and within those fixed areas, the randomization takes place. As an example, in the Hildebrandt et al. (2020) paper, there was concern that geography could play a role (you could be in one of four quadrants of our area, each with different climate and suitability to agriculture), and you could also be in a yam- or non-yam-producing area (yam being the major cash crop in the area). So instead of simply dividing the 100 communities into 50 treatment and 50 control, eight strata were created (each being in one of four geographic quadrants and being in majority yam or not majority yam community). Each of the strata would have approximately 100 communities divided by eight strata so $100/\varepsilon$ or 1/2 or so communities. The randomization into approximately 50% treatment and 50% control would then take place within each strata. Each strata would then have, by construction, a similar number of treatment and control communities. We show the treatment and control villages in our Esoko price alert intervention in Fig. 9.4. The cluster and strata formation is shown in Fig. 9.5.

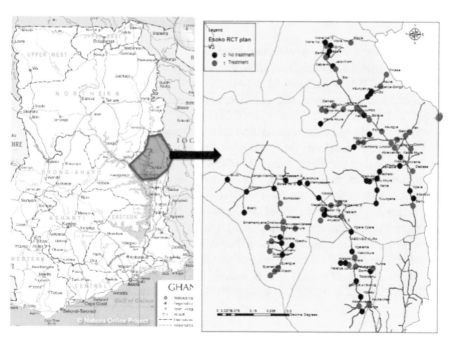

Fig. 9.4 Treatment and control villages in the Esoko price alert intervention

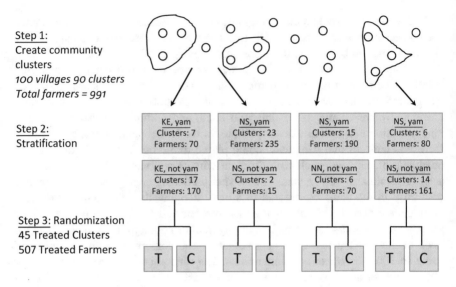

Fig. 9.5 Formation of clusters and strata

The basic econometric model is easily explained as below:

$$p_{iit} = \lambda + \kappa T_i + \psi X'_{ii} + \omega_k + \omega_t + e_{iit} \qquad (9.1)$$

In this equation, here, p_{iit} represents the variable of interest (here, the producer price outcome for farmer λ living in community λ selling in month t). The variable T_i is the treatment size indicator (one if treatment and zero if control), X'_{ii} denotes a set of additional covariates, and ω_i and ω_t denote randomization strata fixed effects whether or not it is in the kth strata) and time period fixed effects, respectively. (The λ and e_{iit} are of course the constant and the error terms.)

Equation 9.1 measures the price alert intervention. One can imagine similar evaluation techniques being used for both the mobile phone trading platforms and the commodity exchange platforms (innovations in Sect. 4.2 and 4.3). For example, in rural areas, some communities could be exposed to the commodity exchange services where others would not be. Assuming lack of communication between the communities (the spillover problem) and either general lack of knowledge of the existence of the commodity exchange or difficulty in accessing it, it would be possible to target the services of the commodity exchange to some communities and not others. The randomized control techniques just described for the price alerts could then be applied to the commodity exchange intervention.[1]

---

[1] The authors, with co-researchers Chris Udry, Lauren Bergquist-Falcao, and Lorenzo Casaburi, are engaged in one such evaluation.

6 Results and Lessons

6.1 Price Alerts Services

Hildebrandt et al. (2020) show that Esoko price alerts had a positive effect on yam prices received by farmers in their study. The initial results of the paper show that there was an increase of 8.73 Ghana cedis (GH₵) per 100 tubers of yam relative to those farmers who had received no Esoko price alerts. This was equivalent to a 5% increase in prices. This initial peak declines steadily over time, making the effect small in magnitude and statistically insignificant; there was a price decline of 0.01 GH₵ per 100 tubers of yam, leading to a 0.8% decrease. This decline is due to a mechanism the paper called "bargaining spillover." In such a landscape, middlemen cannot distinguish between those farmers with price knowledge and those without. So, the whole pool of farmers ends up adjusting their bargaining strategies. Similarly, the traders have to decide if each farmer is an informed or uninformed one. Since traders know that informed farmers will reject low prices, they have to estimate how high they can push the offer without getting a rejection. Therefore, the trader's strategy depends both on the farmer's actual price knowledge but also on how well the trader assesses this. As such, providing information has positive effects even on farmers that did not access price information directly. Prior to the price alert intervention, maize prices were homogenous because bargaining was less prevalent and farmers had a reference "market price." In other words, this study finds positive price effects – obvious effects in the short run and more subtle effects (because all farmers benefit) in the long run. Such results are supported by a similar study by Courtois and Subervie (2015) who found that Ghanaian farmers received about 10% higher prices for maize and groundnuts when they had access to the market information system (MIS). Furthermore, this is consistent with previous evidence which shows that the introduction of information and communication technologies (ICT) reduced price dispersion as agents were able to bargain for better prices, Jensen (2007) being one of the first and classic papers in this recent line of literature and also of Aker (2008). This is further supported by a more recent study on maize farmers in Mozambique, for whom the introduction of mobile phones led them to experience a statistically significant decrease in maize price differentials of 10–13% (Zant, 2019).

The Hildebrandt et al. (2020) paper shows that there could be a big difference between short-run and long-run effects. In the short run, some market participants adjust their behavior (farmers in that paper), while others (traders) adjust their behavior only in the long run. In the long run, all market participants get to adjust their behavior upon introduction of an innovation, which could change the outcomes relative to the short-run effects.

The long-term implications of better access to market price data are not fully clear. Mitra et al. (2018) use an asymmetric bargaining model to study a market price information intervention among potato farmers in West Bengal, India. The authors concluded that access to better price information does not necessarily benefit farmers in their negotiations with middlemen because they don't have access to alternative markets. In Karnataka, India, Levi et al. (2020) evaluated the implementation of the Unified Market Platform (UMP) on market prices and farmers' profitability. They found that the UMP had generated a greater benefit for farmers with high-quality produce, increasing, on average, the prices of maize, groundnuts, and paddy by 5.1%, 3.6%, and 3.5%, respectively. The provision of price information alone might not be enough to facilitate trade among small farmers. While such interventions reduce information asymmetries between traders and farmers, if the market agents do not have outside options for their sales, information will do little to improve their marketing outcomes. On the other hand, if farmers have access to larger markets and have increased bargaining power, more information may represent potential gains, as farmers could potentially access the traders who are ready to pay the higher prices.

While most studies that evaluate the impact of ICT diffusion in the agricultural sector find significant results, a few of them find no evidence of an effect. Futch and McIntosh (2009) investigated the introduction of village phones in Rwanda and found that while the technology did increase the proportion of farmers arranging their own transport to markets, there was no significant increase in the commodity prices that those farmers received. Similarly, Fafchamps and Minten (2012) evaluated the impact of Reuters Market Light (RML), a service that provided farmers with agricultural information through mobile phones in Maharashtra, India. Ultimately, the authors found no differences in average prices for farmers with RML subscriptions. Aker and Fafchamps (2015) analyze the expansion of mobile phones in Niger and find no evidence of increases in farm gate prices as well.

When farmers' knowledge of prices increases, the research has found an effect on both the farmers' production and postharvest decisions. Hildebrandt et al. (2020) show that there is a significant impact on produce prices and production decisions in Ghana. The authors notice that by providing price information of a certain crop, farmers are incentivized to produce more of a particular crop (in this case, yam). They find farmers report growing a new crop or growing more of an existing crop. In addition, price alerts also caused fewer farmers to sell in the local markets and induced them to sell at the farm gate.

In similar studies, Baulch et al. (2018) argue that some price discovery mechanisms might target large traders who are able to sell in large volumes. Thus, only a few small farmers can access these market options through their farming associations. In Central Malawi, Ochieng et al. (2020) find that greater efforts are needed to sensitize the farmers and traders on the quality and quantity requirements of such structured markets, which could result in an increase in the farmer's level of commercialization in such markets. Mitchell (2017) in a study conducted in Gujarat, India, shows that there is an implicit increase in producer prices, leading to an increase in the amount of crops produced.

Additionally, market information systems allow farmers to decrease their postharvest losses. Jensen (2007) finds that the introduction of phones in fisheries in Kerala reduces waste by 4.8%. Fafchamps and Minten (2012) argue that for Indian farmers in the Maharashtra state, the price improvement generated through the price alerts leads them to better agricultural practices and postharvest handling. Finally, Dixie and Jayaraman (2011) show that to avoid postharvest losses, farmers in Zambia used SMS text messages to coordinate among local truckers and enhance product transportation.

6.2   Commodity Exchanges

While price alerts and mobile phone-based trading provide farmers with a market, these solutions present some constraints. Even though farmers are informed about prices – which earns them bargaining power, this does not necessarily translate into better bids: middlemen still have direct access to the markets that the farmers, many times, do not (see Mitra et al., 2018; Mitchell, 2017). Furthermore, most mobile applications for trading, if not all, do not include an algorithm that can account for the difference in crop quality that is offered and demanded by farmers. Including this within the mobile app will present two challenges: comprehensive enforcement and inconsistency in the farmers' ability to grade crops correctly and effectively (see Newman et al., 2018; Levi et al., 2020). Lastly, many of the studies on mobile applications notice that few farmers have smartphones and that many are illiterate, forcing these projects to rely on human interaction (see Aker et al., 2016). As we will see below, a commodity exchange addresses and overcomes these problems by altering the agricultural landscape in multiple fronts: bargaining dynamics, production decisions, and household dynamics.

   Commodity exchanges are modern marketing systems based on warehouse receipts, allowing small farmers to store their surplus safely while they wait for prices to increase. Furthermore, these stored commodities serve as a collateral to secure loans to finance household consumption and investment in the meantime (see Miranda et al., 2019). The effects of using warehouses go beyond the immediate storage facilities and financing opportunities for farmers. For instance, ECX reduced price dispersion between export prices and retail prices and facilitated the tendency of prices of the same commodity to move together (see Andersson et al., 2017). It also led to an increase in the quantity of coffee exported; Minten et al. (2014) showed that exported quantities from Ethiopia were 50% higher in 2012 than 10 years earlier.

   Exchanges have the potential to change farmers' production decisions, as grading systems reveal high- and low-quality grains. Prior to a commodity exchange, almost all crops from the production areas are physically transported to a regional or national central market, usually outdoors, for auction. Without formal regulation, a significant volume of the crops can be adulterated by mixing high-quality grains with low-quality ones. This, in addition to information friction and scattered markets, leads to under-provision of quality crops (Bai, 2018). A commodity exchange allows differentiating between a crop's quality, usually through sorting and inspecting grains, and consequently, the transactions cover every class of the grading given by the commodity exchange warehouses, otherwise a limitation that most mobile trading applications face (Demise et al., 2017). This, in consequence, has the potential to encourage farmers to produce higher-quality products.

7 Conclusion

In this chapter, we have described a development problem facing smallholder farmers in places like sub-Saharan Africa and other similar regions. The problem can be summarized as a lack of effective markets for their goods. We described innovations which have the potential to address parts of this problem. We discussed three innovations, which, in order of increasing complexity, are the mobile phone-based price alerts, electronic trading platforms, and national commodity markets. The three are all similar in the sense that at their core, they are technologies which allow the transmission of information to farmers and those who trade with farmers. We described briefly these innovations and sketched the basic randomized control trial methodology for their evaluation. We indicated the results and lessons learned from these evaluations – both from the authors' own work and from work in the academic literature.

Discussion Questions

Are experiments needed to rigorously test some of the conjectures of this chapter?

There are a number of discussion questions this chapter suggests. The first set of questions have to do with formally proving a number of the conjectures of this chapter. Although there is some current ongoing research, the impacts and benefits attributed to commodity exchanges in the earlier section need to be verified with rigorous field experiments. Designing a nationwide experiment for the commodity exchange intervention has a number of difficulties, which of course are not surmountable.

Could the innovation be harmful? Is price information potentially harmful to farmers?

We need to be mindful of whether the new technologies introduced could inadvertently harm those the innovation is meant to help. One can imagine a situation where the introduction of price information to some farmers could actually hurt those farmers. It is possible that when traders realize that some farmers have superior information, they will shun those farmers and go to other farmers. The farmers with the superior information may then find themselves in the situation where they no longer have any (or as many) traders coming to their farms to trade. In the extreme case where they have no traders coming to them, they may therefore be much worse off than if they did not have the price alerts. So, general equilibrium effects could cause farmers to be hurt. In the Hildebrandt et al. (2020) paper, the authors find that the farmers are not hurt. This potential for harm is not realized in that case. There are, however, at least conceivable situations where technology could be harmful to the intended recipients of that technology.

Can technology cause increased inequality among farmers?

Next, even as we introduce innovations to help farmers, those innovations may increase inequality among the farmers. If the price alerts or the commodity exchange innovations are more likely to help the farmers who are already better off, then inequality may increase among the farmers. In the work of Hildebrandt et al. (2020), for example, it is found that the price alerts benefit primarily those who produce the cash crop yam. If those farmers are also those who are richer to begin with, as it is reasonable to suppose, then the innovation would help the better-off farmers more than the less well-off farmers. It is an open question how much of a concern that should be to us. On the one hand, all of the farmers are poor, so helping some of them via technology, even if only the relatively better-off ones, should be a good thing. On the other hand, our intent may be to help the poorest of the poor, so we may be worried if those farmers are not reached by our technological innovation. This question of course requires more data and more study. We also mentioned earlier the commodity market innovation. One could similarly imagine that the better-off farmers are the ones more likely to engage with the commodity exchange. In that case again, the introduction of the innovation could lead to increased inequality. With all technological innovations, this leads to a pair of discussion questions: will the innovation result in more inequality, and is that inequality harmful (e.g., it does not help those we really care about, like the poorest of the poor), or is it benign (e.g., enough of the farmers benefit that the inequality is unimportant relative to the wider gains the technology enables)?

Building Capacity Through Research
With both the price alert intervention with Esoko and the commodity exchange intervention with GCX, the researchers worked extremely closely with host institutions. After the research, both institutions emerge stronger with greater capacity. With Esoko, insights from the field were transmitted immediately. In one example, researchers noticed that farmers on the Esoko app preferred using local traditional units (bowls or what is called locally "Olonka") rather than the standardized units, which was duly and immediately communicated to Esoko for action. With GCX, our earlier work with farmers encouraged the government to make the final steps in the establishment of the exchange. The earlier work gave the researchers credibility in front of the government when pushing for the exchange.

Mitigating Climate Change

One can think of two reasons why the innovations in technology discussed here would be essential as climate change begins to take its toll and agricultural systems begin to change:

I. As climate changes, supply will inevitably change in unpredictable ways. The need for speedy price discovery and extensive knowledge of prices would therefore be needed for farmers to react to the changes. Markets would become more and not less important in these more volatile climate change-induced environments.

II. As evidenced by the responses to the Covid-19 lockdowns, technology may be important for staying connected and working when there are difficult environmental conditions in existence. There is a slight advantage in not having to travel so much to find market partners when there are environmental challenges. The technological innovations mentioned here enable markets to form with much less physical movement of people.

Role of Gender in Agriculture Markets

While women account for almost 50% of the agricultural labor force in sub-Saharan Africa, they are often constrained in their access to markets and price information, especially where engagement in markets involves travel and searching for customers in faraway villages, often with the risk of crime. Hence, if women themselves were able to access market knowledge through mobile phones and commodity exchanges, they would possibly benefit even more than men would. Hildebrandt et al. (2020), Aker and Ksoll (2016), and Gomez and Vossenberg (2018) focus on gender and show positive impacts on women, for the reasons just mentioned. There are also factors related to within-household bargaining and dynamics which yield benefits to women from technology. One can imagine more peace in the household as the price alerts provide proof of the sales one household member is able to obtain at the market from the household farm. Indeed, some farmers in the Hildebrandt et al. (2020) study mentioned that the Esoko price alerts were useful because with it, they said, "my wife will not cheat me." This was probably a situation where the man worked on the farm, the women did the selling, and there was little trust between the two. As a second example, after the integration of the Agricultural Commodity Exchange for Africa (ACE) in Malawi, 75% of the women interviewed by Gomez and Vossenberg (2018) started keeping books and carrying out some financial planning or budgeting. The warehouse receipt system allowed women to earn more bargaining power, which then benefited their farming opportunities. One of the quotes from that paper is as follows: "ACE helps me to make informed business decisions which are atypical for a woman."

References

Adjognon, G. S., Gassama, A., Guthoff, J. C., et al. (2019). Implementing effective warehouse receipt financing systems: Lessons from a pilot WRS project in the Senegal River valley. *The World Bank, 139445,* 1–28.

Aker, J.C. (2008). *Does digital divide or provide? The impact of cell phones on grain markets in Niger.* Center for Global Development Working Paper 154.

Aker, J. C., & Fafchamps, M. (2015). Mobile phone coverage and producer markets: Evidence from West Africa. *The World Bank Economic Review, 29*(2), 262–292.

Aker, J. C., & Ksoll, C. (2016). Can mobile phones improve agricultural outcomes? Evidence from a randomized experiment in Niger. *Food Policy, 60,* 44–51.

Aker, J. C., Ghosh, I., & Burrell, J. (2016). The promise (and pitfalls) of ICT for agriculture initiatives. *Agricultural Economics, 47*(S1), 35–48.

Allen, T. (2014). Information frictions in trade. *Econometrica, 82*(6), 2041–2083

Andersson, C., Bezabih, M., & Mannberg, A. (2017). The Ethiopian commodity exchange and spatial price dispersion. *Food Policy, 66,* 1–11.

Atkin, D., Khandelwal, A. K., & Osman, A. (2017). Exporting and firm performance: Evidence from a randomized experiment. *The Quarterly Journal of Economics, 132*(2), 551–615.

Bagwell, K. (2007). The economic analysis of advertising. *Handbook of Industrial Organization, 3,* 1701–1844.

Bai, J. (2018). *Melons as lemons: Asymmetric information, consumer learning and quality provision.* Working paper.

Balashova, E., & Sharipova, S. (2018). *Impact of ecosystem services on a sustainable business strategy in urban conditions.* In MATEC web of conferences 170:01012/.

Banerjee, A. V. (2013). Microcredit under the microscope: What have we learned in the past two decades, and what do we need to know? *Annual Review of Economics, 5*(1), 487–519.

Baulch, B., Gross, A., Nkhoma, J.C., et al. (2018). *Commodity exchanges and warehouse receipts in Malawi: Current status and their implications for the development of structured markets.* International Food Policy Research Institute Working Paper 25.

Baumüller, H. (2013). *Enhancing smallholder market participation through mobile phone-enabled services: The case of M-Farm in Kenya.* INFORMATIK 2013–Informatik angepasst an Mensch, Organisation und Umwelt. Bonn.

Bernard, T., de Janvry, A., Mbaye, S., et al. (2017). Expected product market reforms and technology adoption by Senegalese onion producers. *American Journal of Agricultural Economics, 99*(4), 1096–1115.

Courtois, P., & Subervie, J. (2015). Farmer bargaining power and market information services. *American Journal of Agricultural Economics, 97*(3), 953–977.

Demise, T., Natanelov, V., Verbeke, W., et al. (2017). Empirical investigation into spatial integration without direct trade: Comparative analysis before and after the establishment of the Ethiopian commodity exchange. *The Journal of Development Studies, 53*(4), 565–583.

Dillon, B., & Dambro, C. (2017). How competitive are crop markets in sub-Saharan Africa? *American Journal of Agricultural Economics, 99*(5), 1344–1361.

Dixie, G., & Jayaraman, N. (2011). Strengthening agricultural marketing with ICT. *The World Bank Module, 9,* 205–237.

Drott, D.Y., & Svensson, J. (2010). *Tuning in the market signal: The impact of market price information on agricultural outcomes.* Working Paper.

Duflo, E., Glennerster, R., & Kremer, M. (2007). Using randomization in development economics research: A toolkit. *Handbook of Development Economics, 4,* 3895–3962.

Eggleston, K., Jensen, R., Zeckhauser, R. (2002). *Information and communication technologies, markets, and economic development.* The global information technology report 2001–2002: Readiness for the networked world, pp. 62–74.

Esoko (2019). Deliver Information At The Right Time. https://esoko.com/messaging/

Fafchamps, M., & Minten, B. (2012). Impact of SMS-based agricultural information on Indian farmers. *The World Bank Economic Review, 26*(3), 383–414.

FAO. (2019). *The State of Food and Agriculture: Moving forward on food loss and waste reduction.* http://www.fao.org/3/ca6030en/CA6030EN.pdf. Accessed 15 May 2020.

Futch, M. D., & McIntosh, C. T. (2009). Tracking the introduction of the village phone product in Rwanda. *Information Technologies & International Development, 5*(3), 54–81.

Gomez, G., & Vossenberg, S. (2018). Identifying ripple effects from new market institutions to household rules-Malawi's agricultural commodity exchange. *NJAS-Wageningen Journal of Life Sciences, 84*, 41–50.

GSMA Mobile. (2018). *The Mobile Economy-Sub-Saharan Africa 2018.* https://www.gsma.com/mobileeconomy/sub-saharan-africa/. Accessed 15 May 2020.

Goyal, A. (2010). Information, direct access to farmers, and rural market performance in central India. *American Economic Journal: Applied Economics, 2*(3), 22–45.

Harrison, A., & Rodríguez-Clare, A. (2009). Trade, foreign investment, and industrial policy. In D. Rodrik & M. Rosenzweig (Eds.), *Handbook of development economics* (Vol. 5, pp. 4039–4214). Elsevier.

Hildebrandt, N., Nyarko, Y., Romagnoli, G. et al. (2020). *Price information, inter-village networks, and bargaining spillovers: Experimental evidence from Ghana.* https://papers.ssrn.com/sol3/papers.cfm?abstract_id=3694558

Höllinger, F., Rutten, L., Kiriakov, K. (2009). *The use of warehouse receipt finance in agriculture in ECA countries.* Paper presented at the World Grain Forum, St. Petersburg 6–7 June 2009.

Jensen, R. (2007). The digital provide: Information (technology), market performance, and welfare in the South Indian fisheries sector. *The Quarterly Journal of Economics, 122*(3), 879–924.

Kaminski, J., & Christiaensen, L. (2014). *Post-harvest loss in sub-Saharan Africa—What do farmers say?* The World Bank Policy Research Working Paper 6831.

Katunze, M., Kuteesa, A., Mijumbi, T., & Mahebe, D. (2017). Uganda warehousing receipt system: Improving market performance and productivity. *African Development Review, 29*(S2), 135–146.

Levi, R., Rajan, M., Singhvi, S., et al. (2020). The impact of unifying agricultural wholesale markets on prices and farmers' profitability. *Proceedings of the National Academy of Sciences, 117*(5), 2366–2371.

Minten, B., Tamru, S., Kuma, T., Nyarko, Y. (2014). *Structure and performance of Ethiopia s coffee export sector.* Paper presented at the international agricultural trade research Consortium's (IATRC's) 2014 annual meeting: Food, resources and conflict, San Diego, 7–9 December 2014.

Miranda, M. J., Mulangu, F. M., & Kemeze, F. H. (2019). Warehouse receipt financing for smallholders in developing countries: Challenges and limitations. *Agricultural Economics, 50*(5), 629–641.

Mire, M.M. (2019). *Effect of E-commerce on performance in agricultural sector in Kenya: A case of Twiga foods limited.* Dissertation, Doctoral dissertation, United States International University Africa.

Mitchell, T. (2017). Is knowledge power? Information and switching costs in agricultural markets. *American Journal of Agricultural Economics, 99*(5), 1307–1326.

Mitra, S., Mookherjee, D., Torero, M., et al. (2018). Asymmetric information and middleman margins: An experiment with Indian potato farmers. *Review of Economics and Statistics, 100*(1), 1–13.

Nakasone, E. (2013) *The role of price information in agricultural markets: Experimental evidence from rural Peru.* Paper presented at the Northeastern Universities Development Consortium, Boston University, Boston, November 2014.

Nakasone, E., Torero, M., & Minten, B. (2014). The power of information: The ICT revolution in agricultural development. *Annual Review of Resource Economics, 6*(1), 533–550.

Newman, N., Bergquist, L.F., Immorlica, N., Leyton Brown, K., et al. (2018). Designing and evolving an electronic agricultural marketplace in Uganda. In: *Proceedings of the 1st ACM SIGCAS conference on computing and sustainable societies, Menlo Park and San Jose*, June 2018.

Nyarko, Y., Pellegrina, H. (2020). *Commodity exchange markets in Africa*. Working Paper.
Ochieng, D.O., Botha, R., Baulch, B. (2020). *Market information and access to structured markets by small farmers and traders: Evidence from an action research experiment in Central Malawi*. International Food Policy Research Institute Working Paper.
Reda, A., Duong, Q., Alperovich, T., Noble, B., et al. (2010). Robit: An extensible auction-based market platform for challenged environments. In: *Proceedings of the 4th ACM/IEEE international conference on information and communication technologies and development, London*, 13–16 December 2010.
Saenger, C., Torero, M., & Qaim, M. (2014). Impact of third-party contract enforcement in agricultural markets – A field experiment in Vietnam. *American Journal of Agricultural Economics, 96*(4), 1220–1238.
Ssekibuule, R., Quinn, J.A., Leyton-Brown, K. (2013). A mobile market for agricultural trade in Uganda. In: *Proceedings of the 4th annual symposium on computing for development, University of Cape Town, Cape Town*, 6–7 December 2013.
Startz, M. (2016). The value of face-to-face: Search and contracting problems in Nigerian trade. Available at SSRN 3096685.
Svensson, J., & Yanagizawa, D. (2009). Getting prices right: The impact of the market information service in Uganda. *Journal of the European Economic Association, 7*(2–3), 435–445.
Verhoogen, E. A. (2008). Trade, quality upgrading, and wage inequality in the Mexican manufacturing sector. *The Quarterly Journal of Economics, 123*(2), 489–530.
World Food Program. (2010). *Cambodia food market analysis and survey report*. https://www.wfp.org/countries/cambodia. Accessed May 15 2020.
Zant, W. (2019). *Mobile phones and Mozambique traders: What is the size of reduced search costs and who benefits?* Tinbergen Institute working paper TI 19-047/V.

Chapter 10
Fintech for Rural Markets in Sub-Saharan Africa

Jenny C. Aker

1 Development Challenge

Since the 1990s, cash transfer programs have been an important part of social protection policies in low-income countries. As of May 2020, approximately 159 countries had 700 types of social protection programs in place, over 200 of which were cash-based measures (Gentilini et al., 2020). While implementing such programs raises numerous challenges, from targeting, to funding, to choosing the modality, one key issue is distribution. In higher- and middle-income countries, such programs are often implemented electronically, via either bank transfers or prepaid debit cards. Yet in lower-income countries with limited financial infrastructure, social programs often require physically distributing cash in small denominations to remote rural areas. Globally, 1.7 billion adults remain unbanked, without an account at a financial institution or through a mobile money provider (Demirguc-Kunt et al., 2017). This lack of access to financial services not only increases the logistical challenges associated with implementing cash transfer programs but also potentially creates substantial direct and indirect costs for program recipients. This is especially the case in sub-Saharan Africa, where money transfer costs are among the highest in the world (World Bank, 2017). In such an environment, how can cash transfers or salaries be distributed more efficiently using digital technologies? In addition to efficiency, can digital transfer mechanisms improve the welfare of program recipients along other dimensions? Could it benefit or disadvantage particular subgroups? And could public investments in transfer infrastructure lead to spillovers for person-to-person (P2P) transfers, especially in an area of the world where remittances represent 2.5% of GDP (World Bank, 2018)?

J. C. Aker (✉)
Tufts University, Medford, MA, USA

© The Author(s) 2023
T. Madon et al. (eds.), *Introduction to Development Engineering*,
https://doi.org/10.1007/978-3-030-86065-3_10

These questions are at the heart of this case study on Niger, a landlocked country in the Sahelian zone of West Africa and one of the poorest countries in the world. The Sahel is the transition area between the Sahara Desert to the north and the tropical zones to the south, receiving approximately 200–800 millimeters of rainfall per year. The region has witnessed some of its most serious climate-induced food shortages in the 1970s and 1980s, with approximately 250,000 drought-related human fatalities occurring in the 1960s, 1970s, and 1980s. Since that time, Niger has been subject to frequent droughts, the most recent of which occurred in 2018 (OCHA, 2018).

Given that agricultural production in Niger is primarily rainfed, with a unimodal distribution, inter-annual deviations in rainfall are strongly correlated with fluctuations in agricultural output, income, and food security. If financial markets are performing optimally, then households could save or borrow to cope with such shocks. Yet, given the limited access to financial services – in fact, Niger is one of the most financially excluded countries in the world – such strategies are difficult to implement, especially for the rural poor (Collins et al., 2009; Karlan & Morduch, 2010; Dupas & Robinson, 2013; Rutherford, 2000). As a result, rural households often rely upon external assistance – whether remittances or external aid – to cope with idiosyncratic and covariate shocks, both of which require money transfers.

This case study focuses on a particular crisis in Niger, the 2009/2010 drought and corresponding harvest failure, which affected more than 2.7 million people (FEWS NET, 2010). In response to this crisis, governmental and nongovernmental organizations (NGOs) implemented a series of social protection programs, including food aid, vouchers, and cash transfers. While seemingly simple, the context was not: At the time, Niger had one bank for every 100,000 people (Demirguc-Kunt et al., 2017), few paved roads, and small-scale conflicts along the Niger-Mali and Niger-Nigeria borders. Thus, the Nigerien government and NGOs typically distributed cash transfers manually, placing cash into individual envelopes and transporting it with armed security forces into remote rural areas (Aker et al., 2016).

In January 2010, a relatively new technology – mobile money – was introduced into the country by one mobile network operator (MNO), known as "Zap." Similar to M-PESA in Kenya, the mobile money product allowed users to transfer money via a text-based system on their phone and pick up their cash at a local agent. In the context of the food crisis, the technology offered a unique opportunity: Rather than physically distributing cash to thousands of beneficiaries, the government and NGOs could disburse cash transfers electronically via the mobile money system. This digital transfer system could potentially (1) reduce the transfer costs for the implementing agency and program recipients, thereby improving program coverage and outreach; (2) lead to other improvements in program recipients' well-being, primarily due to time savings; and (3) allow program recipients to use mobile money for remittances, an important income source in Niger (Aker et al., 2020a, b; Jack & Suri, 2014). To explore the feasibility, cost, and impact of using mobile money for cash transfer programs, researchers collaborated closely with an NGO, Concern Worldwide, to design and implement a randomized control trial (RCT) across 96 villages in one region of Niger.

Between 2010 and 2020, many of the conditions under which this case study took place have not significantly changed in Niger. Niger is still one of the most financially excluded countries in the world (Demirguc-Kunt et al., 2017), and climate-related shocks and food crises are relatively common. These shocks have been further aggravated by the escalation of armed conflict: As of 2019, there were 4.5 million people displaced and 12.2 million people suffering from food insecurity within the West Africa region. While mobile money exists in Niger, with products offered by three different MNOs, its adoption has not taken off as predicted. M-money adoption was estimated at 9% as of 2017 (Demirguc-Kunt et al., 2017), with relatively lower rates in rural areas, despite mobile ownership rates of over 80% (Aker et al., 2020a, b). Thus, similar challenges as those encountered in 2020 were also apparent in 2018, during a second cash transfer program.

It is easy to dismiss the low rates of mobile money adoption as specific to Niger, with limited relevance to other contexts. Yet, many of these statistics are mirrored in the West Africa region. While average mobile phone adoption in the region is 67% and there are 59 mobile money deployments, mobile money has been slower to take off as compared with East and Southern Africa. For example, despite the fact that there were 163 million accounts in 2019, there is significant heterogeneity in adoption across and within countries (GSMA, 2019). In addition, there is a stark contrast between adoption and usage; out of the total number of registered accounts, 34.6% have shown some activity, with the number of active users ranging from 1% in Niger to 20% in Ivory Coast (Fig. 10.1). This appears to be due, in part, to the limited mobile money agent infrastructure, limited interoperability between MNOs within and across countries, and regulatory frameworks in place (Aker et al. 2020a, b; CGAP, 2016). As a result, mobile money has not yet become the transformative technology in countries such as Niger, nor for some of its neighbors, despite potentially high demand for the service.

Despite these caveats, digital financial services may still offer significant potential for small-scale farmers to save, invest, and smooth consumption in West Africa. Since 2012, a number of "second-generation" digital financial services – namely, digital credit, savings, and insurance – have proliferated in East and Southern Africa. As of 2015, 20% of Kenyans were using Safaricom's M-Shwari digital credit and savings product (Cook & McKay, 2015). There are numerous concerns that have been raised with digital credit products, such as high effective interest rates (CGAP, 2016), as well as high delinquency and default rates. While rigorous evidence of their impact is nascent, early studies in this area suggest that such digital products enabled households to smooth consumption in the face of shocks and encourage short-term savings, although they have not specifically studied agricultural outcomes (Bharadwaj & Suri, 2020; Bharadwaj et al., 2019). While MNOs in West Africa primarily offer first-generation digital financial services – i.e., mobile money – there have been more recent digital credit deployments in countries such as Benin, Ghana, Ivory Coast, and Nigeria. Existing studies in this area promise to shed new light on their impacts in the years to come.

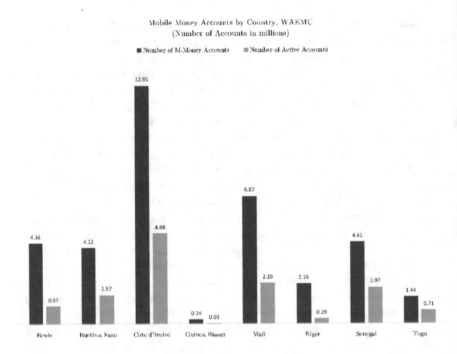

Fig. 10.1 Number of accounts and active accounts per population per country

2 Implementation Context

Niger is one of the largest countries in sub-Saharan Africa, with relatively limited access to roads, financial infrastructure, or electricity. The first mobile money system in Niger was introduced in January 2010. Known as "Zap," the product was developed by one of the MNOs, Zain (later Bharti Airtel). Like most mobile money systems, Zap allowed users to store value in the mobile money account, convert cash in and out of the account, and make transfers by using a set of text messages personal identification numbers (PINs) (Aker & Mbiti, 2010). The cost of making a US$ 45 transfer using Zap was US$ 3 in 2010. Initial coverage, usage, and growth of Zap were limited and geographically focused in the capital city (Niamey) and regional capitals.

Given the context, there were a number of challenges to designing, implementing, and evaluating a mobile money cash transfer program. The first of these was *mobile phone ownership*: While mobile phones were initially introduced in Niger in 2000 and had grown substantially between 2000 and 2010, adoption rates were at 30% by 2010. Although there were high rates of phone sharing within and across households, the nature of the cash transfer program – which targeted vulnerable households within villages and targeted women within the household – meant that mobile phone ownership was a potential constraint to implementation.

Second, beyond the issue of mobile phone ownership, few households in Niger – and specifically in the study region – *knew about (or had used) mobile money*. Since mobile money had only been introduced in January 2010 and the first transfer was scheduled to take place in May 2010, adoption in remote rural areas was less than 1%. This not only meant that households were not registered for mobile money – a process that required some type identification – but they also did not have the special SIM required for the platform.

Third, the mobile money platform was text- and number-based: The program recipient would receive a text notifying her of the transfer with the amount and was required to remember a four-digit PIN number in order to pick up the transfer from the agent. This was therefore the fourth *challenge: Niger had, and still has, some of the lowest literacy rates in the world*, with an average literacy rate of less than 30% and an average of 2 years of completed schooling (Aker & Ksoll, 2019). In our study area, 58% of women had attended some school, but literacy rates among women were less than 15%. This led to challenges in manipulating the mobile phones, as well as recalling PIN codes.

The final challenge was related to the density of *mobile money agents*: As a new product, there were few agents located outside of the capital city, and there were no mobile money agents in the study area. Thus, while the mobile money product had the potential to reduce the costs associated with distributing the cash transfer for the NGO, it also had the potential to increase the costs for the program recipients, essentially shifting the risk to the private sector: the MNOs and the agents.

Designing and implementing the program and corresponding research required close consultation and collaboration with a diverse set of stakeholders: (1) the *NGO*, Concern Worldwide, who was the implementing agency for the cash transfer program; (2) *the 116 villages* who were part of the cash transfer program and were responsible for identifying vulnerable households within the community; (3) *Zain*, the MNO, who was the only mobile money operator at the time; (4) *local traders and retailers*, who were the primary mobile money agents in the region; (5) the local data collection firm, *Sahel Consulting*, and *Tufts University*, who were jointly responsible for designing the research and data collection during the evaluation; and (6) the *Ministry of Social Protection*, who was responsible for overseeing and coordinating diverse cash transfer interventions during the food crisis of 2009–2010.

Close collaboration among these different sets of stakeholders allowed for creative resolution of the four challenges identified above. In order to address the issue of low mobile phone ownership, it was decided to purchase simple mobile phones for program recipients. The team also discussed how the provision of mobile phones might affect program recipients' behavior and hence the research results, which led to the modification of the original research.

In order to address the challenges related to *mobile money awareness and literacy*, Concern Worldwide, the MNO, and researchers collaborated along two key dimensions. First, in addition to the mobile phones, program recipients also received SIM cards that were "Zap-enabled," meaning that that could be registered for (and use) the Zap product, and second, Tufts University and Concern Worldwide developed a training manual and corresponding trainings on how to use mobile

Fig. 10.2 Placeholder

money in a low-literate environment (Fig. 10.2). Building upon research done by Tufts University on a mobile phone literacy project in Niger, Concern Worldwide also distributed a mobile phone poster, which allowed program recipients to find their PIN code on the handset, as well as memorize the number (Fig. 10.3).

Finally, in order to address the issue of *agent network*, Concern Worldwide worked closely with Zain to identify potential agents in the region, in particular by informing Zain of the location of the cash transfer villages. In addition, Concern Worldwide notified Zain of the timing of the cash transfers in advance, so that agents would have sufficient liquidity for program recipients to cash out.

3 Innovate, Implement, and Evaluate

3.1 Innovation

In light of the high costs involved in distributing cash transfers in Niger, the introduction of mobile money offered a new mechanism for disbursing cash transfers to food-insecure households. The starting point for the intervention was

Fig. 10.3 Mobile phone
poster

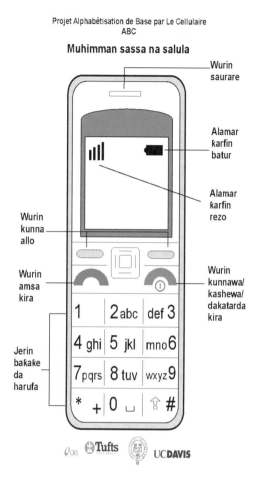

therefore two models of distributing cash transfers – manual and electronic cash (in-person) – and adapting this to the particular context.

The cash transfer intervention in this context was relatively simple: Program recipients among 116 food-insecure villages of the Tahoua region of were provided with a monthly transfer of US$ 45 over a 5-month period, for a total of US$ 225. The transfer was provided on the hungry season (from May to September), in the hopes that this would reduce the likelihood of more severe food insecurity, malnutrition, and the distress sale of assets. While 116 villages were eligible for cash transfers, based upon drought and production data, 20 villages were removed from the evaluation sample, as they either did not have mobile phone coverage (and hence were not eligible for cash transfers via mobile money) or were in highly insecure areas (and hence were not eligible for cash transfers in person). This therefore left a sample of 96 villages for the evaluation.

Eligible households within each village were identified by a village-level vulnerability exercise. Using indicators such as livestock ownership, landholdings, and the number of household members, households were classified into four vulnerability categories (A, B, C, and D), with C and D households selected for participation in the cash transfer program (Aker et al. 2016). The number of program recipient households per village ranged from 12 to 90% of the village population, with an average of 45% (Aker et al., 2016). All targeted households were scheduled to receive the same amount, at about the same time, each month.

Villages were assigned to one of three innovation models:

1. *Manual cash*, whereby program recipients in a given village received the cash transfer in the village or via a nearby village, using the standard model of cash delivery. Program recipients thus received a beneficiary card and were required to travel to the cash delivery point on a specific day of each month. At the cash delivery point, program recipients had to wait in line and have their identity verified before receiving their cash in an envelope.
2. *Manual cash with a Zap-enabled mobile phone*, whereby the program recipients received their cash in a similar mechanism as above but also received a Zap-enabled mobile phone, worth approximately US$ 5. The mobile phone had a Zain SIM, and program recipients could use mobile money if they wished, but they did not receive their transfer via mobile money. They also received the training on how mobile money worked, as explained below.
3. *Zap transfer*, whereby program recipients received the Zap-enabled mobile phone (as was the case in the second model) but received their transfer via the mobile phone. This involved not only distributing the phone to households (with the Zain SIM) but also conducting an interactive training with households to explain how mobile money worked and what they could expect. This model also required collaboration between Concern Worldwide and Zain to create a web-based, password-protected interface with program recipients' phone numbers, transferring money to a bank account connected to the Zap account, identifying and verifying program recipients, uploading an encrypted file onto Zain's system (so that they would not have program recipients' personal identifying information), and sending the cash transfer via SMS to program recipients' Zap accounts.

Among these three models, the primary innovation of interest was the third one. While mobile money could have affected household outcomes in a variety of ways, the primary hypothesis behind this research was related to transfer costs. In other words, by providing the cash transfer via mobile money, it was hypothesized that this would reduce program recipients' costs in obtaining the transfer, in terms of both transport and waiting time. It was further hypothesized that mobile money would affect household outcomes in the following ways:

- The reduction in transfer costs would allow program recipients to invest time in other productive activities during the planting period, as well as change the timing and location of purchases, particularly if they were able to purchase food and nonfood items from agents.

- The introduction of mobile money would allow program recipients to use mobile money for private (person-to-person) transfers, therefore increasing the amount of remittances available from migrants, as well as allowing remittances to arrive when they were needed the most (Jack & Suri, 2014).
- Because the mobile money cash transfer mechanism was more private than the manual cash transfer (as program recipients only received a discreet "beep" letting them know that the transfer had arrived), this could have allowed program recipients – all of whom were women – to have more control over the cash transfer resources.

3.2   Implementation

The design and implementation of the above interventions were developed collaboratively between Concern Worldwide and Tufts University. Initially, Concern Worldwide only wanted to compare two interventions: the manual cash group and the mobile money cash transfer group. When it was realized that mobile phone ownership was only 30% among the target population and that mobile money adoption was essentially zero, the teams quickly realized that improving program recipients' access to mobile phone technology (by providing mobile phones), as well as the mobile money technology (by facilitating registration, SIM cards, and trainings), was required.

The provision of Zap-enabled mobile phones to the mobile money cash transfer recipients therefore required one primary modification to the initial interventions. The first was the addition of the "cash transfer plus mobile phone" intervention group (model no. 2). Since the Zap program recipients were supposed to receive *the mobile phone plus the cash transfer via the mobile phone*, this would imply that there were two differences with the manual cash transfer approach: the mobile phone and the use of mobile money for the cash transfer. If the mobile phone on its own improved program recipients' welfare – either by improving communication on agricultural prices or allowing households to increase access to private transfers – then it would be difficult to disentangle the impacts of the mobile phone from the impacts of the mobile money product. By comparing the manual cash group with the manual cash plus mobile phone group, we were able to answer the question "Conditional on receiving a manual cash transfer, what is the additional impact of the mobile phone?" Then, by comparing the second intervention group with the third, we were able answer the question "What is the additional impact of receiving cash via mobile money?" The addition of the second intervention group created numerous discussions, as this required additional resources (cash to purchase the mobile phones) and trainings. The researchers and Concern Worldwide decided, in consultation, that the primary objective of the research was to measure the impact of the new mobile money transfer technology and that the impact of the mobile phone needed to be netted out.

The second modification was related to the availability of agents in the targeted region. Despite intense work with Zain to encourage them to register agents in the region, the MNO was unable to register a sufficient number of agents by the time of the first cash transfer. As a result, one agent distributed cash to 32 mobile money villages for the first transfer. After additional discussions with Zain, the company was able to register more agents in the region, based upon their own criteria for choosing suitable agents. One key concern, however, was that these agents were equitably distributed across all 96 villages in the evaluation sample, rather than simply focusing on mobile money villages, in order to minimize differences between the mobile money group and the manual cash transfer groups. This was verified during the evaluation stage; the number and density of Zap agents was similar across all groups, without a statistically significant difference between the two. This implied, therefore, that any differences observed between Zap and manual cash villages would not be driven by the differential presence of mobile money agents.

3.3 Evaluation

In order to measure the impact of the mobile money cash transfer innovation on outcomes of interest, we used an RCT at the village level. The 96 evaluation villages were first stratified by administrative division and randomly assigned to one of the three cash transfer innovations, with 32 villages in each group. The primary outcomes measured were those in the original theory of change: (1) transfer costs, both for the implementing agency and for program recipients, including when, where, and how they obtained their cash; (2) uses of the cash transfer, including the different categories; (3) welfare measures associated with the cash transfer, including food security, diet diversity, and nutritional outcomes; and (4) indicators related to mechanisms, in particular related to access to remittances, as well as intra- and inter-household sharing of transfers.

The evaluation collected a wealth of data, including a baseline (May 2010), a midline soon after the transfers (December 2010), and a final round 1 year later (May 2011). The data were a panel dataset, with the primary program recipient as the survey respondent. For each survey found, intensive survey piloting was done; the survey was first written, with the team trained, and then piloted in the field at least three times before being deployed. In addition, before the midline and final survey rounds, qualitative data were also collected before the quantitative surveys, in order to gain insights into impacts that were not initially expected in the initial theory of change. This led to some useful insights about the observability of the transfer within the household (as women wore the mobile phones around their necks and reported that only they knew of its arrival) and a module on intra-household decision-making. In the current context of using pre-analysis plans (PAPs) for rigorous evaluations, these findings may not have been fully integrated into the study or made it into the final paper.

Overall, the evaluation had six key findings:

- The *marginal costs of the mobile money cash transfer were 20% lower than the costs of distributing the cash transfer manually*, but the fixed costs were substantially higher, primarily due to the purchase of the mobile phones. Unlike other studies on digital transfers, we do not find evidence that the m-transfer mechanism had any impacts on leakage (Muralidharan et al., 2016).
- *Mobile money program recipients traveled shorter distances to obtain their transfer* as compared with their manual cash counterparts (both cash only and cash plus mobile). While the manual cash and mobile program recipients traveled an average of 4 km (round-trip) from their village to obtain the transfer, Zap program recipients traveled 2 km to the nearest agent. This is equivalent to a travel time savings of approximately 1 h for each cash transfer or 5 h over the entire program. However, this analysis excludes the cash program recipients' waiting time, which averaged 4 hours per cash transfer, as compared with 30 minutes for Zap recipients, equivalent to a savings of 2.5 days (Fig. 10.4).
- *Mobile money cash transfer recipients that used their cash transfer to buy more diverse types of goods were more likely to purchase protein- and energy-rich foods.* These diverse uses of the transfer resulted in a 9–16% improvement in diet diversity as compared to the cash and mobile groups, primarily associated with the increased consumption of beans and fats. In addition, children in the Zap group consumed an additional one-third of a meal per day (Aker et al., 2016). Yet, Zap households did not reduce their ownership of other durable and

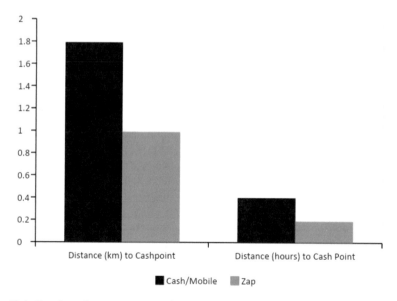

Fig. 10.4 Travel cost by treatment group Source: Aker et al. (2016)

nondurable goods, suggesting that other household members were not decreasing their contribution to household goods as a result of the transfer.

- *These results can be partially explained by two factors: 1) Zap program recipients spent less time on obtaining the transfer* (Fig. 10.4), *and 2) female program recipients in Zap villages had increased bargaining power within the household.* In terms of time savings: While the magnitude was relatively small – approximately 2.5 days over a 5-month period – this is a probably lower bound on actual time savings. In addition, the savings occurred during the planting season, a time when opportunity costs were high, implying that the time savings could have enabled Zap program recipients to engage in other productive activities or invest more time in childcare. There was some suggestive evidence in support of the former channel: m-transfer households were more likely to cultivate marginal cash crops primarily grown by women (Aker et al., 2016). In terms of intra-household decision-making, program recipient reported that mobile money was less observable to other household members, thereby allowing them to temporarily conceal the arrival of the transfer (Aker et al., 2016). This was supported by proxy measures of intra-household decision-making: Zap program recipients were more likely to travel to weekly markets, spend more on children's clothing, and maintain the improved diet diversity results 6 months after the program, well after the cash transfer had been spent (Aker et al., 2016).
- *While program recipient households used mobile money to receive their transfer, they did not use it to receive remittances or to save, and there were no significant differences in the frequency or amount of remittances received.* These results were perhaps not surprising, as the agent network was not widespread at the time, and the mobile money system could not be used for transfers to Nigeria, the destination for a majority of migrants.
- *There were no differences in costs, expenditures, or other measures of well-being between the cash transfer or mobile phone plus cash transfer groups*, suggesting that the mobile phone (on its own) had no additional effect on household welfare.

While these results are promising, there are several limits to the generalizability of these results. First, our case study studied the impact of different transfer mechanisms during a food crisis, when the marginal utility of income can be high. And as a result, uses of the cash transfer could be more diverse and less focused on food items in other contexts. And second, since Niger is one of the poorest countries in the world, with low rates of literacy, financial inclusion, and mobile money adoption, the context might be different from other countries with higher rates of financial inclusion and a more thoroughly developed mobile money infrastructure, especially those in East Africa. Nevertheless, Niger's educational, financial, and mobile money indicators are not vastly different from other Sahelian countries in West Africa, including Burkina Faso, Mali, northern Ghana and northern Ivory Coast, suggesting that these results might be informative for those contexts (GSMA, 2019; Findex, 2017).

3.4   Adaptation

In 2018, the question of how to safely and efficiently distribute cash transfers arose in the context of another project in Niger, one which studied the impact of training and cash transfers on the adoption of an environmental technology (Aker & Jack, 2020). The study took place across 180 villages in the Zinder Region, with 110 villages assigned to receive either conditional or unconditional cash transfers. Overall, 1750 program recipients were supposed to receive a one-time cash transfer between April and June 2018. Located in the far east of the country, the villages were relatively close to the Nigerian border and had relatively high rates of migration and mobile phone ownership (over 60%). The project was a collaborative effort between Tufts University (the research lead), Sahel Consulting (the data collection firm), and the Ministry of Environment, neither of whom had the capacity to distribute cash transfers manually.

Based upon the Zap research in 2010, the team understood that distributing cash transfers via mobile money required substantial investment and that several preconditions were necessary: mobile phone ownership, mobile money usage, and a mobile money agent. Despite high rates of migration and intense demand for money transfers, fewer than 3% of households had used mobile money as of 2017 (Aker et al., 2020c). In addition, there were only four mobile money agents in the entire region. This suggested that neither mobile money infrastructure nor adoption had changed significantly since 2010.

In light of these conditions, the stakeholders generated four options for distributing the cash transfers: (1) the manual distribution of cash transfers by Sahel Consulting, similar to the role of Concern Worldwide in 2010; (2) the electronic distribution of cash transfers via local money transfer providers, similar to Western Union and MoneyGram; (3) the electronic distribution of cash transfers as airtime credit, which program recipients could then convert into cash; or (4) the electronic distribution of cash transfers via mobile money.

Options no. 1 and no. 2 were rejected by the implementation partners. The first option (manual distribution) was rejected as too risky, as it would require transporting US$ 35,000 in cash to remote rural areas, and security would be needed to reduce the likelihood of theft. The second option (electronic distribution via local money transfer companies) was rejected because it was not a relative improvement over mobile money: The location of the money transfer agents was similar to the location of mobile money agents, and the money transfer agents could not guarantee that they would have sufficient liquidity on hand.

As a result, the stakeholders focused their discussion on options no. 3 and no. 4, each of which had advantages and disadvantages. As each stakeholder had different opinions on the relative merits of each option, the team used a technique called "Analytical Hierarchical Process" (AHP) to weight the options and make a decision (Saaty, 1980; Leal, 2020). In essence, this approach involved stating the goal of the exercise (finding the best cash transfer mechanism) and the criteria as to how this decision would be made (viz., the costs to the program recipients, the number of agents, the knowledge of the technology, and the risks to the implementing agency). Each individual on the team assessed each option along each criteria and used this to come up with a ranking. The rankings are all compared, and weights were developed.

Based upon this exercise, option no. 4 scored relatively higher on a number of criteria, in particular the lower risk to the Ministry and Sahel Consulting, as well as the program recipients. As a result, the team decided to implement option no. 4 but with a slight modification. In essence, this involved sending the money via the mobile money to the program recipient's mobile phone number, as this did not need to have a mobile money account (called "envoie code"). The recipient received a code and was able to take this code to the nearest mobile money agent to pick up the money. If the program recipient did not have a mobile phone, then they were asked to provide the number of someone whom they trusted. This modification did not require providing mobile phones, nor registering program recipients on the mobile money platform, as was necessary in 2010. With some additional monitoring – namely, by calling program recipients and the village chief and working closely with the MNO – over 98% of program recipients received their cash transfer and received the full amount of the cash transfer.

4   Results and Lessons Learned

Overall, this case study shows how the introduction of a new digital technology could be harnessed to quickly distribute cash transfers in the context of a slow-onset emergency, where access to financial institutions and money transfer providers is limited. While the original case study took place in 2010, given the fact that over 100 million adults received their cash transfers manually as of 2017 – as well as the growth in cash transfer programs in response to the COVID-19 crisis – the results provide a number of lessons learned for cash transfer programs for the unbanked in urban and remote rural areas. In the 10 years since this study has taken place, there have been a number of other studies using digital platforms to provide cash transfers (Muralidharan et al., 2016; Haushofer & Shapiro, 2016), showing reductions in transfer costs, leakage, and improvements in other welfare measures. Many, yet not all, of these studies have taken place in certain contexts, such as Kenya and India, which have robust and well-developed digital technology systems.

Outside of these areas, several things are clear. First, mobile money offers significant opportunities to distribute cash transfer programs at scale, especially among the poor, who tend to be less likely to have a bank account. Nevertheless, it requires significant increases in the density of mobile money agents, as well as increases in mobile money adoption among the unbanked, which has been a constraint in many countries, especially in West Africa. Increasing the density of mobile money agent may require some innovation on behalf of regulators, banks, and mobile phone operators to register different entities as mobile money agents. If the number of agents cannot be increased in the short term, then distributing such transfers can impose higher costs on cash transfer recipients, as well as added risks associated with a "rush" on mobile money agents. Beyond increasing the number of agents, creative solutions may also be required to encourage mobile money adoption, especially among the urban poor, either by having a more flexible approach to registration or by using a technology that allows a user to send money to a nonmobile money user, similar to our work in 2018. These issues, of course, will need to be balanced with concerns regarding corruption and leakage.

There is significant room for more research on the constraints to the growth of digital financial services (Fintech) in many countries in sub-Saharan Africa, especially in West Africa. While digital credit, savings, and insurance have taken off in countries such as Kenya, Uganda, Rwanda, and Tanzania, many of the digital financial services (DFS) products are still "first generation" in West Africa. This could be due to four related and interconnected issues: (1) the regulatory framework in West Africa, which leads to competition between the "bank-led" and "MNO-led" models of mobile money; (2) the effect of the regulatory framework on MNOs' interest in, and profits from, mobile money; (3) the incentives provided to mobile money agents, which therefore reduces their number and activity; and (4) the lack of interoperability of mobile money products within and across West African countries, which affects the potential for its use for remittances, the main driver of demand in the region.

5 Summary and Interpretive Text Boxes

Gender

The cash transfer program targeted women within the household. The potential assumption of implementing partners was that this targeting criteria might increase women's bargaining power within the household, yet did not consider cultural differences across different ethnic groups in terms of women's access to markets. As a result, a number of proxy measures for intra-household decision-making were developed.

Failure

A clear implicit criteria for failure for the mobile money technology is whether households used the technology beyond the immediate need to "cash out" – ie, either by saving cash on their mobile phone or by using mobile money for P2P transfers. None of these occurred in this context, and mobile money adoption was not sustained in the longer-term, nor was the presence of mobile money agents.

Responsible Research

The clear ethical responsibility in this context was the principle of Do No Harm; recipient households were extremely vulnerable and in the midst of a food crisis. Hence, it was decided from the outset that no pure control group would be included in the experiment; in other words, every intervention village received a cash transfer. In addition, this also meant that every effort was made to limit the time burden of surveys on recipient households. These modifications meant that the research could not answer the question "What is the impact of the cash transfer program?", but rather, what is the impact of this cash transfer mechanism? In addition, the research could not answer questions related to expenditures and consumption, as these data were not collected. This principle was agreed upon at the outset by all involved.

6   Discussion Questions

- Did you think that the current model of distributing cash was a problem in Niger prior to the program? If so, why? If not, why not?
- Do you think that providing mobile phones was necessary in this context?
- What other factors might have been taken into consideration to increase the sustainability of the adoption and usage of mobile money in the medium and long term?
- Should the identification and registration of mobile money agents be left solely to the responsibility of the MNO or in collaboration with the public sector? What is the best way to ensure collaboration?

References

Aker, J. C., & Jack, B. K. (2020). *Harvesting the rain: The adoption of environmental technologies in the Sahel*. Unpublished working paper.

Aker, J. C., & Ksoll, C. J. (2019). Call me educated: Evidence from a mobile phone experiment in Niger. *Economics of Education Review, 72*(2019), 239–257.

Aker, J. C., & Mbiti, I. M. (2010). Mobile phones and economic development in Africa. *Journal of Economic Perspectives, 24*(3), 207–232.

Aker, J. C., Boumnijel, R., McClelland, A., & Tierney, N. (2016). Payment mechanisms and anti-poverty programs: Evidence from a mobile money cash transfer experiment in Niger. *Economic Development and Cultural Change, 65*(1).

Aker, J. C., Prina, S., & Welch, J. (2020a, May). Migration, money transfers and mobile money. *AEA Papers and Proceedings, 110*, 589–593.

Aker, J. C., Sawyer, M., Goldstein, M., O'Sullivan, M., & McConnell, M. (2020b, April). Just a bit of cushion: The role of a simple savings device in planned and unplanned expense events in Rural Niger. *World Development, 128*

Aker, J. C., Prina, S., & Welch, J. (2020c). Migration, money transfers and mobile money: Evidence from Niger. *AEA Papers and Proceedings, 110*, 589–593.

Bharadwaj, P., & Suri, T. (2020). Improving financial inclusion through digital savings and credit. *American Economic Review Papers and Proceedings*.

Bharadwaj, P., Jack, W., & Suri, T. (2019). *Fintech and household resilience to shocks: Evidence from digital loans in Kenya*. Working paper.

CGAP. (2016). *Market system assessment of digital financial services in WAEMU*. World Bank.

Collins, D., Morduch, J., Rutherford, S., & Ruthven, O. (2009). *Portfolios of the poor: How the world's poor live on two dollars a day*. Princeton University Press.

Cook, T., & McKay, C. (2015, April). *How M-Shwari works: The story so far*. Access to Finance Forum, 10.

Demirguc-Kunt, A., Klapper, L., Singer, D., Ansar, S., & Hess, J., (2017). *The Global Findex database 2017: Measuring financial inclusion and the Fintech revolution*. The World Bank.

Dupas, P., & Robinson, J. (2013, June). Why don't the poor save more? Evidence from health savings experiments. *American Economic Review, 103*(4), 1138–1171.

Famine Early Warning Systems Network (FEWS NET). (2010). *Niger Food Security Outlook 2010*. FEWS NET.

Gentilini, U., Almenfi, M., & Dale, P., Demarco, G., & Santos, I. (2020). *Social protection and jobs responses to COVID-19: A real-time review of country measures*. World Bank working paper.

GSM Association. (2019). *State of industry report on mobile money: 2019.*

Haushofer, J., & Shapiro, J. (2016, November). *The Quarterly Journal of Economics, 131*(4), 1973–1204.

Jack, W., & Suri, T. (2014). Risk sharing and transactions costs: evidence from Kenya's mobile money revolution. *American Economic Review, 104*(1), 183–223.

Karlan, D., & Morduch, J. (2010). Access to finance. In *Handbook of development economics.* Elsevier.

Leal, J. E. (2020). A simplified version of the analytical hierarchy method. *Computer Science.*

Muralidharan, K., Niehaus, P., & Sukhtankar, S. (2016). Building state capacity: Evidence from biometric smartcards in India. *American Economic Review, 106*(10), 2895–2929.

OCHA. (2018). *Niger: Food assistance fact sheet.* Washington, D.C..

Rutherford, S. (2000). *The poor and their money.* Oxford University Press.

Saaty, T. L. (1980). Decision-making, scaling and number crunching. *Decision Sciences, 20*(2), 404–409.

World Bank Group. (2017). Remittance prices worldwide. Issue 21.

World Bank. (2018). *Migration and remittances: Recent developments and outlook.* Migration and development brief 29.

Part III
Expanding Human Potential: Technology-Based Solutions for Education and Labor

Temina Madon ⓘ

Introduction

For many people, our time (or labor) is the greatest asset we have to invest. We may operate in informal markets, working as self-employed micro-entrepreneurs, or we may hold jobs in the formal sector. But if we live in poverty, we likely consume whatever we earn, leaving little for savings or household investment.

One path to a better life is social protection: the government policies and programs that address poverty by ensuring a basic standard of living for all. But in countries without the resources to finance safety nets, the most common path out of poverty is to boost human potential, equipping people with the skills to participate in a global, market-based economy. It is therefore unsurprising that many economic development initiatives have focused on education and skills development, alongside job creation and entrepreneurship (World Bank, 2019).

Yet this raises several fundamental questions. What are the most cost-effective ways to invest in people's potential—both during youth and after they have transitioned into adulthood? What tools are needed to connect people with the jobs or enterprises they are prepared for and passionate about, so that society can make productive use of their available labor? And once people have found jobs or entrepreneurial opportunities, how can we ensure the safety and dignity of their work? The case studies in this section will address each of these questions, in turn.

Improving the quality of education is a clear opportunity for innovation. Education is often viewed as a human right; as a result, spending on schooling is a large share of total government expenditure, particularly in low- and middle-income countries. An extreme example is Sierra Leone, which since 2018 has spent a third of total public sector expenditure on education (UNESCO). Despite these investments, classroom resources in the poorest regions are still limited, with policy

T. Madon
Center for Effective Global Action, University of California, Berkeley, Berkeley, CA, USA
e-mail: temina@ocf.berkeley.edu

makers struggling to identify the most cost-effective ways to improve learning outcomes. Public schools are often poorly managed, with limited investment in teacher supervision and professional development, and teachers often fail to instruct students at the right level (Angrist et al., 2020). Yet, it is these government-run schools that serve the most disadvantaged youth.

In spite of these challenges, engineers have made substantive advances in education, through both hardware and software innovations. Indeed "edtech" has grown into a profitable sector in some developing countries. However, the products with greatest adoption (and benefits) often target upper-middle and upper class households. There are fewer solutions tailored for use by teachers within school settings, and where they have been deployed, classroom technologies have a mixed record. There are numerous examples of computer-based learning exacerbating gaps in learning among the poorest children (Evans & Popova, 2016).

In Chap. "Customized E-Learning Platforms," Nicola Pitchford describes an ongoing effort to integrate tablet-based e-learning into classroom instruction in Malawi, operating initially as a pilot and then scaling over time. She finds that integrating the technology into primary school education can significantly improve numeracy, if it is embedded within a country's infrastructure for the delivery of education. There are also important lessons from scaling this class of technologies: how do you move from proof of concept to efficacy trials, and on to effectiveness at scale? Pitchford argues that researchers must not simply test hypotheses, but rather they should participate longer term in the production of education within a country. Ultimately, any researcher's exit from that country should leave capacity for ongoing implementation (with fidelity) as well as ongoing experimentation.

Learning assessment remains a critical issue: how to measure educational outcomes from a cognitive development perspective, particularly when using digital technology? Do current learning metrics focus enough on social and emotional outcomes, and on mental health, or do we need more expansive models of learning and capacity building? How might technology play a role in measurement innovation?

Of course investment in education and improved learning outcomes does not necessarily translate to productive employment. In many countries, there are high rates of unemployment, particularly among young workers and migrant populations. A lack of employment opportunities in the formal sector, accompanied by the youth bulge found in many developing countries (also known as the "demographic dividend"), threatens political instability in many countries. One recent trend in economic development has been to train unemployed youth in business and entrepreneurship (McKenzie, 2021). Yet, small-scale entrepreneurship, on its own, is unlikely to fuel equitable economic growth. Countries must build institutions that support the growth of existing firms, as well as the transition from informal labor and self-employment to employment by growth-oriented firms (Goswami et al., 2019).

A potential solution is the emergence of digital jobs platforms, which address frictions in labor markets by matching employees and firms and reducing search costs in the informal sector. Labor markets in many developing countries suffer from asymmetric and missing information—for example, lack of certification for a job candidate's skills. Is it possible to eliminate some of these market failures, especially in countries with large unemployed youth populations? In Chap. "Digital Networking and the Case of Youth Unemployment in South Africa," Shaw and Wheeler evaluate an intervention to increase use of LinkedIn in South Africa, revealing that users of the platform are indeed more likely to access jobs.

Once workers find profitable employment, are there technologies that can support their safety, welfare, and productivity? In Chap. "Amplifying Worker Voice with Technology and Organizational Incentives", we explore approaches that enhance firm efficiency through improved management practices— namely through worker feedback, incentives for responsive managers, and improvements in workplace environments. The chapter's authors find that capturing workers' voices using technology, and providing confidential feedback to employers, can improve a firms' productivity, through reductions in absenteeism and attrition. The existing systems for this, such as a "suggestions box" on the factory floor, are antiquated and cannot deliver the scale or accountability required for impact. So the researchers have developed, through sequential randomized trials, a set of apps that can help resolve employee grievances across the factory floor.

There is still the larger question of how to grow firms—how to increase the number of jobs available, and improve the profitability of self-employment. Some of these questions are addressed in Part II of this book. In this section, we instead focus on innovations deployed along the arc of a human's development, from the cultivation of abilities (via education and skills development) to the efficient allocation of labor across markets to the creation of safe and meaningful work conditions across local and global supply chains. In each of these areas, there is promise for further innovation to unleash human potential.

References

Angrist, N., Evans, D. K., Filmer, D., Glennerster, R., Rogers, F. H., & Sabarwal, S. (2020). *How to improve education outcomes most efficiently?* World Bank Policy Research WP 9450. World Bank.

Evans, D. K., & Popova, A. (2016). What really works to improve learning in developing countries? An analysis of divergent findings in systematic reviews. *The World Bank Research Observer, 31*(2), 242−270.

Goswami, A. G., Medvedev, D., & Olafsen, E. (2019). *High-growth firms: Facts, fiction, and policy options for emerging economies*. World Bank Publications.

McKenzie, D. (2021). Small business training to improve management practices in developing countries: re-assessing the evidence for 'training doesn't work'. *Oxford Review of Economic Policy, 37*(2), 276−301.

UNESCO Institute for Statistics (UIS) database, http://data.uis.unesco.org, June 29, 2022.

World Bank. (2019). *World development report 2019: The changing nature of work*. Washington, DC: World Bank. https://doi.org/10.1596/978-1-4648-1328-3.

Chapter 11
Customised E-Learning Platforms

Nicola Pitchford

1 Development Challenge

More than 617 million children and adolescents worldwide do not possess the basic reading and mathematics skills required to live a healthy and productive life and contribute towards economic growth (UNESCO, 2017; see Text Box 11.1). This global learning crisis is particularly acute in sub-Saharan countries, where minimal proficiency levels are not being met by 88% of children and adolescents in reading and 84% in mathematics. Across the region, education reform is required to address significant disparities in access, quality and equity. Children living in challenging contexts, such as remote rural locations, urban slums and border cities, are particularly vulnerable, and certain groups of children are disproportionately affected – especially girls, children living in extreme poverty and children at risk of, or diagnosed with, special educational needs and disabilities.

Text Box 11.1: How the Global Learning Crisis Affects Capacity

UNESCO (2018) report that the global learning crisis is costing $129 billion a year.

'Ten per cent of global spending on primary education is being lost on poor quality education that is failing to ensure that children learn. This situation

(continued)

N. Pitchford (✉)
School of Psychology, University of Nottingham, Nottingham, UK
e-mail: nicola.pitchford@nottingham.ac.uk

© The Author(s) 2023
T. Madon et al. (eds.), *Introduction to Development Engineering*,
https://doi.org/10.1007/978-3-030-86065-3_11

Text Box 11.1 (continued)

leaves one in four young people in poor countries unable to read a single sentence and perform basic mathematics'.

UNESCO (2019) warn that the world is off track in achieving the global education goal, SDG 4.

'Even though more and more children are starting school, one in six aged 6–17 will still be excluded by 2030. And of those who are enrolled, many are not learning or will drop out early. By 2030, when all children should be in school, four in ten young people will still not complete secondary education. The European Commission highlighted that by the end of the Agenda for Sustainable Development, about 800 million young people, half of them girls, will not possess basic skills. This is a critical global issue, as many adolescents and young people will not move into the job market with the right skills needed for relevant employment that responds to the challenges faced by their countries. Exacerbated by complex global threats like climate change, the situation is most acute in low-income countries and holds back economic growth and political stability, which would help bring people out of poverty and foster economic growth and social wellbeing'.

The United Nations has recognised the significance of this global crisis in education in the 2030 Sustainable Development Goals (SDGs). All member states have agreed to work towards addressing SDG 4: Quality Education, which aims to *ensure inclusive and equitable quality education and promote lifelong learning opportunities for all.*

Radical solutions are required to eliminate existing barriers to quality education for all children, anywhere in the world. Whilst the vast majority of children are now in primary school globally, their learning attainment is highly uneven and often persistently low (Hubber et al., 2016). This is not simply a matter of efficiency but one of human rights and social justice (Barrett et al., 2015).

SDG4 has focused attention on the quality of learning outcomes. Evidence suggests that technology-enhanced learning materials could be utilised by developing countries to promote significant and cost-effective improvements in learning (e.g. Mitra et al., 2005). However, realising these benefits across different school contexts remains a major challenge. For example, the One Laptop Per Child project had limited success in the United States, Peru, Rwanda and Tanzania, mainly due to poor implementation (e.g. limited Internet access in schools, poor support for repairs, non-child-directed delivery) and lack of teacher training (see Hubber et al., 2016). Evidence from the Second Information Technology in Education Study (SITES) indicates that significant investment has been made in technology-enhanced learning initiatives across the developing world, but this investment has not yet resulted in learning gains (Law et al., 2008). One reason may be that teachers and schools implement such initiatives in ways that reinforce rather than transform existing practice (Ruthven, 2009). This suggests there may be considerable potential in initiatives that improve learning but which have a close match to existing pedagogical practice.

Unlocking Talent The initiative described in this case study, Unlocking Talent, was established in 2013 to address the global learning crisis by harnessing technology to deliver high-quality education. It takes an inclusive system approach to embedding evidence-based e-Learning platforms within schools, with the aim of improving early-grade learning outcomes in reading and mathematics. Unlocking Talent is a growing, global initiative made up of an alliance of partners that focus on putting children and their educational needs first. At its core, the project uses e-Learning to help overcome the challenges that hold learners back. These include, but are not limited to, lack of trained teachers, overcrowded classrooms, lack of learning resources and high dropout rates. For context, in Malawi, where Unlocking Talent began, the pupil to teacher ratio is 77:1, and the dropout rate at primary age is over 50% (UNICEF Malawi, 2019).

Whilst e-Learning platforms alone will not solve these problems, good educational technology is key to raising learning standards and could complement other anti-poverty interventions, such as cash transfers. For example, a study by Kilburn et al. (2017) showed that financial constraints are a major barrier to school attendance in Malawi, because the cost of uniforms and books is beyond what many families can afford. Cash transfers can increase attendance and reduce dropout, but this does not change learning outcomes. High-quality e-Learning platforms can increase learning outcomes, although children need to attend school on a regular basis to benefit from this technology. Hence, to raise learning outcomes in low-income contexts, such as Malawi, multiple synergistic interventions are required.

Why Malawi?

In 2013, the Unlocking Talent alliance started to introduce a customised e-Learning platform within state primary schools in Malawi, to explore if technology could provide a means of delivering high-quality education to marginalised children and raising learning outcomes. Malawi was chosen because of the extreme challenges facing education within the country. Whilst most innovators pilot their education technology in small private schools, the Unlocking Talent alliance focused on state schools from the onset, so as to reach the most marginalised children.

Malawi is one of the poorest countries in Africa (UNESCO-IBE, 2010) and has some of the largest educational challenges globally, ranking second lowest in the Southern and Eastern Africa Consortium for Monitoring Educational Quality (SACMEQ) III report (Hungi et al., 2010). Whilst primary education is free for all children aged 6–13 years, the education system is in poor condition, fraught with chronic overcrowding, limited resources and inadequately trained teachers (Milner et al., 2011). As a consequence, many of Malawi's children do not attend or fail to complete primary education.

Allied to very poor retention rates, very few pupils meet minimum learning standards. Less than 10% of poor rural girls meet such standards, and attainment of basic reading and mathematics is among the lowest in the region (World Bank, 2010). As a result, adult (15+) rates of reading and mathematics proficiency are just 66% (UNESCO Institute for Statistics, 2016).

Additionally, structural inequalities exist that disadvantage certain groups (Yates, 2008; see Text Box 11.2). This has significant consequences for Malawi's potential for long-term economic growth. Poor reading and mathematics skills mean many people in Malawi struggle with basic day-to-day tasks, such as being able to read the newspaper, understand forms and buy and sell goods at the local market, resulting in long-term dependency and inability to secure and sustain employment and actively participate in society. As discussed by our research team (Pitchford et al., 2019), levels of reading attainment are linked to a country's economic growth and to an individual's health, nutrition, rate of fertility and mortality and income potential (Verner, 2005); however, mathematics attainment has been shown to have a stronger link to an individual's income potential (Crawford & Cribb, 2013; Dickerson et al., 2015; Geary, 2004). Findings from several African countries indicate an increase in mathematics test scores of only 0.1 standard deviations is associated with an increase in income of between 2% and 6.5% (Dickerson et al., 2015).

Text Box 11.2: How Structural Inequalities Within Malawi Education Affect Gender and Other Vulnerable Groups
There has been strong international focus on disparities in learning attainment within countries. To illustrate how particular groups of children are disproportionately affected by education in Malawi, data on numeracy levels are summarised below from the SACMEQ-III report (Spaull, 2012).

Gender: 64% of grade 6 girls were classified as functionally innumerate compared to 56% of boys.

Wealth: 61% of pupils from the poorest 25% of society were classified as functionally innumerate compared to 58% of pupils from the richest 25% of society.

Location: Higher functional innumeracy rates were reported for grade 6 pupils in rural (63%) than urban (51%) regions.

Disability: 98% of children with special education needs and disabilities in Malawi are excluded from quality education (MacDonnell Chilemba, 2013).

Wealth and location interact as, in general, poverty is higher in rural than urban regions, but within urban regions, there are pockets of extreme poverty. Rapid urbanisation has resulted in large informal settlements in major cities, often with very poor education facilities, which can overturn the general urban advantage (Chimombo, 2009).

In addition, children with home languages other than Chichewa, such as those residing in the north of Malawi, can be further disadvantaged by receiving instruction in school in a language (Chichewa) other than their home language (Essien et al., 2016).

2 Implementation Context

The context in which we have implemented a customised e-Learning platform in Malawi is extremely challenging, due to limitations in key components of the country's integrated system of learning attainment. These include challenges with school infrastructure (e.g. space, grid supply of electricity, Internet connectivity), capacity and capability of teachers (most of whom have limited digital literacy skills and little time for learning new skills), short school days (restricting available time to introduce a new intervention without missing out on other key areas of the curriculum) and overcrowded and mixed-ability classes. With an average of 77+ children in a class and often more, all of different abilities, it is difficult to ensure all children have sufficient access to the technology at the required dosage to be effective.

In addition, there are concerns about theft of technology and the greater impact this has on school security and community beliefs about the value of technology for learning. Malawi has a seriously underfunded education sector that struggles to provide even the most basic learning materials, let alone expensive technology. This all adds to the complexity of implementing an e-Learning platform in a low-income country such as Malawi.

With these challenges, it was necessary to work with many stakeholders to introduce an e-Learning platform within the Malawi primary education system, with the view to scale nationally (see Fig. 11.1). Every country and context are different, so having partners with the right skills, processes and support needed to implement education technology is vital for success. From the outset, the Unlocking Talent alliance has been partnering with the Government of Malawi to ensure the project is fully embedded within the Ministry of Education, Science and Technology (MoEST) and aligns with the national primary education strategy. Voluntary Service Overseas (VSO), an international development organisation, leads the Unlocking Talent initiative and is responsible for implementing the project in Malawian primary schools. VSO has developed a close working relationship with the MoEST in Malawi over the last 50 years, which has enabled them to have an active project management unit working with all relevant departments and the senior management team in the MoEST. The other core partners of the Unlocking Talent project are the software developer – onebillion – and our research team at the University of Nottingham.

Innovative, interactive, child-centred apps, designed and developed by onebillion, are the centre of this e-Learning platform, which is delivered through handheld tablets. This case study describes nearly a decade of research evaluating the effectiveness of the onebillion apps – and VSO's implementation of the platform – within Malawian primary schools, providing the critical evidence base for scaling the initiative.

To ensure the research is culturally sensitive and to draw on specialist in-country knowledge of education practice and policy, the research team has partnered with academics at the University of Malawi. The in-country academic partners

Funders
Including Royal Norwegian Embassy, KFW, DfID, UNICEF,
Comic Relief, Cisco, Airtel

Publish   **Practice**   **Policy**

onebillion

Research & Evaluation
University of Nottingham & University of Malawi
Postdoctoral researchers, doctoral researchers, Master's students

Fig. 11.1 Partnerships for scaling a customised e-Learning platform in Malawi adopted by the Unlocking Talent alliance

operationalise the research by assembling, training and managing teams of local evaluators who carry out assessments with children and schools in the local language. This partnership also builds research capacity through supervision of doctoral and masters students and early career researchers at the University of Malawi. To promote scaling of the project, in line with the research evidence, the alliance has engaged funding partners including the Norwegian Embassy; Kreditanstalt für Wiederaufbau (KFW); Comic Relief; the UK Foreign, Commonwealth and Development Office; UNICEF; Cisco; and Airtel.

Other critical stakeholders are end users of the technology, including teachers, school leaders, primary education advisors, community leaders, parents and children themselves. In developing their customised e-Learning apps, onebillion has gone to lengths to ensure that they meet the needs of children, through rigorous piloting of their software and hardware with children in different settings. When implementing the technology, VSO engages with parents, community leaders, teachers and school leaders, to promote cultural acceptance and sensitisation and to address any concerns they might have, by holding sensitisation meetings at local schools. In addition, there is a steering committee for the Unlocking Talent project in Malawi, comprising the national secretary for education, district-level education officers, representatives from school management (head teachers) and class teachers, to ensure their voices are heard and inform contextually sensitive shaping of the project as it scales.

3   Innovate, Implement, Evaluate and Adapt

3.1   Innovation: Customised E-Learning Platforms

Customised e-Learning platforms have potential to address some of the major challenges related to the global learning crisis facing many low-income countries in sub-Saharan Africa and elsewhere around the world, especially when based on the seven principles of universal design: (1) equitable use, (2) flexibility in use, (3) simple and intuitive use, (4) perceptible information, (5) tolerance for error, (6) low physical effort and (7) size and space for approach and use (The Center for Universal Design, 1997). Key software features include a wide array of multisensory representations of information (such as pictures, sound, video and animation), interactive tasks of varying degrees of difficulty, clear goals and rules, learner control, response feedback and repetition. These features promote an individualised learning environment that places the child in active control of their learning, with little or no adult support (Condie & Munro, 2007; Rose et al., 2005). The apps are delivered through touch-screen tablets, which are mobile and lightweight, eliminate the need for extra dexterity-reliant devices (e.g. keyboard and mouse) and have the capacity to store multiple child-friendly apps (Kucirkova, 2014).

We have found that customised e-Learning platforms have potential for wide-reaching acceptance and uptake by primary school teachers. The platforms can address the broad range of individual abilities typically represented within a large class and provide high-quality tailored instruction that is consistent for all children. However, certain alignments at a country level, such as language of instruction, are required for e-Learning platforms to be optimally effective. For example, Outhwaite et al. (2020) have shown that proficiency in language of instruction correlates positively with progress through the onebillion software. In addition, teachers are more likely to adopt e-Learning platforms that align closely with existing curriculum and pedagogical practice, as they supplement what is being taught in class and therefore can be used as a remedial resource for children that are struggling to learn with standard classroom instruction (Pitchford et al., 2018).

E-Learning Platform Used in the Unlocking Talent Project
As noted earlier, the software adopted by the Unlocking Talent initiative has been developed by onebillion – joint winners of the Global Learning XPRIZE. The software provides a comprehensive course of hundreds of learning units designed to support the acquisition of basic reading and mathematics from scratch, with little or no adult support. The course content is grounded in the national primary curriculum of Malawi and covers topics that are essential for the development of basic reading and mathematics, such as identification of letter sounds, syllable and phonological

Fig. 11.2 Examples of activities within the interactive apps developed by onebillion to support acquisition of basic reading and mathematics

awareness, reading aloud, comprehending single words and sentences, counting to 100, adding, subtracting and telling the time (see Fig. 11.2).

The course content is modular, guiding children through a series of learning units, and is delivered in their local language, Chichewa, by a friendly virtual teacher. The software offers child-centred tuition through interactive picture, audio and animation formats with clear objectives, instructions and immediate formative feedback, consistent for all users. These app features align with the principles of active, engaged, meaningful and socially interactive learning (Hirsh-Pasek et al., 2015) and are characteristic of direct instruction and retrieval-based learning, within a self-paced, individualised and inclusive learning environment, all of which are known to be important for raising learning outcomes (Grimaldi & Karpicke, 2014; Kirschner et al., 2006; Slavin & Lake, 2008). For example, the onebillion software enables individualised reward and feedback on each interaction with the software. Children are rewarded with a bright yellow tick and a high-pitched sound when completing a task successfully, but if they perform an inappropriate action, the software alerts this incorrect response by giving a low-pitched sound. Individualised immediate feedback is known to motivate children to stay on task, especially children with attention difficulties (e.g. Marx et al., 2018), yet this is scarcely available from teachers in a crowded classroom (see also Hirshleifer (2017) for another example of how educational technology can increase student rewards).

When used in formal education settings, such as primary schools, a remote monitoring feature within the software records when children are using the apps and logs their progress through the app content. This information is fed back to teachers, which enables teachers to direct attention to children that become halted on a particular topic and assist children who are making slow progress. Other stakeholders can access this information at a school level, so they can monitor how frequently schools are using the technology and how fast children are progressing.

The latest version of the software includes a customisation algorithm which is designed to rapidly assess a child's skill level each time they open the app. The software then places the child at an appropriate skill level within the app content. This adaptive learning feature has been designed especially for non-formal education contexts, such as out-of-school outreach programmes, refugee settings and direct-to-community approaches for children living in remote contexts where access to formal education settings is scarce.

Whilst onebillion designs and develops the software content and features, VSO and the research team provide feedback on usability within schools, which can be used to enhance the software through an ongoing iterative redesign process. Likewise, class observations by the research team of the intervention's delivery by teachers in schools are fed back to VSO to improve ease and fidelity of implementation.

3.2   Implementation Approach

The main approach for implementing the e-Learning platform that is currently being used in Malawian primary schools involves small groups of children (typically 30–60 children depending on the size of the school) attending a 40-minute session at a learning centre, a specially designed classroom equipped with solar power to enable use throughout the day, even in remote rural regions that are off-grid (see Fig. 11.3a). Individual children use a handheld tablet connected to a set of headphones to access the interactive, child-centred apps developed by onebillion (see Fig. 11.3b). Teachers help, guide and praise children as they learn with the app. They also use a dedicated teacher tablet to run the session. One teacher, who is selected as the learning centre coordinator, is responsible for the proper functioning and effective use of the learning centre. VSO members train teachers in how to use the technology and how to lead sessions using the software. This in-service training involves two 4-hour sessions which are delivered by VSO to teachers in the school. This implementation approach has been subjected to formal evaluation and has been shown to be effective at raising learning outcomes (Pitchford, 2015; Pitchford et al., 2019). It does, however, pose some challenges for schools, as it involves small groups of children leaving their class at any one time throughout the school day to

(a) A Learning Centre in Malawi where the onebillion software is delivered to small groups of children facilitated by a class teacher.

(b) Individualised learning approach with the onebillion software by a child in Malawi.

(c) Whole class implementation approach of the onebillion software by a class teacher in Malawi using a solar-powered projector.

Fig. 11.3 Illustration of different implementation approaches of the onebillion e-Learning platform in Malawian primary schools: (**a**) and (**b**) show the main implementation approach for individualised learning within a learning centre, whereas (**c**) shows a whole class implementation approach using a solar-powered projector facilitated by the class teacher

attend a session in the learning centre, which can be disruptive. In addition, with large class sizes and short school days, it is often difficult to provide all children with access to the learning centre on a regular basis.

To address these issues, VSO is currently exploring other implementation methods, such as paired learning using headphone splitters, to increase regular usage of the technology within primary schools. An alternative implementation approach, which could have potential to reach all children within a targeted ability group, involves connecting a tablet to a solar-powered projector which beams the onebillion app content to a large screen (e.g. a white sheet) attached to the classroom wall (see Fig. 11.3c). In this dynamic classroom environment, the teacher is able to guide learning of the whole class simultaneously, actively engaging children with the visual and audio content in the software. The teacher actively scaffolds learning by navigating the digital content, moving around the class and selecting individual children to complete in-app tasks, either by responding directly on the tablet or by indicating their response on the large screen. Class teachers are equipped with a rich source of high-quality learning materials and a large suite of interactive activities that they can utilise to support their pedagogical practice. However, these alternative approaches do not capitalise on the customised learning features that are key to the original e-Learning platform and, to date, have not been subjected to formal evaluation, so it is currently unknown if they will be effective at accelerating learning.

3.3   Evaluation: Research Approach

Since 2013, a research team from the University of Nottingham has been evaluating the effectiveness of the Unlocking Talent project at raising basic reading and mathematics of primary school children in Malawi and other countries, including the United Kingdom, Brazil, South Africa, Tanzania, Kenya and Ethiopia. Other independent evaluations have also taken place by Imagine Worldwide in Malawi and the Education Endowment Foundation in the United Kingdom.

This case study describes a staged and systematic approach to evaluating the effectiveness of the Unlocking Talent project in different countries, as illustrated in Fig. 11.4. This starts with piloting the technology within a particular context to assess its feasibility at raising learning outcomes and then builds evidence across increasingly large-scale randomised control trials (RCTs) which compare groups of children either learning with the customised e-Learning platform (intervention group) or learning through standard classroom instruction (control group). Performance of individual children is measured at baseline, prior to the delivery of the intervention, then again at endline, after the intervention has ceased, and differences in attainment over time are compared between groups. This measures the impact of the intervention in terms of learning gains. In addition, process evaluations are conducted to assess the fidelity of the intervention's implementation against intended use. A process evaluation involves conducting class observations with a

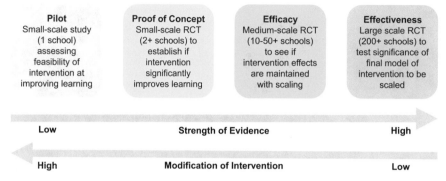

Fig. 11.4 Staged and systematic research approach to evaluating the effectiveness of the customised e-Learning platform adopted by the Unlocking Talent project. A series of successfully larger RCTs helps to de-risk the intervention, effectively creating 'checkpoints' as the project scales

structured checklist, interviewing teachers and surveying school and district education managers on the drivers thought to influence the fidelity of the intervention. Drivers of implementation fidelity in this context include child-level features (e.g. engagement with task), class-level features (e.g. trained teaching staff, dedicated space, calm environment), facilitator features (e.g. technical support, pedagogical support, behaviour management) and school leadership features (e.g. support for the intervention, provision of time within a daily regime to deliver the intervention). This combination of quantitative and qualitative research methodologies provides rich contextual information about the extent to which the intervention is effective at accelerating learning and how modifications might improve its effectiveness. In scaling the project, the strength of evidence accumulates, and the scope for modifying the intervention reduces, such that by the end of a large-scale effectiveness trial, the final model of the intervention – one that will be scaled nationally – should be identified.

Given the multisector approach to implementing a customised e-Learning platform within primary schools across Malawi, a holistic research approach is warranted that investigates different components of an integrated system that influences educational attainment. Key components of this integrated system include the technology (e.g. software and hardware), schools (e.g. teachers' capacity and capabilities, school leadership, availability of time within the school day, class size), communities (e.g. parent and community attitudes and support towards the introduction of technology in primary schools) and government commendation and advocacy for the project, as well as what the child brings to the learning task (e.g. prior ability in reading and mathematics, attendance at school, development of associated cognitive and motor skills and any associated health issues that may impact propensity to learn). All of these components have potential to influence how well children engage, interact and learn with this technology-based intervention to require investigation.

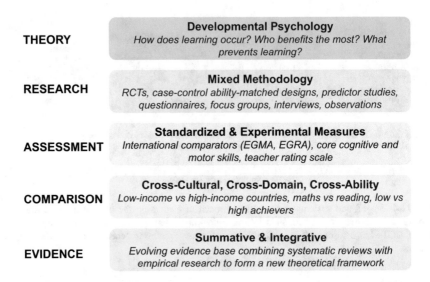

THEORY
Developmental Psychology
How does learning occur? Who benefits the most? What prevents learning?

RESEARCH
Mixed Methodology
RCTs, case-control ability-matched designs, predictor studies, questionnaires, focus groups, interviews, observations

ASSESSMENT
Standardized & Experimental Measures
International comparators (EGMA, EGRA), core cognitive and motor skills, teacher rating scale

COMPARISON
Cross-Cultural, Cross-Domain, Cross-Ability
Low-income vs high-income countries, maths vs reading, low vs high achievers

EVIDENCE
Summative & Integrative
Evolving evidence base combining systematic reviews with empirical research to form a new theoretical framework

Fig. 11.5 Summary of the TRACE research framework for evaluating technology-based education interventions

This holistic approach to evaluating the Unlocking Talent project is captured in the Theory, Research, Assessment, Comparison, Evidence (TRACE) research framework, summarised in Fig. 11.5. It is grounded within the theoretical framework of developmental psychology, especially theories of reading and mathematics acquisition. It involves examining how children acquire these foundational skills with the assistance of this customised e-Learning platform, establishing which children benefit from this intervention and identifying factors that prevent or facilitate learning. The breadth of the research questions necessitates mixed-method research, involving both quantitative and qualitative methodologies, including RCTs, case-control ability-matched designs, predictor studies, questionnaires, focus groups, interviews and observations.

Reliable and valid assessment of children's abilities is a critical aspect of this research. Yet typically, assessment is inadequate in low-income countries, such as Malawi, due to (1) a paucity of sensitive tools, normed for a particular country, that can measure different aspects of child development and scholastic attainment; (2) a scarcity of trained evaluators that speak the local language; and (3) the poor quality of national- or school-level tests, which renders them unsuitable as outcome variables. Governments and funders often require programmes to utilise tests of attainment that allow for international comparisons to be made, such as the Early Grade Reading Assessment (EGRA; USAID, 2010) and the Early Grade Mathematics Assessment (EGMA; USAID, 2011), yet these are often crude measures of attainment as they typically span a large age range. Measures of core cognitive and motor skills that are associated with reading and mathematics attainment, which are suitable for use in the early primary years, are sparse in Malawi, to say the least.

To address this lack of suitable measures, a digital assessment was developed by the Nottingham research team that can be administered to groups of children, thus increasing efficiency in the assessment process, and has been shown to be a reliable and valid cross-cultural measure of core abilities (Pitchford & Outhwaite, 2016b). Paucity of suitable measurement tools is particularly problematic when assessing the extent of functionality in children with special educational needs and disabilities across domains that may impact their ability to interact with digital technologies. Again, to address this need, a teacher-rating scale was developed by the Nottingham research team for measuring teacher perceptions of extent of difficulty in individual children across five key areas of functioning: mobility, hearing, vision, language and learning. This enabled a quantitative measure of extent of disability; this was shown to be predictive of children's progress in mathematics with the onebillion app (Pitchford et al., 2018).

To gain a comprehensive understanding of the potential reach of the Unlocking Talent project, different comparisons must be conducted. For example, the onebillion software covers content in reading and mathematics, so comparing learning gains (i.e. difference in performance at baseline and endline on measures of attainment) across these domains highlights weaknesses in the software at supporting acquisition of particular skills. These can then be addressed by the developers. Similarly, comparisons of learning gains between low-achieving and high-achieving children at baseline indicate the scope of the software at supporting learning for all children. In addition, profiling different groups of learners in terms of their cognitive and motor skills can provide insight into why some children struggle to learn whilst others excel at learning with this technology. This can inform whether the intervention should be targeted towards particular groups of children.

Furthermore, comparisons of learning reading and mathematics with this technology across primary school children in low-income and high-income countries, such as Malawi and the United Kingdom, enhance understanding of the universality of this customised e-Learning platform, which is important when considering where and how to scale to other countries.

Synthesising evidence from empirical research and systematic literature reviews of other technology-based learning programmes produces an evolving evidence base, which can be augmented over time. This generates new theoretical and implementation frameworks on which informed practice and policy decisions can be based.

3.4 Adaptation: Scaling Approach

Within Malawi, the scaling approach for the Unlocking Talent project currently focuses on key areas of activity.

Teacher Training In preparation for scaling the Unlocking Talent project nationally across Malawi and fully embedding it within the primary education sector, VSO is working alongside the MoEST to incorporate use of digital education technology into the training courses delivered at government-owned teacher training colleges. This builds capacity within the teaching profession to maintain effective use of technology-based education interventions as they are introduced within primary schools and supplements the in-service teacher training that VSO delivers as part of their current implementation process.

Direct-to-Community Approaches To increase the reach of the Unlocking Talent project to the many children in Malawi that do not attend primary schools, direct-to-community approaches are being piloted. To address the learning needs of out-of-school children, onebillion has developed 'onetab' – a robust, solar-charged, low-cost tablet pre-installed with their adaptive reading and mathematics software. The onetab device has been designed to work in challenging environments, with a protective bumper, bright screen, loudspeaker and strong solar-charging connector to ensure it can stand up to daily use by children in the home. The adaptive learning algorithm that onebillion has incorporated into their latest version of the software, which is installed on the onetab device, enables a direct-to-community approach, as children can access reading and mathematics instruction at an appropriate level for their ability, independently of a class teacher. This innovative adaptation extends the current implementation approach of the Unlocking Talent project within formal education settings to non-formal education settings, such as out-of-school outreach programmes and refugee camps in Malawi. It can also provide uninterrupted education during unexpected school closures, for example, in response to global pandemics such as COVID-19, as learning can continue to take place within the home.

Implementation Toolkits As the Unlocking Talent project scales within Malawi and to other countries, it is critical to maintain a high level of implementation fidelity, to ensure that the customised e-Learning platform used in the project continues to be implemented as intended. A common consequence of scaling innovative education initiatives is 'pollution' of implementation, as different organisations take on delivery in other contexts and make alterations to the programme, which can lead to dilution of previously observed intervention effects (Mihalic, 2009; Samara & Clements, 2017). In an attempt to prevent wash-out and maintain high implementation fidelity of the Unlocking Project as it scales, VSO, the University of Nottingham and Imagine Worldwide have recently produced and published a series of toolkits that provide guidance to organisations on designing, planning, launching and monitoring an education technology-based project. The toolkits are designed for non-governmental organisations (NGOs) and governments that are thinking about designing and implementing tablet-based learning programmes. They are targeted at programme directors, project managers, software developers, site leaders, staff and facilitators, including teachers, parents and community leaders. The toolkits are modular, so organisations can pick and choose sections depending on their needs. To date, the toolkits have not been rigorously tested, but were designed to reinforce fidelity across implementation contexts (see also Outhwaite et al., 2019).

Localisation The Unlocking Talent alliance is also being scaled by introducing this customised e-Learning platform to other countries that are in need of education reform. Localising the software is an important step in this process, and onebillion is responsible for this. For each new language they introduce, onebillion works with universities, linguists, authors, book editors, teachers, parents, children and pedagogy experts to ensure the software is localised appropriately to the needs of the child. Further information on how onebillion localise their software is summarised in Text Box 11.3.

Text Box 11.3: Process of Localising the Onebillion Software to Different Languages
When introducing the onebillion software to different communities, it is important to localise the software to ensure it meets the language needs of children. The process adopted by onebillion to localise their software to different languages involves the following:

1. **Partners.** onebillion works with partners to discuss when to localise their software into target language(s). This is a complex task that needs to be adopted comprehensively, so it is important that partners understand what is involved.
2. **Scoping.** onebillion works with local linguistic experts to (i) build a description of the culture and language, including phonetic structure and high-frequency words; (ii) determine the order that letters should be introduced to the child in order to meet syllables and words as soon as possible; (iii) identify, create and test any new learning units needed; (iv) check the bank of words and images against the language and culture, working with graphic artists, to ensure they further the child's understanding; and (v) creatively translate all stories, keeping them relevant and engaging for the child.
3. **Recording.** onebillion works with voice-over actors who are native speakers in the new language area, to ensure a familiar accent for the child. They record everything in-house to ensure their high standard is maintained – this involves approximately 160 stories; 185,000 words; and 32,000 audio files.
4. **Sequencing.** Course designers at onebillion sequence the course, combining existing curricula and their own experience and carefully selecting a variety of activities for the child. This is an iterative process based on feedback from children in the new context that the software is being localised for.
5. **Quality assurance.** Onebillion fully test all 4000+ units, multiple times, for functionality and language. This happens both in-house and with native speakers, in the target countries. Only then is the course ready for the child to start learning.

(continued)

Text Box 11.3 (continued)

6. **Tools**. The localisation process is supported by onebillion's in-house software and the structure of their software, which enable them to work with experts all over the world and to tailor the course for each new language.

Funding Model Scaling requires funding from a diverse portfolio of donors. The funding approach is agile, with the Unlocking Talent alliance responding to opportunities as they arise, as well as being proactive in seeking funding from donors operating in countries where the alliance wishes to expand. As VSO is committed to evidence-based practice, evaluation research is cost into their implementation proposals to donors. Dedicated funding is also available for global challenges and education research from some organisations. For example, the Education Endowment Foundation funded a large-scale efficacy trial of the onebillion software with young children in the United Kingdom who were struggling to learn basic mathematics.

4 Results/Lessons Learned

Evaluation Findings

Using a mixed-method approach, as outlined above, our research team has demonstrated that children and teachers enjoy using the customised e-Learning platform with onebillion software adopted by the Unlocking Talent project and that this technology significantly raises learning outcomes compared to standard classroom practice (Pitchford, 2015). The randomised evaluation incorporated a placebo group that used the handheld tablets in the learning centre to engage with a simple design software that required similar drag and drop movements as the onebillion maths software. This group did not show significantly enhanced learning in maths above the level shown by the control group. As a result, the observed learning gains can be attributed to the software's implementation, rather than novelty effects arising from interacting with a new technology (see Table 11.1).

Furthermore, the benefits of learning with this customised e-Learning platform are sustained even as the programme scales (Pitchford et al., 2019). When learning with the onebillion software, gains in ability have been shown to be consistent over time, across different cohorts of children, between different countries, and assessed by different groups of researchers. This demonstrates that consistent and reliable learning gains can be achieved with this education technology regardless of context. In general, this amounts to a 3+ month advantage for children learning basic mathematics and a 4+ month advantage for children learning basic reading with the onebillion software compared to standard classroom instruction (Education

Table 11.1 From Pitchford (2015): group performance (mean (SD), min–max) at pretest and post-test and percentage gain for the tests of mathematical concepts (MC; maximum score 48) and maths curriculum knowledge (CK; maximum score 50)

Assessment	Intervention group (maths tablet)	Placebo group (non-maths tablet)	Control group (standard practice)
Standard 1	$n = 22$	NA	$n = 20$
MC pretest	2.0 (1.7), 0–5	3.4 (4.3), 0–13	
MC post-test	5.1 (4.6), 0–17	4.5 (4.3), 0–13	
% gain	**6.5**	**2.3**	
CK pretest	2.4 (2.7), 0–12	2.7 (1.7), 0–6	
CK post-test	7.7 (6.3), 0–24	5.9 (5.3), 0–20	
% gain	**10.6**	**6.4**	
Standard 2	$n = 38$	$n = 35$	$n = 37$
MC pretest	8.6 (6.5), 0–20	10.1 (5.7), 0–21	8.4 (6.3), 0–19
MC post-test	14.6 (6.6), 0–24	12.5 (7.3), 0–23	10.5 (6.5), 0–19
% gain	**12.5**	**5.0**	**4.4**
CK pretest	6.4 (6.3), 0–24	8.3 (8.1), 0–28	5.5 (5.3), 0–21
CK post-test	20.7 (10.3), 0–37	15.1 (8.5), 2–31	10.8 (7.4), 0–26
% gain	**28.6**	**13.6**	**10.6**
Standard 3	$n = 44$	$n = 44$	$n = 43$
MC pretest	14.9 (5.7), 0–23	15.4 (5.6), 0–22	14.9 (5.2), 0–22
MC post-test	19.5 (5.2), 5–27	19.3 (5.1), 4–26	18.6 (6.0), 0–26
% gain	**9.6**	**8.1**	**7.7**
CK pretest	13.5 (9.3), 0–34	14.8 (10.1), 1–37	11.0 (10.3), 0–33
CK post-test	35.2 (7.0), 19–46	24.9 (8.0), 3–36	23.4 (6.9), 7–36
% gain	**43.4**	**20.2**	**24.8**

Endowment Foundation, 2019; Imagine Worldwide, 2020; Outhwaite et al., 2020; Pitchford et al., 2019).

Our research in Malawi, in particular, shows that after interacting with the onebillion software on a daily basis at primary school for eight consecutive weeks, additional benefits are observed: children's attentional skills improved, enabling them to concentrate better in class (Pitchford & Outhwaite, 2019). Importantly, as shown in Table 11.2, girls learn just as well as boys with this technology, and when implemented at the start of primary education, the Unlocking Talent intervention can prevent gender-based attainment gaps in mathematics that typically emerge over the first year of schooling in Malawi through standard classroom instruction (Pitchford et al., 2019).

The ease of use and flexible structure of this customised e-Learning platform make it suitable for all children, including those with special educational needs and disabilities. We have shown that a group of children attending a special needs unit attached to two Malawian state primary schools, who presented with a diverse range of difficulties across the group, could interact and engage with the onebillion software and made progress through the app, albeit at a slower pace than mainstream

Table 11.2 From Pitchford et al. (2019): learning gains (mean %) and effect sizes (Cohen's d) for boys and girls in mathematics (EGMA) and reading (EGRA) after intervention with the customised onebillion e-Learning platform compared to standard practice. For mathematics, standard practice promotes a gender advantage for boys over girls, whereas girls learn just as well as boys with the technology-based intervention

	Intervention group		Control group	
Assessment	Boys	Girls	Boys	Girls
EGMA % gain M (SD)	19.95 (15.11)	20.86 (16.14)	17.05 (16.12)	11.90 (12.61)
Cohen's d	**0.06**		**0.36**	
EGRA % gain M (SD)	8.32 (15.74)	10.64 (15.11)	4.29 (7.73)	5.09 (7.33)
Cohen's d	**0.15**		**0.11**	

peers (Pitchford et al., 2018). However, extent of disability, as measured by teacher ratings, correlated negatively with progress through the app, indicating that children with profound difficulties struggled to learn with this technology. Our research suggests that whilst customised e-Learning platforms might be suitable for some children with mild-to-moderate special educational needs and disabilities, children with profound difficulties require a different type of educational support.

Similar learning gains with this customised e-Learning platform have also been shown in the United Kingdom, demonstrating this technology is equally effective across countries differing vastly in income status and primary education provision (Education Endowment Foundation, 2019; Outhwaite et al., 2017; Outhwaite et al., 2018; Pitchford & Outhwaite, 2016a). Furthermore, our research in UK primary schools has highlighted the importance of implementation in securing improved learning outcomes for basic mathematics with the onebillion app (Outhwaite et al., 2019). Qualitative research conducted through a process evaluation of using the onebillion software showed that when schools established a consistent routine for embedding the apps into their daily classroom schedule, children learnt more basic mathematics than when schools varied the implementation routine. Results showed that 'established routine' predicted 41% of the variance in children's learning outcomes with the apps. Established routine involved schools implementing the intervention at a consistent time each day, having a dedicated member of staff whose responsibility it was to implement the intervention, having well-organised equipment (e.g. colour coding the tablet devices so that they were easily identifiable by children) and having a dedicated space within the classroom and having a seating plan where children used the onebillion apps. In another study, the effectiveness of the onebillion software in supporting acquisition of basic mathematics by early-grade children attending a bilingual immersion school in Brazil was evaluated, and proficiency in language of instruction was shown to be positively associated with progress through the apps (Outhwaite et al., 2020). These findings are important as they show that factors related to the implementation of this intervention in primary schools are critical for securing improved learning outcomes – seemingly, the customised e-Learning platform and interactive apps alone are not the only factors driving success.

Overall, this emerging evidence base, built upon a robust scientific methodology and psychological theories of reading and mathematics development, has generated confidence that the Unlocking Talent project can be implemented successfully at scale in Malawi. The solution significantly raises attainment of core foundational skills, and it can address some of the structural inequalities that are inherent to the country's education system.

Challenges and benefits of working with partners.

Global challenges, such as providing quality education for all, require genuine cooperation and collaboration from multiple sectors to enable policy-makers to make informed decisions on the best available evidence. This is a difficult landscape to navigate and one that a traditional academic career does not necessarily prepare you for. Working with partners across a range of sectors has its challenges, but these are often offset by the benefits it affords. External partners can bring insights into the research design and process, based on their wealth of knowledge from working on the ground. This adds value to the quality of the research and provides a direct pathway to apply the research findings to people that matter.

When initiating global challenges research, it is useful to map who the partner organisations are at the outset of the design process (e.g. Fig. 11.1) as this will be critical to the sustained success of the project. Partnerships take time to build and require constant nurturing to maintain engagement and foster trust. This is easier to achieve when partners share a mutual goal and appreciation of the varying demands that different organisations may have, which at times might clash with what you are trying to achieve. If difficulties arise, it is best to focus on the overarching goal you and your partners are trying to achieve and move towards a way of working that will facilitate you reaching this.

Working in partnership with other sectors also requires a flexible and fast way of operating which is not aligned to the academic year. This is especially relevant to technology-based fields, where adaptations and innovations are rapid and frequently precede the evidence base. There is often an urgency in addressing global challenges. Yet, despite the time it takes to accrue a strong evidence base, we need to know that what we are doing is going to be effective and of benefit to the end users – without incurring harm. Forging ahead with adaptations and innovations to a product that has preliminary evidence of being effective, before developing a full understanding of factors that determine its effectiveness, can result in product developers altering the components that make it a success.

In addition, the language used by different groups of partners can often hold distinct meanings, so clarity of expression is key to successful communication. An example is the term 'significance'. When researchers use this term, they are referring to statistical significance – the probability that a given result occurs by chance – however, partners from other sectors frequently use this term to imply importance. This can change the interpretation of research results when partners are communicating research findings, so it is important to engender a mutual understanding of shared terminology.

Addressing global challenges also requires large-scale investment. Academics that are new to this field may underestimate the resources needed to be able to conduct high-quality research at a level that will have global impact. It is also a competitive field in which the funding structures supporting this work are often short-lived, requiring a quick turnaround of activities, which does not promote long-term understanding of the impact the work can have. Funding opportunities that are supported through the government can change rapidly in a volatile political environment. This makes planning for long-term projects difficult. Whilst agile funding streams can address global challenges as they happen, such as in response to the COVID-19 pandemic, it can make it difficult to sustain funding for large-scale projects that require long-term impact evaluations, such as the effectiveness of technology-based education programmes on improving learning outcomes.

Despite the challenges of working in this field, the rewards can substantially outweigh the difficulties. Working collaboratively with the Unlocking Talent alliance towards a shared goal and working with humility are what makes for long-term sustainable change.

Implications for practice and policy. To support future work by practitioners, education software developers and policy-makers, here, we provide several key implications of the implementation and evaluations of the Unlocking Talent project in Malawi and other countries.

Practitioners
- Schools should look for customised e-Learning platforms that are localised to the language of instruction, are easy to use and have a flexible structure, as these are accessible to learners of varying abilities, even children with special educational needs and disabilities.
- For children with special educational needs and disabilities, teachers and parents should assess the extent of disability before using interactive educational apps, as this type of technology might not be suitable for children with profound difficulties.
- When implementing customised e-Learning platforms within formal education settings, teachers should maintain a well-established classroom routine to optimise effectiveness.
- School leadership should monitor implementation fidelity of customised e-Learning platforms within classrooms on a regular basis, to ensure teachers are using the technology as intended.
- Teachers and parents wishing to use interactive apps with their children should consider children's proficiency in the languages spoken within the software, as children need to have a sufficient language proficiency to access curriculum content and to respond to instruction given in educational apps. Look for apps that have been carefully localised to the language of instruction.

Software Developers
- Software developers should consider accessibility of their apps for different groups of children, especially those with special educational needs and disabilities, and build in adaptations to maximise accessibility for all learners.
- Software developers need to consider the role of children's language development in their app design features and content and ensure that the language used in their apps is appropriate for the age of the intended users.

- Software developers should consider adapting their technology on the basis of user feedback and research findings to increase uptake and improve quality of their products.
- Building assessment features into software could facilitate monitoring of children's progress by teachers and researchers. This might involve in-app assessment of skills to be taught in a learning activity prior to app-based instruction, followed by an end-of-activity in-app quiz which requires children to apply the knowledge learnt to new materials. Progress charts built into software can display this graphically over time, providing information on the rate of learning by individual children.
- Monitoring of time on task, number of correct and incorrect responses per learning activity, sequence of responses per learning activity and time to complete a learning activity can also be valuable information to teachers and researchers; developers can consider how to build these metrics into their software. This instrumentation can provide rich information as to how children are interacting with the software content.

Policy-Makers

- Governments should consider introducing customised e-Learning platforms that encompass features of universal design into their primary education sector, as this technology has been shown to significantly raise learning outcomes compared to standard classroom practice across a range of low-, medium-, and high-income contexts.
- Governments considering introducing customised e-Learning platforms within their education system should assess the country's readiness to implement this technology effectively, including scoping the infrastructure, teacher capacity and capabilities and availability of time within the school day required to support successful implementation.
- When introducing e-Learning platforms within an education system with the view to scale, key partners (especially within government) need to be engaged from the outset to ensure sustainability, cultural sensitivity and accountability. Maintaining budget lines for e-Learning within the education budget will also foster commitment to long-term sustainability.
- Interactive apps that have been shown to mitigate gender differences should be implemented at the start of formal education to enable girls and boys to learn at a similar rate before gender disparities become entrenched.
- Governments could consider introducing customised e-Learning platforms direct to communities, to reach children who cannot attend school and to provide continued education during unexpected periods of school closures, such as in response to the COVID-19 pandemic.
- Governments, schools and parents should seek professional guidance in choosing customised e-Learning platforms to improve children's learning outcomes, to ensure selected technologies are supported by a rigorous scientific evidence base demonstrating effectiveness.

Discussion Questions

1. Within limited funds, which children should be prioritised to receive the customised e-Learning platform currently being used by the Unlocking Talent project in Malawi? Provide a rationale for your answer.
2. How might the e-Learning platform currently used by the Unlocking Talent project in Malawi be adapted to enable more children to access the technology on a regular basis?
3. What adaptations could be made to e-Learning platforms to increase accessibility and learning for children with special educational needs and disabilities?
4. Which factors might influence learning outcomes with customised e-Learning platforms when implemented outside the school system, such as through direct-to-community approaches? How might these be addressed through policy, practice or technological advances?
5. How might other educational technologies, such as radio or TV programmes, be used in low-income contexts to provide high-quality education for children not attending school? What features would need to be incorporated into these types of technologies to maximise learning in non-formal education settings?

References

Barrett, A., Sayed, Y., Schweisfurth, M., & Tikly, L. (2015). Learning, pedagogy and the post-2015 education and development agenda. *International Journal of Educational Development, 40*, 231–236.

Chimombo, J. (2009). Changing patterns of access to basic education in Malawi: A story of a mixed bag? *Comparative Education, 45*(2), 297–312.

Condie, R., & Munro, B. (2007). *The impact of ICT in schools – A landscape review*. Retrieved 19 9 2013, from http://dera.ioe.ac.uk/1627/

Crawford, C., & Cribb, J. (2013). Reading and maths skills at age 10 and earnings in later life: A brief analysis using the British Cohort Study. *Centre for Analysis of Youth Transitions*. 2013. Report No. 3.

Dickerson, A., McIntosh, S., & Valente, C. (2015). Do the maths: An analysis of the gender gap in mathematics in Africa. *Economics of Education Review, 46*, 1–22.

Education Endowment Foundation. (2019). *onebillion: App-based maths learning*.https://educationendowmentfoundation.org.uk/projects-and-evaluation/projects/onebillion-app-based-maths-learning/

Essien, A. A., Chitera, N., & Planas, N. (2016). Language diversity in mathematics teacher education: Challenges across three countries. In R. Barwell, P. Clarkson, A. Halai, M. Kazima, J. N. Moschkovich, N. Planas, M. Phakeng, P. Valero, & M. Villavicencio Ubillús (Eds.), *Mathematics education and language diversity* (pp. 103–119). Springer.

Geary, D. C. (2004). Mathematics and learning disabilities. *Journal of Learning Disabilities, 37*(1), 4–15. https://doi.org/10.1177/00222194040370010201

Grimaldi, P. J., & Karpicke, J. D. (2014). Guided retrieval practice of educational materials using automated scoring. *Journal of Educational Psychology, 106*, 58–68.

Hirshleifer, S.R. (2017). *Incentives for effort or outputs? A field experiment to improve student performance*. Unpublished manuscript available at https://economics.ucr.edu/repec/ucr/wpaper/201701.pdf

Hirsh-Pasek, K., Zosh, J. M., Golinkoff, R. M., Gray, J. H., Robb, M. B., & Kaufman, J. (2015). Putting education in "educational" apps: Lessons from the science of learning. *Psychological Science, 16*(1), 3–34.

Hubber, P., Outhwaite, L. A., Chigeda, A., McGrath, S., Hodgen, J., & Pitchford, N. J. (2016). Should touch screen tablets be used to improve educational outcomes in primary school children in developing countries? *Frontiers in Psychology, 7*, 839.

Hungi, N., Makuwa, D., Ross, K., Saito, M., Dolata, S., van Cappelle, F., Paviot, L., & Vellien, J. (2010). *SACMEQ III project results: Pupil achievement levels in reading and mathematics.* Southern and Eastern Africa consortium for monitoring educational quality, working document, number 1.

Imagine Worldwide. (2020). *Tablet-based Learning for Foundational Literacy and Math: An 8-month RCT in Malawi.*https://www.imagineworldwide.org/resource/tablet-based-learning-for-foundational-literacy-and-math-an-8-month-rct-in-malawi-executive-su

Kilburn, K., Handa, S., Angeles, G., Mvula, P., & Tsoka, M. (2017). Short-term impacts of an unconditional cash transfer program on child schooling: Experimental evidence from Malawi. *Economics of Education Review, 59*, 63–80.

Kirschner, P. A., Sweller, J., & Clark, R. E. (2006). Why minimal guidance during instruction does not work: An analysis of the failure of constructivist, discovery, problem-based, experiential, and inquiry-based teaching. *Educational Psychologist, 41*(2), 75–86.

Kucirkova, N. (2014). iPads in early education: Separating assumptions and evidence. *Frontiers in Psychology, 5*(715), 1e3.

Law, N., Pelgrum, W. J., & Plomp, T. (2008). *Pedagogy and ICT use in schools around the world: Findings from the IEA SITES 2006 study.* CERC-Springer.

MacDonnell Chilemba, E. (2013). The right to primary education of children with disabilities in Malawi: A diagnosis of the conceptual approach and implementation. *The African Disability Rights Yearbook, 1*(1).

Marx, I., Hacker, T., Yu, X., Cortese, S., & Sonuga-Barke, E. (2018). ADHD and the choice of small immediate over larger delayed rewards: A comparative meta-analysis of performance on simple choice-delay and temporal discounting paradigms. *Journal of Attention Disorders.* https://doi.org/10.1177/1087054718772138

Mihalic, S. (2009). *Implementation fidelity.* Unpublished manuscript available from http://citeseerx.ist.psu.edu/viewdoc/download?doi=10.1.1.180.9133&rep=rep1&type=pdf

Milner, G., Mulera, D., Banda, T., Matale, E., & Chimombo, J. (2011). *Trends in achievement levels of grade 6 learners in Malawi.* SACMEQ.

Mitra, S., Dangwal, R., Chatterjee, S., Jha, S., Bisht, R. S., & Kapur, P. (2005). Acquisition of computing literacy on shared public computers: Children and the "hole in the wall". *Australasian Journal of Educational Technology, 21*, 407–426.

Outhwaite, L. A., Gulliford, A., & Pitchford, N. J. (2017). Closing the gap: Efficacy of a tablet intervention to support the development of early mathematical skills in UK primary school children. *Computers and Education, 108*, 43–58.

Outhwaite, L. A., Faulder, M., Gulliford, A., & Pitchford, N. J. (2018). Raising early achievement in math with interactive apps: A randomized control trial. *Journal of Educational Psychology.* https://doi.org/10.1037/edu0000286

Outhwaite, L. A., Gulliford, A., & Pitchford, N. J. (2019). A new methodological approach for evaluating the impact of educational intervention implementation on learning outcomes. *International Journal of Research and Method in Education.* https://doi.org/10.1080/1743727X.2019.1657081

Outhwaite, L. A., Gulliford, A., & Pitchford, N. J. (2020). Language counts when learning mathematics with interactive apps. *British Journal of Educational Technology.* https://doi.org/10.1111/bjet.12912

Pitchford, N. J. (2015). Development of early mathematical skills with a tablet intervention: A randomized control trial in Malawi. *Frontiers in Psychology, 6*, 485.

Pitchford, N. J., & Outhwaite, L. (2016a). Apps teaching early maths skills. In N. Kucirkova & G. Falloon (Eds.), *Apps, technology and younger learners: International evidence for teaching.* Routledge Press.

Pitchford, N. J., & Outhwaite, L. A. (2016b). Can touch screen tablets be used to assess cognitive and motor skills in early years primary school children? A cross-cultural study. *Frontiers in Psychology, 7*, 1666.

Pitchford, N. J., & Outhwaite, L. A. (2019). Secondary benefits to attentional processing through intervention with an interactive maths app. *Frontiers in Psychology.* https://doi.org/10.3389/fpsyg.2019.02633

Pitchford, N. J., Kamchedzera, E., Hubber, P. J., & Chigeda, A. (2018). Interactive apps promote learning in children with special educational needs. *Frontiers in Psychology, 9*, 262.

Pitchford, N. J., Chigeda, A., & Hubber, P. J. (2019). Interactive apps prevent gender discrepancies in early grade mathematics in a low-income country in sub-Sahara Africa. *Developmental Science, 22*, e12864. https://doi.org/10.1111/desc.12864

Rose, D. H., Meyer, A., & Hitchcock, C. (2005). *The universally designed classroom: Accessible curriculum and digital technologies.* Harvard University Press.

Ruthven, K. (2009). Towards a naturalistic conceptualisation of technology integration in classroom practice: The example of school mathematics. *Education & Didactique, 3*(1), 131–152.

Samara, J., & Clements, D. H. (2017). Interventions in early mathematics: Avoiding pollution and dilution. *Advances in Child Development and Behavior, 53*, 95–126.

Slavin, R. E., & Lake, C. (2008). Effective programs in elementary mathematics: A best-evidence synthesis. *Review of Educational Research, 78*(3), 427–515.

Spaull, N. (2012). Malawi at a glance. SACMEQ at a glance series. *Research on Socio-economic Policy (RESEP).* Available http://resep.sun.ac.za/index.php/projects/

The Center for Universal Design. (1997). *The principles of universal design, version 2.0.* North Carolina State University.

UNESCO. (2017). *More than More Than One-Half of Children and Adolescents Are Not Learning Worldwide.* Fact Sheet No. 46, September 2017 UIS/FS/2017/ED/46. UNESCO Institute for Statistics.

UNESCO. (2018). *Global education monitoring report 2019: Migration, displacement and education – Building bridges, not walls.* UNESCO.

UNESCO (2019). https://en.unesco.org/sites/default/files/the_world_is_off_track_in_reaching_the_global_education_goal.pdf

UNESCO Institute for Statistics. (2016). *Results of the 2016 UIS education survey.* http://uis.unesco.org/en/news/results-2016-uis-education-survey-now-available

UNESCO-IBE. (2010). *World data on education report.* Paris, France.

UNICEF Malawi. (2019). *2018/2019 education budget brief.*

USAID. (2010). *Malawi 2010 early grade reading assessment: National baseline report.*

USAID. (2011). *Malawi early grade mathematics assessment: National baseline report 2010.*

Verner, D. (2005). *What factors influence world literacy? Is Africa different?* (Policy research working paper no. 3496). World Bank. https://doi.org/10.1596/1813-9450-3496

World Bank. (2010). *The education system in Malawi* (World Bank working paper 182). World Bank.

Yates, C. (2008). *Keeping children in school: A review of open education policies in Lesotho and Malawi* (SOFIE opening up access series no. 5). Institute of Education, University of London.

Chapter 12
Digital Networking and the Case of Youth Unemployment in South Africa

Patrick Shaw and Laurel Wheeler

1 Development Challenge

In developing countries, the system of paid labor is characterized by a duality of informal and formal jobs. Work in the formal sector is tied to employers that are registered with the government and pay taxes. In contrast, work in the largely unregulated informal sector often involves self-employment or employment with family-owned enterprises. For example, small-scale street vendors, taxi drivers, and freelancers are typically part of the informal economy. Although wages tend to be higher and less volatile in the formal sector, the informal sector comprises 60% of all nonagricultural employment in most countries in the Middle East, North Africa, Latin America, Asia, and sub-Saharan Africa (ILO, 2018; Kingdon & Knight, 2001). Relative to other developing countries, South Africa has a large formal sector and a high rate of unemployment (Statistics South Africa, 2019).[1] This case study examines the application of existing technology to mitigate barriers between young South African workseekers and formal employment opportunities with government-registered enterprises.

---

[1] According to Statistics South Africa (2019), the informal sector comprises only 20% of total nonagricultural employment in South Africa.

P. Shaw (✉)
Brown University, Providence, RI, USA

RTI International, Research Triangle Park, NC, USA
e-mail: patrick_shaw@brown.edu

L. Wheeler (✉)
Department of Economics, University of Alberta, Edmonton, AB, Canada
e-mail: lewheele@ualberta.ca

In the first quarter of 2019, approximately 28% of South Africans were unemployed (Statistics SA, 2019). South Africa's high aggregate unemployment is due to a combination of factors, including spatial segregation between workers and firms, labor regulations, a weak education system, racial discrimination due to the legacy of apartheid, and restrictions on informal enterprises (Banerjee et al., 2008). The spatial mismatch between where jobs are located and where workseekers live increases the time and monetary cost of submitting applications in person. These costs are high enough to impede the job search process of some workseekers (Kerr, 2017). Employment regulations in South Africa are unusually difficult to navigate, making legal disputes in hiring commonplace (Magruder, 2010; Rankin & Roberts, 2011). As a result, employers are often reluctant to hire someone without referrals or significant work experience. And South African schools assess student achievement inaccurately and inconsistently, which limits employers' ability to use educational qualifications to differentiate between job candidates (Lam et al., 2011; Taylor et al., 2011).

In this context, transitions into formal employment have been particularly difficult for young workseekers. South Africa currently has one of the highest rates of youth unemployment in the world, with more than 55% of South Africans aged 15–24 years not in employment, education, or training (Statistics SA, 2019). This is twice the unemployment rate of adults aged 25–64 in the same time period. Figure 12.1 shows that the youth-adult unemployment gap has been high and relatively stable for many years (Mosomi & Wittenberg, 2020). Furthermore, large numbers of young people in South Africa – particularly young men – have stopped searching

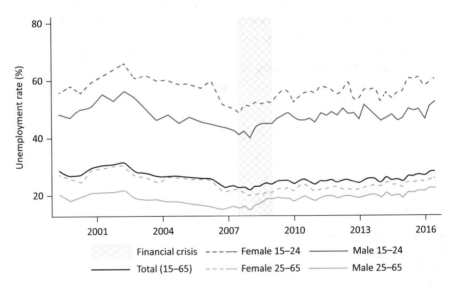

Fig. 12.1 South African unemployment by age and gender
Note: This figure is reproduced from page 4 of Mosomi and Wittenberg (2020). Mosomi and Wittenberg generated the figure using data from Post-Apartheid Labour Market Surveys (PALMS)

For work-seekers...	For employers...	What is the result?
In the frictionless world... Young people have no obstacles to education. They receive the optimal level of education based on their skills and their expectations about future wages. Their educational certifications accurately reflect their potential productivity.	Employers widely advertise job openings, reaching all work-seekers.	Employers hire the most qualified workseekers out of the pool of prospective workseekers. Unemployment duration is short and retention is high.
Work-seekers have full information about all job openings.	Employers have full information on all work-seekers, based on educational certificates, resumes, and interviews.	
Applying to jobs is costless.	Interviewing job candidates is costless.	
In the real world... Young people may have limited opportunities to develop skills or experiences, either due to cost or due to their age.	Employers may be more inclined to higher older workseekers who have had more on-the-job work experience or more educational opportunities.	
In countries like South Africa, oftentimes educational certifications are weak signals of ability.	Employers may not perceive certifications to be valuable information about worker productivity, making them rely more heavily on other means of getting information about workseekers.	Employers are more likely to hire older workseekers than younger workseekers with the same skills and qualifications. Unemployment durations for young workseekers may be high and match quality may be low.
Due to their age and relative lack of work experience, young people tend to have smaller professional networks, meaning that they have less information about job openings and their fit for different types of jobs.	When prospective employers lack credible information about worker productivity, they often rely on referrals based on professional networks, disadvantaging young workseekers.	

Fig. 12.2 Youth employment frictions

for work altogether, likely due to discouragement resulting from persistent high unemployment.

Although youth unemployment is particularly stark in South Africa, young people in many parts of the world contend with high rates of unemployment, underemployment, and unstable employment (ILO, 2017). More than 70 million people between the ages of 15 and 24 are unemployed worldwide, representing a youth unemployment rate three times the adult unemployment rate (ILO, 2017). Young workseekers tend to have less formal education and on-the-job training than adults, which could suggest to prospective employers that they are less qualified. Youth with limited prior work experience may also be disadvantaged by their lack of references from past employers and the small size of their professional networks. As outlined in Fig. 12.2, these "frictions" that interfere with the hiring process could explain the disproportionately negative employment and earnings outcomes of young people.

Past research has established that pervasive youth unemployment has wide-ranging and long-lasting implications for out-of-work young people and for the countries where they live. Long unemployment spells in youth increase the probability of unemployment into adulthood and may have lasting effects on a range of other schooling and economic outcomes (Schmillen & Umkehrer, 2017; Mroz & Savage, 2006; Bell & Blanchflower, 2011). Research out of Europe suggests that, in the absence of early policy interventions, even short periods of high youth unemployment may become a regular feature of a country's economy (Caporale & Gil-Alana, 2014). Entrenched youth unemployment is associated with economic instability, crime, and political unrest in many countries (Fougere et al., 2009; Okafor, 2011; Ajaegbu, 2012).

This case study describes how researchers adapted an existing technology to target the disconnect between young workseekers and employers in South Africa. While young people in South Africa experience particularly high rates of unemployment, they also demonstrate relatively high levels of digital proficiency (Pew Research Center, 2018), suggesting they may constitute a demographic likely to benefit from technological solutions to formal sector employment challenges. Researchers taught groups of young workseekers to use the world's largest online professional networking platform, LinkedIn. In principle, there are many reasons why an online networking tool such as LinkedIn could increase employment in South Africa. The South African workforce ranks among the highest and most diversified digital skills of any developing country, and digital proficiency continues to improve with Internet infrastructure improvements (Choi et al., 2020). Workseekers can use the online platform to acquire information about job openings, to apply for jobs, or to elicit referrals from other users. At the same time, employers can use the platform to acquire information about applicants or to post job openings. Results of the evaluation suggest that training young workseekers to use LinkedIn speeds up their transition into formal employment.

Addressing the youth unemployment crisis is imperative to South Africa's continued development as an economic leader in sub-Saharan Africa. Today, the nonprofit, private, and public sectors in South Africa each play a role in the removal of barriers to youth employment. The public sector has led the way, devoting myriad resources to aiding workers and employers in the job search process. Shortly after becoming a democracy in 1994, the Government of South Africa developed a multipronged strategy for improving young workseekers' access to jobs, focusing on improving formal education and expanding vocational training opportunities, creating public employment programs, and funding job placement programs. Especially since the mid-2000s, skills and training programs have proliferated in a variety of forms throughout the country. One of the most high-profile initiatives targeted at employers is the Employment Tax Initiative (ETI), a wage subsidy launched by the South African government in 2014 that effectively reduces the cost of hiring young South African workseekers. Despite evidence that many of these interventions have had some success (e.g., Ebrahim et al., 2017), levels of youth unemployment remain high, which has spurred the growth of private and nonprofit solutions. This case study suggests an important role for software development in tackling the challenge of youth unemployment.

2 Project Formation

This case study details how an international team of researchers from Duke University and RTI International, in partnership with a local nonprofit organization and a leading professional networking service, set out to tackle the problem of youth unemployment in South Africa. Between 2015 and 2019, the project team adapted an existing technological solution to job search barriers, introduced the innovation to young South African workseekers, and conducted an evaluation to assess the effectiveness of the innovation. Specifically, the team studied whether the online digital professional networking platform, LinkedIn, could be used to improve the employment outcomes of young, economically disadvantaged workseekers in South Africa. The evaluation was a randomized control trial (RCT) that randomly assigned some job training cohorts to receive training on LinkedIn platform use. Wheeler et al. (2019) detail the results of the evaluation.[2] This section of the case study describes how the research team devised the study with several practical factors in mind, including the existence of a reliable implementation partner and high levels of digital connectivity in the South African population.

2.1 Tailoring a Solution to the Local Context

LinkedIn is a social media site built to facilitate professional networking and development. Users of the site create public profiles containing information about their skills, qualifications, and employment experience. Profiles may also include endorsements written by former supervisors or colleagues. Workseekers and prospective employers engage with LinkedIn in several ways that may help connect workseekers to jobs. To name a few, workseekers can use LinkedIn to learn about job openings, to form virtual professional networks by adding other users as connections, and to submit job applications costlessly online. Employers can use LinkedIn to advertise job openings and to screen applicants based on information contained in their LinkedIn profiles. Survey data suggest that both workers and firms commonly use LinkedIn to find jobs (Collmus et al., 2016). Hiring managers reportedly use LinkedIn at both the recruitment stage and the interview stage to fill in information gaps about workseekers (Caers & Castelyns, 2010; Roulin & Levashina, 2018). Other types of digital networking have been proven effective in connecting local employers, mostly in high-income contexts, to nonlocal contract workers, mostly in low-income contexts (Agrawal et al., 2015). But research into the effectiveness of digital technologies as employment solutions remains thin. The study conducted by the research team is the first to experimentally evaluate the employment effects of training workseekers to join and use an online professional networking platform.

More than 75% of LinkedIn users come from outside the United States, including more than 7 million users based in South Africa (LinkedIn, 2020). South Africa is LinkedIn's largest market in Africa. But only approximately 20% of LinkedIn users in South Africa are between 18 and 24, reportedly due to the perception that

---

[2] Since the time of writing, the paper has been published in the American Economic Journal: Applied Economics. It can be located using the following citation: Wheeler et al. (2022). LinkedIn(to) Job Opportunities: Experimental Evidence from Job Readiness Training. *American Economic Journal: Applied Economics*, 14(2), 101–25.

professional networking platforms are geared toward workers with more experience (Barbarasa et al., 2017). The underrepresentation of the South African youth population on the platform may represent an opportunity for growth. Once they sign up to join LinkedIn, young people in low- and middle-income countries like South Africa engage with the platform more actively and add connections faster than their older counterparts. LinkedIn is widely used by South African firms as well.[3] Although LinkedIn may be particularly relevant for high-skilled positions in the labor market, firms in South Africa also actively use LinkedIn to recruit workseekers for the types of entry-level positions that most economically disadvantaged youth would occupy. Approximately 50% of positions advertised in South Africa are entry level.

Relative to other developing countries – especially other countries in sub-Saharan Africa – digital connectivity in South Africa is high. More than 50% of South Africans own a mobile device that can connect to the Internet and applications, and nearly 60% of the population is able to access the Internet by any means (Pew Research Center, 2018). Smartphone ownership among South Africans aged 18–29 is at least ten percentage points higher than the overall mean in South Africa. With the caveat that Internet access is not uniform across the entire South African youth population (Oyedemi, 2015), these statistics suggest that there may be scope to expand the use of digital professional networking platforms like LinkedIn within the population of unemployed South African youths.[4] At baseline, 89% of the participants in this case study reported having an account on at least one of the following social media platforms: Facebook, Twitter, LinkedIn, or Mxit.

2.2 Identifying a Local Partner

The Harambee Youth Employment Accelerator (Harambee) is perhaps one of the most well-known nonprofit organizations focused on helping young people find jobs in South Africa. Harambee provides job training and matching services to thousands of economically disadvantaged young workseekers in South Africa every year, and their organization continues to grow. As of 2018, they also operate in Kigali, Rwanda. Harambee offers several different training programs ranging in duration from 3 days to 8 weeks. The programs are designed to help workseekers acquire the skills needed to be successful in jobs in financial services, sales, logistics, and operations. At the end of some of the training programs for high performers,

---

[3] There were more than 250,000 active job postings on the LinkedIn platform at the end of the intervention.

[4] A study of students at a South African university found that students were online for 16 hours of the day (Uys et al., 2012). Only a small minority of the participants in the case study have a college degree, suggesting that the samples are not entirely comparable. Income inequality is one factor that seems to determine inequality in Internet access. Individuals living in rural South African communities may be significantly less likely to have and use a smartphone than those in urban areas (Dalvit et al., 2014).

Harambee also arranges interviews between workseekers and prospective employers belonging to the network of firms Harambee has built over the years.[5] In its first 9 years of operation, more than 700,000 workseekers completed Harambee's training programs, materializing into more than 160,000 jobs and work experiences with more than 500 employers.

Harambee admits workseekers aged 18–29 from low-income households with little to no work experience. Only a small minority possesses more than a high school education. The research team identified LinkedIn as a possible tool to overcome the specific set of constraints facing the workseekers participating in Harambee's job training programs. For many of the reasons listed in the introduction, traditional job search strategies are often unsuccessful for this sample of workseekers. The young people eligible for Harambee's services tend to be those who lack professional referrals, know relatively little about specific job vacancies, and struggle to afford job application costs and fees. Harambee job training participants belong to a demographic that may be particularly likely to see returns to nontraditional job search strategies. Among the top 20 firms with a history of interviewing and hiring Harambee workseekers, at least 12 had active job postings on LinkedIn for positions that require only high school education at the time of the intervention. Firms using LinkedIn to recruit for entry-level positions report high demand for many of the skills Harambee seeks to hone during their training programs, including sales, marketing, and soft skills (Barbarasa et al., 2017).

Participants in the research study included 1,638 Harambee "candidates" enrolled in 30 of Harambee's "bridging programs" across South Africa. The research team was able to involve Harambee candidates in the study due to a long-standing partnership between Harambee and several of the scholars on the research team. Investments in relationships with implementation partners or government agencies are often important inputs into the research process. In this case, the success of the research project hinged on the opportunity to partner with Harambee. The research team capitalized on Harambee's job readiness training model as well as on their experience with research.

Harambee attributes its success in large part to its reliance on research. The organization administers a series of longitudinal surveys to the workseekers who go through their training program, both at the start of the program, when the trainees are "candidates," and years after the completion of the program, when they are "graduates." With a data store of millions of psychometric and non-psychometric assessments as well as longer-term employment outcomes, Harambee draws on rich analytics to make adjustments to their training programs over time. In addition, Harambee has a history of partnering with researchers from academic institutions to evaluate the efficacy of various employment interventions. Due to

---

[5] Although workseekers eligible for Harambee's services tend to face less promising job prospects than the average South African workseeker, the individuals invited to join the "bridging programs" are those receiving the highest scores on the cognitive, communication, and numeracy assessments administered by Harambee.

past involvement in large-scale randomized experiments, Harambee possesses a great deal of institutional knowledge about how to conduct rigorous experimental research. With Harambee as a partner, the intervention would be led on the ground by those with an understanding of important elements of the research design. For instance, Harambee training managers understood the need to keep control and treatment cohorts uninformed of their treatment status, also known as blinding, to mitigate bias that may result from participants behaving differently or reporting their outcomes differently based on knowledge of being treated.

Perspectives on the Researcher-Implementation Partner Relationship

Patrick Shaw and Laurel Wheeler: "Developing a good relationship with implementation partners is essential. Without the perspective and experience of researchers in the project space, the study is limited to second-hand expertise. Edwin Lehoahoa, a Harambee manager in public/private partnerships and burgeoning South African researcher, was particularly integral to the success of the research project. Edwin helped lead the implementation of the study and provided invaluable insights at several stages of the project. Case in point, we consulted with Edwin on a weekly basis during the writing process for this case study."

Edwin Lehoahoa: "Our journey towards making professional social networks more widely accessible and engaging depends on not only our ability to design them to attract the younger generation but our ability to break the data cost barrier. Being an organization that is demand-led, when Harambee was afforded the opportunity to learn about the linkages that platforms in general can create between workseekers and opportunity holders, we were thrilled. Especially the idea of learning how LinkedIn, a platform traditionally used as a networking platform by professionals and highly experienced individuals can be evolved in order to attract and link more marginalised entry level work seekers with opportunities that are less formal and require lower experience and qualification levels. Data is our biggest asset at Harambee and this study afforded us the ability to use some of the outcomes to shape our journey towards becoming a platform organization that promotes inclusion and drives social change at scale."

Best practices for conducting research in developing countries point to the importance of relying on the knowledge and experience of local implementation partners. The most productive relationships are built on the foundation of mutual respect for the other party's expertise.

To select the sample for the study, the research team relied on Harambee's existing process of recruiting and vetting workseekers. Harambee provided their administrative data on workseeker characteristics, including the data from cognitive and noncognitive assessments administered during the onboarding process. After sample recruitment, Harambee's 6- to 8-week job training programs for high-performing workseekers, known within the organization as "bridging programs," constituted the perfect environment within which to randomly introduce a curriculum based on LinkedIn.

For the most part, the incentives of the research team aligned with the incentives of the organization. From Harambee's perspective, the intervention would be low cost and potentially high reward. Time spent on the intervention would displace only 4 hours of the standard 6- to 8-week job training programs. Recruitment, screening, and other activities off of which the research study piggybacked were already taking place, thereby representing zero additional cost. The organization's familiarity with research and randomized experimentation also minimized the cost of educating the job readiness trainers responsible for implementation. At the same time, Harambee was invested in learning the outcome of the research. As an organization devoted to understanding the youth unemployment crisis in South Africa, the results of the research could be invaluable in terms of the design of future training programs or in terms of advocacy work between Harambee and the government and private sectors.

2.3   Defining and Refining the Problem Boundary

The research team cemented their interest in studying digital networking technologies as solutions to youth unemployment when they answered a call for proposals issued by LinkedIn's social impact sector. Before finalizing the research design, however, researchers iterated through multiple sets of research questions in the early phases of the project. The original proposal focused on the power of networking technology to elevate workseeker expectations and aspirations. The underlying theory was that young workseekers may hold inaccurate and pessimistic beliefs about their interconnectedness to prospective employers and other workers. The research team hypothesized that providing young workseekers with a visual mapping of their professional connections would improve their employment outcomes.

Ultimately, the set of research questions addressed by the intervention did not include questions related to data visualization. In practice, the research evolved largely with the team's understanding of data availability. For reasons discussed in subsequent sections, LinkedIn did not provide the research team with access to all the data collected and maintained by their organization. While LinkedIn shared coarse measures of user network size, they did not share detailed data on the connections themselves. For example, the provided data would indicate that workseeker A had N LinkedIn connections 6 months after the Harambee training program, but not that workseeker A connected to workseeker B on date T. The lack of granularity in the networks data forced researchers to redefine the problem boundary. In addition to information about the number and type of network connections, LinkedIn shared data on account opening, profile completion, and search activity such as how often participants used LinkedIn to view and apply for jobs. These measures enabled the research team to assess whether introducing young workseekers to LinkedIn increases the probability of opening an account (the extensive margin) and increases account usage (the intensive margin).

Data limitations also precluded researchers from conducting an in-depth exploration into the ways in which employers engage with LinkedIn in hiring. Job search is a two-way street. Just as workseekers may use the platform to collect information about job openings and prospective employers, employers may use the platform to collect information about job candidates. LinkedIn provided data on the number of times a study participant's LinkedIn profile was viewed in the final month of the Harambee training program; however, LinkedIn did not provide detailed data linking employer views to candidate profiles. To help contextualize and interpret the results of the evaluation and to fill in the gaps created by data availability, the research team administered a short end-of-evaluation survey to hiring managers affiliated with employers that regularly partner with Harambee.[6]

The final problem boundary was also shaped by the backgrounds and incentives of the research team, made up of a combination of economists, education researchers, and data scientists. The consensus view was that the research design should allow for measurement of LinkedIn usage as well as short- and long-run employment outcomes. But the mix of researchers generated interest in a variety of causal mechanisms, ranging from economic channels (e.g., screening and signaling) to behavioral channels (e.g., aspirations and educational engagement). In other words, researchers were interested not only in evaluating whether the intervention improved employment outcomes but also in understanding why. As part of the brainstorming stage of the research design, the team sketched a simple theory of change graphic outlining some of the hypothesized channels through which LinkedIn could affect employment outcomes. That sketch, which is reproduced in Fig. 12.3, guided the research design by helping researchers identify a set of measurable outcomes that may be related to LinkedIn use.

The original theory of change groups mechanisms into five categories: labor market information, networks, signaling/screening, psychosocial change, and educational change. In terms of the **labor market information channel**, LinkedIn could provide workseekers with information about job vacancies. It could also act as a centralized database of industries and employers, improving workseekers' understanding of how their particular skills and qualifications match up to opportunities. Establishing or expanding a **professional network** allows workseekers to connect with individuals in the workforce who may have information about available job opportunities. Professional networks may also prove to be important sources of referrals. LinkedIn could provide workseekers with improved technology to **signal their ability** to prospective employers who would then use this information to **screen prospective hires**. Traditional signals may be particularly weak for young people with limited work experience and educational attainment. Finally, LinkedIn

---

[6] The survey elicited responses from the human resources staff at three large firms that jointly employ 20% of the study participants. None of the hiring managers reported using LinkedIn in hiring.

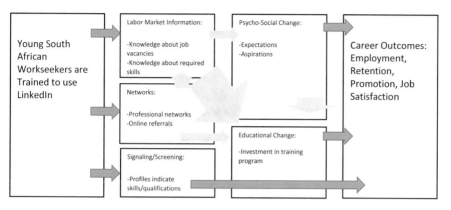

Fig. 12.3 Original theory of change

use signaling may be responsible for different types of behavioral change, including **educational investment** and **psychosocial change**. In terms of educational investment, LinkedIn usage could either increase workseeker enthusiasm for the job readiness training program and increase their effort or lead to discouragement and decrease effort. In terms of self-beliefs, LinkedIn may alter a workseeker's perception of his/her fit for certain jobs through exposure to role models, among other things. Through these mechanisms, LinkedIn may ultimately impact both the propensity to be employed and the quality of the match between the young person and a particular job.

At the start of the study, the research team was broadly interested in how workseekers and employers alike could utilize digital networking to overcome frictions in the job search process. The initial belief was that digital networking presents young workseekers with an opportunity to gain job search skills and learn about suitable job opportunities; it presents prospective employers with an opportunity to search online for suitable workers. By the time the research design was finalized, the focus had narrowed to a set of testable hypotheses shaped by the local context, the partnerships with Harambee and LinkedIn, data availability, and the composition of the research team.

3 Innovate, Implement, Iterate, Evaluate, and Adapt

3.1 Innovation

The act of finding a job has evolved vastly with the rise of technology. Throughout the mid-to-late twentieth century, the standard job search strategy involved handing out resumes in person or placing or answering print job ads. Respondents to the 1967 Current Population Survey (CPS) administered in the United States could select from five job search categories: checking with a public employment agency, checking with a private employment agency, checking with an employer directly, checking with friends or relatives, or placing or answering ads. In the early 1990s, "checking with the employer directly" was the modal search strategy, and "checking with private employment agencies" was the most successful (Bortnick & Ports, 1992).

As Internet access expanded, listings about open positions previously posted in the classified sections of local newspapers began to move online (e.g., Career-Builder.com, Monster.com). This worked both ways, as workseekers were increasingly able to post resumes and have employers do the searching. By 1998, 15% of workseekers in the United States were using the Internet to search for jobs (Kuhn & Skuterud, 2000). As the online market became flooded with resumes, referrals and professional networking became increasingly important.[7] LinkedIn was founded in 2002 as a platform that could simultaneously provide information about job openings, serve as a form of professional credentialing, and expand job search networks. Since then, demand for online platforms such as LinkedIn has continued to grow with the proliferation of smartphone usage.

LinkedIn has become a staple in many parts of the world, particularly in Organisation for Economic Co-operation and Development (OECD) countries. But LinkedIn tends to have lower visibility among workseekers and employers searching within developing countries. In some regions of the developing world, more localized platforms serve a function similar to LinkedIn. For example, Rozee.pk provides digital professional networking services in Pakistan (https://www.rozee.pk/), Babajob in India (http://www.babajob.com/), Silatech in the Middle East and North Africa (https://silatech.org/), and Shortlist in Africa and India (https://www.shortlist.net/). LinkedIn therefore exists within an ecosystem of online networking platforms that cropped up in response to the evolution of job search strategies.

This research project involved three innovations based on the existing LinkedIn technology. First, promoting the use of LinkedIn within the South African market is an innovation in application. Survey evidence from OECD countries shows that workseekers use LinkedIn to learn about prospective employers (Sharone, 2017). In addition, firms in OECD countries report that they collect information about job applicants from online media (Stamper, 2010; Shepherd, 2013; Kluemper et al., 2016). But there existed little research into how workseekers and firms use LinkedIn in the South African context.

Second, the research team collaborated with LinkedIn and Harambee to develop a curriculum that would provide *real* access to this technology – not just getting users online and connected, but training them on how digital networking could help them gain employment. After co-developing the curriculum with a member of the Harambee staff, the team disseminated the LinkedIn training curriculum

---

[7] Although Internet job search may not have been the most effective tool when it was first introduced (Kuhn & Skuterud, 2004), later research suggests that it has since contributed to reducing workseekers' unemployment durations (Kuhn & Mansour, 2011).

to Harambee job readiness trainers and taught them how to properly implement the LinkedIn training. The curriculum entailed a 1-hour presentation in the first week of the job training program plus additional coaching and discussion sessions in later weeks that covered building successful profiles, joining networks for targeted occupations, and on-platform searching for openings and companies. The curriculum is available in the paper's supplementary materials from the American Economic Journal: Applied Economics.[8] An example of the curriculum appears in the appendix.

The third innovation was to encourage the use of LinkedIn within an existing local learning environment. The rationale was that introducing the technology within an active job readiness training program with proven success would not only serve to augment digital literacy but also ensure sustainability. Other features contributing to sustainability of the technology include the stability of LinkedIn and the low cost of the intervention.[9]

3.2   Design and Implementation

Before introducing the LinkedIn training curriculum to Harambee workseekers, the research team spent approximately one year designing the intervention. As depicted in Fig. 12.4, this was the first phase of many that would take place between 2015 and 2019. The evaluation was designed as a randomized control trial, relying on randomized treatment assignment to expose the causal relationship between LinkedIn and employment outcomes.

Researchers exploited much of Harambee's preexisting infrastructure to streamline sample selection. As part of their regular process, Harambee recruits and screens workseekers and administers cognitive and noncognitive assessments during onboarding. These activities provided the sampling frame and data on population characteristics. Prior to randomization, the research team conducted a series of power calculations to determine how many workseekers would need to be included in the experiment. Using information about the average size of Harambee's job readiness training programs, as well as other information about design features and parameter estimates from South African survey data, researchers decided to include 30 training cohorts in the study. To illustrate this process, a power calculation exercise appears at the end of this chapter. Next, researchers conducted pairwise randomization of Harambee's next 30 job readiness training cohorts across its South Africa locations, effectively creating matched treatment-control pairs by location. All of these design decisions were logged and registered in the American Economic Association's Randomized Control Trial (RCT) Registry before implementation.

---

[8] https://doi.org/10.1257/app.20200025
[9] Cost-benefit calculations are given by the exercise at the end of the chapter.

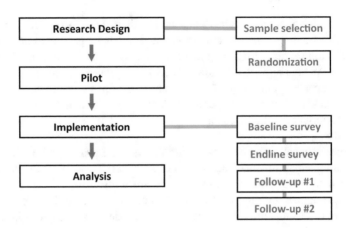

Fig. 12.4 Project phases

In the second phase of the project, the research team gathered in South Africa to run a pilot study of the feasibility of the research design. As part of the pilot activities, the team devoted substantial time to writing and revising the survey instruments. Focus group interviews with Harambee job readiness training program participants helped refine survey questions and tailor the LinkedIn training curriculum to the sample population. Another outcome of the pilot was the development of a short informational session directed at job readiness trainers to improve their understanding about the experimental nature of the intervention.

Approximately one month after the pilot study, the main intervention kicked off with the start of a job readiness training cohort in Johannesburg assigned to the control group. Two days later, the second study cohort began their training in Cape Town. Scaling up from the pilot study was relatively seamless, in large part thanks to Harambee. Researchers positioned the intervention directly within Harambee's existing training programs and used a portion of the job readiness training time to administer both the LinkedIn training and the baseline and endline surveys. All Harambee training centers provide workseekers with access to computers that they otherwise may not have. Without free Internet access, some workseekers may have been unwilling or unable to sign up for and use LinkedIn. In addition, administering the surveys during the Harambee training program ensured high response rates without the need to offer cash incentives for completing surveys. This costless survey administration is in contrast to the 6-month and 12-month follow-up surveys, which were costly in terms of hiring the third-party professional survey firm as well as in terms of the airtime incentives provided to study participants to increase response rates.

Because Duke and RTI researchers were based in different parts of the world, Harambee played an important role in overseeing the day-to-day execution of the intervention. Harambee's job readiness trainers administered the LinkedIn treatment. Each cohort was led by one job readiness trainer responsible for overseeing the entirety of the training for the 6–8-week program. Although some training cohorts included in the study did overlap temporally and geographically, participants of different training cohorts were kept largely separate, minimizing the opportunity for treatment spillovers. Trainers were familiarized with the LinkedIn training curriculum prior to the start of the intervention, but they were not notified of their cohort's treatment status until the first day of each program. Study participants would never be made aware of their treatment status, and the only difference between control and treatment training programs would be the addition of the LinkedIn training curriculum in the treatment cohorts.

Between 2016 and 2018, while control and treatment training programs took place across Johannesburg, Pretoria, Cape Town, and Durban, the research team focused their energy on data collection, cleaning, and storage. The last training program included in the study concluded in early 2018, but data collection efforts did not conclude until one year later. The impact evaluation ultimately consisted of a sample of 1,638 workseekers from 30 different training programs across four South African locations. The project produced a longitudinal dataset that followed workseekers from the start of their training at Harambee to 12 months after the completion of their training program.

3.3 Implementation Challenges and Iterations

The partnership with Harambee enhanced implementation efficiency. Nonetheless, the intervention was not spared the occasional implementation challenge. The challenges can be grouped into three broad categories: treatment administration, survey administration, and general data collection and cleaning.

(a) Treatment Administration

The intervention experienced several unanticipated lengthy delays between training programs. Harambee's job readiness training programs included in the study had scheduled start dates ranging from May of 2016 to November of 2017. Many of these programs overlapped with holiday seasons, which impacted enrollment and survey response rates. At times, four different programs would start in a 3-week span, and then no program would start for months at a time. Toward the end of the study, the programs were delayed significantly due to lack of enrollment. The research team expected to complete the intervention by mid-2017 but in reality did not finish until early 2018.[10]

In addition, a few technological glitches interfered with treatment administration at the individual level. During the LinkedIn training, some users were inadvertently locked out of their LinkedIn accounts. This issue, which affected five study

---

[10] In analysis, researchers control for timing to minimize the impact of delays and seasonality on estimates of treatment effects.

cohorts, was resolved with assistance from LinkedIn. A robustness check treating the affected cohorts as "non-compliant" shows that the lockout issues did not meaningfully change the results of the study. The cause of the issue was never identified.

The research design ceded much of the control over implementation to the trainers responsible for running Harambee's job readiness training programs. Early in the intervention, the research team learned the importance of instilling within these trainers a sense of accountability for the LinkedIn component of the training. One of the early training cohorts assigned to the treatment group did not receive the LinkedIn training because the trainer responsible for that cohort neglected to administer the curriculum. The research team was not able to oversee the implementation in person, and because the cohort was not based out of the main offices in Johannesburg, the research partners at Harambee were also unable to observe. Although the research team had built a strong relationship with Harambee as an organization, they had not at the onset built buy-in with each of the trainers nor implemented a trainer feedback system.

For the majority of the job readiness cohorts included in the intervention, implementation challenges were minimal. The research team informally received feedback from Harambee regarding points of confusion or difficulty for both the implementers and the workseekers. For example, an implementer from Johannesburg opined that "some of (the workseekers) hadn't spent a lot of time on the platform because they were intimidated by the platform layout," likely because they "didn't know where to find things or where to begin." The team addressed these concerns with updates to the LinkedIn training. As expected, implementer comfort with the curriculum progressed naturally throughout the study. The team also developed a short educational session for trainers, and Harambee further socialized them to the expectations of the project through one-on-one outreach. This alleviated many of the initial struggles for the program trainers.

(b) Survey Administration

Survey implementation at scale is a large coordination effort. The study had a baseline survey administered in the first week of the program, an endline survey administered in the last week, and 6- and 12-month follow-up surveys administered after the end of the training. Researchers sought to collect data from each study participant to maximize the representativeness of survey responses, but some amount of nonresponse was unavoidable. The baseline survey missed study participants who were absent on the day of the survey or who enrolled late in the job readiness training program, the endline survey missed some participants who found employment before the end of the program or who left the program for other reasons, and the follow-up surveys missed participants who were not reachable by the survey firm after the training program, often due to changing phone numbers. Nonresponse was under 1% for the end-of-program employment measures, 32% in the 6-month surveys, and 40% in the 12-month surveys. Wheeler et al. (2022) show that nonresponse does not differ by treatment status and is only weakly related to the baseline characteristics of study participants.

Survey administration presented additional challenges that proved costly, from either a time or pecuniary standpoint. The research team invested countless hours in cleaning survey data. The web-based baseline and endline surveys required respondents to enter their own identifying information in open text questions. The identifying information, which facilitated the merging of data from different sources, often did not match across surveys due to data entry errors. To resolve the mismatches, selected members of the research team collected and cross-referenced multiple identifying pieces of information, including the respondents' Harambee identification number, full name, date of birth, location, gender, and email address.[11] These data required an immense amount of cleaning. Furthermore, in some instances, study participants started the baseline survey, got kicked offline or inadvertently closed the window, and then started a new survey. Although rare, these responses showed up on the back end, forcing the research team to sift through duplicate responses as a regular part of the data cleaning process. The research team also invested time in coaching trainers on how to facilitate the web-based baseline and endline surveys. The research team explained when and how trainers could offer assistance without biasing responses. Finally, loss of power interrupted a few of the web-based surveys. Although not an unusual occurrence in South Africa, these power outages necessitated the readministration of some of the web-based surveys.

The follow-up survey administration proved extraordinarily costly from both a time and pecuniary standpoint. Because a companion study found low rates of response to both web- and SMS-based surveys in the same setting (Lau et al. 2018), the research team hired a professional survey firm to administer the follow-up surveys by telephone, which had a large impact on the budget. To maximize response rates, the survey firm called participants up to 30 times until successfully contacting them, which was both time-consuming and expensive. The call center averaged nine calls per respondent in the 6-month follow-up with 14% of respondents requiring 20 or more call attempts. Participants were even more difficult to reach at the 12-month follow-up, when the survey firm averaged twelve call attempts per respondent, with 24.5% of respondents requiring 20 or more call attempts. The final cost of the follow-up survey effort was US$93,000, which amounts to approximately US$30 per study participant per follow-up survey or US$44 per completed survey.

(c) Other Data Collection and Cleaning

Data collection and cleaning challenges measured larger than any other implementation challenge. In addition to the baseline, endline, and follow-up surveys administered specifically for this research project, the evaluation relied on LinkedIn data and Harambee administrative and performance data. The data provided by these third-party sources presented standard challenges of data availability and data

---

[11] The Duke researchers did not have IRB permission to access the identifying information of the study participants. RTI researchers collected those data and anonymized the data before sharing with Duke researchers.

quality. At the research design phase, researchers did not know with certainty which measures would be provided by various sources, so some of the planned analysis was not feasible. In addition, the Harambee in-training performance measures that were prespecified as outcomes of interest in reality contained too many missing values to include in analysis.

Sourcing timely LinkedIn data proved to be one challenge to data collection. Delays were due in part to the European Union's General Data Protection Regulation (GDPR), which was introduced in the middle of the study. For a short period of time, LinkedIn paused data sharing as the organization ensured compliance with the new regulations. Researchers addressed this issue statistically to account for the possibility of both of time-varying data collection effects as well as time-invariant geography effects. LinkedIn data collection presented another challenge in the form of missing data. The data shared with researchers contained missing values for a small number of study participants who had LinkedIn accounts and should have been included in the dataset. The research team investigated the problem by manually cross-referencing the names of the study participants with the names and work experiences of LinkedIn users. According to LinkedIn, these active accounts were not found by automated scraping procedures due to the existence of different emails, misspellings or typing errors, or variations on given names.

Data issues persisted throughout the study, so it was necessary to iterate data collection methods as the study progressed to ensure data reliability and validity. The product of the data collection efforts was a large panel dataset that combined microdata from more than ten different sources and tracked study participants from the start of the intervention to 12 months post intervention. The final dataset contained web-based and phone-based survey data, LinkedIn data, and Harambee administrative data. Simply connecting all of these pieces was difficult, and the result required extensive cleaning.

A final set of data challenges arose as a byproduct of inter-institution collaboration. Different institutions impose different standards for data storage. In accordance with Internal Review Board (IRB) standards at different institutions, the research team kept two datasets: one containing identifying information and a random identifier for each subject and an anonymized dataset containing the random identifier plus all substantive information from the surveys. The two datasets were stored separately and hosted on a cloud-based storage service that can be encrypted. No identifiable individual-level data was shared with Harambee or any other organization. In compliance with data sharing agreements between RTI and LinkedIn, no LinkedIn data was given to Duke researchers or stored on Duke servers.

3.4   Evaluation

The evaluation compared the experiences of study participants exposed to the LinkedIn training (the "treated" participants) to the experiences of study participants not exposed (the "control" participants). The research team discovered that, at the end of Harambee's job readiness training program, treated participants were 10% more likely to find immediate employment than control participants. Higher levels of employment persisted for at least 12 months following the intervention. Furthermore, treatment-control differences in employment appear to be driven by treatment-control differences in LinkedIn use. The research team found that exposure to the LinkedIn training increased the probability that a participant had an active LinkedIn account and also increased the intensity with which the participant engaged with LinkedIn. The results suggest that, at least in this specific context, training young workseekers to use LinkedIn speeds up their transition into employment.

Because the intervention was designed to be a solution to the challenge of youth unemployment in South Africa, researchers identified short-run and long-run employment as the primary outcomes of interest. Harambee provided a measure of short-run employment, a binary indicator of whether the study participant was employed immediately out of the job readiness training program. The research team administered surveys to collect data on longer-run employment 6 and 12 months post training. In addition to the binary employment information, the surveys solicited information about the quality of the worker-job match, measured by job retention, promotion, and permanency. In practice, researchers found little evidence of treatment impacting match quality.

Treatment-control differences in current employment suggest that something about the intervention was responsible for improving employment outcomes. In order to attribute those differences to LinkedIn use, the team analyzed whether and how study participants engaged with the LinkedIn platform. Extensive margin measures of LinkedIn use – such as whether study participants opened an account during the intervention – were examined alongside intensive margin measures of LinkedIn use, such as profile completion, number of profiles viewed, number and types of connections made, and number of job applications submitted through the platform. LinkedIn provided these measures at three points in time, roughly corresponding to the end of the training program and 6 and 12 months post training. LinkedIn designed their own tools for tracking user behavior and sent the research team the usage variables they had already constructed. In addition to using LinkedIn-constructed measures, researchers employed statistical techniques to devise alternative measures of LinkedIn usage. For instance, researchers used the first principal component of a set of intensive and extensive margin measures to reflect an aggregate measure of LinkedIn use. The research team found evidence of compliance with treatment, based on the majority of these measures. The LinkedIn treatment increased LinkedIn use both by encouraging study participants to open accounts and by increasing the intensity with which they engaged with their accounts.

Researchers designed the evaluation to include a host of other measures that would be used to shed light on the role of the hypothesized mechanisms laid out in the original theory of change (see Fig. 12.3). In terms of information provision, researchers found that treatment significantly increased the number of profiles viewed and the number of jobs viewed on the LinkedIn platform, suggesting that study participants may have used the platform to acquire information. The average number of profile and job views was nevertheless quite low. LinkedIn provided researchers with a measure of the number of articles read on the platform, but the vast majority of participants read zero articles.

In terms of network formation, LinkedIn provided the research team with four different variables describing a study participant's professional network on the platform: number of connections, number of connections with a bachelor's degree, number of connections in a managerial position, and the average power of connections.[12] Treatment had a positive impact on all network measures. LinkedIn measures did not include information about the ways in which participants interacted with their connections or about the identities of the network connections.

Treated participants were significantly more likely to have complete LinkedIn profiles, suggesting a role for the signaling of workseeker ability. Although the research design did not provide researchers with the opportunity to observe the ways in which employers used and interpreted LinkedIn profiles, several measured changes are consistent with the signaling/screening channel. First, researchers observe that the treatment employment rate rises by the end of the job readiness program, suggesting a quick mechanism like LinkedIn profiles helping workseekers pass employer screening. Second, treatment almost triples the number of times a workseeker's profile is viewed by other LinkedIn users in the month of program completion. This may reflect prospective employers viewing profiles during hiring. However, limited information from the short firm survey conducted ex post suggests that this mechanism likely was not important in determining the results. None of the hiring managers who responded to the survey reported that they viewed LinkedIn profiles during hiring.

Finally, to measure change in educational investment, researchers collected self-reported measures of interest in the job training program and studied job readiness trainer reports describing things like participant energy and intellectual curiosity. Treatment did not have an impact on these measures. To test the psychosocial channel, researchers included in their survey instruments questions about study participants' expectations and aspirations. Similarly, treatment had almost no impact on these measures. Although researchers only measured a subset of the possible types of self-belief, they found no evidence to suggest that the LinkedIn training impacted employment through its impact on behavioral change.

The research was not designed to test mechanisms directly but rather to provide suggestive evidence on the channels through which the LinkedIn training may affect the main outcome of interest: employment. The evaluation shows that LinkedIn training was responsible for increasing post-program employment from 70% among control participants to 77% among treated participants. This 10% increase is

---

[12] The average power of connections is calculated by LinkedIn based on the number of connections as well as the education and skills of connections. It is a measure of network quality that is constructed by LinkedIn.

comparable to the average effect of other active labor market interventions targeted toward long-term unemployed workseekers (Card et al., 2017). While recognizing that this intervention took place within a unique context, a case may be made for adopting similar technologies to solve youth unemployment challenges in other countries. This is particularly true in light of the high benefit-to-cost ratio associated with the intervention. The research team calculated that the cost of the intervention totaled approximately US$48 per study participant.[13] That is relative to the treatment-induced increase in earnings of approximately US$417 over the course of 1 year. Cost-benefit calculations are given by the exercise at the end of the chapter.

3.5 Adaptation

Based on the benefit-to-cost ratio alone, a case may be made for exploring the use of LinkedIn training in other countries. High youth unemployment is an issue that affects many developing countries, in particular countries in the Middle East and North Africa. But there are a few unique features of the study context that may limit generalizability. First, the study participants were disadvantaged but high achieving, suggesting higher-than-average technological familiarity for their demographic. Second, South Africa has high rates of LinkedIn use by workseekers and employers. Without a strong LinkedIn presence in a country, take-up would be low. Third, this case study describes an intervention that takes place within an existing job readiness training program. In the absence of an organization like Harambee, it is unclear whether the LinkedIn curriculum would be effective.

Scaling up from this intervention more broadly even throughout South Africa requires some consideration of the obstacles to digital networking. Platform accessibility and inclusivity are imperative for widespread take-up of digital networking tools. More than nine in ten Harambee workseekers use Internet-enabled phones (though not necessarily smartphones), but South Africa has some of the highest data costs in the Southern African Development Community. Thus, lowering costs through subsidies or incentives or expanding free public Internet access would be a big step in scaling. An additional option, which Facebook has recently tested, provides a stripped down application version available at little to no data cost and allows youth from marginalized communities and those without smartphones to engage with the network. This also provides an opportunity for marginalized young people to have a voice in the development of these types of innovations.

Another aspect of scalability pertains to the quality of other operators in the employment ecosystem. Harambee has found success in its workflow model, but the organization cannot reach every individual that needs training and support.

---

[13] The estimated cost is US$48 using purchasing power parity adjustments but US$21 at the nominal exchange rate.

The structure of the LinkedIn intervention can be easily replicated in a similar "facilitating" environment and in other sectors. Existing vocational training and job readiness programs across sub-Saharan Africa do not necessarily simulate Harambee's environment.

4 Lessons Learned

Five years after devising a research question about whether LinkedIn could increase youth employment in South Africa, the research team has produced the first experimental evidence that training participants in a job readiness training program to use an online professional networking platform improves employment outcomes. The LinkedIn training curriculum increased immediate employment by 10%, an effect that persisted for at least a year after job readiness training. The research team estimates that higher rates of employment among participants who received the LinkedIn training were largely driven by enhanced LinkedIn use.

Given the fairly distinct characteristics of the study population and employment context, these findings suggest several potential avenues for future research. Are there specific aspects of digital networking that drive the employment results? Can these tools be scaled or used at all outside of the job readiness environment? Would digital networking be as effective in a different country or with a different population of workseekers?

Even if the findings of the evaluation apply only to the employment conditions of a specific population, the lessons learned from the research project are wide reaching, including lessons about policy design, lessons about study implementation, and lessons about future technology design. Policy recommendations are rather straightforward, such as reducing data costs and encouraging the use of digital networking platforms in vocational and other job training programs. These insights into policy design have already informed Harambee's advocacy with the government and private sectors. The results of the research study have strengthened Harambee's call for a reduction in data costs to enable young people to access digital professional platforms.

In terms of study implementation, the research project underscored the importance of investing in redundant backup strategies to account for losses in Internet or power, such as laptops or tablets that can be used offline for training content delivery and survey data collection. In addition, for researchers working with a remote implementation partner, careful monitoring for fidelity of implementation, at least toward the beginning of the study, may alleviate some of the issues this research team experienced in delivering the intervention.

Finally, for future developers of digital job platforms, this research suggests a need for strategies to generate greater take-up. For populations that are economically disadvantaged, digital platform developers can design applications with low data demands. For populations that are not technology savvy, digital platform developers can build supportive curricula and training materials into their platforms, to help users onboard and interact with the platform in a deeper way.

A.1 Appendix

(i) LinkedIn Training Curriculum

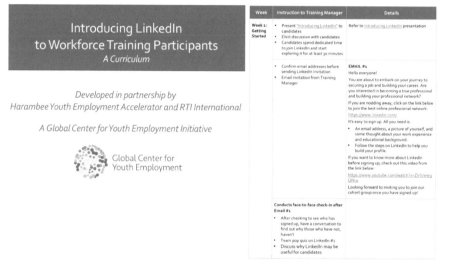

This image depicts the first part of the LinkedIn training curriculum administered by job readiness training managers. The full curriculum is accessible with the published version of the paper from the American Economic Journal: Applied Economics sited at https://doi.org/10.1257/app.20200025.

(ii) Power Calculation Exercise

For most randomized evaluations, power calculations are an important step in the research design. Researchers design their experiments such that they are *powered* to detect post-intervention differences between control and treated participants. The power of the research design is the probability of rejecting the null hypothesis that treatment does not change outcomes when the null hypothesis is in reality false. One concern is that, if effect sizes are small, researchers may be *underpowered* to detect them and will fail to reject the null hypothesis when in reality the intervention was responsible for small changes in outcomes.

At the beginning of the intervention, researchers must consider trade-offs between the sample size they'll use in their study and the minimum effect size they want to be able to detect, known as the *minimum detectable effect size* (*MDE*). As they increase the sample size, they decrease the MDE. In other words, with more data, the researchers are able to detect increasingly small responses to treatment.

In reality, time and resources are limited, and the study sample cannot include the entire population of interest. This is where power calculations come into play.

Let's illustrate the basic principles of power calculations by setting the parameters and calculating the minimum detectable effect size.[13] The relationship between the MDE and the design parameters is given by Eqs. (12.1) and (12.2), with the parameters defined below:

$$MDE = \left(t_{(1-\kappa)} + t_{\alpha/2}\right) * \sqrt{\frac{1}{P\left(1-P\right)J}} * \sqrt{\rho + \frac{1-\rho}{n}} * \sqrt{\sigma^2} \qquad (12.1)$$

$$\sigma^2 = \frac{(1-\tau) + \tau * M}{M} \qquad (12.2)$$

The relevant parameters include the following:

1. **Test size critical value** $(t_{\alpha/2})$: The value of this parameter is obtained from a standard t distribution. The researcher selects a significance level α and uses a t-table to identify the critical value associated with a two-sided test:

   (a) Assume: The level of statistical significance you aim to achieve is 5%
   (b) Question: What is the value of $t_{\alpha/2}$?
   (c) Solution: $t_{\alpha/2} = 1.96$.

2. **Test power critical value** $(t_{(1-\kappa)})$: The value of this parameter is also given by a t-table based on the level of power specified by the researcher:

   (a) Given: Researchers are typically satisfied with power of 80%.
   (b) Question: What is the value of $t_{(1-\kappa)}$?
   (c) Solution: $t_{(1-\kappa)} = 0.84$.

3. **Number of clusters** (J): This study involved randomization of training groups rather than individuals, and treatment was administered at the group level. The number of clusters reflects the number of training groups. The researcher will solve for the number of clusters needed to obtain a large enough sample size for a given MDE:

   (a) Question: Does the MDE increase or decrease as the number of clusters increases?
   (b) Solution: The MDE decreases as the number of clusters increases, corresponding to an improvement in the study's power.

---

[13] Note that these parameters and calculations are based on the design factors specific to the study at hand. They differ across studies based on differences in terms of the number of treatment arms, the level of randomization, the degree of compliance, etc. For more information, Duflo, Gennerster, and Kremer (2007) provide an excellent treatment on the principles of power calculations.

4. **Number of units within each cluster** (n): The number of units within each cluster reflects the expected number of individuals within each training group:
(a) Assume: The number of units within each cluster is identical
(b) Question: Does the MDE increase or decrease as the number of units within each cluster increases?
(c) Solution: The MDE decreases. Again, increasing the overall sample size improves the power of the study (decreases the MDE).

5. **Intracluster correlation coefficient** (ρ): This parameter accounts for the correlation of outcomes within groups. The intracluster correlation coefficient is the share of variance explained by within group variance. Estimates of ρ could come from pilot data or from national surveys of representative samples:

(a) Given: Based on data on the baseline characteristics of the first few groups in the study, researchers estimate $\rho = 0.0165$.[14]
(b) Tip: For any nonzero ρ, increasing J improves precision more than increasing n.

6. **Fraction treated** (P): As part of the research design, the researcher determines the fraction of the sample that receives treatment. Unless the treatment is expensive, it is generally optimal to treat half the sample because $P = 0.5$ minimizes the MDE:

(a) Given: $P = 0.5$.

7. **Number of posttreatment measures** (M): Some outcomes are measured at multiple points in time after the intervention. This parameter is determined by the research design:

(a) Given: The researchers plan to measure the outcome of interest at 6 months and 12 months post-intervention, so $M = 2$.
(b) Question: How does the number of posttreatment measures affect the power of the study?
(c) Solution: The number of posttreatment measures factors into the residual variation. Increasing M decreases residual variation, which increases the power of the study.

8. **Intertemporal correlation coefficient** (τ): This parameter accounts for the correlation of outcomes over time. It is the share of the variance explained by variance over time. This parameter only becomes important if outcomes are measured at multiple points in time. Estimates of τ often come from longitudinal surveys of representative samples:

(a) Given: Based on multiple waves of data from that National Income Dynamics Study, researchers estimate that $\tau = 0.312$.

---

[14] Note that it's best to use a range of estimates. Point estimates are given here for the sake of clarity.

9. **Residual variation (σ^2):** This parameter is calculated based on the values of τ and M in a relationship given by Eq. (12.2):

   (a) Question: What is the value of σ^2 based on the information given in this example?

   (b) Solution: $\sigma^2 = 0.656$.

   (c) Question: Does the MDE increase or decrease with σ^2?

   (d) Solution: The MDE increases with σ^2, indicating that decreasing the residual variation improves the study's power.

Putting it all together:

Question (1): Given the above parameter values, what is the minimum detectable effect size associated with 30 groups of 70 participants per group?

Question (2): Interpret this.

Question (3): How do you think researchers would determine if that's an acceptable MDE for the study?

Solution (1): 0.1447.

Solution (2): Given the above parameter values, a sample of 2,100 study participants evenly divided into 30 groups would allow researchers to reject the null hypothesis with statistical certainty only if the treatment effect is greater than 0.1447.

Solution (3): Power calculations are not an exact science. Researchers typically base this determination on evidence from previous research and intuition. Based on past studies, what effect size would you expect to see? Would you want to be able to detect smaller effect sizes if it's feasible from a cost standpoint?

(iii) Cost-Benefit Calculation Exercise

Given:

1. Harambee's average cost per candidate enrolled in the 8-week version of the job readiness training program is US$3,833 purchasing power parity (PPP).
2. The job readiness training program meets for 40 hours per week for each of the 8 weeks.
3. Harambee allocates approximately 4 hours to LinkedIn training per job readiness program.
4. The job readiness training program costs cover staff time for training, administration, etc.; facility rental; IT costs; and participant stipends.
5. The intervention increases employment by 6.9 percentage points in the sample of 890 treated candidates.
6. The South African national minimum hourly wage is approximately US$3 PPP or US$6,050 annual earnings.

Question 1: Assuming that the LinkedIn training does not increase any of Harambee's fixed or variable operational costs, what is the average cost per candidate of the LinkedIn intervention?

Question 2: What is the cost of the LinkedIn intervention per each additional candidate employed as a result of the intervention?
Question 3: Calculate the lower bound for benefit of the treatment in terms of earnings. Assume no one earns more than the minimum wage. How much does treatment increase the average candidate's annual earnings?
Question 4: What is the benefit-cost ratio based on your answers to the above?

Answer 1: US$48 PPP.
(Answers are rounded to the nearest dollar.)
Answer 2: US$694 PPP.
Answer 3: US$417 PPP.
Answer 4: 8.7:1.

References

Agrawal, A., Horton, J., Lacetera, N., & Lyons, E. (2015). Digitization and the contract labor market: A research agenda. In A. Goldfarb, S. Greenstein, & C. Tucker (Eds.), *Economic analysis of the digital economy* (pp. 219–250). University of Chicago Press.
Ajaegbu, O. O. (2012). Rising youth unemployment and violent crime in Nigeria. *American Journal of Social Issues & Humanities, 2*(5), 315–321.
Banerjee, A., et al. (2008). Why has unemployment risen in the new South Africa? *The Economics of Transition, 16*(4), 715–740.
Barbarasa, E., Barrett, J., & Goldin, N. (2017). *Skills gap or signaling gap?: insights from LinkedIn in emerging markets of Brazil, India, Indonesia, and South Africa Solutions for Youth Employment, Washington D.C.* LinkedIn.
Bell, D. N. F., & Blanchflower, D. G. (2011). *Youth unemployment in Europe and the United States*. IZA Discussion Paper No. 5873, April 2011
Bortnick, S. M., & Ports, M. H. (1992). Job search methods and results: Tracking the unemployed, 1991. *Monthly Labor Review, 115*(12), 29.
Caers, R., & Castelyns, V. (2010). LinkedIn and Facebook in Belgium: The influences and biases of social network sites in recruitment and selection procedures. *Social Science Computer Review, 29*(4), 437–448.
Caporale, G. M., & Gil-Alana, L. (2014). Youth unemployment in Europe: Persistence and macroeconomic determinants. *Comparative Economic Studies, 56,* 581–591.
Card, D., Kluve, J., & Weber, A. (2017). What works? A meta analysis of recent active labor market program evaluations. *Journal of the European Economic Association, 16,* 894–931.
Choi J, Dutz M A, Usman Z (2020) The future of work in Africa: Harnessing the potential of digital technologies for all.. The World Bank Group
Collmus, A. B., Armstrong, M. B., & Landers, R. N. (2016). Game-thinking within social media to recruit and select job candidates. *Social media in employee selection and recruitment: Theory, practice, and current challenges,* 103–124.
Dalvit, L., Kromberg, S., Miya, M. (2014). The data divide in a South African rural community: A survey of mobile phone use in Keiskammahoek. *Proceedings of the e-Skills for Knowledge Production and Innovation Conference 2014,* Cape Town, South Africa: 87-100
Duflo, E., Glennerster, R., Kremer, M. (2007). Using randomization in development economics research: A toolkit. CEPR Discussion Paper No. 6059
Ebrahim, A., Leibbrandt, M., Ranchhod, V. (2017). The effects of the Employment Tax Incentive on South African employment. WIDER Working Paper 2017/5

Fougere, D., Kramarz, F., & Pouget, J. (2009). Youth unemployment and crime in France. *Journal of the European Economic Association, 7*(5), 909–938.

International Labour Organization. (2017). *Global employment trends for youth 2017: Paths to a better working future.* International Labour Office.

International Labour Organization. (2018). *Women and men in the informal economy: A statistical picture.* International Labour Office.

Kerr, A. (2017). Tax(i)ing the poor? Commuting costs in South African cities. *South African Journal of Economics, 85*, 321–340.

Kingdon, G., Knight, J. (2001). *Why high open unemployment and small informal sector in South Africa? Centre for the Study of African Economies, Economics Department,* University of Oxford.

Kluemper, D., Mitra, A., & Wang, S. (2016). Social media use in HRM. In *Research in personnel and human resources management* (pp. 153–207). Emerald Group Publishing Limited.

Kuhn, P., Mansour, H. (2011). Is Internet job search still ineffective? IZA Discussion Paper No. 5955

Kuhn, P., & Skuterud, M. (2000). Job search methods: Internet versus traditional. *Monthly Labor Review, 3*, 3–11.

Kuhn, P., & Skuterud, M. (2004). Internet job search and unemployment durations. *American Economic Review, 94*(1), 218–232.

Lam, D., Ardington, C., & Leibbrandt, M. (2011). Schooling as a lottery: Racial differences in school advancement in urban South Africa. *Journal of Development Economics, 95*, 121–136.

Lau, C. Q., Johnson, E., Amaya, A., LeBaron, P., & Sanders, H. (2018). High stakes, low resources: what mode(s) should youth employment training programs use to track alumni? Evidence from South Africa. *Journal of International Development, 30*(7), 1166–1185.

LinkedIn. (2020). The LinkedIn register. https://news.linkedin.com/about-us#1. Accessed 12 August 2020

Magruder, J. (2010). Intergenerational networks, unemployment, and persistent inequality. *South Africa American Economic Journal: Applied Economics, 2*, 62–85.

Mosomi, J., & Wittenberg, M. (2020). The labor market in South Africa, 2000–2017. *IZA World of Labor, 2020*, 475.

Mroz, T., & Savage, T. H. (2006). The long-term effects of youth unemployment. *Journal of Human Resources, XLI*(2), 259–293.

Okafor, E. E. (2011). Youth unemployment and implications for stability of democracy in Nigeria. *Journal of Sustainable Development in Africa, 13*(1).

Oyedemi, T. (2015). Participation, citizenship and internet use among South African youth. *Telematics and Informatics, 32*(1), 11–22.

Pew Research Center. (2018). Internet connectivity seen as having positive impact on life in sub-Saharan Africa

Rankin, N., & Roberts, G. (2011). Youth unemployment, firm size and reservation wages in South Africa. *South African Journal of Economics, 79*, 128–145.

Roulin, N., & Levashina, J. (2018). LinkedIn as a new selection method: Psychometric properties and assessment approach. *Personnel Psychology, 72*(2), 187–211.

Schmillen, A., & Umkehrer, M. (2017). The scars of youth: Effects of early-career unemployment on future unemployment experience. *International Labour Review, 156*(3-4), 465–494.

Sharone, O. (2017). LinkedIn or LinkedOut? How social networking sites are reshaping the labor market. In *Emerging conceptions of work, management, and the labor market.* Emerald Publishing Limited.

Shepherd, B. (2013). Social recruiting: Referrals. *Workforce Management, 5*(18).

Stamper, C. (2010). Common mistakes companies make using social media tools in recruiting efforts. *CMA Management, 12*, 13–22.

Statistics South Africa. (2019). *Quarterly Labour Force Survey (QLFS), 1st quarter 2019.* Statistics South Africa.

Taylor, S., Van Der Berg, S., Reddy, V., Janse Van Rensburg, D. (2011). How well do South African schools convert grade 8 achievement into matric outcomes? Tech. Rep. 13/11, Stellenbosch Economic Working Papers

Uys, W., et al. (2012). Smartphone application usage amongst students at a South African university. IST-Africa 2012 Conference Proceedings: 1–11

Wheeler, L.E., et al. (2019). LinkedIn(to) job opportunities: Experimental evidence from job readiness training. University of Alberta Working Paper No. 2019-14

Wheeler, L.E., et al. (2022). LinkedIn(to) Job Opportunities: Experimental Evidence from Job Readiness Training. *American Economic Journal: Applied Economics, 14*(2), 101–25.

Chapter 13
Amplifying Worker Voice with Technology and Organizational Incentives

Achyuta Adhvaryu, Smit Gade, Piyush Gandhi, Lavanya Garg, Mansi Kabra, Ankita Nanda, Anant Nyshadham, Arvind Patil, and Mamta Pimoli

1 Development Challenge

Across the globe, the apparel industry operates through a vast value chain of both formal and informal workers. This includes farmers that grow cotton, factories that turn fabrics into finished garments, and retail stores that sell these garments to consumers. While retail stores are found in many different parts of the world, most production and manufacturing processes are concentrated in low-income countries. This establishes a certain degree of separation between the retailer and manufacturer and even more between the consumer and the frontline worker.

This disconnect, even if unintentional, can lead to business practices that do not internalize the well-being of workers, and the consequences of this disconnect can be drastic. An extreme example is the Rana Plaza tragedy in 2013, in which over 1000 workers died in a commercial building collapse in Bangladesh, caused largely by persistent neglect of working conditions and disregard for workers' concerns around building safety.

This case study describes how technology can be used to increase connection and engagement among stakeholders on the factory floor. The solutions we describe hold the potential to increase transparency and accountability for vulnerable workers

A. Adhvaryu (✉)
William Davidson Institute, Ross School of Business, University of Michigan, Michigan, MI, USA

Good Business Lab, New Delhi, Delhi, India
e-mail: nyshadha@umich.edu

S. Gade · P. Gandhi · L. Garg · M. Kabra · A. Nanda · A. Patil · M. Pimoli
Good Business Lab, New Delhi, Delhi, India

A. Nyshadham
Ross School of Business, University of Michigan, Michigan, MI, USA

© The Author(s) 2023
T. Madon et al. (eds.), *Introduction to Development Engineering*,
https://doi.org/10.1007/978-3-030-86065-3_13

323

across South Asia's garment industry. In the status quo, relationships between workers and their supervisors are rife with frictions, owing to production pressures, gender dynamics, low wages, weak work motivation, low self-esteem, and feelings of unbelonging (particularly for migrant workers). While the relationship between managers and workers – and the extent to which workers feel valued – is a key determinant of firm performance (Adhvaryu et al., 2019; Ashraf & Bandiera, 2018; Bandiera et al., 2009; Hoffman & Tadelis, 2018), these issues are too easily deprioritized by business owners, who often fail to perceive their role in firms' profitability and long-term survival.

The development challenge of creating high-performing firms that can retain their workers – while providing quality jobs to low-income women – is central to economic growth. We address this challenge through technology that enables responsive and supportive interpersonal relationships among workers and supervisors. Our approach, like many innovations described in this textbook, will combine a software platform coupled with a set of novel iteratively designed organizational management strategies. The result is a solution that is highly adapted across developing country contexts – and one that can achieve impact at scale.

1.1 Establishing Connection: Workers in the Garment Industry

The garment manufacturing industry in many low-income country contexts is highly labor-intensive. It spans both the formal and informal sectors of the economy. Production for multinational retail brands in almost all cases happens in the formal setting where firms hire their employees through contracts and often provide workers with basic benefits such as earned leave and government-mandated (in India) health insurance coverage.

In India alone, the textile and apparel industry employs over 45 million individuals directly, making it the second largest sector in terms of employment after agriculture. The industry has been an integral part of the Indian economy since the seventeenth century, reaching a size of $140 billion in 2018. While COVID has dampened its growth, along with India's economy as a whole, according to data released by the Confederation of Indian Textile Industry (CITI), the recovery for the domestic market is expected to be quite steep after the pandemic. The domestic market alone is estimated to reach USD 120 billion by 2024. The sector contributes 2.3% to India's GDP, 7% of the country's manufacturing production, and 13% of the country's export earnings.

Box 13.1: Protections for Formal Sector Workers in India
In the formal setup of the apparel industry, workers are employed in a setting that requires compliance with both legal regulations and retailer standards for

(continued)

> **Box 13.1** (continued)
>
> worker safety. Most workers are hired not on a short-term contract basis, but as full-time employees with workplace benefits such as retirement plans and employee health services. While short-term contracts and informal work do form a part of the industry, this case study focuses on formal workers.
>
> In India, wages for frontline workers are benchmarked to government minimum wage policy, which is largely determined at the state level. The minimum wage consists of two parts – a "basic" portion and a "dearness allowance," which is intended to allow for cost of living adjustments. The dearness allowance is adjusted every year for inflation, while adjustments to the basic wage level are made roughly every 5 years by the Government of India and commonly result in larger increases than the more frequent inflation adjustments.
>
> Wages are fixed; however, workers are entitled to receive additional compensation if they work overtime. Collective bargaining is allowed, and salaries are negotiated according to company policies and local labor laws. There are several trade unions working for garment factory workers, for example, the Karnataka Garment Workers Union (KOOGU) in the southern state of Karnataka.

Our research team has worked for the past decade with Shahi Exports Pvt. Ltd., India's largest apparel export house. Shahi operates close to 60 large garment factories in India, employing nearly 120,000 workers. Here, we describe the day-to-day work conditions for Shahi employees.

Frontline workers typically earn minimum wage, which in India is USD 105–135 per month. Salaries are deposited to workers' bank accounts around the seventh day of every month, usually without delays in disbursement. Working hours and holidays are in accord with national labor laws: work shifts are 9–10 hours long, including a lunch break for 30–45 minutes (depending on the factory). The job is 6 days a week, with a day off on Sundays. Approximately 80% of the total workforce are migrants from rural India, and they send remittances of up to USD 40–55 per month on average to family members in their villages, leaving little opportunity for personal savings.

The majority of the garment manufacturing workforce comprises young female workers who stay on the job for an average of 1 year. Based on data analyzed from our partner firm, almost 50% of migrant workers drop out of the workforce within the first 6 months after joining. The industry's direct and indirect costs of turnover are high, especially in India. Direct costs are incurred by the continuous need to hire and train new cohorts of workers; indirect costs include regularly disrupted production, reduced productivity due to high absenteeism, and low worker morale.

A typical migrant worker comes to the manufacturing area from a rural district in or outside of their residing state. They often face difficulty in acclimatizing to new language, culture, food, co-workers, and cohabitants – all while being separated from their families and social networks back home. In this unusually high-pressure environment, greater exercise of worker voice could improve well-being and sustain young women's labor market participation.

In fact, larger garment manufacturing firms that contract with international buyers actually do invest in employee well-being and development. For example, the industry has scaled soft-skills training among workers, and well-being measures have become an integral part of management practice. Some firms also provide medical facilities, child care, and welfare schemes for workers. However, a majority of small garment factories, especially those producing for domestic buyers (or subcontracting for medium and large firms), lack transparency and accountability with regard to working conditions and employment contracts.

Facilities provided by the firm discussed in this case study include free child care in every unit with full-time caretakers, support staff, and nutritious free food; medical center in every unit, equipped with free medication, ambulance, and first-aid services; bank accounts for every employee and secure ATMs on site; support in applying for relevant government benefits; merit-based scholarships for employees' children; human resource team at all units trained in basic counseling skills including behavior change, working with families, and group counseling; specialized counseling cells in 15 units, with professional counselors; and regular employee engagement programs, cultural events, and festival celebrations.

1.2 Attrition, Voice, and Worker Well-Being

Despite recent improvements in employee benefits, there are significant issues facing Indian garment workers and firms. International buyers expect manufacturers to complete regular social audits, conducted by third-party organizations, yet ironically, this process discourages factory management from keeping records of reported grievances. It also disincentivizes managers from taking strict action on complaints, for fear of unfavorable audit reports (which harm the factory's business). Operating in a high-pressure manufacturing environment means that worker concerns are overlooked, particularly when taking action compromises production.

Most Shahi workers – about 80% – are women working as frontline machine operators (see Table 13.1). Supervisory roles remain male-dominated. This systemic gender imbalance leads to persistent power asymmetry. In addition, most entry-level workers come from low-income and disadvantaged backgrounds, exacerbating their lack of agency and voice. They have few alternative job opportunities (see Box 13.2) and are therefore hesitant to report legitimate issues for fear of retaliation, which in the extreme can lead to job loss.

Anecdotally, we find that systematic gender imbalance, coupled with an environment of pressure and fear, manifests in the form of shouting, abuse, and harassment on the factory floor. These issues often go unreported and unresolved. Predictably,

Table 13.1 Snapshot of workforce at Shahi Exports

Parameter	Description
Gender	Mostly women (70–80%)
Origin	Migrant (20–30%) Local (70–80%)
Age group	Migrant (18–25 years) Local (18–45 years)
Marital status	Migrant (mostly unmarried) Local (unmarried and married)
Living with/at	Migrants (hostels and rented houses) Local (rented houses, parents or in-laws)
Mobile phones	Migrants (mostly smartphone after they come to the city) Local (mostly feature phone, sometimes it belongs to their parents/husband)
Work experience	Migrant (1–2 years) Local (5–6 years)

worker absenteeism and attrition in the garment industry are remarkably high, each averaging between 8% and 10% every month at our partner firm.

Box 13.2: Indian Garment Workers Have Few Outside Options
In an Outside Employment Opportunities Survey conducted at 12 factories of Shahi Exports, we surveyed more than 2300 workers about their wage expectations and about outside job opportunities. The primary outside option was another garment factory job (listed by 37% of workers); for most workers (55%), there is no outside option.

1.3   A Lab for Good Business

Albert Hirschman's seminal work – *Exit, Voice, and Loyalty* (1970) – posits that worker voice and exit, or attrition, are intimately related. Voice is defined as follows:

> Any attempt at all to change, rather than to escape from, an objectionable state of affairs, whether through individual or collective petition to the management directly in charge, through appeal to a higher authority with the intention of forcing a change in management, or through various types of actions and protests, including those that are meant to mobilize public opinion.

In this case study, we explore how technology can be used to establish a connection between factory management and workers, in the pursuit of long-term worker well-being. Can technology, combined with novel managerial improvements, empower workers to voice their concerns, grievances, and suggestions?

Answering these questions requires us to define worker voice and identify how it can be enabled. We must also explore how to build responsive management practices, how to sustain buy-in for worker voice among factory stakeholders, and how to instill bilateral trust in a method for employer-employee communication.

There are multiple existing means of amplifying workers' voice in our context – ranging from a worker helpline and suggestion boxes to HR outreach and unionization. Although these mechanisms are at times valuable to workers, they are not often associated with trust; transparency; and, importantly, *anonymity*. The lack of a reliable, transparent, and anonymous mechanism for enabling voice leaves workers feeling unheard and neglected, leading to attrition. On the other hand, managers in this context lose touch with the pulse of their factories, which in the long run allows minor issues to snowball into larger, more systemic problems.

These observations and questions started us on a journey to develop technological solutions informed by human-centered design to promote voice among garment workers with a buy-in from the management. In the first leg of this journey, we evaluated the impact of an employee satisfaction survey, an elementary form of worker voice, on worker satisfaction and attrition. In a second experiment, we leveraged a digital SMS-based worker grievance redressal platform to automate the process. Learning from these efforts and analyzing other worker voice management tools available in the market, we documented how existing technologies fail to fully address needs in the Indian and wider South Asian garment manufacturing context. Finally, we designed a novel, lower-cost solution that leverages digital technology to enable worker voice coupled with managerial incentive programs to promote buy-in from factory staff.

Our resulting worker communication solution, branded as *Inache*, enables workers to communicate their suggestions and grievances via SMS or voice call, and mobilizes managers to listen and help via system-level incentives. We are currently evaluating the impact of this solution on management outcomes, worker satisfaction, and ultimately worker attrition and productivity.

This journey has also spurred the creation of a nonprofit organization, called Good Business Lab (GBL), which is registered in the United States and India. GBL continues to design, evaluate, and proliferate new workplace technologies and programs aimed at a myriad of persistent workplace issues. GBL also produces insights about the adoption, effectiveness, and paths to scale of these innovations. This chapter describes the evidence generated by GBL for the case of worker voice, as well as documents the multistaged process we follow when engaging in user-centered software design. The journey is far from complete, but as you read, we hope you see the potential impact of this deliberate and innovative approach.

2 Context for the Innovation

2.1 Existing Mechanisms for Worker Voice

The stakeholders that influence worker voice the most are the manufacturers that employ workers and the brands that source from these manufacturers. Typically, corporate brands mandate third-party "audit" of manufacturing factories for compliance with social, environmental, health, and occupational safety requirements. Providing an avenue for workers to voice their concerns, unfortunately, doesn't take priority in this list. In the rare instance that it does, the requirement is limited to having suggestion boxes on the factory floor, an age-old traditional practice sans a robust proof of concept in the current context. We envision providing such avenues to voice concerns, beyond compliance. However, an unwelcome outcome of the auditing culture has been the general fear of transparency among factories, which gravely affects their willingness to adopt worker voice measures.

To better understand the context, we first undertook a design audit. This involved understanding the organization's sociocultural fabric and the prevalent norms and rituals – both explicit and implicit. We looked at existing mechanisms to report grievances or suggestions, or ask questions, that a worker had access to, at work. We found that there were several communication channels in place at Shahi (see Table 13.2), and we gathered data on these channels for 40+ factories.

We also studied existing processes and organizational hierarchies to understand factory dynamics. This was done by undertaking focus group discussions with top management leaders, human resources (HR) offices, and production staff members and conducting semi-structured interviews with more than 30 workers, representing

Table 13.2 Employer-employee communication channels

Direct communication	Worker directly approaches the management* to communicate their concerns
	Specifically human resources (HR), which is responsible for grievance redressal in the factories
Announcements	Management makes general announcements for all workers related to holidays, work shifts, etc. in the factories
Suggestions	Suggestion and complaint boxes are located in the factory premises that allow workers to anonymously submit their concerns
Committees	Multiple worker-management committees have been set up to handle different kinds of grievances. Example: internal complaints committee, which handles grievances related to sexual harassment
Helpline	A phone number, handled by the management, that workers can call at, to raise any issues (records usually not digitized)
Handbook	A booklet that dispenses information to workers on factory-related policies and practices
Help desk	Counter operated by HR executives in large factories during lunch hours, where workers can reach out and get their queries resolved

different personas (e.g., migrants and locals, married and unmarried, novices and experienced, different job functions). Moreover, we made observations on the factory floor through short ethnographic studies.

We discovered shortcomings with the mechanisms outlined in Table 13.2, including a lack of a formal go-to process, accountability, and transparency in grievance registration. Often, issues raised by workers through direct communication with HR staff were not recorded anywhere. In addition, turnaround time for response varied – from a few hours for direct communication with HR on the shop floor to several months for suggestion boxes and worker committee meetings. Besides suggestion boxes, it was impossible to maintain anonymity through any other mechanisms. This restricted workers from sharing sensitive grievances, like cases related to sexual harassment, verbal abuse, and bullying.

While helplines seemed to be an effective method for registering cases, many workers opted to call after office hours (sometimes at 10 PM or 11 PM, a practice discouraged by firms). Finally, while announcements and handbooks were effective in disseminating information to a mass audience, carrying the handbook and paying attention to announcements during working hours were a pain point, from workers' perspective. Like suggestion boxes, they also failed to support two-way communication. Help desks were effective in large factories with ample staff (personnel) but ended up being futile in smaller factories with limited staff who felt burdened with excess responsibilities.

2.2 Modeling the Factory Environment

To further build an understanding of people's interaction with internal factory systems and how they together enable and impact worker voice, we created a model of the factory's environment. We modeled the "macro" environment, focusing on Shahi's organizational structure, and the "micro" environment, focusing on the worker's universe of experiences.

Macro-environment First, we explored how factory management perceived the value and threat of enabling worker voice. We mapped stakeholders according to their incentives, motivations, and commitment to worker voice. By studying the administrative records produced by stakeholders, we discovered that the factory HR and organizational development (OD) teams were working in tandem to manage grievances. We were able to formulate the roles and responsibilities of each unit, resulting in an organizational structure detailed in Fig. 13.1.

Each factory's HR team works as an interface between the workers and the management. The HR team is responsible for employment-related procedures, such as salary, attendance, and onboarding workers. The organizational development (OD) team is responsible for implementing worker well-being programs on the ground, for example, maintaining hostel facilities (room and board) for migrant workers, running rural training centers and upskilling programs, and implementing

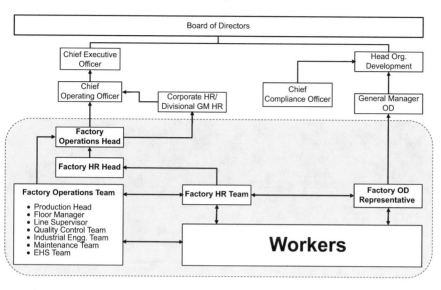

Fig. 13.1 Organizational hierarchies within the factory

other social compliance programs in the factories. The core team for grievance management in the factory comprises representatives from the HR and OD departments.

After interviewing Shahi's staff, we were able to map interest in worker voice tools and trace various stakeholders' levels of influence within the factory. This exercise allowed us to identify "paths of least resistance," i.e., easiest ways to align stakeholders' interests with our reforms to management practice.

Microenvironment To understand the worker's perspective, we first mapped all of the actors a worker interacts with (see Fig. 13.2), including supervisors, doctors, security guards, hostel wardens, and management staff. We also mapped the nature of workers' relationship (positive, negative, or neutral) with each stakeholder. In any manufacturing setting, the workplace hierarchy is rigid, and it influences and drives stakeholders in different ways. We tried to reinterpret the hierarchy, incorporating the frequency with which workers interact with each stakeholder. We identified two important loops that exist in this ecosystem: (1) conflict loop and (2) peace loop. The conflict loop (highlighted in red) consists of workers, line supervisors, and other production staff. There exist predictable reasons for conflict, for example, low efficiency, the production pressure of meeting targets, inferior production quality, and skewed gender ratio (most of the workers are female, while their supervisors/managers are men). On the other hand, the peace loop (highlighted in green) consists of workers, HR executives, OD representatives, and medical/healthcare staff. Workers tend to reach out to one of these stakeholders for help or assistance, including mustering emotional support in a conflict or distress situation.

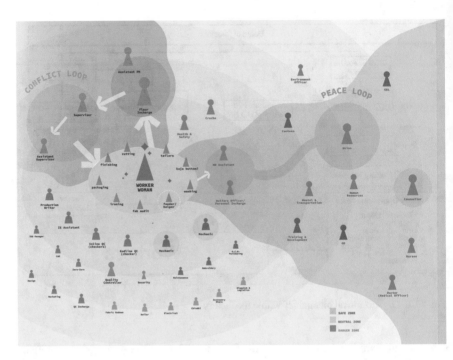

Fig. 13.2 The microenvironment of workers

The rest of the factory's stakeholders don't interact directly with the workers very often and can be categorized as a neutral group. Naturally, on the shop floor, there is a balancing act between the conflict loop and peace loop.

Apart from the internal stakeholders, there are key external partners – like NGOs, trade unions, government agencies, and retailers – that impact workers in various ways. NGOs and trade unions usually have strong local connections with workers, nurtured over a sustained period of time. They have a reputation of caring deeply about workers' interests. This lends them credibility, not only among workers but also suppliers and retailers. Government agencies and departments inform state- and national-level policy which affects workers' day-to-day lives (e.g., how much they earn, regulating factory working conditions such as mandating the requirement of a child care center in factories in India). They have far-reaching networks to undertake widespread information dissemination. Retailers or brands that do business with the supplier often leverage their influence to demand various business practices and measures from suppliers, which directly impact workers, for example, requiring third-party social audits and mandating training programs.

We dug deeper to study a day in a worker's life – to understand their high and low points – through in-depth interviews with approximately 30+ workers. We used empathy maps, a method used to understand users from what they say, how they feel throughout a day, and how they respond to stressful situations (see

Fig. 13.3). This process surfaced the cognitive load imposed by financial strain, especially from instances of delayed wage payments. Such stress, coupled with day-to-day issues, has a tendency to pile up and become a cause of great distress for a worker. We also learned that in this population, very few workers owned a smartphone (approximately 45%). In fact, most workers owned and used feature phones, and only half were formally educated. Of those with formal education, most completed high school only. These observations would impact our tool development significantly.

3 Innovate, Evaluate, and Scale

3.1 Experiment No. 1: A Study of Worker Voice

Firms like Shahi revise their wages frequently. They can choose to raise wages by more than is required by minimum wage, although this is rare. In general, workers face substantial uncertainty about the size of annual wage increases, with unpredictable government and firm decision-making. Anecdotal evidence suggests that worker dissatisfaction is especially high after annual firm-wide wage increases – a fact that may be explained, in part, by the disappointment brought about by wage-related uncertainty.

In our first voice experiment, we investigate how this disappointment might lead to higher quit rates. The firm-level wage hike in this case took effect in April 2016, the same year when the *basic* component of the minimum wage was revised by the government, which takes effect once every 5 years.

Through a randomized controlled trial, we tested the impact of voice on worker satisfaction. Employee feedback channels are an important form of voice, with both intrinsic and instrumental value. The intrinsic value comes from the ability and opportunity to express one's opinion about the workplace or workplace practices. The instrumental value is derived from the changes in the workplace resulting from feedback.

We wanted to test whether feedback operates primarily through the intrinsic channel, by providing workers an option to voice their (dis)satisfaction. So, we deployed a simple employee satisfaction survey in the months immediately following the 2016 wage hike.

We selected a sample of 2000 workers, spanning 12 factories, to participate in the study. Approximately half were randomly selected to receive the survey (treatment group), and the other half made up the control group. The survey was anonymized and recorded (1) worker satisfaction with the job, supervisor, wage, and workplace environment and (2) opinions about supervisor quality (e.g., whether mistakes are held against workers, whether it is difficult to ask others for help, whether supervisors encourage learning, and whether workers can trust their supervisors to advocate for them, listen to them, and help solve their problems). For all 12

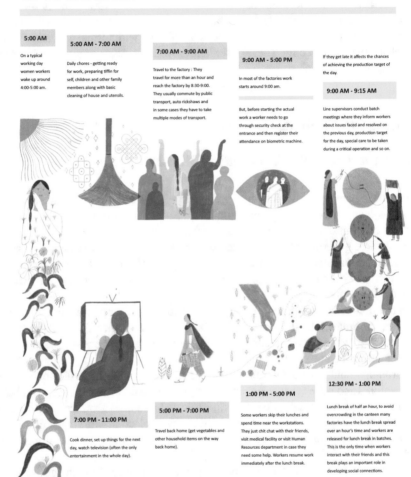

A DAY IN THE LIFE OF A WORKER

We extensively studied the life of workers keeping working and non working days in mind. Most of the women workers are engaged in two shifts.

FIRST SHIFT:
Working in the factory

SECOND SHIFT:
Fulfilling household duties, which extends into the weekend

This routine continues even if they are on their period - breaks are hard to come by.

Often being the sole caregiver of the family puts a lot of pressure on them.

5:00 AM

On a typical working day women workers wake up around 4:00-5:00 am.

5:00 AM - 7:00 AM

Daily chores - getting ready for work, preparing tiffin for self, children and other family members along with basic cleaning of house and utensils.

7:00 AM - 9:00 AM

Travel to the factory : They travel for more than an hour and reach the factory by 8:30-9:00. They usually commute by public transport, auto rickshaws and in some cases they have to take multiple modes of transport.

9:00 AM - 5:00 PM

In most of the factories work starts around 9:00 am.

But, before starting the actual work a worker needs to go through security check at the entrance and then register their attendance on biometric machine.

If they get late it affects the chances of achieving the production target of the day.

9:00 AM - 9:15 AM

Line supervisors conduct batch meetings where they inform workers about issues faced and resolved on the previous day, production target for the day, special care to be taken during a critical operation and so on.

12:30 PM - 1:00 PM

Lunch break of half an hour, to avoid overcrowding in the canteen many factories have the lunch break spread over an hour's time and workers are released for lunch break in batches. This is the only time when workers interact with their friends and this break plays an important role in developing social connections.

1:00 PM - 5:00 PM

Some workers skip their lunches and spend time near the workstations. They just chit chat with their friends, visit medical facility or visit Human Resources department in case they need some help. Workers resume work immediately after the lunch break.

5:00 PM - 7:00 PM

Travel back home (get vegetables and other household items on the way back home).

7:00 PM - 11:00 PM

Cook dinner, set up things for the next day, watch television (often the only entertainment in the whole day).

11:00 PM

They go to bed, the interesting aspect to explore would be the dreams that they see with eyes closed and how they try to fulfil those dreams by taking responsibility of family and ownership at work in the factory.

....and this is a 365/366 days routine for them!

Fig. 13.3 Day in the life of a worker

factories, we also obtained routine administrative data collected by the firm. These included retention rates and personnel data including gender, education, hometown, department, and job type.

Our findings were compelling: many workers used the survey to express dissatisfaction with various aspects of their jobs. As shown in Fig. 13.4, over 20% of workers agreed or strongly agreed with the first two statements: that mistakes were held against them and that asking for help was difficult. Smaller proportions (between 6 and 16%) provided negative evaluations of their supervisor, indicating their supervisors were not encouraging, not someone they could trust, or indifferent about helping solve problems. Combining responses, over 50% of the sample responded negatively to at least one of the six statements.

Reviewing workers' satisfaction levels, we also gained interesting insights. Though average satisfaction with the job, supervisor, and workplace environment was quite high (over half reported being extremely satisfied), satisfaction with wage levels were much lower. More than half were either somewhat or extremely dissatisfied.

We learned that workers' expectations were substantially higher than the realized wage hike: they expected a hike that was roughly three times the size of the actual increase. On average, workers expected to earn about USD 17 (16% of total salary) more per month than their realized wages.

Further analysis, combining administrative data on retention, revealed that the effects of the voice intervention were strongest among the most disappointed. Individuals who were disappointed by the wage hike were more likely to quit, but the voice intervention was particularly able to lower quit rates among them. At the average level of wage disappointment (USD 17), those who received the voice intervention were 19% less likely to quit than the control group. For those who were not disappointed at all, the intervention had no statistically significant effect. This set of results (see Fig. 13.5) suggests that the survey voice intervention worked primarily by mitigating disappointment.

3.2 Experiment No. 2: An Off-the-Shelf Software Solution

Encouraging results from the first experiment led us to a partnership with the Children's Place, a clothing company, to test an existing worker engagement technology platform, called WOVO by Labor Solutions. Through interviews with various stakeholders, we learned that worker voice was undermined by the negative repercussions of raising complaints and the weak accountability for complaint resolution by management. These trends demotivated workers. In addition, existing grievance mechanisms at Shahi Exports (such as suggestion boxes and worker committees) either lacked anonymity or offered no mechanism for ensuring feedback to workers on the status of complaints. WOVO provided both anonymity and accountability.

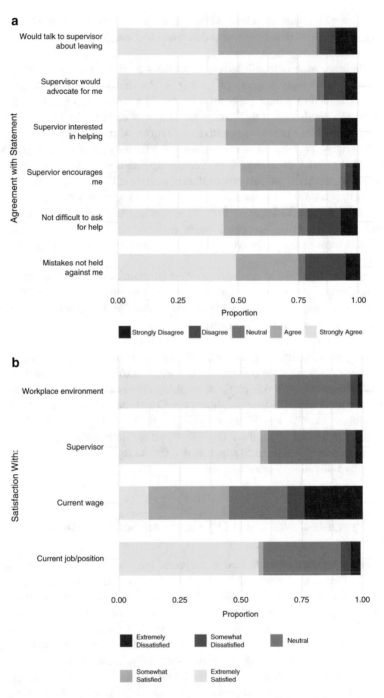

Fig. 13.4 Employment satisfaction survey. (**a**) Plot 1: evaluation of job conditions and supervisor characteristics. (**b**) Plot 2: satisfaction levels

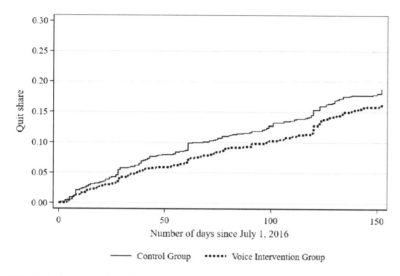

Fig. 13.5 Quit share over time for treatment and control groups

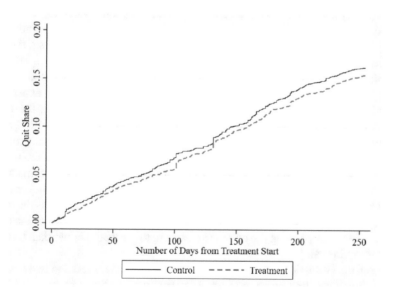

Fig. 13.6 Attrition outcomes

The platform is built as an SMS-based worker grievance redressal tool, which 1) allows workers to anonymously submit messages (questions, grievances, or suggestions) via SMS (or the WOVO smartphone application) and 2) provides management with a dashboard to respond to manage and gather data about such communication. Our experiment was again designed as a randomized trial, randomizing 7,500 workers in two Shahi factories into either treatment or control.

The treatment group was formally registered on the WOVO platform and received user training. The control group had access to the tool and could submit grievances. However, they were neither formally registered on the platform nor did they receive user training.

This study was designed to minimize the threat of spillovers between treatment and control groups. Manufacturing in the garment industry is organized in assembly lines, consisting of 50–70 workers arranged in sequence. Workers are often shuffled between lines. Together, these conditions would have created resentment if some workers were provided with access to a service like WOVO, and others had the service withheld (and only learned about it indirectly). To address this, we sent the treatment group private reminders via SMS, encouraging them to use the tool. No reminders were sent to the control group. This drove variation in tool usage between treatment and control group and helped estimate impact.

To study the impact of WOVO on business and social outcomes, we collected survey data and analyzed administrative records. Enumerator-led surveys were used to capture self-reported measures of worker well-being. More than 2,600 workers were randomly selected for survey follow-up, across treatment and control groups. The survey modules captured worker demographics, satisfaction level, mental health challenges, types of grievances and their redressal, and phone availability and usage. We did two rounds of surveying: at baseline, before the start of the intervention, and at end line, 7 months after the start of the WOVO program. Administrative data included measures of worker productivity, retention, and absenteeism.

In the first experiment, every study participant got an opportunity to answer the survey, which can be interpreted as 100% utilization; in the second experiment, the utilization rate for the technology was just 4.69%. Nevertheless, the results from the second experiment show reduction in absenteeism and attrition across the study population. Absenteeism was reduced by 5% in the treatment group, on a base monthly absenteeism rate of 9.6% in the control group (see Table 13.1). We find that the attrition was 6–10% lower in the treatment group compared to the control group, though these results are not precisely estimated (they are only significant at 10% level for one of the specifications). We interpret this as the impact of access to the tool, irrespective of use. This suggests that the impact of the tool is largely delivered by having the option (i.e., access to feedback) rather than by actual grievance redressal.

During the evaluation period, a total of 354 cases were registered in two factories, representing a ninefold increase in utilization compared with the suggestion boxes posted in factories (which had a utilization rate of 0.52% in the same period). Treatment group workers used the SMS tool about four times more compared to control group workers.

Worker surveys recorded significantly higher trust in the tool among the treatment group. Among treated workers, 84% were aware of the tool, and 92% of all cases registered through the tool were from this group. In comparison, only 55% of workers in the control group were aware of the tool, and only 8% of all cases received were sent by this group. User training and (SMS) reminders administered

Table 13.3 Effects of treatment on absenteeism

	Absenteeism	
	(1)	(2)
Treatment	-0.0046^{**}	-0.0044^{**}
	(0.0021)	(0.0021)
Observations	51943	51943
Control Mean	0.096	0.096
Individual Controls	No	Yes

Notes: Standard errors (clustered at worker level) in parentheses. $* \ p < 0.1$, $** \ p < 0.05$, and $*** \ p < 0.01$. "Individual controls" include fixed effects for age, tenure, and education categories (including a "missing" category). "Control Mean" is the mean of the dependent variable among control group workers.

Columns 1–2: The unit of observation is the worker-month, restricting to months during the experimental period. Both regressions control for the worker's average pre-treatment absenteeism, strata fixed effects, and month-year fixed effects (Fig. 13.6).

Table 13.4 Analysis of case types

Consolidated Suggestions of all Divisions of (January–December 2018)					
Types of complaints/Divisions	KNITS	KPD	LSD	MNB	Total
Workplace Harassment/Verbal Abuse	143	18	47	108	316
Sexual Harassment	23	0	12	11	46
Salary / OT cases	2	1	0	12	15
In-disciplined Behaviour	10	0	11	12	33
Suggestions	7	1	0	14	22
Infrastructure related issues	2	0	2	1	5
Total	187	20	72	158	437

to the treatment group were able to strengthen trust in the tool by 2%, driving a fourfold increase in utilization.

A summary of the different types of cases registered on WOVO during the study is shown in Table 13.3. The data have been aggregated from WOVO's raw output and cleaned to remove any test cases by Labor Solutions and Shahi. The greatest number of cases was about provident funds and banking, which are considered to be questions rather than grievances and are not usually considered to be urgent. This suggests that workers leveraged the tool to share routine issues. However, some of these issues require workers to physically visit HR and are out of the purview of HR's control, e.g., receiving checkbooks or linking mobile numbers to their bank accounts. Nevertheless, WOVO is a useful way to record and track these issues to build a collective case for action (Table 13.4).

GBL invested heavily in qualitative data collection in this study. In a series of interactions with HR managers in the two factories, the team captured rich details on the experience of implementing and using WOVO. Fifteen HR team members were involved in this study; they found the dashboard relatively easy to use and were comfortable using SMS to communicate with workers, ranking it at 4 on a scale of 1–5 (5 being easiest). One-third of the HR staff felt that the type of complaints was different than they were used to, suggesting that workers might be using WOVO to report issues that they didn't report earlier. Around one-third felt that the total number of complaints through offline mechanisms (like suggestion boxes) had also increased, while 55% felt no change in the number of complaints they had to resolve. More than half felt an increase in their workload due to the tool, since it involved cases that could have been resolved more efficiently if the worker came to speak with them directly (e.g., workers using the tool to enquire about their provident fund may not have provided enough details, like employee codes, in the SMS; this created a communication lag of several days, which delayed processing; when workers drop by the HR office, these issues can be resolved within minutes).

In terms of interpersonal relations within the HR team, 26% felt that relations with their superiors improved, mainly due to transparency of the process. However, 13% felt that due to the expectation of resolving grievances quickly, relations worsened. Over half felt that the tool brought about a positive change in the HR culture, one-third felt that work culture remained the same, and 6% felt it had deteriorated the culture. They felt an increase in the need to actively solve grievances and experienced a better connection with the workers.

With respect to utilization of the service, we realized that restricting communication to only SMS service, as permissible on the WOVO platform, was suboptimal, given the limited (digital) literacy of the workforce. In our baseline survey, we found that only 45% of workers owned smartphones, and only 54% had completed high school. Most workers owned and used feature phones. Strictly, SMS-based communication further affected each step of the user journey. For example, workers use Roman and regional language script depending upon the configuration of their feature phones. As per tech specifications, 1 SMS = 150 Roman alphabets = 90 regional language alphabets. Thus, each grievance could take up to 2–3 SMS, and each SMS could cost 0.5–1 INR. The costs can double quickly if back and forth is needed to resolve a complaint, discouraging workers from reporting sufficiently or reporting at all.

The consensus among managers was that existing job roles of floor managers needed to be leveraged to troubleshoot and increase the speed of the redressal process. For example, if a manager is responsible for handling canteen-related complaints, then that manager needs to be involved in the resolution process for related complaints. This decentralized troubleshooting would enable smoother and quicker grievance redressal and seemed preferable than having a single centralized "super administrator" for a factory.

Usage climbed once workers developed trust in the service. Historically, traditional grievance management systems together brought in 400–500 cases in 3–4 months across all 60+ factories. With WOVO, across just two factories, the increase in registered cases was sizable, with over 300 cases lodged in 7 months. In collaboration with ex-labor officers and legal experts, GBL designed and delivered capacity-building workshops to the management, in order to resolve and manage the increased caseload efficiently and effectively.

Managers felt that the tool brought workers closer to the management, yet they underscored the need for an incentive system to encourage managers to execute the tool properly, including taking action on incoming grievances. To this end, they suggested introducing nonmonetary incentives, alongside the tool.

3.3   Experiment No. 3: A Custom Solution

After two (impactful) experiments with our research team and with intensifying dialogue about worker engagement, Shahi became committed to the idea of transparent, digital worker management communications. However, the implementation of WOVO highlighted key areas where the product fell short of Shahi's needs, including incompatibility with voice call features and the high per-worker cost of subscription to the software platform. With Shahi's scale (employing more than 100,000 workers), a subscription service would be a large recurring cost for the business and likely unsustainable. Thus, while there was buy-in from management to implement digital tools, it made sense for Shahi to develop a custom software solution for its 60 factories, tailored to the realities of the Indian garment industry. Shahi asked our design team to develop a worker voice solution from the ground up.

We undertook several qualitative methods to approach the design and functionality of the tool and its implementation strategy. The first step in this process was to understand user behavior, for both workers and managers. We dug deeper into Shahi's macro-environment to map stakeholders' incentives, motivations, and commitment to worker voice. We looked at benefits and limitations of the existing mechanisms a worker has access to, to report grievances or suggestions or ask questions at work. We used design thinking methods to learn about the problems workers face at the workplace and understand what holds them back from sharing their grievances. We mapped a day in a worker's life, especially the highs and lows in a typical day, conducted in in-depth one-on-one and group interviews; tried to gauge how they feel through empathy maps; and assess how they respond to stressful situations (described in detail under Sect. 2).

To provide access to more workers and in turn drive utilization, there was consensus to build a voice call feature, along with messaging feature, to report grievances. There were, however, two options to operationalize it. First, through interactive voice response (IVR) by staffing a call center, centrally or in an individual factory, that can receive and respond to grievance reporters (workers) in real time. Having a person at the other end to record grievances would ensure a better experience for workers and clear reporting. A central call center would be less costly and efficient than a factory call center, but a factory call center would be preferred with regard to language, context, and access to sensitive data.

The second option was to have separate phone lines for different factories through which each grievance is recorded as voicemail which will be transcribed by a designated staff member during office hours. With this option, workers will still be able to send queries 24×7, but management would only be able to respond during business hours.

We decided to go with the second option because, if IVR was leveraged, a worker would be instructed by a computer-generated audio to register their complaint under the right category and subcategory; for example, press 1 to register a complaint regarding the Canteen, and press 2 to select subcategory. Given that we have eight case categories and 25+ subcategories, from a behavioral perspective, we deemed this could create frustration among workers who are incidentally in a distressed state when they call. In addition to that, the attrition rate among production workers stands at 8–10% per month, so technically, after a year, there is a new cohort of workers on the shop floor. After leaving the factory, most workers are supposed to receive their provident fund[1] (PF) and other savings, implying that a huge number of cases that a factory receives are related to PF, gratuity, etc., and these cases are most efficiently handled by direct factory representatives. Moreover, workers frequently change their phone numbers making it harder to track their complaints. Thus, the decision to provide a separate phone number (option B) to each factory would provide a single point of (familiar) contact in the factory to each worker and would facilitate an understanding of the pattern of such cases in each factory. This choice facilitated accurate mapping and tracking of factory-specific cases, allowed us to design a pertinent performance incentive model for HR (covered presently), and led to devising a decentralized grievance management, which meant that individual factory management would be responsible for managing complaints received from its facility, in effect empowering and encouraging management in the respective factories.

Keeping in mind the low worker literacy, including digital literacy, and minority ownership of smartphones among workers, we realized that the tool needed to be accessible outside of an Android or iOS ecosystem. Multiple regional languages are spoken across different states in India. For example, in Karnataka, 58% of the workers speak Kannada, 15% speak Hindi, 9% speak Oriya, 5% speak Tamil, and 8% speak Telugu. This meant that the tool needed to be compatible with all local languages spoken by workers in a factory.

To improve uptake and buy-in from factory managers, we worked with them to design the dashboard for our tool. We leveraged their familiarity with online platforms such as Facebook and G Suite to build a familiar and easier dashboard user experience. For example, the user interface is inspired from the habits of the software user on social media platforms, e.g., uploading a document in the evidence section and ability to comment.

Software or digital tools are revered for bringing transparency and accountability in the reporting process; however, an important factor in the success of any grievance

---

[1] Provident fund (PF) is an investment fund contributed to by employees, employers, and (sometimes) the state, out of which a lump sum is provided to each employee on retirement. Many workers leave their jobs after 5 years of service, at which point they are entitled to PF and gratuity benefits. Recent guidelines by the Government of India suggest linking of bank and PF account number to Aadhar Card, PAN Card, and mobile phone number. Sometimes, this process proves to be a pain point for workers. Hence, we consider issues related to PF and other schemes as major severity cases.

management system is building trust through timely case resolution. To facilitate this, we made sure that grievance handlers are notified, via email, as soon as a new case is received on the portal. To bolster accountability and action, regular performance review reports are generated and shared with factory heads, overseeing grievance redressal. This ensures that cases are resolved within a stipulated time and workers get regular updates on the status of their complaints.

To further tackle concerns regarding timely resolution and accountability among managers, we designed performance-based incentives for grievance handlers, an idea that was suggested by the HR itself in the qualitative findings from our second experiment. The incentives were nonmonetary in nature because (a) there are multiple programs that run in these factories at any given time, now and in the future, and providing monetary incentives for one particular program might affect expectations of future programs as well, and (b) we wanted to develop a sustainable model that can be easily replicated in other factories. In fact, literature supports the idea of nonmonetary incentives as being effective in increasing performance among employees as discussed by Sonawane (2008) and Sorauren (2000).

Factory representatives shared that recognition and acknowledgment of their efforts by peers, workers, and senior management were a credible motivator. Suggested nonmonetary incentives were achievement certificates, free dinner for family, a recognition email by factory heads, etc.

Finally, we came up with a nonmonetary individual performance review system, which was, however, ceded for team incentives that the management found to work better as it was a team effort (Bandiera et al., 2009; Hamilton, 2003). Moreover, healthy team dynamics was favorable to remove the problem of free ridership and ensure within motivation rather than it being enforced on anyone by their superiors.

Another critical element to build trust in the process was to ensure that the quality of grievance redressal was up to the mark. To this effect, we introduced certain parameters like tone of conversations with the workers and quality of evidence uploaded on the tool. This reduced subjectivity in quality testing and enabled quality checkers to decide whether the case was resolved in a satisfactory manner or not. If not, they had the option to reopen that case, which would subsequently affect factory's performance incentives. For example, if the number of cases reopened for a particular factory in a month exceeds 10% of total cases, the factory loses out on incentives for that month.

All these shared insights from the design exercise, along with the learnings from the prior experiments, guided the development of our homegrown worker voice solution, *Inache* (Fig. 13.7).

Inache is a two-way communication tool for workers to anonymously communicate grievances and suggestions via SMS or call to the management. Incoming grievances are managed on the Inache dashboard by a management representative group in a factory, called the grievance handling team or the resolution team. This team is made up of case reporter (CR), case manager (CM), and case troubleshooter (CT). CR is responsible for transcribing and interpreting the voice note or SMS received on the dashboard; this process generates a case incidence report (CIR).

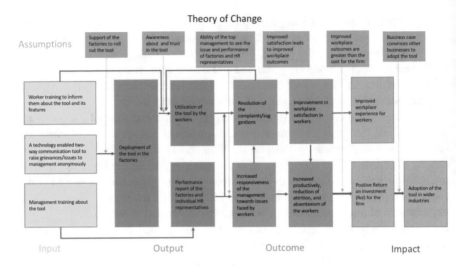

Fig. 13.7 ToC worker voice tool

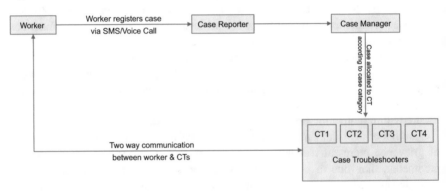

Fig. 13.8 Journey of a case in a factory

Once a CIR is prepared, the case is transferred to the CM who then assigns the case to a CT, well positioned to resolve the grievance (Fig. 13.8).

To ensure timely resolution and accountability among managers, we have designed team performance–based incentives. Workers and managers are given user training and related information on *Inache* during induction sessions preceding its rollout. We have set parameters in place to test the quality of grievance redressal. In fact, the software automatically generates performance reports in real time; it also alerts if any predefined tolerance levels of quality are breached.

3.4 Limitations and Future Developments of Inache

Currently, the software (HR dashboard) is only supported on the web and optimized for a desktop computer. Most users like case managers and troubleshooters, however, spend the bulk of their time on the shop floor, preventing them from engaging with the tool when they are not in front of their computers. Hence, there is a need for a mobile optimized version, which simultaneously calls for looking into data security for when the user is not in factory premises or the handheld device is not in the user's possession.

Further iterations could include an artificial intelligence-based chatbot, which is capable of understanding different accents and of resolving simple, day-to-day grievances, like how many number of leaves are pending for a worker.

Having interactive kiosks for workers in the factory can drive engagement, providing a friendly touch point for workers, since they are familiar with ATM kiosks. These kiosks can be placed near medical facilities, HR offices, canteens, etc. While having kiosks in the premises takes away from the perception of complete anonymity, we believe that workers will be willing to use them to register those grievances, which do not jeopardize their safety, for example, banking- and salary-related cases, which cannot be solved without details of the worker, and cases related to canteen, maintenance, etc. Furthermore, having kiosks in the premises can serve as a reminder for the existence of Inache and can signal management's commitment to worker voice. Having said that, we will work toward designing best practices, placement of the kiosks, and user interaction to provide maximum anonymity.

We piloted Inache in three factories for 4 months before the final rollout, in which we tracked (a) number of cases registered daily; (b) average time taken for the case to travel from one team member to another, i.e., from case reporter (CR), to case manager (CM), to case troubleshooter (CT); (c) average time taken to respond to the worker; and (d) average time taken to resolve the issue and close the case in the dashboard.

The decentralized grievance management required 4–10 people to be a part of the resolution team in a factory. Faced with paucity of manpower, we decided to roll out the main intervention in a phased manner. The first phase involved those factories where sufficient staff existed to handle the tool rollout and where no new hires would be necessary. The second phase involved the remaining factories where new hires would be needed.

Real-time feedback from the users – case reporter (CR), case manager (CM), and case troubleshooter (CT) – in pilot factories helped strategize the composition of grievance handling teams in each factory, depending on its size (number of workers) and design user (CR, CM, CT) training modules. To select the respective grievance handling teams, we created a core committee of senior members from the OD team, allocating each one of them a cluster of factories. This committee was then responsible for nominating suitable candidates from each factory to work on case resolution and giving them user training.

Hackathon

During phase two rollout, we organized a *2-day hackathon* with all the stakeholders involved in the implementation of the tool to highlight the issues they were facing and discuss potential solutions. The goal of this exercise was to learn about their day-to-day usage of the tool and identify the need to refine features and develop a standard operating protocol for the users. We learned the following:

- Workers were testing out the tool by sending in greetings like "hi," "hello," "hey," and "good morning." This made it difficult for grievance handlers to interpret it or decide if they needed to respond. Not responding could breed distrust among workers.
- While using the voice note feature to send in grievances, some workers misunderstood the feature to be a customer support-type service and kept waiting for someone to assist them on the other end.

These issues were overcome by giving live demos to around 35,000 workers and explicitly mentioning that the voice note feature was simply another way of sending a message, not a customer support service.

To measure the social and economic impacts of our technical innovation (*Inache*) and the implementation strategy, we designed a randomized controlled trial. This experiment spans 42 factories which will be divided into the following groups: (1) pure control group: this cluster of factories maintains the status quo, i.e., they do not get the tool, user training, or performance incentives for the managers; (2) grievance tool–only group: this treatment group gets access to the tool and user training for workers; and (3) grievance tool and incentives group: this treatment group gets the whole suite of services, i.e., tool, user training, and performance incentives for the managers. The intervention is at the factory level, in which each worker in a factory gets access to the treatment/control assigned to that factory.

To measure worker well-being outcomes, we will survey randomly selected 36 workers from each factory on themes, such as loyalty toward the company, feedback on their supervisors, workplace satisfaction, trust on redressal mechanisms, and availability of outside job options. To measure business outcomes, administrative data on productivity, attrition, and absenteeism will be collected and tracked.

Challenges with Worker Surveys

Our worker well-being outcomes are primarily informed by one-on-one surveys with workers. In prior experiments, we have found that the majority of responses by workers have been tipped toward a positive/favorable scale.

(continued)

To understand what might be leading up to this, we piloted various versions of the survey on the field – ranging options from five-point Likert scale to three-point Likert scale and from human face emojis to WhatsApp emojis – to elicit honest responses. Out of these piloted versions, we have found the three-point scale WhatsApp emojis to work the best:

The physical survey environment also played a major role in this decision – most surveys are conducted inside the factory, albeit in a separate room away from their supervisors and fellow workers, and we found that it was easier for workers to point to the appropriate emotion rather than saying the option out loud, especially if they had an unfavorable opinion about their workplace, wages, supervisors, etc. Additionally, we kept the questionnaire short (about 15–20 min) to make sure we minimize survey fatigue among workers and surveyors.

We trained surveyors in this new style of questioning by using sample questions/scenarios so that the surveyors knew how to present various options to the worker.

The experiment is still underway at the time of preparation of this chapter. Inache has completed 7 months in the treatment factories, and while we cannot talk about its impact on workers and business right now, as of October 2020, the cumulative utilization rate (total cases/total workers*100) is at 3.19%, and more than 2000 total cases have been received from a total of over 60,000 workers.

At this stage, it is useful to bring the reader's attention to the lesser absolute number of the utilization rate in this experiment compared to the second experiment (utilization rate of 4.69%). While the present number is smaller, it is not necessarily indicative of lesser impact. Firstly, the context of the two experiments is vastly different; the second experiment was situated in the factories in a northern Indian state, while the current one is situated in the southern states, implying innumerable differences in the types of issues faced at the workplace, gender and social norms and language and ethnicity, among others. Next, until early February, only phase one factories, i.e., a total of 14 factories and 38,000 workers, had access to the tool, so the utilization rate takes the denominator 38,000. As shown in the graph, between the start and early February, the utilization rate went up at first with the access to a better tool but later came down as one-time or irregular issues were resolved satisfactorily. Post February, the tool was rolled out in phase two factories also, and the denominator for calculating the utilization rate took the value of 65,000 (all workers). The second wave of drop in utilization rates around March–April was seen due to a COVID-induced nationwide lockdown, and since then, however, the utilization has been increasing.

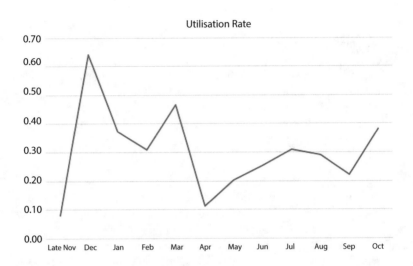

The pandemic has certainly heightened worker anxieties by introducing numerous uncertainties, and it poses a big challenge for our work. It might impact the outcomes of our study both directly (attrition due to reverse migration) and indirectly (anxious workers with low productivities). Even barring COVID, we need to investigate how to justifiably calculate the impact of the tool – which questions to pose, which metrics to track, etc. We need to probe how to interpret the utilization numbers in terms of tool effectiveness and which matrix to use for measuring the effectiveness.

The theory of change suggests that deploying an anonymous grievance redressal tool will encourage more workers to submit suggestions, queries, and complaints. Assuming Shahi continuously monitors the implementation and managers' incentives are well aligned, higher usage by workers could drive responsiveness and redressal by managers, thereby increasing workplace satisfaction among workers. Moreover, the results from the earlier experiments suggest favorable business impact of voice, like reduced attrition (Adhvaryu et al., 2019) and reduction in absenteeism (Boroff & Lewin, 1997). If proven, this project has the potential to provide a positive return on investments for Shahi and will be a testament for adoption in wider industries as well.

4   Lessons Learned

It is valuable for businesses to provide its workers an opportunity to voice, and technology can make this possible by bringing transparency, anonymity, and speed. We witnessed this firsthand when comparing the analogue heavy-handed "employee satisfaction survey" from the first experiment to a digital solution employed in the latter experiments.

During the second experiment, we realized that without contextual understanding, it is challenging to choose an appropriate technology suitable for workers in the low-income setting. In addition to that, scaling existing worker voice solutions in the market was not feasible because of high subscription fees or the technology not best suited to our context. So, we decided to develop the solution in-house. Given the scale of Shahi, employing upward of 100,000 workers in India, the costs of developing its own solution seem justifiable. But we couldn't have ruled out the existing solutions or undertaken a successful innovative feat if we had not engaged in rigorous, early-stage experimentation.

This endeavor led us to ask important questions, which a solution must answer to be able to fulfill the needs of the workers and the business. What's in it for them? Value proposition for each stakeholder is different and needs to be prepared and communicated aptly. Our bottom-up approach to development, which touched stakeholders from all levels, yielded a sense of ownership among them, playing an important, mostly favorable, role in case resolution and Inache's sustained use. We observed the constraints workers face and how their reporting behavior changes in the advent of worker voice technology, and we carefully studied the role of the employer, specifically HR, in influencing this behavior. In the process, we also realized that simply developing a technical solution would not adequately address the more fundamental issues around trust, accountability, quality, and incentives. Therefore, our offering extends to other processes of grievance management than just reporting, recording, and tracking architecture, with, namely, team-based performance incentives for the management and quality control benchmarks.

Gender

The garment industry in countries like India and Bangladesh faces a particular conundrum – even though the majority of the workforce it employs is women, one rarely sees women in higher managerial positions. This trend is a consequence of prevailing gender norms and patriarchal inertia in such regions. Most of the women start and end up as frontline tailors. This disproportionate gender representation at managerial levels affects a factory's culture – it dampens career aspirations for women and often perpetuates a culture of discrimination in the industry. In such a gendered environment, the role of a communication tool like Inache is to give voice to these women workers. The ability to express themselves, air out their grievances, and hold their supervisors accountable is an effort to address the challenges that larger structural hierarchies pose. Inache, as a worker voice communication tool, becomes an enabler for workers to access systems that are designed to provide aid and support them.

Pivots

Two-way communication and anonymity are the biggest incentives for workers to use worker voice technologies as an alternative to traditional grievance redressal mechanisms like suggestion boxes. There exist several worker voice technological tools; one such tool, WOVO, was evaluated by GBL. This evaluation established the necessity and importance of worker voice interventions in a factory setting. However, the evaluation also highlighted the limitations such technology poses because it was not developed keeping a factory setting in mind. The evaluation brought forth the challenge of limited penetration of technological literacy among the workers. In our baseline survey, we found that only 45% of workers owned a smartphone and 54% of workers had a high school diploma. Another limitation of such tools is that their expense is often not sustainable for a company of Shahi's size and capacity. The only way to move forward and inculcate the use of worker voice technologies within factory settings is to design a tool, keeping the factory ecosystem in mind. To this end, a voice call feature was included for workers who did not feel comfortable sending SMSs. Additionally, the low-level technological design of Inache meant that it could be used on basic feature phones as well.

A Day in the Life: Stories from Two Garment Factory Workers

Financial strain is often felt more by those under the pressure of time. Devi (name changed to preserve anonymity) called Inache helpline number with one such grievance over her provident fund (PF) loan. She recounts in disbelief how fast she received a confirmation message replete with a case number to identify her issue, sans any lag. For workers like Devi, getting used to procedural delays is almost a survival tactic, but its absence, she candidly shares, brought solace, long before the issue was resolved. Within 2 days of registering her grievance, she received a message asking her to share further details. Soon after, she was requested by the HR department to meet the accounts department who resolved all her queries on the pending PF Loan. Paperwork followed, and within a week, her need, urgent at the time, was resolved on priority.

Day-to-day issues, though minor, can pile up and become a cause of great distress. Sheila (name changed to preserve anonymity) shares how she requested for an extra trolley in the dispatch department through the Inache helpline number. The HR department followed up the next day, understood her requirements, and made the necessary arrangements within a few hours. Sheila was elated with the timely resolution – it made her work easier,

(continued)

increased her trust and reliability in the system, and improved her appreciation for the employer.

Changing Technology and Political Landscape

The mobile technology landscape in India is under rapid transformation today. While some changes affect us favorably, others could necessitate overhaul of our service offerings. Hereunder, a few key macro-level developments are mentioned that are currently underway. It remains to be seen how they will affect the development of our new marketable versions:

- In India, the telecom market is a red ocean, where different players try to outcompete with each other through price. Every month, telecom operators come up with new data and voice call plans. This is especially attractive to low-income populations that are highly price-sensitive, thus switching from one operator to another in a short span of time. This is a major issue in maintaining a stable worker master data, which maintains workers' phone numbers that are registered in the Inache platform. When a worker instead chooses to use a new, unregistered phone number to file a complaint, it could create confusion about the authenticity of the cases. We tend to receive many cases from unregistered phone numbers. To overcome this, we try to frequently update the worker master data. We also try to keep workers informed about the importance of maintaining the same phone numbers for longer periods during training, and we even do short refresher training frequently to this effect.
- Today, prices of feature phones and smartphones are comparable. Google has introduced low-cost Android Go-powered smartphones for emerging markets like India. Despite this, smartphone adoption rate among women remains low in India because the decision to buy (or not buy) a smartphone still rests with the head, primarily husband or father, of the family who perceive it as an item of leisure. Low smartphone penetration prevents us from leveraging mobile applications, which could be instrumental in creating a more engaging platform for workers. If during the RCT, which is underway, a majority of Shahi's workforce switches to owning a smartphone, we will look into developing a mobile app for Inache.
- Recent developments in the telecom industry in India, for example, the rise of Jio and merger of Idea-Vodafone, have affected the quality of phone services in general. Telecom is moving from 2G/3G to 4G, while the adoption rate of 4G phones is really low among our targeted population. This means that most workers are using 2G/3G phones with teleservices that are 4G phone compatible. This adversely affects the SMS delivery rate,

(continued)

with approximately 70% success. While there is not much we are able to do to tackle this right now, we are looking into other network providers, which have better success rates.

- The Telecom Regulatory Authority of India (TRAI) regulates the telecommunications sector in India. According to which, an enterprise is not permitted to receive and send an SMS from the same phone number. This means that the phone number, which the worker uses to report grievances, cannot be used by the factory to contact the worker. Instead, the factory has to use a different phone number to respond, which we eventually did, and this can lead to confusion and apprehension among workers. We try to inform workers of this peculiarity, via demonstrations, during the training sessions.

Discussion Questions

1. Why does the problem of trusted communication mechanisms on factory floors still exist? What other solutions or mechanisms could accentuate Inache's intended outcomes?
2. What are the pros and cons of decentralized grievance management mechanism?
3. Suggest some innovative ideas to utilize technology in low tech literacy contexts.
4. Discuss monetary vs nonmonetary incentives for case managers.
5. Discuss additional hypotheses/treatment arms which can be tested in a similar setting.
6. Can you draw any parallels between enabling voice on the factory floor and enabling voice in your own university/work setting? What has/has not worked for you?
7. In the system of factory communication, can you identify negative and positive loops that either curb or enable worker voice?
8. Beyond business returns highlighted in this case study, can you identify some inherent moral reasons for enabling worker voice?
9. What could be a better approach to introduce digital tools to this population (workers and managers)?
10. Can you imagine any spillover effects on other areas, not already mentioned, of using Inache in this setting?

References

Adhvaryu, A., Nyshadham, A., & Tamayo, J. A. (2019). *Managerial quality and productivity dynamics*. No. w25852. National Bureau of Economic Research.

Ashraf, N., & Bandiera, O. (2018). Social incentives in organizations. *Annual Review of Economics, 10*, 439–463.

Bandiera, O., Barankay, I., & Rasul, I. (2009). Social connections and incentives in the workplace: Evidence from personnel data. *Econometrica, 77*(4), 1047–1094.

Hoffman, M., & Tadelis, S. (2018). *People management skills, employee attrition, and manager rewards: An empirical analysis*. No. w24360. National Bureau of Economic Research.

Expectations, Wage Hikes, and Worker Voice: Evidence from a Field Experiment-Achyuta Adhvaryu, Teresa Molina, Anant Nyshadham (2019). https://www.nber.org/papers/w25866.pdf

Boroff, K., & Lewin, D. (1997). Loyalty, voice, and intent to exit a union firm: A conceptual and empirical analysis. *Industrial and Labor Relations Review, 51*(1), 50–63.

Sonawane, P. (2008). Non-monetary rewards: Employee choices & organizational practices. *Indian Journal of Industrial Relations, 44*(2), 256–271.

Sorauren, I. (2000). Non-monetary incentives: Do people work only for money? *Business Ethics Quarterly, 10*(4), 925–944.

Part IV
Water, Health, and Sanitation

Ashok J. Gadgil ⓘ

Introduction

Water, Health, and Sanitation are central to human well-being. All three were historically seen as *outcomes* of economic development, rather than as *drivers*. So, the earlier thinking was that when communities, countries, or populations become (economically) prosperous, they will have enough surplus to pay for good quantity and quality of water and obtain and maintain good health and good sanitation. In a more modern view, good-quality water-health-sanitation are seen as drivers, not outcomes, to get people out of poverty and ill health, and focused efforts on these services are seen as an important part of global efforts to help communities escape absolute poverty (World Bank, 2017; United Nations, 2004).

Furthermore, technology and innovation are disruptors that can sometimes directly improve water-health-sanitation access, without waiting for communities to first become prosperous. Directed creative technical efforts to improve water, health, and sanitation can bring improvements in these services far quicker than the time it might take for first improving general economic prosperity and then obtaining its eventual results.

Of course, it doesn't always work that way, and the efforts of the best minds and the most motivated teams can be sometimes unexpectedly undercut by aspects of the real world that were not considered relevant. Practice of the emerging discipline of Development Engineering is full of such examples. I would argue that there are two benefits that nevertheless come from trying: (1) so long as we are willing to analyze, think through, and share the findings about the causes of failures, we all can learn together, and our success rate will improve over time and (2) better to try and fail, rather than get scared and not try at all. So long as we don't make the same

A. J. Gadgil
Department of Civil and Environmental Engineering, University of California, Berkeley, Berkeley, CA, USA
e-mail: ajgadgil@berkeley.edu

mistakes again and again, we can hope to make fewer and fewer new mistakes and reach at least partial, or even full, success.

The four chapters in this section provide first-hand examples from four different decade-long projects in the Water-Health-Sanitation area. The narratives are complex and rich in content and have many lessons to take away, based on one's prior knowledge, perspective, and background.

The Chap. 14 by Gadgil et al. takes us along the invention, development, and field testing of a novel technology, "ECAR," from its beginnings in 2005 to its demonstration plant that became operational and financially viable in 2016. Along the journey, we learn how the authors recover from major and unexpected setbacks and overcame several technical obstacles. They were also very lucky to successfully find a funding sponsor to support the design, building, and commissioning of the full-scale demonstration plant over a period of 3 years, from 2012 to 2015. The chapter ends with an interesting twist. Although the multi-institutional consortium sorted out the implementation hurdles, managed community engagement and hand-off to industry, and successfully obtained the needed regulatory and business permits, until 2021, till full 5 years of its successful operation were completed (both technically successful and financially viable), the demonstration plant remained the only plant of its kind in India, where almost 100 million rural people are estimated to be drinking arsenic-poisoned water. So, the chapter ends with a bit of a puzzle – what work remains, and what is still missing so that this and similar technical solutions can rapidly go to scale?

The chapter by Daniel Wilson on measuring adoption of fuel-efficient cook-stoves with digitally enabled sensors has several important lessons and remarkable backstories compressed in its storyline. The most important lesson to me is the relentless effort, resourcefulness, innovation, and patient hard work to get the deceptively simple project in remote war-torn Darfur from A to B. The effort would be only a an academic-year of effort for one or two students in a well-resourced and peaceful setting; however, for this team, it extended into being a multiyear effort for a team that required more than a dozen dedicated and talented team members. What are the overarching takeaways from this chapter? Different people will have different takeaways. For the author of this section introduction, there are three. First, the chapter illustrates a core principle of good experimental science: measure all key parameters in more than just one way if at all possible, so your measured data does not fool you. In measuring adoption, for decades and decades, field researchers have relied on social surveys. Only recently compact non-intrusive sensors and data loggers became available and affordable. The team used both and compared the results, and then they discovered that the social survey methods gave hugely and consistently inaccurate data reflecting the biases of the respondents. The physics-based automated sensors gave quite different answers to the same questions. Second, the team compared social survey data collection via (1) laboriously writing down answers by hand, and later transcribing these into a spreadsheet, versus (2) having a smart handheld device on which the answers could be entered directly during the interview to be transmitted wirelessly to a computer. And how did these two methods compare? Despite being more tedious, tiring, and slow, the former method yielded the same (erroneous) results as the handheld smart device. So, here technology did not improve, or yield any advantage over, the tedious and diligent data collection and data entry by hand. And third, I note with admiration the resilience, the resourcefulness, and the breadth of creativity energy brought to bear on sensor design, training of field workers, and software-based automation in the analysis of massive amounts of collected data! Here technical ingenuity shines and yields a completely unexpected insight – how a very slight "nudge" from a pre-announced second visit by a field worker to a non-adopting household greatly improved adoption of the stoves! Here, the massively rich data, its automated analysis, and human behavioral insights in adoption all show their interplay!

The chapter by William Tarpeh et al. covers a period from about 2012 to the present (2020, at the time of writing the chapter). It provides, at its very starting pages, an excellent exposition of how to select and define a problem that is possibly amenable to technology invention and innovation and will have globally significant impact, particularly in resource-poor farming communities, on their agricultural productivity. It goes on to describe how the team developed and explored two competing pathways for capturing nitrogen from urine. These pathways were then assessed for their cost and feasibility under real-world constraints in a developing country context, for agricultural applications, in parallel providing local economic activity and also addressing the problem of environmentally harmful uncontrolled disposal of urine on the community scale. This compact but broad-ranging chapter covers not just the progress of the work through the 8 years but also offers lessons on building partnerships, teaches the relevant chemistry and chemical engineering, presents a business model canvas for the product, and discusses paths for scale-up. All in all, an exciting and exhilarating ride! For the truly outstanding aspects of the journey (brought out in detail in the chapter narrative and also in the summarizing bullet points at the end of the chapter) reflect the key strengths: a learning mind-set, nimbleness, ability and willingness to pivot, resilience and flexibility in plans, and opening up and exploring multiple pathways to get to the next stage so as to not hit a blind alley in the journey! It is an inspiring chapter to read and an approach that could be emulated at a high level in attacking some entirely different problem.

As Hyun et al. note in their chapter on NextDrop, the technology development started with noticing a problem with intermittent piped water supply almost universally experienced in urban India: the customers received the water only a few hours per week but had no clear idea of when the water would start flowing through their faucets. Thus, for vast majority of customers in this situation (i.e., those who could not afford an overhead water tank in their loft or above the roof or lived in flimsy housing that structurally could not support one), this meant waiting and waiting for water to arrive. The smart solution thought of by NextDrop engineers was to save their customers the uncertainty and the wait, by a collaboration with the water supply utility workmen. In this model, NextDrop would send an SMS message to tell the customers when the water was going to arrive. This worked well in the small city of Hubli-Dharwad. However, when the company expanded operations into the large metropolis of Bengaluru, a carefully conducted field evaluation by independent researchers showed that the system did not work.

The authors (who are part of the evaluation research team) take a deeper look to understand why. They explain to us the relevant institutional linkages; organizational behavior; personal motivations of the water utility staff; the limitations experienced by low-income households that often had only one cell phone per family, cluttered with irrelevant marketing SMS messages; and the hassles of the wage earner, who was daily traveling long commute distances, making it totally impractical to rush back to catch the water as it arrived at the home faucet. The chapter offers impressive lesson in how the best technical minds can miss serious but non-technical roadblocks to their success and also how a careful evaluation of the overall effort from an independent third-party research group can unravel the mystery of an unexpected failure after an initial success. Given the complexity of challenges in development engineering projects, analyzing and sharing failures and their causes is key to learning by those who follow. The standard of publishing in technical literature commonly is that success leads to a publication and failure is written off to some bad luck, or market failure, and commonly remains not only unpublished, but also poorly analyzed by the research team, and discarded. When that happens, extremely valuable lessons are missed. It is a pleasure to see that is not the case here.

References

United Nations. (2004). *Safe water, sanitation fundamental for poverty reduction, commission on sustainable development told.* Press Release April 27, 2004. Available at: https://www.un.org/press/en/2004/envdev772.doc.htm. Accessed on 14 Dec 2020.

World Bank. (2017). *Millions around the world held back by poor sanitation and lack of access to clean water.* Press Release August 28, 2017. Available at: https://www.worldbank.org/en/news/press-release/2017/08/28/millions-around-the-world-held-back-by-poor-sanitation-and-lack-of-access-to-clean-water. Accessed on 14 Dec 2020.

Chapter 14
Stopping Arsenic Poisoning in India

Ashok J. Gadgil ⓘ, Susan Amrose, and Dana Hernandez

1 Introduction and Overview

The World Bank defines absolute (i.e., near-destitute) poor as those who earned less than USD 1.90 per day in purchasing power parity (PPP), in 2015 (Katayama & Wadhwa, 2019). At that time, there were 736 million people on the planet in absolute poverty. Of these, 24% lived in India (Sanchez-Paramo, 2020).

It is imperative to address problems that affect the poor by inventing, developing, testing, and scaling up new technologies (elaborated further **in the first four chapters of this book**). Often, these problems are unique or especially intense to those in poverty. Efforts to solve the problems with novel technologies, unless scalable and unless they take root in the local social system, are doomed to remain small one-off examples (see an excellent critique of such approaches in Rybczynski, 1991). The technologies to address many of these problems are typically not spillover technologies (such as solar photovoltaics (PV) or mobile phones), but rather those which are targeted to their specific needs. If the world fails to respond to these needs, new technologies may continue to increase the divide between the rich and poor in the coming years, as suggested by the Human Development Report (Conceição, 2019).

A. J. Gadgil (✉) · D. Hernandez
Department of Civil and Environmental Engineering, University of California at Berkeley, Berkeley, CA, USA
e-mail: ajgadgil@berkeley.edu; danaah@berkeley.edu

S. Amrose
Department of Mechanical Engineering, Massachusetts Institute of Technology, Cambridge, MA, USA
e-mail: samrose@mit.edu

© The Author(s) 2023
T. Madon et al. (eds.), *Introduction to Development Engineering*,
https://doi.org/10.1007/978-3-030-86065-3_14

An effort to solve one urgent problem afflicting hundreds of millions of these people is described in this chapter. The period covered is from its beginning in 2000 up to its status in 2019.

This technology intervention is aimed at helping to end the arsenic poisoning of about 200 million people who drink groundwater contaminated with naturally occurring arsenic, at concentrations above its maximum contaminant level (MCL) in drinking water. MCL levels for various contaminants in drinking water are recommended by the World Health Organization (WHO) and set by national governments. The WHO-recommended MCL for arsenic in drinking water is 10 parts per billion (ppb; for water contaminants, 1 ppb is 1 μg/L). Millions of people are consuming water with arsenic at 10, 30, or even 60 times that level, resulting in disastrous health consequences due to chronic arsenic poisoning.

The trajectory of this technology intervention has many threads, and a simple way to organize the developing story is to present all the threads in parallel, chronologically. However, the multidisciplinary nature of the intervention and the parallel nature of the threads make a simple chronological organization difficult to follow. The authors have therefore tried to balance two perspectives to view the trajectory more comprehensively: chronological and disciplinary. It is imperfect, and we the authors hope the reader will forgive us for that imperfection.

About 250 individuals and about 20 organizations have made vitally important contributions to this effort. This chapter is written as viewed through the eyes of only three of these individuals, and we are aware that the other views may be somewhat different, emphasizing different aspects of the work and with different perspectives. We take responsibility for our inevitable blind spots and distortions that result from our limited viewpoints.

Beyond this chapter, the three most relevant papers for further reading are Gadgil et al., 2012; Amrose et al., 2013; and Hernandez et al., 2019.

2 Problem Definition

2.1 Arsenic as Development Challenge

The right to clean drinking water was recognized as a basic human right by the United Nations (UN) General Assembly Resolution 64/292, in 2010 (Wikipedia, 2020a). However, UNICEF and the WHO recently noted that much more progress is certainly needed, given that 2.2 billion people globally still lack access to safe drinking water (WHO Newsletter, 2019). As part of the Sustainable Development Goals (SDGs) and specifically SDG 6, the UN has set the target of providing affordable, accessible, and safe drinking water for all by 2030 (UN, 2020).

An estimated 200 million people worldwide are exposed to arsenic-contaminated drinking water (Naujokas et al., 2013; Minatel et al., 2018). In the 1980s, much of rural India and Bangladesh switched from surface water, which was causing a high incidence of diarrheal disease, especially among children, to groundwater extracted with hand pumps. To prevent gastrointestinal illness, tens of millions of hand pumps were installed (Nordstrom, 2000).

However, prior to widespread tube well installations, the groundwater had not been tested for the presence of arsenic. Signs of chronic arsenic poisoning were first discovered in the 1980s in West Bengal, which led to the first diagnosis of arsenicosis in Bangladesh in 1993 (Ghosh & Singh, 2009). Further investigations and measurements in the 1990s led to the discovery that the groundwater in Bangladesh and adjacent parts of India contained concentrations of arsenic far above acceptable levels. In 2000, the WHO called the situation the "largest mass poisoning of a population in history" (Smith et al., 2000). Recent studies have pointed to unacceptably high levels of arsenic in groundwaters in many parts of the world, particularly the USA, Mexico, Argentina, Nepal, Vietnam, Cambodia, the Philippines, and China (Wang & Wai, 2004; Karagas et al., 2015). Researchers at the Swiss Federal Institute of Aquatic Science and Technology (EAWAG) have published a global map highlighting regions of the world with high levels of geogenic arsenic in groundwater based on statistical modeling (Amini et al., 2008).

Chronic exposure to toxic levels of arsenic soon became apparent in the relevant populations as hyperkeratosis, leading to skin lesions, ulcers, gangrene and amputations, neuropathy, and internal cancers, as well as low IQ in exposed children (Wasserman et al., 2007; Ahmad and Bhattacharya, 2019). Tens of millions of the rural poor in afflicted regions have no viable alternative but to drink groundwater with toxic levels of arsenic.

Arsenic is tasteless, colorless, and odorless, and testing for arsenic is expensive and cumbersome. Arsenic chemistry is complex – in contaminated waters, arsenic occurs in two different oxidation states (As(III) and As(V)), with each oxidation state behaving differently (Smedley & Kinniburgh, 2002). As(III) is more mobile and harder to remove through adsorption mechanisms because of its neutral charge at circumneutral pH. Furthermore, co-contaminant ions commonly found in groundwater (e.g., phosphate) compete for the same binding sites (Roberts et al., 2004) that aim to capture As(III) and As(V), further complicating arsenic removal, particularly because the WHO's recommended MCL of arsenic in drinking water is only 10 ppb. Arsenic concentrations of 250, 500, and even 1200 ppb are found in groundwaters of Bangladesh and Northern India.

Cancer risk assessment models predict that for a population of 100,000 people drinking water at 10 ppb arsenic over their lifetime, there will be 700 cases of excess internal cancers. Internal cancer risk rises linearly with arsenic concentration at these values. For example, consuming 250 ppb arsenic in drinking water will result in a 25-fold increase in excess internal cancers than consuming drinking water with 10 ppb arsenic (National Research Council, 2001; Smith et al., 2000, 2002).

Although internal cancers eventually lead to death, there are also high economic costs to households of those who lose arms or legs to amputations from gangrene. This cost is due to loss of their income, and the costs of hospitalizations and medical costs. There are also high social and psychological costs for the affected individuals who were considered respected breadwinners but are then perceived as a burden on their family or socially ostracized (Chakrabarti, 2017; Rahman et al., 2018). An analysis of economic benefits of arsenic removal from groundwater used for drinking was presented for the Indian state of West Bengal by Prof. Joyashree Roy in 2008 (Roy, 2008).

Even for a community exposed to arsenic in their drinking water, the health burden of arsenic falls disproportionately on the poor. This is due to their lack of access to timely health care, poor nutritional status leading to a higher susceptibility to negative health impacts, lower likelihood to afford treatment, and lower likelihood to be educated about arsenic and its risks. Furthermore, women and girls are the most highly impacted, with instances of women and girls who show signs of arsenicosis being socially ostracized, rejected by their communities, and even rejected by their families (Brinkel et al., 2009; Das & Roy, 2013).

Technologies that can remove arsenic from drinking water are indeed available. However, resource-poor communities have inadequate technical, managerial, and financial capacity ("TMF Capacity"; see US EPA, 2017). Historically, well-intentioned groups, often funded by charitable donations, have installed household- or community-scale arsenic remediation technologies in rural communities. In India, these are called arsenic removal units (ARUs). In almost all cases, the ARUs are installed for free. However, in most cases, the installers leave the community without a clear social organization of responsibility for maintenance and repairs, or, in some cases, without even leaving clear instructions. Many times, the consumables and replacement parts were too expensive for the community. Thus, the existence of effective technologies has not translated to solving the arsenic problem. This suggests the need for an effective and affordable arsenic removal technology that is robust and integrates with successful social embedding for long-term sustainability. This vision challenged a group of researchers in Berkeley to create an innovation that sustains over the long term and meets the above criteria.

The Murshidabad district of West Bengal, located on the Indian side of the border with Bangladesh, is one of the poorest and worst arsenic-affected poor regions of West Bengal. In his doctoral research in applied economics, Abhijit Das studied the fate of recently installed ARUs in rural parts of Murshidabad district. Das visited and tracked the ARUs as they were being installed in Murshidabad villages. There was some expectation that some of the ARUs weren't working well in the field – that is why Das selected this to be his research topic. However, Das found that an astonishingly high fraction (nearly 95%) of ARUs in his study sample failed within 1 year of installation (Das et al., 2016) (Fig. 14.1). All of the ARUs studied had incorporated sound arsenic removal technologies that were effective at removing arsenic in laboratory settings. However, the failure rate in the field was astonishingly (and embarrassingly) high! Another study, this time focused on the functioning ARU units, found that only 50% were regularly being used by the local population (Inauen et al., 2013).

The technologies did not really fail. The *technologists* had failed. The implementing agencies had considered the problem solved as soon as the ARUs were installed (Das & Roy, 2013). The high risks of failure in a purely top-down approach to arsenic mitigation are discussed and analyzed in detail by Chakrabarti (2017). The systems were unsustainable: financially nonviable, not embedded in the societal context, without incentives or social structures for their continued maintenance and repair, or without knowledge transfer to local community stakeholders about how vital the ARUs were for protecting their health. People didn't know how they worked or why it was important to keep them working. Toward the end of his

Fig. 14.1 Arsenic removal units (ARUs) in Murshidabad district of West Bengal, India. Panel (**a**) a working ARU (typical for 5% of installed units at 1 year after installation). Panel (**b**) a defunct ARU (typical of 95% of units at 1 year after installation)

doctoral research, Das documented and photographed ARUs – most were used to tie cattle or goats and store hay, and had fallen in disrepair within 1 year. Thus, the fact remains that as more and more of the population in India comes to depend on groundwater aquifers for their drinking water supply and more areas of India are discovered to have arsenic in their groundwater, the arsenic poisoning of tens of millions of Indians and Bangladeshis continues, almost two generations after the discovery of arsenicosis in populations of rural West Bengal.

3 Approach

3.1 Theory of Change

The theory of change describes the eventual long-term goal and how the proposed activity (with its required inputs and outputs) may lead to outcomes, which in turn lead to the desired impact goals, under certain assumptions and favorable conditions.

The theory of change is an excellent way to draw out the assumptions and favorable conditions that may be different for different actors in the "activity" and which may not reflect the real-world conditions (out of ignorance, wishful thinking, just hope, etc.). Spelling out the theory of change reduces the risk that these assumptions and conditions remain hidden and challenges the team to come together with a sharper clarity about which assumptions and favorable conditions

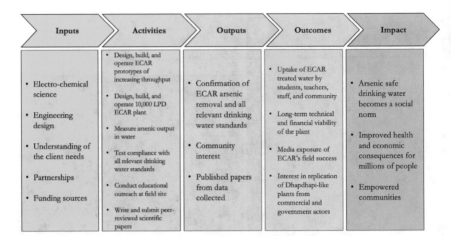

Fig. 14.2 The theory of change (ToC) that the ECAR team formulated for the project. The inputs began in 2005 at UC Berkeley, and by 2019, the team was continuing to put in efforts to see the project outcomes

are necessary to make the desired impact. It also clarifies which activities might be irrelevant and which could potentially be undertaken to reduce the risk of failure in reaching the ultimate desired goal. The theory of change is a large domain of intellectual inquiry, primarily from the fields of project evaluation and monitoring. More can be found on the Wikipedia page and the references cited there (Wikipedia, 2020b). For brevity, Fig. 14.2 summarizes the theory of change for the project described in this chapter.

3.2 Overview of Project Progression from Lab Bench to a Full-Scale Pilot

This section describes the coalition of partners, also called "the team," that is working with an intention to create a sustainable solution to the arsenic crisis, planning from the start for the solution to have the capacity to achieve a scale commensurate with the immense scope of the problem. From the start, the members that formed the team believed and internalized the idea that this problem is so vast and complicated, that it is outside the abilities of any single narrowly focused group to solve, and that it would require a coalition of partners with high mutual trust and complementary subject-matter strengths. While technology innovation was seen as essential, it had to be undertaken within the narrow parameter space allowed by the requirement of financial viability and scalability of a sustainable solution – it was critical for the coalition to have a deep understanding of the past failure stories and the cultural, institutional, economic, and historical background of the ultimate customers and local community of users of the technology. Throughout this chapter, the coalition's funding support is described along with the other aspects of the work. This is because, without funding, it is impossible to carry out technology innovation and cover expenses of travel, instruments, measurements, lab experiments, field surveys, building prototypes, and conducting field tests. Here, only funding support that was successfully obtained is described. The team was always on the lookout for funding, and funding is often thin for impact-driven, highly applied research focused on solving problems of the poor – and the success rate of proposals or approaches to funders was less than 20% overall. So, 80% of persistence and patient effort is not noted here and will remain undescribed further.

The initial work on an electrochemical approach to solve this problem focused on science, technology, and innovation. That effort was undertaken by a research team at Lawrence Berkeley National Lab and the Civil and Environmental Engineering Department at the University of California, Berkeley (respectively abbreviated as LBNL and UCB, both at Berkeley, CA, USA), led by Prof. Ashok Gadgil. That period covered the first 5 years of research, from 2005 to 2010. In these early years, Dr. Susan Amrose, first as a doctoral student and then as a project scientist (now a research scientist at MIT), worked closely with Gadgil to test and develop electrochemical approaches to arsenic mitigation. Initially, the team started with a 200 mL glass beaker on a bench top, removing only As(V) spiked in clean deionized water. The effort focused on deepening our understanding systematically until the Berkeley team had designed a 100 L batch reactor and tested it successfully with a mixture of equal parts of As(III) and As(V) in a synthetic Bangladesh groundwater matrix, at an arsenic concentration approaching 1000 ppb. Early research publications show how the Berkeley team gained knowledge little by little as they started (Gadgil et al., 2012; Li et al., 2012; van Genuchten et al., 2012; Amrose et al., 2013).

Funding for this effort first came from LBNL's Laboratory Directed Research and Development ("LDRD") program which provides small limited-duration innovation grants. The Dow Chemical company had established a funding program at the business school of UC Berkeley called Sustainable Products and Solutions, which funded this work from 2009 to 2012. The Berkeley team also won funding from phase one and phase two of the US EPA's program called "P3" (for People, Planet, and Prosperity). Gaps and shortfalls were filled with Gadgil's Rudd Chair funds from UC Berkeley campus. These resources allowed the Berkeley team to take the necessary technical risks that led to a steady increase in knowledge and confidence, to build up to the 100 L reactor.

Current PhD student Dana Hernandez is playing a major role in its ongoing development, along with postdoc and former PhD student Siva Rama Satyam Bandaru, who has been engaged in this project since 2011. Numerous other doctoral students have been involved in the effort at Berkeley, and six of them (Amrose, van Genuchten, Delaire, Bandaru, Glade, and Hernandez) have made this work central to their dissertations, each making a very substantial contribution to advancing our knowledge. NSF doctoral fellowships which funded three of the six doctoral students have been of tremendous financial help in this effort.

The strong focus on creating a sustainable solution that goes to scale led Gadgil to approach Prof. Joyashree Roy, the founder and coordinator of the Global Change Programme at Jadavpur University (GCP-JU; Kolkata, India). Gadgil and Roy had known each other since 1997–1998 when Roy spent a year at LBNL as a Ford Foundation Postdoctoral Fellow in environmental economics. GCP-JU is an interdisciplinary organizational unit in Jadavpur University and explores global change as it manifests itself in India, through the lens of applied social sciences, particularly, applied economics and sociology. In this coalition, UCB contributed technology expertise, while GCP-JU provided critical insight into the sociocultural, economic, and political factors surrounding arsenic mitigation in West Bengal. GCP-JU also provided access to a network of local stakeholders and influential leaders that allowed us to translate our research into impact. Last, but not the least important, GCP-JU brought credibility to the coalition effort, because JU is a highly respected university in India, officially recognized to be in the top ten public institutions of eminence, and is seen as a highly credible source of knowledge by the general population. In hindsight, the coalition members acknowledge that this multidisciplinary partnership was essential for the progress that was later achieved.

During 2011–2013, the team conducted a field trial of the 100 L reactor in West Bengal, which led to their increased confidence to design, build, and test an even larger prototype. This larger 600 L reactor was tested at a nearby school called Dhapdhapi High School. These field trials were successful, even as new technical and social-embedding problems surfaced and were successfully dealt with (more on these later). The Berkeley side of the coalition won funding support for this work from UC Berkeley's Blum Center for Developing Economies and the United States Agency for International Development (USAID)-funded large multicampus multi-year program, called the Development Impact Lab, which is also led from the Blum Center of UC Berkeley. During 2011–2013, the team was forced to change its field site and start all over (as described later on) owing to unexpected external events. The team's technical and social research in these years is reflected in the corresponding publications (Amrose et al., 2014; van Genuchten et al., 2014; Gadgil et al., 2014).

By 2014, the development of ElectroChemical Arsenic Remediation (ECAR) had reached a key fork in the road. The team had shown successful performance of the ECAR technology at the 600 L scale, both in the lab and in the field at the Dhapdhapi school. However, they had not yet built or demonstrated the long-term technical effectiveness, commercial viability, and social acceptance of ECAR technology to the scale of a fully functioning demonstration plant. Doing that required more funding support. Most innovative technologies for the poor in the developing world wither on the vine and become obscure memories or dusty curiosities at this point in their development.

The list of the team's failed funding efforts will be skipped to focus on the successful one. The US and Indian governments jointly established in 2000 an autonomous foundation in India IUSSTF (Indo-US Science and Technology Forum) which funds and promotes technology transfer and commercialization from the US inventors and innovators to the sectors that India considers of critical importance. Each proposal requires participation from an Indian industrial partner of credible depth of technical, marketing, and management expertise. The Berkeley and JU team searched for many months in vain for such a partner through multiple channels, which even included Gadgil giving lectures in India to Indian industry representatives with the help of the Confederation of Indian Industry (CII), all to no avail.

In January 2012, Gadgil won the Zayed Sustainability Prize, for Lifetime Achievement in the individual category. At the awards ceremony, he met Mr. Jean-Pascal Tricoire, the CEO of Schneider Electric, who was there to receive the Zayed Sustainability Prize that Schneider Electric had won in the large corporate recognition category. Schneider Electric had just purchased a majority share in an Indian company, Luminous Power Technologies, (now called Livpure) founded by Mr. Rakesh Malhotra. During their interaction, Mr. Tricoire offered to introduce Gadgil to Rakesh Malhotra who might be looking to license a novel breakthrough technology to solve a key drinking water problem. It further helped that a VP of Schneider Electric in India, Dr. Satish Kumar, also personally knew Gadgil for over a decade. This entirely fortuitous circumstance led to an introduction of Livpure to the ECAR team. Within the year, in December 2012, Livpure, JU, and Berkeley jointly applied to IUSSTF for funding.

A lot of effort followed the application to refine the proposal to communicate the exact scope of work for meeting the due diligence by IUSSTF. Finally, in July 2014, the collaboration of the three parties was granted the IUSSTF award for a plant that would scale up and demonstrate the technology, to be completed within just 2 years. This changed the trajectory of ECAR from what otherwise would have been just a dead exhibit in the dusty museum of ignored technical curiosities.

In the meantime, in 2013, the Berkeley-JU team, led by Gadgil, won the top honor of the Prince Sultan Bin Abdulaziz International Prize for Water (www.PSIPW.org) – the Creativity Prize. Winning this top international prize for the ECAR technology innovation increased the team's credibility. Ultimately, the team believes that the recognition was pivotal in raising funds from the IUSSTF to build a demonstration water treatment and distribution unit at a rural school in West Bengal.

Since the team had only 2 years to design, build, and commission the demonstration plant, the team went back to the same fabricator on whom they had relied for the 600 L smaller project. This fabricator, Shri Hari Industries, is a small engineering company outside Mumbai, and their chief design engineer, with several years of design experience, had been personally interested in contributing to engineering innovation (Fig. 14.3). In close partnership with Berkeley researchers, Shri Hari Industries staff were able to rapidly iterate the design of the ECAR reactors with the Berkeley team to improve ease of manufacture, shipping, and operation, thereby significantly improving reliability and reducing capital costs.

Figure 14.4 sketches the progression of the technology over a 12-year span. From 2016 onward, the ECAR full-scale pilot plant has been operating at Dhapdhapi

Fig. 14.3 Ashok Gadgil, Narendra Shenoy (chief design engineer of Shri Hari Industries), and Susan Amrose in November 2014, discussing the design details of the water distribution manifold for the two ECAR reactors then under construction

Fig. 14.4 The progression of ECAR, beginning as a 200 mL beaker, with its development to 100 L and 600 L prototypes, and finally scaled as two ECAR reactors of 1000 L each. The 2005 photo shows Ashok Gadgil and Susan Amrose at the Berkeley Lab. The 2011 photo shows Ashok Gadgil and Siva Bandaru inspecting the 100 L prototype at Amirabad High School. The 2012 photo shows Siva Rama Satyam Bandaru, Susan Amrose, Caroline Delaire, Ashok Gadgil, Paramitha Chaudhuri, Sudipta Ghosh, Suman Chakraborty, Chandan Bose, Amit Dutta, Peter Kuin, and Sayantan Sarkar with the 600 L prototype at Jadavpur University. The 2017 photo shows Susan Amrose standing next to one of the two reactor tanks at Dhapdhapi High School

High School in West Bengal, India. An automated water dispensing system daily offers 1 L of arsenic-safe drinking water to each of 2500 students and 400 staff and teachers free of cost and offers the excess water (about 4000 L) for sale to the surrounding community at a locally affordable price of less than 1 US cent per L.

The Berkeley-JU team was excited to partner with Livpure. Livpure was a large and vibrant water-treatment company, ranking as one of the top three in the booming market of under-the-sink reverse osmosis (RO) units for urban Indian households. Livpure had an excellent understanding of the Indian business and regulatory environment and a strong marketing team, with thousands of distributors supporting their RO products. The Berkeley-JU team hoped and expected that Livpure would advise and influence the direction of development and maturation of the technology during the demonstration plant stage. If the demonstration plant was successful, then they expected that Livpure would bring capital, sophisticated marketing, and engineering resources to further improve the engineering design and implement a wide-scale solution to the arsenic poisoning problem. This was based on Gadgil's experience with WaterHealth International (more in the following section). Conversely, Livpure expected a prompt flow of funds from the government for building more ECAR plants. For various reasons, these expectations turned out to be misaligned, and the subsequent technology diffusion did not occur as originally hoped.

The outcome was that technology maturation was primarily driven by members of the Berkeley-JU team, whose members poured all their efforts into ensuring that the system worked well technically, economically, and socially and were able to demonstrate its robustness, effectiveness, and market acceptance. However, there has been little interest from the industrial partner in expanding plant capacity, increasing the sales with outreach to the community, or replicating this plant to other locations. Yet, they did not abandon the plant and continue to operate and manage it, test the water monthly, and conduct the water sales. They have continued to get the product water tested monthly by a trusted third-party lab and have posted the results outside the door of the ECAR plant for all interested parties to see. This was vitally important to maintain the trust of the community members that come daily to purchase water from the plant. At community's insistence, the plant was kept operational even during the COVID pandemic that required a shut down of all non-essential businesses.

4 Implementing the Approach

This section describes the team's process of first understanding the arsenic problem in West Bengal, and then creating the ECAR innovation to address it, iteratively implementing the innovation, and adapting the innovation to scale and for other contexts. The iterative nature of the invention innovation process unfolds in primarily three different ways: (1) the technical solution itself is iterated to make it work better, rejecting some aspects of earlier designs; (2) the understanding and framing of the problem and its boundary conditions is adjusted through interactions with potential customers, iteratively improving the problem definition; and lastly, (3) the demands placed on translating the scientific understanding into a specific hardware and operating protocols also become better understood and are iteratively improved during the field experiments and field testing. In all this effort, the core strength of the team comes from integrity, candor, and an attitude of willingness to learn – a learning mindset.

4.1   Prior Experience

4.1.1   WaterHealth International

The team's conceptualization of the arsenic problem was shaped, from the start, by Gadgil's experience in microbial drinking water treatment, particularly in India. In 1995, Gadgil had invented UV Waterworks, a device that used UV light to disinfect drinking water at a rate of 15 L per minute, at a cost of USD 0.05/1000 L, suitable for small community-scale deployment. Although the technology was low cost, robust, and low maintenance and had a high positive impact on health, creating real impact had required a market-based business model in which clean water was sold at a locally affordable price.

A start-up company, WaterHealth International (WHI), was able to leverage the extremely low cost of water production from the technology to maintain a locally affordable price point while making a profit. Other good practices of WHI included ensuring good quality control, spending the necessary funds for media outreach, marketing, and getting support of local village-level nonprofit organizations for local health education about diarrheal disease and microbial pathogens in water.

Furthermore, achieving a high volume of sale of the purified drinking water was tied to WHI's financial viability. Therefore, the company was incentivized to continue with relevant public health education and to continue to gain the public's trust that the water was high quality (through monthly third-party testing of the water they sold and publicly posting and disseminating the resulting certified data). Gadgil knew from this experience in India that if someone's livelihood and profit depended on the continuous provision of arsenic-safe drinking water on a day-to-day basis, then that person (or organization) would not let the technology fail for lack of proper maintenance. This shaped the team's framing of the technology innovation to focus, not just on something that could produce arsenic-safe water, but on something that could do so inexpensively enough to generate a small profit while selling the arsenic-safe water at a locally affordable price.

4.1.2   Arsenic in Bangladesh

Gadgil had started research on affordably removing arsenic from drinking water at Berkeley in 2000. The team initially focused on an innovative approach to build a core-and-shell particulate adsorbent for arsenic, called Arsenic Removal Using Bottom Ash (ARUBA). The ARUBA's core was based on extremely low-cost bottom ash particles (10 micron diameter) from coal-fired power plants, which Gadgil obtained from India and invented a process to coat them with ferric hydroxide. The powder did remove arsenic successfully from contaminated groundwater, and the waste passed the US EPA's Toxic Characteristic Leachate Protocol (TCLP; a test to determine if waste is acceptable for disposal in municipal landfills in the USA). However, the adsorbent proved to be inexpensive only when

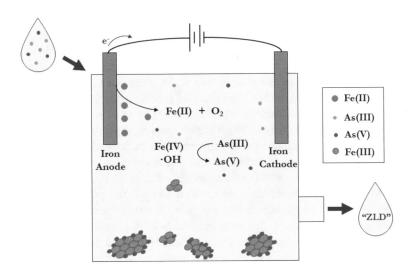

Fig. 14.5 Schematic of ECAR process with iron electrodes. Anodic dissolution of Fe(0) from the iron anode releases Fe(II) in the bulk solution. This Fe(II) reacts with dissolved oxygen (O$_2$) to form insoluble Fe(III) and simultaneously generates reactive intermediates (like Fe(IV) and *OH) that oxidize arsenite, As(III), to arsenate As(V). The Fe(III) captures the As(V), and then, the aggregated precipitates settle out of solution. The letters "ZLD" on the outlet water highlight that this process is zero liquid discharge, a coveted goal of process engineering

it was deployed at large scales, on the order of tons per day. In 2005, the team had started to explore another technology pathway that could be affordable and profitable at both small and large scales while enabling a company to survive and grow through the expected slow growth of the arsenic-safe water market.

The ARUBA research and field tests continued through 2008, and the team later published their findings (Mathieu et al., 2010). Switching directions from ARUBA to a new electrochemical technology was the first pivot in the research direction. This new technology, called ElectroChemical Arsenic Remediation (ECAR) (Fig. 14.5), used a small amount of electricity to remove arsenic from diverse groundwaters in a way that is highly effective and low cost, produces little arsenic-laden waste, and requires minimal supply chain development of readily available materials, among other attractive factors.

There was another important shift in the team's thinking in the period from 2008 to 2010. Throughout the development and field testing of ARUBA, it had seemed logical to conduct field tests in the country most severely afflicted with arsenic: Bangladesh. The team members reached out to the premier engineering university in (Bangladesh University of Engineering and Technology – BUET) which continues to offer excellent education in engineering. Team members met several times with UNICEF officers in Dhaka who oversaw UNICEF's programs to improve water quality and with various officers of BRAC, the world's largest NGO, famous for its breadth and depth of work in Bangladesh. The team also met with the senior echelons of BRAC University and BRAC Bank. The most important and most consistent takeaway from all of these meetings was that the arsenic problem in Bangladesh was intensely politicized. External agencies (the UN, World Bank, British Geological Survey, and others) were blamed for the arsenic disaster, and they were expected to pay for arsenic-free water. There was a highly politicized division of public opinion, with a large group of opinion leaders who were hostile to any suggestion that arsenic remediation should ever be part of a commercially viable enterprise. Thus, the only way to reach scale would be to work through a delivery model fully supported by either the government or an external aid agency. The team tried many times to obtain this support or to find alternatives, but it was clear that this could not be achieved with the time and resources at the team's disposal. Right across the border, in the Indian state of West Bengal, the problem was equally grievous, but no angry political battle was being fought over who would pay and no blame game being played. The team decided in 2010 to move all future work to India so it could fight on one front – against arsenic – and did not also have to fight on a second front, the political front, to get things done. Team members believe this was another key and important pivot that led to the success at Dhapdhapi in West Bengal six years later, in 2016. They believe this could not have been achieved in the highly contentious political atmosphere of Bangladesh.

4.2   The Amirabad Experience

With leadership from JU, the Berkeley-JU team reached out to numerous key stakeholders in India and West Bengal, in particular, to inform them of their planned work and to obtain advice and feedback. This included multiple levels of officeholders and experts in numerous organizations. Two key lessons the team learned through the process were (1) the critical importance of proper handling and disposal of arsenic-bearing waste and (2) the absolute necessity of reaching out to the local public through multiple channels of communication and contact over an extended period, until members of the public became familiar with and socially "normalized" to using the technology.

In 2010, the team selected a field site with the help of a local college teacher and his NGO contacts from the JU team, in the small town of Amirabad, in the district (roughly equivalent to a county in the USA) of Murshidabad. This was the same district where, earlier, Das had conducted his doctoral research. Amirabad is a six-hour drive from JU, so short trips and day visits were impractical. Nevertheless, via a small local NGO known to the JU team members, the team leaders approached the local high school (actually, the Amirabad High Madrassa or High Religious School) and were offered a classroom with electricity for conducting the field tests of the 100 L ECAR reactor, with plans for testing successively larger ECAR reactors. The 100 L reactor was first designed, built, and tested in Berkeley, then duplicated and tested in JU, and finally, transported to Amirabad. For local outreach, the team held public community meetings as well as multiple meetings and discussions with school teachers explaining to them what the team was doing and why. This included explaining how mature technologies emerge from science (the progression in technology readiness level (TRL)), since technology maturation research is an unfamiliar activity to most people (even in the USA), and expectations of speedy progress continued to run far ahead of what the team could humanly deliver.

As a good faith effort to provide near-term relief, the team paid for a US-based NGO ("Project Well") to dig a protected dug well on the school grounds. Project Well had given assurance that the well would be a good interim solution. It turned out that very quickly the well ran dry, and no operational problem was addressed by the NGO or its local representatives. However, that was only the beginning of the team's troubles at Amirabad.

Although the team had moved the field research work from Bangladesh to West Bengal in favor of finding a less politicized situation, the team discovered the truism that "politics is everywhere – once there are more than two people involved!" In 2011, elections were imminent. Various political factions and leaders looking for opportunities to strengthen support from their base are often eager to promise a quick-fix technological solution to win the adulation and endorsement of their local constituencies. So, while the team was field testing just a prototype, the leaders and communities expected immediate continuous water service delivered to them by the research team. To counter these high expectations, the research team held various public meetings and clearly stated their short-term goals. They also met with multiple administrators to discuss the way forward for a long-term solution. However, these efforts had limited impact. Given the past experience of multiple failed technology trials, the community was looking for quick and assured solutions. Meeting this expectation was not only scientifically unfounded but also beyond the resources and capacity of the research team.

By the fall of 2011, the field trials of the 100 L ECAR reactor at the Amirabad school had been a technical success (the background work and field test results are described, along with cost estimates, by Amrose et al. (2013)). Work was soon completed in Berkeley on the design of the next-iteration larger reactor, now 600 L, based on the technical lessons learned. By the spring of 2012, the 600 L reactor, a 600 L settling tank, a gantry (a small mobile crane), and all related equipment had been moved to JU and tested successfully. It was time to find a large indoor space in the Amirabad school to accommodate all the equipment.

Initially, the fieldwork was conducted with encouragement and support from the Amirabad school's administration led by the headmaster, who was carefully neutral, pro-science, and pro-research. Additionally, the headmaster lived in Amirabad and saw the beneficial implications for the community if the novel ECAR technology was to be successful. Unexpectedly (for the team), the Department of Education instituted their normal "rotation" of headmasters from one school to another, which happens every few years, and the supportive headmaster was rotated out. The new headmaster lived out of town in an area with safe drinking water. He was, from the start, a bit equivocal about allowing ECAR fieldwork in the school's classroom. He was inclined to support the faction in the school teachers that began to vocally oppose the presence of the team on the school grounds in the absence of the team offering a guarantee for providing safe drinking water supply. With increasing virulence, this group attacked the team's honesty of purpose and suggested that the team represented outsiders collaborating with others from a wealthy western superpower (the USA), who were "exploiting" the school for their research. Some insight of this activity came, but too late, from the local NGO and their contacts at the JU side of the team. In hindsight, the team feels that the large distance (almost 7 hours each way by car from JU) led to inadequate communication from the team to dispel rumors. This all ended badly. In an extremely tense meeting in the summer of 2012 with the school headmaster and a group of teachers whose screaming accusations he acquiesced in, the team announced its decision to pull out, leave the school site, restore the classroom to its original state, and return it to the school. The team felt disappointed and angry, including at themselves. Much work to build social and political capital had been lost.

This led to two immediate consequences for the team.

The first and immediate consequence was the beginning of an internal process of reflection, discussion, and review to try to understand what went wrong, what the team could have done differently, and what safeguards could have been put in place to avoid such a major setback. Team members were sure that given the intensity of the verbal exchange at the meeting and the political situation at the school, the damage was beyond repair and pulling out was absolutely the right decision. In a few short years, after news of the success at Dhapdhapi reached the Amirabad school, the headmaster and NGO there telephoned the team leader at JU several times asking the team to restart the work at Amirabad. However, given his complete lack of support for the field test in 2012, the team asked for the invitation to be made in writing before it would be even considered (he never wrote).

4.2.1 Assessing Failure

One reason the team's effort fell victim to the local sociopolitical division and acrimony was the inadequate two-way formal communication between the team, the community, and the school's opinion leaders. This communication was hampered by the long distance from JU to Amirabad and absence of a formal mechanism for two-way communication. As a result, the impressions and expectations from the opposing faction of teachers did not surface and get corrected promptly. They operated under the growing belief that the research team was performing the work for some future financial gain from unseemly business profit and that the school should see a mature technology delivering safe drinking water to the school within weeks or, at most, months from the start of the field test. This comprehension led to two major shifts in the team's thinking: (1) the team must formalize agreements with field site owners by creating a signed memorandum of understanding (MOU) that (although legally nonbinding) spells out the roles, responsibilities, and mutual expectations from all sides, and (2) the team needed to pay a lot more attention and invest more effort in managing the expectations of the field site influencers and opinion leaders. When MOUs were created, a copy of the MOU would be given to each signee so that it could be referenced in future discussions, and if further adjustments were to be made, these would also be in writing and signed by all parties.

A second consequence of the blowup at Amirabad was the urgent requirement to find another field site to continue the work. The team decided to be more systematic, rather than just finding friends of friends of friends, to find the next field site. A more rational and systematic approach was needed and implemented. The team followed the example of Harvey MacKay searching for a bank lender for his start-up business, described in his well-known book *Swim with the Sharks Without Being Eaten Alive* (Mackay, 2005). Two team members were charged with visiting every school within concentric circles drawn with the JU campus at its center; starting with the shortest travel radius (30 minutes), they kept on increasing the radius of their search after exhausting the schools in circles of successively increasing travel-time radii (i.e., 30, 60, 90, 120 minutes, etc.). Their goal was to come back with a list of three different potential schools that each met the following criteria:

1. Strong interest and support from the school administration in the effort.
2. Availability of infrastructure at the school (e.g., room, water, electricity).
3. Presence of arsenic in the local water being used at the school.
4. Reasonable travel time from JU to the school.

In about a month, the team (of Das and Bandaru) brought a list of three strong candidate schools (Table 14.1), which the team discussed and ranked. The highest weight was given to strong support from the school administration. The top ranked school, Dhapdhapi High School, was a mere ~2-hour drive from JU (Fig. 14.6), well connected also by railways/public buses, and had 2500 students and about 250 ppb arsenic in the local groundwater that supplied the school. A small municipal pipe delivered a small amount of potable water intermittently, but the daily delivered volume was completely inadequate for the daily needs of 2500 students (in the age group of 11–19 years) and 400 teachers and staff. Another nearby school ranked equally well; however, it was a girls-only school. After burning their fingers at Amirabad with accusations of unethical motivation, the team favored the coed school at Dhapdhapi. There was less risk of another messy accusation, particularly because the field engineers who would spend many months at the school would include both men and women. Thus, after following a systematic and rational approach, the team found the second field site within just a few weeks, and the work was able to proceed.

4.3 The 600 L Prototype

Prof. Joyashree Roy, leader of the JU team, reached out to the Dhapdhapi higher secondary school administration and was warmly received by the headmaster. This time, the team took care to have a clearly written MOU with the school administration, which was countersigned and approved by the secretary of the school's

Table 14.1 The data for the top three schools used to select a site for the demonstration of the ECAR technology

	School 1: Co-ed primary school	School 2: Co-ed Dhadhapi high school	School 3: Girls only high school
Arsenic concentration (ppb)	~100–200	~225	~200
Water source	Hand pump, shallow tube well	Hand pump, shallow tube well	Shallow tube well attached with submersible pump
Number of students	200–250	2500	510
Suitable space	Adequate	Excellent	Excellent
Travel time from JU	2 hours 15 minutes	2 hours	45 minutes

governing board. This MOU proved to be a lifesaver during the few occasions of misunderstandings and differing mutual expectations that inevitably arose. This is an important point. In a community-scale technology trial, not only the official hosting agency, but also the community becomes involved, often with incomplete or incorrect information. Therefore, careful attention to managing community expectations becomes very important. The team went through the process of holding community outreach meetings, including open-mike meetings with school teachers and staff, as well as meetings with a pair of student representatives from each class. In parallel, the Berkeley team had completed the design of the 600 L reactor.

After successful testing at JU's Environmental Engineering Lab in the summer of 2012, the 600 L system (which was previously meant to go to Amirabad) was moved to Dhapdhapi High School, where the administration provided a dedicated unused classroom for the field test (Fig. 14.7). While 600 L was too small to serve an entire community or even the school, it was a good step-up from the small 100 L device (about the size of a large trash can) previously tested in Amirabad. The next step would be a device that would deliver approximately 6,000–10,000 L/day. This capacity was seen as optimal, based on prior experience and calculations, to obtain good economies of scale while meeting the (projected) demand of the local community members, who would have to walk to the plant to pick up the water. For larger throughputs, the coverage territory becomes too large, and potential customers at the perimeter of the cover area are understandably generally unwilling to walk the long distances back to their households carrying water. This was another lesson from Gadgil's WaterHealth work.

The field test of the 600 L device at Dhapdhapi ended successfully (results were published as a journal paper (Amrose et al., 2014)). Soon, the team received the exciting news that the IUSSTF grant was being released in July 2014 that would enable demonstration of a full-scale pilot of a community ECAR plant.

The team took a calculated risk by significantly changing the design of the 2000 L reactors compared to the 600 L reactor without going through another pilot study at the 600 L scale with the new design. The 2000 L reactors incorporated completely different hydraulics and a more sophisticated train for processing the product water.

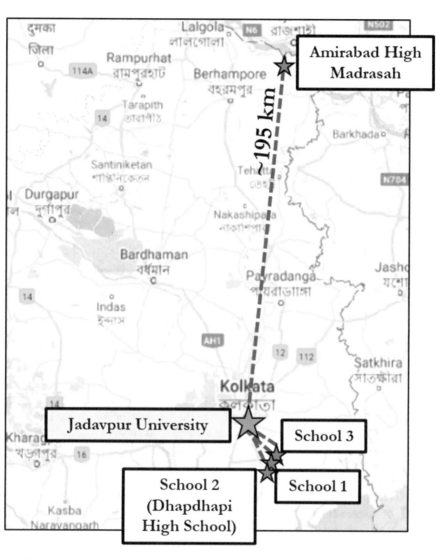

Fig. 14.6 Map shows the location of Jadavpur University and the distances to promising school sites. The straight-line distance from JU to Amirabad (the first school partner) is 195 km, but time to travel is 6 hours and 45 minutes by car. The travel time to Dhapdhapi is only about 2 hours and 15 minutes

This was based on the team's confidence in their scientific understanding and also the very limited window of available time (2 years).

Fig. 14.7 Professor Gadgil discussing the chemistry of ECAR with Siva Bandaru, at that time a field engineer in the Berkeley team. Dr. Susan Amrose is on the right. This is in the classroom of Dhapdhapi High School that was allocated for the 600 L prototype trial in 2012

4.4 Scaling Up and Commissioning the Demonstration Plant

Learning from weaknesses observed during the long-term performance of the 600 L reactor, the electrode layout and hydraulics of the reactors for the 10,000 L per day (LPD) demonstration plant changed, though the core chemistry remained the same. The reactors were again designed in Berkeley and built in Mumbai by the same trusted small fabricator, Shri Hari Industries. The Berkeley team placed two of their engineers in Mumbai for daily collaboration and participation during the fabrication. The 10,000 LPD design comprises two identical reactors, each capable of treating 1000 L of arsenic-contaminated groundwater per batch. The transfer of water from each reactor tank into the settling phase occurs alternately, operating as a semi-batch process (Fig. 14.8). As the water in the second tank is transferred, raw water fills the first reactor tank for the next batch of treatment. There was no possibility of testing the reactors in Mumbai or at JU due to their size. However, by then, the Berkeley team had developed high confidence in their understanding of ECAR reactor design, so the reactors were transported by truck directly to Dhapdhapi.

In March 2015, the two 1000 L reactors arrived and were installed at Dhapdhapi High School. Civil work was also completed by Livpure on an external shed, built immediately adjacent to the school, to house part of the water treatment system. Electrical upgrades were also completed for the school power supply to receive an additional 10 kW power to help supply the ECAR plant.

The team worked with officers of the State Pollution Control Board and a reputed hazardous waste collection company to establish and get approval of a plan for safe

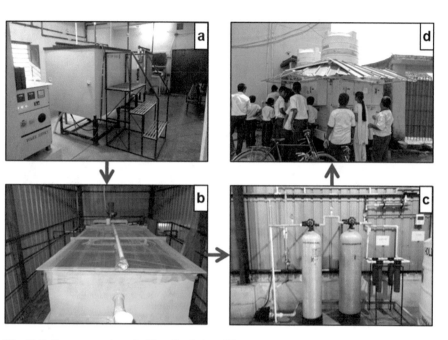

Fig. 14.8   Treatment train at the Dhapdhapi plant. Water treatment begins at (**a**) the ECAR reactors where electrolysis occurs, followed by (**b**) flocculation and settling and (**c**) tertiary treatment, consisting of a rapid sand filter, several micron filters, and a UV light. An activated carbon filter is also seen in panel 3, but is not in use (nor necessary to ensure safe-drinking water, but was part of the pilot testing phase). Water is then distributed to students, teachers, and staff through (**d**) the water kiosk that contains four automatic dispensing units

sludge management and disposal. Livpure and the team coordinated to complete wiring, plumbing, and tertiary treatment systems for the full plant integration, as well as to conduct initial testing and troubleshooting. The first local operator was hired and trained. Maintenance equipment for the electrode plates was fabricated and tested.

Clean, arsenic-safe drinking water was first produced intermittently in mid-2015. However, no water was allowed for human consumption until the team had developed very high confidence in the plant performance. This confidence emerged by mid-September 2016, 6 months after plant operations began on a daily basis. During that time, the produced arsenic-safe water was discarded into a soak pit, much to the distress of some of the local stakeholders, who were left drinking arsenic-contaminated water while the team completed all testing.

The Government of India had certified several commercial labs (under the accreditation scheme "National Accreditation Board Labs" (NABL)) for conducting water quality tests – but not all NABL labs were equally reliable, as the team discovered by paying several of them to test blind samples of calibrated arsenic solutions of known strengths. Throughout this period, raw and finished product water samples were air shipped to Berkeley for analysis, in addition to water quality

Fig. 14.9 Plant inauguration on July 8, 2016. The picture shows teachers and staff of Dhapdhapi High School, as well as researchers from UCB and GCP-JU. This picture was taken right after a meeting with teachers and staff where Prof. Gadgil and Prof. Roy explained project efforts and answered questions from students, teachers, and staff. Sustained support and advice from the headmaster, Mr. Biswas (in center, in white shirt sleeves) was crucial for overall project success

analysis being conducted by a local Indian NABL-certified lab, which the team had independently verified to be trustworthy. Sending the samples all the way to the USA was a way to overcome some of the limitations of the technical environment faced in the team's field research. Another research group in JU had a highly accurate instrument (inductively coupled plasma-mass spectrometry, or ICP-MS) for measuring water quality, including arsenic. However, that ICP-MS remained inaccessible to the team. Another sensitive instrument at JU had been functioning, but the essential supply of argon for its operation had run out, and the empty cylinder of argon was not being replaced or refilled over many months. These are some small examples of the numerous difficulties of conducting high-quality scientific research in resource-constrained areas without a steady flow of adequate funds to support research infrastructure.

The leaders of the Berkeley team visited the completed pilot plant for commissioning (Fig. 14.9) and met with the leadership of Livpure, relevant officers of the Ministry of Drinking Water and Sanitation, and USAID staff in Delhi, to discuss progress and steps forward. Progress at the demonstration site was reported in India's prominent daily newspaper the Times of India.

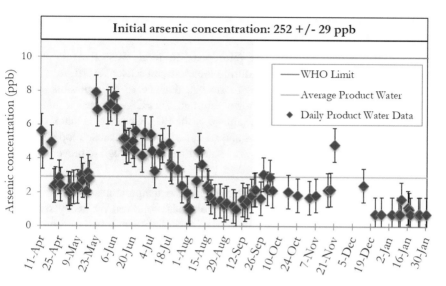

Fig. 14.10 ECAR product water results from April 11, 2016, to January 30, 2017, demonstrate that the Dhapdhapi plant is highly effective in reducing high arsenic concentrations (average initial: 252 +/− 29 ppb) to concentrations much below the WHO limit of 10 ppb (average final: 2.9 +/− 1 ppb). The gray-shaded region (left half) depicts the time period for which water was treated and arsenic level was carefully monitored but the water was not allowed for consumption. The white region (right half) depicts the time period during which water was made available for consumption by students, teachers, and staff. This figure is adapted from Hernandez et al. (2019)

After the technical challenges (described in Sect. 4.5) were identified and overcome, water production from the full treatment plant (Fig. 14.10) began to reliably meet all chemical and biological aspects of the Indian drinking water quality standard (IS:10,500:2012), including exceeding the international standards for arsenic corresponding to the WHO-recommended MCL; the initial arsenic concentration of approximately 250 ppb was reliably reduced to <5 ppb (Fig. 14.10).

4.5   Technical Challenges

While the deployment of ECAR as a large-scale plant was ultimately successful, there were a number of unanticipated technical challenges that surfaced and were overcome along the way. One of the primary reasons for engaging in field work is to uncover and resolve these kinds of contextual challenges. The major benefit of operating in field-relevant environmental conditions is the opportunity to see effects that may simply not be anticipated or cannot be replicated in the laboratory. These specific challenges resulted from the field conditions, multi-stakeholder management, and maintenance requirements. It is worthwhile to give a few examples, since the devil is always hiding in the details. As another way to put it, "In theory, there is no difference between theory and practice, but in practice, there is a world of difference between theory and practice."

Intermittent Power Supply The pilot plant in rural West Bengal, relies on intermittent grid power. Especially during the monsoon season (April–July), power outages occurred almost daily due to storms, damage to transmission lines, or substation breakdowns, etc. When power returned, the ECAR process would simply resume where it had left off (one advantage of the ECAR process is that it does not lose efficiency when operation stops and starts). However, because it operated as a semi-batch process, power outages were sometimes long enough to prevent a single daily batch of 2000 L from being treated.

Hot and Humid Climate The summer daytime temperatures in West Bengal reach as high as 40 degrees Celsius in the shade, followed by heavy rains and high humidity (average relative humidity of 84% between July and September). Such extreme environmental conditions led to the rapid corrosion of bare electrical contacts (due to high humidity), which in turn led to heat buildup that was not dissipated under high ambient temperatures, leading to melting and charring of plastic insulation. In addition, powdery white precipitates (later determined to be calcium carbonate, or calcite) appeared in the product water during summer and the water delivered to consumers in intense summer heat was unpleasantly hot. With the onset of winter, the team also observed unexplained significant deterioration in the performance of the tube settler process. After various troubleshooting efforts, the team realized that the groundwater being pumped for treatment in the ECAR tanks was warmer than the water held in the tube settler, which had equilibrated to colder overnight ambient temperatures. As such, the iron flocs were short circuiting the settling path in the tube settler, causing turbidity levels of up to 50 NTU at the outlet of the tube settler. To solve this seasonal problem, it was decided to pump up and fill the ECAR tanks with raw groundwater at the end of each workday and store it in the ECAR tanks overnight, so that water too would equilibrate with ambient cold night temperature by the morning. This was an adequate solution, and the tube settler functioned properly throughout the winter months.

4.6 Implementation Challenges

Training of Local Operators At the start of the field trial, there were no written instructions, either in English or Bengali, on plant operation. Thus, training was "on the job" and based on (sometimes inconsistent) verbal instructions from various team members and partners. Locally trained plant operators heard different instructions from different stakeholders, leading to confusion. This eventually led to the development of an operations manual to remove inconsistencies and ensure clarity. However, that did not always ensure common-sense operation. In one case, the arsenic concentration strangely spiked in the product water above the target level (though it remained below the recommended WHO MCL of 10 ppb). This is seen in Fig. 14.10 for the period starting at 23rd of May to about 18th of July 2016. With a lot of effort, the engineering science team traced it to the recirculation pumps in the ECAR reactors being shut off, mid-processing, by the plant operator. It turned out that the plant operator's action was due to the pump noise interfering with his personal cell phone calls. After identifying the cause, the research team explained the importance of keeping the recirculation pumps on and that it was an essential and nonnegotiable part of his job as an ECAR operator.

Ensuring Continuity of Knowledge Within the Team As this project involved many researchers, interns, and partners, at certain phases, there was insufficient time allocated to training the next incoming team members. The training process requires expenditure of precious time in a time-bound project. However, turnover of some team members is inevitable (e.g., as some graduate) in such multi-year projects. This leads to discontinuity in knowledge if documentation was incomplete, overlooked, or was felt to be only supplemental information. The team's stumbling from this problem led to a partial solution – as far as possible, the team had field staff work in pairs, so if one of them rotated out, the other in the pair would have detailed field knowledge that could be useful to train the replacement hire. This patchwork solution seems to have worked well enough.

Revisiting and Correcting Early Mistakes Some of the initial actions and decisions during the actual construction were suboptimal for the long-term operation of the demonstration plant. This was the result of different priorities between the implementing partners; the industry partner prioritized cost saving via shortcuts and cheaper components and parts, while the research partners preferred the plant to remain robust and resilient (in their future absence) with higher-quality parts and with a layout facilitating easy troubleshooting, should problems develop later. Some problems emerged very soon – owing to the initial shortcuts and cheap components right after the pilot plant was commissioned in April 2015. Fortunately, the research partners had committed to support two field engineers at the plant for many months even after commissioning to diagnose and overcome unexpected problems.

First, there were misdirected water jets into one of ECAR reactor tanks. During ECAR treatment, it is important to continuously replenish the dissolved oxygen in the water, which is being depleted by the chemical reactions in the ECAR process. This was done through cleverly engineered water jets integrated into a water recirculation system, which also helped maintain chemical homogeneity in the bulk solution. Without understanding the underlying chemistry, the installers had misdirected these water jets, preventing the adequate recharge of dissolved oxygen. On discovering this mistake, the field engineers inserted a metal spacer in the reactor and ensured that the jets impinged the water surface as intended.

Second, to cut costs, the industrial partner had purchased inappropriately rated and falsely labeled electrical wires, which were additionally also poorly installed electrical wires. This caused overheating and melting of the plastic insulation on the wire. Furthermore, these wires were buried in a shallow trench in the floor and covered with concrete. The field engineers, responsible to commission the plant, had to reinstall and replace all of the faulty wiring and move it to overhead cable trays for future easy access. In addition, the electrical contacts between the wires and the steel plate electrodes also needed to be replaced and reinstalled. The galvanized-iron (GI) connectors corroded and were replaced with tin-coated connectors which were coated with electrical grease as a preventative measure against corrosion and to improve lifetime

Finally, there were large energy losses and voltage drops at the power supply. The research team had ordered a custom low-voltage DC power supply that would deliver 80 Amps of DC current to each reactor at a voltage that could be adjusted manually between 5 volts and 15 volts. They discovered that the busbar used for the power supply's output electrical distribution was not electrical-grade copper, but just plain aluminum (a poorer and cheaper conductor of electricity). The poorer conductivity could have been compensated with a larger cross section of the busbar, but was not. So, the reactors kept receiving lower voltage than desired from the power supply because of the significant voltage (and energy) loss in the aluminum busbar. A similar problem arose from not using electrical-grade copper in other parts of current distribution within the ECAR reactors. It took some time and effort to diagnose these problems. The reactor performance improved after the field engineers replaced the busbars of aluminum and poor-grade copper with busbars of electrical-grade copper.

Additional Engineering Work Needed During Commissioning The demonstration plant included many components (Fig. 14.8) in addition to the ECAR reactor tanks. It was impractical to test and fine-tune the performance of the large pieces of equipment (e.g., for particle separation) in a controlled laboratory setting. As a result, the researchers worked on these aspects of the demonstration plant on-site for the first many months. This included tests to calibrate the correct dose of particle coagulant and flocculation treatment duration and observe and diagnose any fluctuations in the overall performance of the particle separation equipment. Only then could a maintenance routine be established to ensure reliable long-term plant performance. For each of the treatment steps, alterations or maintenance requirements were identified. Two examples follow.

The actual implemented tube settler proved much less effective than expected. It was then discovered that the plastic lamella that form the core of the tube settler was incorrectly installed by the overconfident installation expert (who remained in denial!). The misassembled lamella assembly was damaged beyond repair and had to be discarded. New lamellae were purchased and correctly installed by Berkeley field engineers. The field engineers that identified and rectified the problem learned on the job and found the solution by searching on the Internet and watching YouTube videos about how to correctly assemble and install the lamellae. After the retrofit, the tube settler reliably delivered water with low turbidity (<1 Nephelometric Turbidity Units (NTU)).

Originally, the finished product water was pumped up to the top of the (three story) school building and stored in a dark blue plastic tank. In summer, the tank became quite hot making the water unpleasant to drink, and furthermore, the team

Fig. 14.11 Handoff of Dhapdhapi plant from UCB and GCP-JU to Livpure on January 31, 2017. From left to right: Joydeep Bhattacharjee (Livpure), Pratik Mukherjee (Livpure), Joyashree Roy (GCP-JU), and Ashok Gadgil (UCB). All manuals were handed over to Livpure after the signing. From this day onward, Livpure has taken over the responsibility for operating and maintaining the plant while giving access for the researchers to the plant, on an as-needed basis

engineers figured out that this high temperature was the cause of precipitation of calcite dust floating on the water. The tanks' location was moved to a shaded spot (on top of the water dispensing unit), and the tanks were changed to white color. This kept the water cool and avoided calcite precipitates. The reduced height also decreased the excessively high rate of water flow at the dispensing unit.

The team had planned the project timeline such that two technical persons (either field engineers or researchers) were present and engaged full time at the site for the first 9 months of the plant operation. They were invaluable in identifying, diagnosing, and solving numerous technical problems and transferred the knowledge to the industry partner. The field researchers also developed manuals for maintenance, engineering, and plant operation that were reviewed by the project team lead and handed over to Livpure, the industrial partner, on January 30, 2017 (Fig. 14.11).

Until commissioning and for the first several months of regular operation, all the sludge produced was needed by researchers at JU conducting research on safe encapsulation of the sludge in structural concrete blocks (Roy et al., 2019). Doctoral research along similar lines was also conducted earlier by Tara Clancy at University of Michigan, Ann Arbor. See Clancy et al., (2013) for a good review of arsenic-bearing waste management. The team arranged for safe handling and storage of the sludge on site (Fig. 14.12) and its periodic pickup and disposal at the hazardous chemicals waste disposal facility in Haldia, by a licensed chemical waste disposal company, Ramky.

Fig. 14.12 Rathin, one of the locally trained plant operators, is safely storing sludge into the on-site cage, using proper personal protective equipment (PPE)

5 Reaching Scale: Opportunities and Challenges

5.1 Lessons Learned from Innovation, Implementation, Evaluation, and Adaptation

Based on the extensive experience of past efforts to take new technologies to scale, the team came to define the concept of the "Critical Effort Zone" (introduced in the published literature by Hernandez et al. (2019)). The Critical Effort Zone (Fig. 14.13) is introduced as a part of the commonly defined innovation chain, along with schematic cash flow for a financially sustainable innovation. As stated in Hernandez et al. (2019), "the Critical Effort Zone matches the period in the innovation chain for which the expected cash flow of a project reaches its largest negative value. In the social embedding process of technology maturation, this zone requires intense efforts for trust building with key social actors, and ultimately, for acceptance of the technology by the society that will use the innovation. These efforts require understanding of human behavior, strategic planning, and deep social science understanding of the local social context. If efforts during this period are unsuccessful, the innovation can stagnate or remain unused and perish. In contrast, if the efforts are successful, the technology may progress toward successful commercialization and scale-up."

During this critical zone, GCP-JU served as the locally reputed and trusted scientific partner that helped deploy capacity-building strategies beyond the large prototype phase, aiming for long-term sustainability of the project beyond the pilot

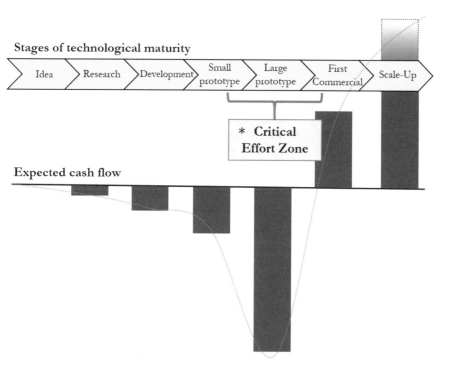

Fig. 14.13 A schematic of the Critical Effort Zone for the ECAR project. This zone encompasses the 600 L prototype at Dhapdhapi High School, its first demonstration plant, and short duration after the handoff to the industry partner. The figure highlights that the greatest financial expenditure falls within the Critical Effort Zone, and if the team does not succeed here, scale-up will not be possible. Note that the vertical axis for the cash-flow bar charts is not to scale. This figure is adapted from Hernandez et al. (2019)

phase. The duration of the Critical Effort Zone is different for each project. For the ECAR project, it took about four years (Dec. 2012 to Jan. 2017), comprising part of the small prototype, the entire large prototype, and the initial commercial stages of distributing ECAR drinking water to the community. The Critical Effort Zone consists of successful operation of the field pilot, well-designed communication to the public by the scientists in field operations, and flexibility toward any technology redesign needed to fit cultural practices. The actions taken during this zone must ensure a good fit of the technology to the market needs or "Product-Market Fit" (Andreessen, 2007), and must build the technical, management, and financial capacity for a transition to commercialization. During the Critical Effort Zone, the project team deployed and socially embedded the first ever large-scale ECAR pilot plant and delivered arsenic-safe drinking water to the first few thousand people daily for long-term health security.

5.2   ECAR Business Model

Although a technology is capable of delivering arsenic-safe water, only a technology combined with a viable business model is capable of delivering arsenic-safe water sustainably, day after day for many years. The ECAR team was very aware that many grant-funded projects fail to create impact once the grant resources are spent. Thus, the team worked to build a business model into their early conceptual framing of the long-term sustainable operation of ECAR technology. Although this chapter has focused largely on the technical efforts, parallel efforts led largely by GCP-JU and also by Berkeley focused on understanding what kind of business model and behavior change strategies would be most appropriate for the arsenic-affected communities of West Bengal. Previously, Prof. Joyashree Roy had published on the household spending on safe drinking water in Kolkata (Roy et al., 2004). A doctoral student in the Gadgil group, Caroline Delaire, also investigated, through rigorous in-person surveys of over 500 households in Murshidabad district of rural West Bengal, the factors that influence consumer's decisions about purchasing arsenic-safe drinking water and alternatives (Delaire et al., 2017). The Berkeley-JU team believed that a technically successful demonstration plant that was also financially viable would reduce the overall risk perceived by the industry partner and provide it with higher confidence in the financial reward from further scale-up. All this fitted well with the vision of the funding agency (IUSSTF) for the demonstration plant. Partnering with a school such as Dhapdhapi High School would provide a familiar central location and trusted source of public health information and education, critical to help change behavior.

6   Results and Lessons Learned

Multi-stakeholder Management Requires Attention   Partnerships are built and matured at all stages of the technology development and deployment process and require close and consistent attention. This signifies continuously explaining and defining roles among all stakeholders and keeping communication as open as possible. There will be challenges in the project management among partners when hiring new staff members, seeing trained engineers leave, and instructing field researchers from abroad during nature of international efforts separated by time zones and varying levels of resources.

The design of the two 1000 L reactors required the involvement of all three partners: UCB, GCP-JU, and Livpure. UCB provided adequate training about ECAR fundamentals and operation to GCP-JU and Livpure, answering all technical

Fig. 14.14 Official signing, by the headmaster, Shubhendu Biswas (center), of the document giving "Consent to Establish" and "Consent to Operate" to the ECAR water treatment plant at Dhapdhapi High School (November 2016). The "Consent to Operate" grants Livpure 10 years of operation of the plant following yearly performance review of an advisory and a working group. Also in the photograph are Abhijit Das (left) and Joyashree Roy (right)

and scientific questions to the full extent of their knowledge. Livpure pushed to have a design that delivered 10,000 LPD.

Thorough and Transparent Documentation Ensures Continuity Despite Unexpected Changes in Stakeholder Leadership   This large-scale, multi-year, international project involved many stakeholders. Unexpected changes in leadership can disrupt the timeline and the anticipated date of completing milestones, causing the cooperation to be halted or completely altered. A formal documentation signed by key project members (Fig. 14.14) reduces the risk of the new leader, like a new school headmaster, questioning the effort and commitment of the organization to the project. This holds clear accountable expectations among the parties.

Technology Design and Debugging Require Your Boots on the Ground   In the laboratory, the innovation of ECAR became well understood by the team. In the field, ECAR also reliably removed arsenic, exactly as designed. However, the remaining aspects of scaled-up processes of sludge separation and elements of the water distribution technology encountered unexpected challenges that could be identified, diagnosed, and resolved only because of the full involvement of research personnel. Thus, the research team strongly recommends that the first field trials remain research environments, under realistic conditions, in which the system is improved upon after feedback and iterations.

Trust Building Is the Foundation for Social Acceptance of a Technology The concept of social capital becomes evident in development engineering efforts. Local communities have seen many good-willed researchers come and fail and go away, never to return. They will have doubts in their minds about why this work is being carried out. They may keep these doubts among themselves or raise questions publicly about the motives of the team. Conscious efforts toward transparency and building mutual trust are a worthwhile and critical investment of time and effort. These efforts can include periodic public meetings with community leaders and open-mike sessions with the community and making analysis and data fully accessible.

Schools Can Be Effective Locations for Addressing a Public Health Concern A local public school provides an excellent means of transferring information to the students, teachers, staff, and parents and families of the students. The ECAR team held informational sessions and large inauguration events at the school, inviting questions from community members. They assisted with science fair entries and a student poster competition about arsenic in drinking water. The water debit cards used to dispense water from the automatic dispensing units carried messaging on the importance of drinking arsenic-safe water. These activities allowed for the diffusion of public health awareness on arsenic, as well as better understanding and acceptance by the community of the team's efforts in providing arsenic safe drinking water.

Plans Will Change in the Field Researchers will face technical problems in the field in resource-poor settings. It is also likely that researchers will not have immediate access to the principal investigators (PIs) or other supervisors and thus will need to either work within their capacities or hold off on certain action items. These experts are also likely to be remote and will be pressed to diagnose the root cause of the observed problem from incomplete information with very limited or no analytical instruments on-site. Thus, the entire team needs to be resourceful in low-income, rural settings and be prepared for much slower progress with the same hard effort they may make in a technologically better-resourced environment.

6.1 Cost of Reducing Cancer Risk by Removing Arsenic from Drinking Water

The lifetime excess risk of internal cancer from drinking water with arsenic at 250 ppb is 18 per 100 (Fig. 14.15). Reducing arsenic concentration to 3 ppb means that the lifetime risk is lowered to 0.23 per 100, signifying an 80-fold reduction in lifetime cancer risk for those drinking water from the Dhapdhapi plant. The safe water sells at 1 US cent per L (or INR six per 10 L), serving about 8000 people. For a simple estimate (without discounting, etc.), assume 1 L (i.e., eight cups) of drinking water consumption per person per day and a 70-year lifespan. So,

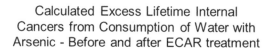

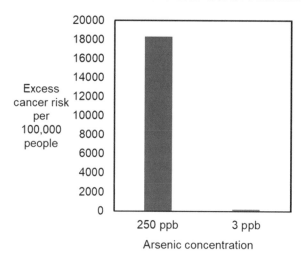

Fig. 14.15 The lifetime excess risk of internal cancer from drinking water with arsenic at 250 ppb, which is typical of groundwater in arsenic-affected areas, is about 18,360 per 100,000 people. The lifetime excess cancer risk of consuming arsenic at 3 ppb is about 230 per 100,000 people. By drinking ECAR water, which sells at 1 US cent per L, there is an 80-fold reduction in risk. These numbers are calculated using the EPA risk model (U.S. EPA 2010, which is based on numerous studies in the scientific literature)

each person spends a sum of USD 255.50 over a 70-year period. When 100 people spend this much (i.e., spend USD 25,550), 17 of their lives are saved from internal cancers. This works out to be USD 1500, without discounting, per life saved. In this calculation, the benefits from avoidance of skin ulcers, gangrene, amputations, neuropathy, diabetes, heart disease, and social ostracization are ignored.

5.2   *Looking Back and Adapting to New Contexts*

It is worthwhile to briefly revisit the theory of change, and where the team ended in India after 14 years of effort (from 2005 to 2019). What worked, what didn't, and why? The team can say with high confidence that ECAR performs technically well, is energy efficient, has zero liquid discharge (ZLD is a coveted goal in sustainability engineering design), and removes arsenic easily to concentrations below 10 ppb. ECAR is also economically attractive, financially viable, and socially acceptable to the community at Dhapdhapi. So, the team fully met three of the four items in the "Outcomes" (Fig. 14.2). The outcome partially met so far is the interest from companies and from public agencies at the state and central government levels in replicating the success at Dhapdhapi. The second-ever ECAR plant is now in operation in India since early 2021, delivering 5000 L per day of arsenic-safe drinking water. The plant is located at Bahraich in Uttar Pradesh, adjacent to the Nepal border. This plant is fully solar-PV powered with roof-top solar panels and has an internal water storage capacity of 15,000 L to provide for three days of water supply in case of power interruption during heavily clouded skies. With this second iteration, the layout has improved to include a spacious interior and redesigned ECAR reactors. This is a great achievement by our industry partner, and our team hopes that other companies and government agencies will support the implementation of ECAR plants across arsenic-affected parts of India.

We note that the Central Government of India took the highly unusual step in February 2016 to release INR 800 crores (INR 8 billion, about USD 120 million) to the states affected by arsenic and fluoride in their groundwater (P.T.I. 2017). These are grant funds for capital expenditures to be used by the State Public Health Engineering Departments (PHEDs) only for community-scale safe drinking water systems. So, now, the problem, the solution, and the money are all there. Furthermore, the ECAR team spent a very significant time and effort to register ECAR at the website of the Ministry of Drinking Water and Sanitation (MDWS) to invite testing, verification, and certification of the implementation and performance of the ECAR technology at the operating plant at Dhapdhapi. After such testing and visits by independent government-appointed experts, the Mashelkar Committee of the MDWS listed ECAR on its website in August 2019, as an approved technology for arsenic remediation of groundwater for drinking (eJalshakti, 2019). With this approval (posted on the website of MDWS), no individual State PHED needs to undertake its own testing and verification of the ECAR technology, removing another barrier to the technology's dissemination.

The estimated number of Indians drinking water above 10 ppb continues to grow, as more testing data become available. Prof. Chander Kumar Singh and his research group at TERI University, Delhi, estimate that the number of Indians exposed to arsenic above 10 ppb in their water exceeds 100 million (Singh 2020).

The IUSSTF project ended in 2016. Beyond IUSSTF, what is the funding landscape for further work on ECAR? And where is it headed? A large multicampus, multi-year project led by UCB on water-energy research, funded by the US Department of Energy, has supported applications of advanced versions of ECAR to remediate arsenic-bearing wastewater from coal-fired power plants in the USA. The Philippines has a groundwater arsenic problem in some regions. Since 2018, Gadgil has collaborated with faculty at their top engineering school, the University of the Philippines Diliman, to transfer ECAR science and technology from Berkeley to UP Diliman, supported by the Government of the Philippines. In 2019, the UCB research group won 3 years of research support from the State of California's program to reduce cancer incidence in California populations. This project supports the UCB group to develop an advanced form of ECAR that must work well in the California socioeconomic and regulatory environment and to conduct a short field test in an affected region of rural California.

The requirement of using ECAR chemistry in rural California has led researchers in the Gadgil Lab to invent two more-advanced versions of ECAR. These two versions are able to treat a much higher flow of water for arsenic removal than ECAR at only marginally higher cost but with a much smaller footprint and labor content. As they rely on the same ECAR principles, therefore the team's confidence is high that they will likely function well in the field. Both the advanced technologies have tested well in the laboratory and in very limited field tests in rural California. The first of them, called Air Cathode Assisted Iron Electrocoagulation (ACAIE), has been disclosed and published (Bandaru et al. 2020), and the second, Fe-EC with External Oxidizer (FOX), is being prepared for publication at the time of this writing (August 2020).

7 Summary of Key Actions as Viewed by the Team

- Stay focused from the start of technology invention on strong science-based effectiveness, affordability, environmental safety of removing arsenic, and technology's fit to a business model.
- Revisit the engineering design often, as the science understanding gets deeper.
- Keep a learning mindset – stay curious about learning everything that might impact the technology, its implementation, and eventual scale-up.
- Actively seek to identify the team's weaknesses ("holes" in combined organizational skills and goals), and always look to fill them by reaching out to potential partners (e.g., unsuccessful outreach to all major industrial players in India and finally finding Livpure serendipitously through a Zayed Prize connection!).
- Undertake conscious efforts to keep relevant authorities at various levels of the government informed so the team is not treated as strangers, upstarts, or fly-by-night operators.
- Ensure, as far as possible, transparency and redundancy of skills, so the work does not collapse owing to one single individual getting removed from the team.
- Aim for written signed MOU agreements from all parties on whom the team critically depend for their cooperation to be successful. MOUs are not legally enforceable, but they avoid later accusations of who promised what, and help in expectation management.
- Undertake outreach effort to the community (e.g., teachers, staff, students, and nearby community) through public meetings every 6 months or so. Report progress or setbacks, manage expectations, and answer questions publicly and honestly to suppress rumors and speculations.
- Work to understand all the stakeholder's needs and constraints, which may change throughout the project, and address the issues rationally and through transparent communication.

8 Interpretive Text Boxes

- **A critical role for academia.** Academia brings to the table not just an unusual depth of knowledge, but is also seen as neutral in the sense of not having a long-term financial or political interest in the new social arrangements arising from introduction of the new technology. Thus, academia could act as a mediator and coordinator of stakeholders (e.g., UCB's help to resolve issues between Shri Hari and Livpure or JU's frequent coordination between Livpure and the community/school/government).
- **Importance of field work.** Apart from the technical aspect of iteratively testing in the field and improving the prototypes, field work greatly helps in developing and strengthening relationships with partners who will be essential to scale-up and sustainable implementation.
- **Expectation management.** It is very important (and difficult) to manage expectations of stakeholders so they are aligned and timelines are matched. It is important to repeatedly emphasize and convey the purpose of research. This is related to outreach efforts to the community, such as removing rumors, inflated expectations, and misunderstandings.
- **Doing Field Work Early and Often.** Field work can shape the research questions being asked. Our field work led to a focus on designing a large-scale rapid settling stage (based on learning, this was the limiting step in 100 L trials), robust waste management (based on feedback from the community), steps to overcome the passivation of electrode plates over long-term use (after 100 L, 600 L, and 2000 L trials), and the need to look at bacterial contamination in addition to arsenic, among other questions.
- **Development engineering research groups are unique.** Intense impact-oriented research is facilitated by operating in a slightly different way than how most academic research groups operate. The UCB part of the team included project scientists and staff engineers who worked as equals in the group, along with the PI, graduate and undergraduate students, and a part-time administrative support.

Questions for Discussion

1. Was it right for the team leadership to have a protected dug well installed at the Amirabad school to offer immediate relief (although it did not work out)? Why or why not? What was the team's responsibility, having installed it, to ensure its successful and continued operation? Was this an example of mission creep?
2. Was the team leadership right in moving their earlier efforts from Bangladesh to West Bengal, India, for the reasons stated? Why or why not?
3. As noted in the narrative, when the team was testing various minor issues in the treatment train after arsenic removal from the 1000 L reactors, they kept on discarding all treated water into a soak pit for 6 months, while local people had to continue drinking the local arsenic-contaminated water as was the past practice. Discuss the ethical and legal pros and cons of this decision. Was that the right decision? Why or why not? What would you do?

4. The narrative mentions rejecting a girls-only school for the field trial (after the Amirabad blowup) and selecting a coed school to avoid another risk of controversy. Was that the right decision? Why or why not? Are there options you think the team should have explored before abandoning the girls-only school for the site of the field trial? The girls-only school is only 45 minutes by car, compared to ~2 hours to Dhapdhapi.
5. Discuss the experience described overall, of attempting technology transfer to what seems like a passive industrial partner. What could have been done differently for ensuring a more energetic commitment and commercial takeoff, particularly given that the industrial partner is a nonexclusive licensee of the technology?

Acknowledgments We would like to thank Professor Joyashree Roy and Dr. Siva Rama Satyam Bandaru for their invaluable roles in the ECAR project, as well as their careful review of this book chapter for accuracy of the narrative.

We gratefully acknowledge LBNL's LDRD program which funded the initial lab research and the Sustainable Products and Solutions program established at UC Berkeley by the Dow Chemical company. We also acknowledge phase one and phase two funding from the US EPA's P3 program. We also acknowledge the Big Ideas program at UCB, the Blum Center, and USAID's Development Impact Lab (DIL) led by UCB. We acknowledge the NSF Graduate Research Fellowship Program (GRFP) support to Case van Genuchten and Dana Hernandez, in addition to support from National Science Foundation's program for Innovation at the Nexus of Food Energy Water Systems (InFEWS). We also acknowledge support from University of California Office of the President (UCOP) for Siva Bandaru's first visit to UCB. We acknowledge JU-GCP's funds which supported their participation in this effort. We acknowledge gift funding from Siemens Stiftung, Reed Elsevier Foundation, National Collegiate Inventors and Innovators Alliance (NCIIA), and Jewish Teen Foundation, Marin County. We are grateful to IUSSTF in our scale-up and commercialization efforts in the last 5 years of the project timeline. Gaps and shortfalls were filled with Gadgil's Rudd Chair funds from UC Berkeley campus. These resources allowed the Berkeley team to take the necessary technical risks that led to a steady increase in knowledge and confidence, to build up to the 1000 L reactor.

References

Ahmad, A., & Bhattacharya, P. (2019). Arsenic in drinking water: Is 10 μg/L a safe limit? *Current Pollution Reports, 5*(1–3), 1137. https://doi.org/10.1007/s40726-019-0102-7

Amini, M., Abbaspour, K. C., Berg, M., et al. (2008). Statistical modeling of global geogenic arsenic contamination in groundwater. *Environmental Science & Technology, 42*, 3669–3675. https://doi.org/10.1021/es702859e

Amrose, S., Gadgil, A., Srinivasan, V., et al. (2013). Arsenic removal from groundwater using iron electrocoagulation: Effect of charge dosage rate. *Journal of Environmental Science and Health, Part A, 48*, 1019–1030. https://doi.org/10.1080/10934529.2013.773215

Amrose, S. E., Bandaru, S. R., Delaire, C., et al. (2014). Electro-chemical arsenic remediation: Field trials in West Bengal. *Science of the Total Environment, 488*, 539–546. https://doi.org/10.1016/j.scitotenv.2013.11.074

Andreessen, M. (2007). *Product market fit.* Stanford University. https://pmarchive.com/guide_to_startups_part4.html. Accessed Jan. 2021.

Bandaru, S. R., van Genuchten, C. M., Kumar, A., et al. (2020). Rapid and efficient arsenic removal by iron electrocoagulation enabled with in situ generation of hydrogen peroxide. *Environmental Science & Technology, 54*, 6094–6103. https://doi.org/10.1021/acs.est.0c00012

Brinkel, J., Khan, M. H., & Kraemer, A. (2009). A systematic review of arsenic exposure and its social and mental health effects with special reference to Bangladesh. *International Journal of Environmental Research and Public Health, 6*, 1609–1619. https://doi.org/10.3390/ijerph6051609

Chakrabarti, K. B. (2017). *Arsenic Contamination in Bengal Basin: Reinventing Mitigation through Participatory Social Innovations.* Dissertation. Washington State University. (Accessible via ProQuest Dissertation Database.)

Clancy, T. M., Hayes, K. F., & Raskin, L. (2013). Arsenic waste management: A critical review of testing and disposal of arsenic-bearing solid wastes generated during arsenic removal from drinking water. *Environmental Science & Technology, 47*, 10799–10812. https://doi.org/10.1021/es401749b

Conceição, P. (2019). Human development report 2019: Beyond income, beyond averages, beyond today: Inequalities in human development in the 21st century.

Das, A., & Roy, J. (2013). Socio-economic fallout of arsenicosis in West Bengal: A case study in Murshidabad District. *Journal of the Indian Society of Agricultural Statistics, 67*, 267–278.

Das, A., Roy, J., & Chakraborti, S. (2016). *Socio-economic analysis of arsenic contamination of groundwater in West Bengal.* Springer.

Delaire, C., Das, A., Amrose, S., et al. (2017). Determinants of the use of alternatives to arsenic-contaminated shallow groundwater: An exploratory study in rural West Bengal, India. *Journal of Water and Health, 15*, 799–812. https://doi.org/10.2166/wh.2017.321

eJalshakti (2019) List of recommended technologies. In: Department of Drinking Water and Sanitation Government of India. https://ejalshakti.gov.in/MISC/InnovationAccrMC_Rep.aspx. Accessed 11 Aug 2020

Gadgil, A., Roy, J., Addy, S., et al. (2012). Addressing arsenic poisoning in South Asia. *Solutions, 5*, 40–45.

Gadgil, A., Amrose, S., Bandaru, S., et al. (2014). Addressing arsenic mass poisoning in South Asia with electrochemical arsenic remediation. In S. Ahuja (Ed.), *Water reclamation and sustainability* (pp. 115–154). Elsevier. https://doi.org/10.1016/B978-0-12-411645-0.00006-7

Ghosh, N., Singh, R. (2009). Groundwater arsenic contamination in India: Vulnerability and scope for remedy.

Hernandez, D., Boden, K., Paul, P., et al. (2019). Strategies for successful field deployment in a resource-poor region: Arsenic remediation technology for drinking water. *Development Engineering, 4*, 100045. https://doi.org/10.1016/j.deveng.2019.100045

Inauen, J., Hossain, M. M., Johnston, R. B., & Mosler, H.-J. (2013). Acceptance and use of eight arsenic-safe drinking water options in Bangladesh. *PLoS One, 8*, e53640. https://doi.org/10.1371/journal.pone.0053640

Katayama, R., Wadhwa, D. (2019). Half of the world's poor live in just 5 countries. https://blogs.worldbank.org/opendata/half-world-s-poor-live-just-5-countries. Accessed 30 Aug 2020.

Karagas, M. R., et al. (2015). Drinking water arsenic contamination, skin lesions, and malignancies: A systematic review of the global evidence. *Current Environmental Health Reports, 2*(1), 52–68. https://doi.org/10.1007/s40572-014-0040-x

Li, L., van Genuchten, C. M., Addy, S. E., et al. (2012). Modeling As (III) oxidation and removal with iron electrocoagulation in groundwater. *Environmental Science & Technology, 46*, 12038–12045. https://doi.org/10.1021/es302456b

Mackay, H. B. (2005). Swim with the sharks without being eaten alive: Outsell, outmanage, outmotivate, and outnegotiate your competition. *Harper Business.*

Mathieu, J. L., Gadgil, A. J., Addy, S. E. A., & Kowolik, K. (2010). Arsenic remediation of drinking water using iron-oxide coated coal bottom ash. *Journal of Environmental Science and Health, Part A, 45*, 1446–1460. https://doi.org/10.1080/10934529.2010.500940

Minatel, B. C., et al. (2018). Environmental arsenic exposure: From genetic susceptibility to pathogenesis. *Environment International, 112*, 183–197. https://doi.org/10.1016/j.envint.2017.12.017

National Research Council. (2001). *Arsenic in drinking water: 2001 update.* The National Academies Press.

Naujokas, M. F., et al. (2013). The broad scope of health effects from chronic arsenic exposure: Update on a worldwide public health problem. *Environmental Health Perspectives, 121*(3), 295. https://doi.org/10.1289/ehp.1205875

Nordstrom, D. K. (2000). An overview of arsenic mass poisoning in Bangladesh and West Bengal. *India. Minor elements*, 21–30.

P.T.I. Press Trust of India (2017). Arsenic affected drinking water in 66,663 habitations: Govt. In: India news, breaking news, entertainment news | India.com. https://www.india.com/news/agencies/arsenic-affected-drinking-water-in-66663-habitations-govt-1815142/. Accessed 11 Aug 2020.

Rahman, M. A., Rahman, A., Khan, M. Z. K., & Renzaho, A. M. (2018). Human health risks and socio-economic perspectives of arsenic exposure in Bangladesh: A scoping review. *Ecotoxicology and Environmental Safety, 150*, 335–343. https://doi.org/10.1016/j.ecoenv.2017.12.032

Roberts, L. C., Hug, S. J., Ruettimann, T., et al. (2004). Arsenic removal with iron(II) and iron(III) in waters with high silicate and phosphate concentrations. *Environmental Science & Technology, 38*, 307–315. https://doi.org/10.1021/es0343205

Roy, J. (2008). Economic benefits of arsenic removal from ground water — A case study from West Bengal, India. *Science of the Total Environment, 397*, 1–12. https://doi.org/10.1016/j.scitotenv.2008.02.007

Roy, J., Chattopadhyay, S., Mukherjee, S., et al. (2004). An economic analysis of demand for water quality: Case of Kolkata. *Economic and Political Weekly, 39*, 186–192.

Roy, A., van Genuchten, C. M., Mookherjee, I., et al. (2019). Concrete stabilization of arsenic-bearing iron sludge generated from an electrochemical arsenic remediation plant. *Journal of Environmental Management, 233*, 141–150. https://doi.org/10.1016/j.jenvman.2018.11.062

Rybczynski, W. (1991). *Paper Heroes: Appropriate Technology: Panacea or Pipe Dream?* Penguin.

Sanchez-Paramo, C. (2020). Countdown to 2030: A race against time to end extreme poverty. In: *World Bank Blogs.* https://blogs.worldbank.org/voices/countdown-2030-race-against-time-end-extreme-poverty. Accessed 30 Jun 2020.

Singh, C.K. (2020) Personal communication to Ashok Gadgil.

Smedley, P. L., & Kinniburgh, D. G. (2002). A review of the source, behaviour and distribution of arsenic in natural waters. *Applied Geochemistry, 17*, 517–568.

Smith, A. H., Lingas, E. O., & Rahman, M. (2000). Contamination of drinking-water by arsenic in Bangladesh: A public health emergency. *Bulletin of the World Health Organization, 78*, 1093–1103.

Smith, A. H., Lopipero, P. A., Bates, M. N., & Steinmaus, C. M. (2002). *Arsenic epidemiology and drinking water standards.* American Association for the Advancement of Science.

U.S. EPA (2010). Toxicological Review of Inorganic Arsenic. Report EPA/635/R-10/001. Accessed from EPA website on June 20, 2017

UN (2020). https://sdgs.un.org/goals. Accessed August 11, 2020.

van Genuchten, C. M., Addy, S. E., Peña, J., & Gadgil, A. J. (2012). Removing arsenic from synthetic groundwater with iron electrocoagulation: An Fe and As K-edge EXAFS study. *Environmental Science & Technology, 46*, 986–994. https://doi.org/10.1021/es201913a

van Genuchten, C. M., Pena, J., Amrose, S. E., & Gadgil, A. J. (2014). Structure of Fe (III) precipitates generated by the electrolytic dissolution of Fe (0) in the presence of groundwater ions. *Geochimica et Cosmochimica Acta, 127*, 285–304. https://doi.org/10.1016/j.gca.2013.11.044

Wang, J. S., & Wai, C. M. (2004). Arsenic in drinking water—A global environmental problem. *Journal of Chemical Education, 81*(2), 207. https://doi.org/10.1021/ed081p207

Wasserman, G. A., Liu, X., Parvez, F., et al. (2007). Water arsenic exposure and intellectual function in 6-year-old children in Araihazar. *Bangladesh. Environmental Health Perspectives, 115*. https://doi.org/10.1289/ehp.9501
Wikipedia   (2020a).   https://en.wikipedia.org/wiki/Human_right_to_water_and_sanitation. Accessed August 11, 2020.
Wikipedia. (2020b). https://en.wikipedia.org/wiki/Theory_of_change. Accessed August 11, 2020.
WHO Newsroom (2019) 1 in 3 people globally do not have access to safe drinking water. https://www.who.int/news-room/detail/18-06-2019-1-in-3-people-globally-do-not-have-access-to-safe-drinking-water-unicef-who (Accessed: 15 July 2020).

Chapter 15
Sensing Change: Measuring Cookstove Adoption with Internet-of-Things Sensors

Daniel L. Wilson

1   The Challenge of Measuring Impact

Practitioners of development engineering aspire to create new technologies, and the goal of these new technologies is to have a positive impact for people in need. Perhaps the technology affords cleaner water, better governance, faster medical diagnosis, or easier access to public services. Regardless of the technology, we became development engineers because we want our work to have positive impacts.

In this chapter, I'll discuss cleaner cooking appliances that were intended to reduce exposure to air pollution and the drudgery associated with traditional wood-fired cooking methods. I'll discuss the trials and tribulations of creating new kinds of sensor technologies to make it easier to measure cookstove adoption. This process spans the course of about 8 years, three major technology variants, and thousands of households in dozens of countries. In the beginning, I'll discuss the challenges of collecting sensor data using clunky industrial data loggers in Sudan. Then, we will move onto a first round of custom-built data loggers deployed on advanced cookstoves in India. Then, in a four-country trial with thousands of households participating, we will explore how an entirely custom Internet of Things (IoT) sensor, survey, and analytics platform were used to address many of the issues that arose in the first two experiments. Finally, I'll share how this cookstove sensor technology was spun off into a startup sensor company. But first, why all this fuss about measuring cookstoves?

Cooking is a critical global development challenge because roughly three billion people rely on traditional wood- or other solid-fueled fires to cook their daily meals (The World Bank, 2011). The smoke from this practice is one of today's greatest

D. L. Wilson (✉)
Geocene Inc., Berkeley, CA, USA
e-mail: danny@geocene.com

© The Author(s) 2023   399
T. Madon et al. (eds.), *Introduction to Development Engineering*,
https://doi.org/10.1007/978-3-030-86065-3_15

environmental health risk factors and is responsible for some 1–4 million premature deaths per year (Forouzanfar et al., 2015; Lim et al., 2013). Cleaner cookstove technologies could positively affect this situation, but we will discuss that more in later sections.

What is impact, and how do we measure it? This is a major topic that development engineers are grappling with today (Thomas & Brown, 2020). Suppose you have developed a new novel cookstove technology that reduces emissions by 95%. If that cookstove is only used by a few dozen farmers as part of your research project, did it have an impact? Alternatively, suppose you make a tiny improvement to a traditional earthen cookstove; your innovation drives down emissions by 5%, and you estimate that 10 million cookstoves are built per year using your technology. Did your small improvement on a traditional technology have an impact?

I would like to propose a framework to think about impacts throughout this case study. That is, impact is proportional to the product of performance, adoption, and scale:

$$Impact \propto Performance \times Adoption \times Scale$$
$$I \propto PAS$$

This specific framework, which is pronounced "IPAS" (people are better at remembering acronyms that have to do with beer), is useful for remembering how multiple, often conflicting, considerations must be taken into account when measuring impact. For the purposes of this chapter, we can think of the definition of each of these variables as follows:

- **Impact** is the realized positive global utility that exists as a result of the existence of the technology.
- **Performance** is the marginal difference in outcome between the intervention state, on an individual product or person level, and the status quo. For example, for cooking, a good metric of performance might be the difference in personal relative risk of disease between breathing smoke emissions cooked on a traditional and cleaner cookstove.
- **Adoption** is the average degree to which the user of an individual unit of technological intervention (e.g., a single cookstove) utilizes that intervention per unit time. For example, this could be the average number of cleaner cookstove uses per week. The study of adoption is part of a broader category of development practice activities often referred to as "monitoring and evaluation."
- **Scale** is the number of users of your intervention.

Of course this model is a gross oversimplification. It is important to remember that, despite the linear simplicity of the IPAS framework, many real-world systems have nonlinear relationships. For example, if a cookstove that reduces emissions by 50% will not reduce negative health impacts by 50% in the same way that cutting a smoking habit from two packs per day to a single pack per day will not reduce lung cancer risk by 50% (Burnett et al., 2014). The IPAS model also ignores important considerations such as disadoption of harmful baseline practices, impacts of training and time on performance, temporal changes in scale, and second-order effects such as how adoption of one technology may lead to virtuous second-order adoption of more beneficial technologies (adopting solar panels leads to adopting lighting that leads to better educational outcomes) or negative second-order effects (adopting microwave cooking leads to consumption of less healthy foods) (Pillarisetti et al., 2014).

Despite the drawbacks of this simplified model, it does emphasize two aspects of technological interventions that engineers are notorious for downplaying: adoption and scale. As engineers, many of us are only trained to improve and measure performance, not adoption or scale. Whether designing a cookstove, water filter, generator, or other intervention technology, we tend to be very diligent about quantifying the performance of our interventions. The desire to measure and report metrics such as efficiency, power, and throughput comes naturally to many engineers. However, it is important to remember that a great-performing intervention that people do not like or that just a few people use will not have a significant impact. In fact, sometimes, higher performance reduces impact because higher performance is often correlated with factors that drive down adoption and scale such as cost, durability, multipurpose use, and difficulty of distribution (Jetter et al., 2012).

So, measuring impact is critically important. We know that measuring impact means more than just measuring performance. We also need to measure adoption (to what extent an average user utilizes a particular innovation) and scale (the quantity of users). For the purposes of this chapter, we will focus exclusively on a case study of how to measure the adoption of cookstoves using Internet of Things (IoT) sensors. This case study will span my PhD research and then the company that grew out of that research. So, let's start by thinking about how to measure adoption.

1.1 Surveys Collect Bad Data

Although most practitioners of development engineering are motivated by good intentions, our positive intentions do not necessarily lead to positive impacts. When we create a new technology, we cannot just hope (or expect) that it will improve lives. Before we can scale up or make any claims about the impacts of technological interventions, we must carefully measure if and how customers use technological interventions. We call these patterns of individual customer interaction with a new technology "adoption."

Historically, surveys have been the most common "instrument" for monitoring adoption (Burwen, 2011; Lewis & Pattanayak, 2012). At first glance, survey instruments seem like an attractive option. After all, most intervention programs are interested in some basic understanding of questions like the following: Do customers use our intervention? Do they like it? Would they buy it? Would they recommend it to a friend? Is the intervention improving their lives in some. meaningful way?

Most of these questions feel like they could be reasonably answered in a thoughtful discussion with a customer. Surveys can be implemented with as little material as a clipboard and pen, training staff to enumerate a survey is relatively straightforward, and respondents can give rich contextual color, anecdotes, and insights in their answers which can be difficult to gather otherwise. These factors can make surveys seem like a low-cost and high-reward instrument for collecting data about technological impacts.

However, surveys suffer from two critically important problems: recall errors and social-desirability bias (Das et al., 2012; Edwards, 1957; Methodology, n.d.; Thomas et al., 2013; Zwane et al., n.d.). Recall errors result from the difficulty surveyees have in recalling facts or events, even if the surveyee is trying in earnest to respond truthfully. In research, my colleagues and I have been interested in understanding if, why, and to what extent people used their traditional and intervention cooking appliances.

However, getting quantified metrics about cookstove use through surveys can be challenging. For example, let me ask you, the reader, how many times and for how many total minutes did you operate your microwave last month? Well? This is a difficult question for anyone to answer. Most of us don't think much about our personal microwave utilization stats, so you might respond to me by just guessing a number that "feels right," or if you're good with mental math, you might do some quick thinking to try and estimate an accurate response. Or, more likely, maybe you won't say much at all. When I ask this question to students in university lectures, most students just freeze and say "I . . . I don't know." In terms of measuring quantified impact, these kinds of answers are not very useful.

While recall errors are problematic, they do not necessarily introduce systematic bias to survey results. In answering the microwave question, we might expect roughly the same number and degree of overestimated and underestimated answers. By contrast, the other main drawback of surveys, social desirability bias, does indeed create problems with systematic bias. Social desirability bias is the tendency for research subjects to offer the normative or desirable response. This kind of normative response, where the respondent tells the survey enumerator the "right" answer that the enumerator "wants to hear," tends to overstate "positive" behaviors and understate "negative" behaviors (Nunnally & Bernstein, 1994).

In our microwave example, imagine I had sent that microwave to you as a gift a few months back. It's a weird gift, for sure, but then, I'm a weird guy who is obsessed with cooking appliances. Today, I sit you down with a pad of paper and a pencil, and I put you on the spot. "How often do you use that microwave?" I ask. "Well," you might be inclined to say, "I use that microwave all the time. Every morning in fact. It's perfect for oatmeal, which I just love by the way, so I use it every morning. Yep."

Nope. You didn't even take it out of the box because you already have a perfectly good stove. Also, you hate oatmeal.

Unfortunately, this is usually the social dynamic in development engineering field studies. Intervention products like pumps, lights, cookstoves, water filters, and mosquito nets are donated or heavily subsidized by the research project. Then, after some time, study participants are asked to self-evaluate their adoption of the technology (e.g., "how many nights last week did you sleep under the mosquito net?") and other qualitative aspects of the technology's appropriateness (e.g., "do you feel like this mosquito net does a good job keeping your family safe from malaria?"). Multiple research studies have demonstrated that responses to these kinds of questions are weakly correlated (or not correlated at all) with actual user behavior (Wilson et al., 2015, 2016a; Wilson et al., 2018). So, how do we measure actual user behavior?

1.2   Sensors Collect Good Data, But It's Hard to Do It Right

Sensors are a great choice for collecting cold unbiased data about the physical world. Sensors can be used to objectively measure facts about the environment such as the heat from a cookstove, the flow of water through a pump, the flip of a switch, or the opening and closing of a door (Clasen et al., 2012; Thomas et al., 2020; Turman-Bryant et al., 2020). When sensors are designed well and implemented in a thoughtful manner, they can collect data streams that offer insights about users' behaviors that could not be realized through surveys alone (Iribagiza et al., 2020; Thomas et al., 2020; Wilson et al., 2017).

As a quick side note: sensors are not immune to bias issues either. Research has shown that, when users know their behavior is being observed, even by sensors, the way they behave changes (e.g., they might use an intervention cookstove more). This effect, where users change behavior when they are aware of being observed, is called the "Hawthorne Effect" (Landsberger, 1958; McCarney et al., 2007; Methodology, n.d.; Thomas et al., 2016).

However, sensors alone do not deliver insights or even intelligible data. There is a wide gap between the raw data collected by data-logging sensors and the insights most of us expect from sensor-based data collection systems (Kipf et al., 2015; Wilson et al., 2020). When it comes to interpreting the data from sensor systems, we have been spoiled by exceptionally vertically integrated and user-friendly sensor systems, especially in the Internet of Things (IoT) age. Take a modern IoT pedometer like Fitbit. At its core, a pedometer is a multi-axis accelerometer measuring acceleration and thousands of samples per second. In the case of Fitbit, millions of person-hours have gone into turning that raw data stream of thousands of acceleration samples per second into a beautiful, intuitive, and comprehensible hardware product and mobile application that allows anyone to collect and visualize metrics about the number of steps a user has taken of the course of time.

So, where's the Fitbit of cookstove adoption tracking? Therein lies the problem. Development engineering practitioners want to ask questions about adoption of novel technologies, and usually, there are no off-the-shelf sensor systems or IoT products available to simplify the data collection and interpretation process. As development engineering practitioners, we find that we're usually not in a situation where we can buy an elegant hardware and software solution to solve our adoption

Table 15.1 Example data
logger data

Datalogger_id	Timestamp	Temperature
Logger-314	2020-05-01 1:00:00	20.0 °C
Logger-314	2020-05-01 1:01:00	20.2 °C
Logger-314	2020-05-01 1:02:00	21.5 °C
Logger-314	2020-05-01 1:03:00	22.1 °C
...	...	...

monitoring and evaluation needs. Using the pedometer analogy, we're back in the situation where we need to collect the thousands of accelerations reading per person per second and then make sense of the data later on. This means collecting raw sensor data about flow, temperature, humidity, acceleration, voltage, power, pH, or whatever other environmental variable correlates well with adoption of the technology you care about.

The tool of the trade for collecting raw sensor data streams is called a "data logger." A sensor is a device that turns some property of the physical world, such as temperature, into a machine- or human-readable format such as voltage or the position of a dial. By contrast, a data logger is a device that records the values of a sensor to memory for later transmission to a computer. Data loggers typically collect and store sensor data into large flat archives such as comma-separated value files (Table 15.1).

In the cookstove example, temperature data loggers could be used to collect large quantities of temperature time series data. Essentially, the output of a temperature data logger would look like this.

Just raw data, data point after data point and file after file, typically for hundreds of data loggers and millions of observations. For example, imagine a 100-household cookstove adoption monitoring and evaluation program. This program monitors two cookstoves per household for 3 months using data loggers that record temperature once per minute. This monitoring campaign would collect roughly 26 million individual data points.

Managing the collection and analysis of this data looks like this can be a major headache. First, as you can probably surmise from the small collection of top Google-ranked data loggers in Fig. 15.1, the data logger industry is industrial and clunky. These devices usually do not come with slick mobile apps and intuitive fleet management tools. Instead, data from data loggers is typically downloaded by hand using proprietary cables and dongles using a piece of industrial software which often only runs on a legacy version of Microsoft Windows as shown in Fig. 15.2. Training field staff to reliably operate this kind of software has been a significant challenge in my research.

Successful projects using traditional data loggers must ensure quality across many critical steps in the data chain. Figure 15.3 shows 12 important steps in the sensor data chain spanning the time period before data collection begins, while data collection is taking place and after collection of data has been completed. At any step along this path, significant data quality issues can be introduced which can dramatically impact the success of a project. Here are just a few real-world examples

Fig. 15.1 The first images for the Google Image Search term "data logger." Accessed on May 30, 2020

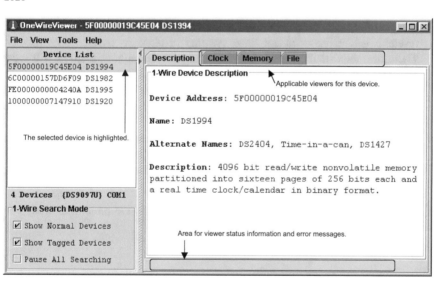

Fig. 15.2 Example of a typical data logger software interface from Maxim OneWireViewer User's Guide, version 1.3

of small mistakes that create significant problems in sensor-based data collection campaigns:

1. Incorrect dates on staff laptops leading to confusion about when data were actually collected. The data logger gets its clock time from the computer that

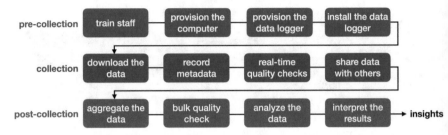

Fig. 15.3 The data chain for monitoring and evaluation of technology adoption with data-logging sensors

provisioned it, and the staff's computers had the wrong time (or even the wrong date) for some unknown amount of time.

2. Staff incorrectly record metadata when downloading data from data loggers. For example, data is downloaded, and the file from this data is mistakenly named household-124.csv instead of household-123.csv; there is no easy way to even know there was a mistake, let alone correctly know where this data was downloaded from.

3. Staff do not have adequate training to interpret data quality in real-time, and research leadership does not have systems in place to oversee data quality. Therefore, data quality problems persist unseen throughout a study. We have observed several examples where research staff diligently visit households every few weeks for months on end to repeatedly download garbage data from a data logger with a broken sensor. This problem is not noticed until months after the data collection part of the study is complete.

4. Graduate students with little to no data science experience are tasked with analyzing huge datasets from these kinds of experiments. It takes months or years for the students to complete the analysis, and often, the analysis suffers from statistical or analytical errors.

If there is any broken link in the data chain, an expensive multi-year study can be significantly damaged or even ruined entirely.

2   Monitoring Cookstove Adoption with Sensors

2.1   Darfur

2.1.1   Implementation Context

In 2012, the war in Darfur, Sudan, had been ongoing for 9 years. As a result of the conflict, millions of people from rural Darfur had concentrated into camps of internally displaced persons (IDPs). People in these camps traditionally cooked on

Fig. 15.4   Darfuri women carrying fuelwood back to IDP camp. Photo credited to Ashok Gadgil

firewood that was freely collected in the rural areas. However, these new large-population centers had put enormous pressure on the woody biomass resource in and around the camps. Over the years, the radius of complete biomass denudation increased to the point where, in 2005, it was estimated that women had to walk a 7-hour round trip from the camp to collect enough firewood to cook for just 2 or 3 days (Galitsky et al., 2006) (Fig. 15.4).

These trips were difficult and dangerous with many reports of sexual violence occurring while women were outside the relative safety of the camps. In many camps, particularly in the arid North Darfur, fuelwood had become out of reach—and collection trips on foot had become impossible. The situation had become desperate enough that many women reported that they had begun to trade their relief food rations to shady businessmen who would truck in firewood from rural Darfur and Chad. Women would use the firewood they bartered for to cook what food they had remaining for their families.

Since 2005, Professor Ashok Gadgil and team at UC Berkeley and Lawrence Berkeley National Laboratory had been developing and distributing an improved cookstove called the Berkeley-Darfur Stove (Fig. 15.5). This cookstove had been demonstrated to reduce fuel use by roughly 50% (Jetter et al., 2012; Rapp et al., 2016). The hope was this reduction in fuel use would lead to a commensurate reduction in time, cost, and risk spent collecting or purchasing fuel wood. By 2012, the work of distributing this cookstove had been transferred to a Berkeley-based nonprofit organization called Potential Energy. Potential Energy was interested in quantifying the impact of the 50,000 or so cookstoves that had been distributed to

Fig. 15.5 Berkeley-Darfur
Stove

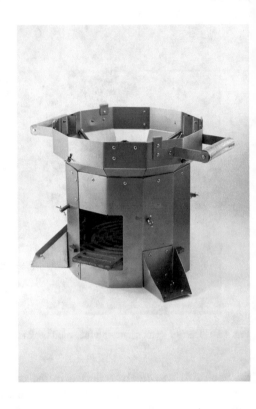

date. At that time, Potential Energy had just changed its name away from the Darfur Stoves Project. Potential Energy's executive and assistant directors, Andree Sosler and Debra Stein, knew me through my relationship with my PhD advisor, Potential Energy board member, and Berkeley-Darfur Stove inventor Ashok Gadgil.

Our goal was to assess the adoption of about 150 Berkeley-Darfur Stoves just after they had been distributed for free in the IDP camp near the town of Al Fashir in North Darfur. The plan was to compare adoption of the cookstove measured by sensors to self-reported adoption data gathered through surveys. With these two data sources, our hope was that we could build some sort of regression in order to make sense of some previously collected survey data about adoption. We hoped to run this study in two contexts: internally displaced peoples' camps near Al Fashir as well as unorganized rural settlements further away from Al Fashir.

Part of Potential Energy's motivation to perform this adoption study was to build a case for carbon financing (Wilson et al., 2016b). At the time this study was conceived, Potential Energy was already working with an organization, Impact Carbon, that validated carbon credits. Still, additional validation about cookstove adoption could have supported Potential Energy's case that the fuel savings of the Berkeley-Darfur Stove helped to offset anthropogenic CO_2 emissions. To measure the adoption of the cookstoves, Potential Energy was interested in using a sensor-based stove usage monitoring system (SUMS). Seminal work on SUMS had been

Fig. 15.6 El Haj Adam of SAG in a sea of Berkeley-Darfur Stoves outside Al Fashir assembly workshop

performed by Ilse Ruiz-Mercado during her doctoral research at UC Berkeley just a few years earlier, and Potential Energy decided that they would like to pursue this approach (Ruiz-Mercado et al., 2008, 2011, 2012, 2013).

Our plan was to work with a local Sudanese nongovernmental organization (NGO) partner, Sustainable Action Group (SAG), to run the study. SAG would offer up one of their staff as the study leader and coordinate all of the day-to-day activities and management of the field staff related to the study. SAG was already intimately involved with the Berkeley-Darfur Stove Project because they were the NGO responsible for the local assembly and distribution of cookstoves after the cookstoves arrived as flat kits to Darfur (Fig. 15.6).

Following in the footsteps of Ilse Ruiz-Mercado, our team had decided to use Maxim iButton temperature loggers as our stove usage monitoring system (SUMS). I'll refer to this term as "SUMS" throughout this chapter whenever discussing a sensing device that is specifically employed to track cookstove adoption. The Maxim iButton is a self-powered data logger with all its electronics, temperature sensor, and battery contained in a metal button of the size and shape of a watch battery. Some models could withstand temperatures as high as 140 °C. The idea was that we could install the iButton SUMS on the outside of the cookstoves, and as the temperature of the cookstove rose, these temperatures would be recorded by the iButtons, and we could later correlate spikes in temperature with cooking events.

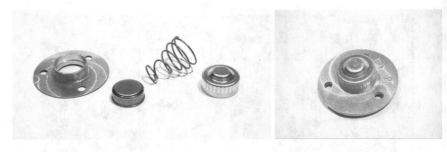

Fig. 15.7 Custom iButton case. The iButton data logger itself is the object that looks like a coin cell battery, second from the left

However, these iButton data loggers were never designed to be used on cookstoves. They had some major weaknesses including a short battery life, and when the battery died, all of the data was lost. Maxim iButtons are most often used in the food industry, mounted to the side of huge milk containers inside refrigerated trucks or inside buffet service stations to ensure food remains cold or hot enough to be safe for consumption. To adapt the iButtons for use on a cookstove, we designed an aluminum case that would hold the iButton and keep it firmly pressed against the surface of the cookstove. This aluminum case also allowed us to clearly stamp a large alphanumeric code onto the case of the SUMS, for example, "A-12," which made it possible to track which data came from which data logger (e.g., "2013-08-12 A-12" was data from A-12 downloaded on August 12, 2013) (Figs. 15.7 and 15.8).

Throughout late 2012, our team attempted to obtain a visa to travel to Darfur to kick off this project with SAG, but due to political turmoil in Sudan and restrictive policies about visitors from the United States, our team was not able to secure visas by early 2013. Instead, our team decided that in January 2013, Angeli Kirk and I would fly to Addis Ababa, Ethiopia, where we would meet up with a representative of SAG, Abdel Rahman, to do a multiday training session. Angeli, a research colleague, PhD student, and friend from Berkeley's Agriculture and Resource Economics program, would meet up for the latter half of the training session (and to assess Potential Energy's expansion opportunities in Ethiopia).

During this training session, we would familiarize Abdel with the design of the experiment as well as how to use the sensors we would need to implement the study. I had brought all of the equipment with me to transfer to Abdel who would then take them into Sudan (Fig. 15.9). We planned to wait until the summer to start the study, and we were still hopeful that we would be able to secure a future visa to perform an in-person training with field staff in Darfur sometime in the late spring of 2013. Therefore, this training served as a kind of "kickoff" where we would get familiar with the sensors and identify the personnel, facility, and equipment resources that SAG would need to complete the study.

Fig. 15.8 Areidy Beltran, undergraduate research assistant extraordinaire, assembles iButton SUMS in the mechanical engineering machine shop during finals week, December 2012

2.2   Innovate, Implement, Evaluate, and Adapt

2.2.1   The Lead-Up

As the early months of 2013 wore on, it became clear that our Berkeley-based team was never going to be able to visit Darfur. In late February, we got an email from Jan Maes, a consultant who worked for Potential Energy. He let us know that a member of the Impact Carbon team, Ellie Gomez, had been abducted from her hotel in Darfur by armed assailants. This was the same hotel I was planning to stay at. Miraculously, Ellie escaped her kidnappers just minutes after the abduction. This close encounter happened in the context of frequent stories about aid workers and NGO employees being kidnapped, held for ransom, and sometimes murdered in Darfur.

The abduction incident threw cold water on our whole study. We had originally planned to do some studies in the rural settlements surrounding the IDP camps, but considering the security situation, these plans were scrubbed. Also, we decided we

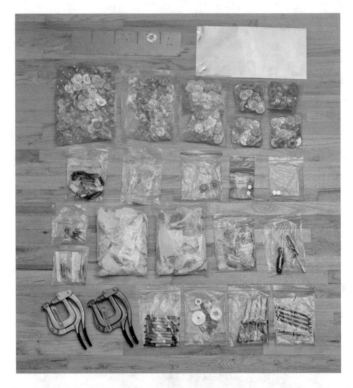

Fig. 15.9 Supplies brought from Berkeley to Ethiopia in January 2013 to be transferred to Darfur

could not even visit individual domiciles in the IDP camps; instead, we would have to ask women to congregate at an IDP Women's Center.

These meetings at the Women's Center were partly necessary to administer qualitative surveys, but they were also a result of the way iButton sensor data needed to be downloaded: to access the data, our field staff needed to carry a laptop computer, a set of cables and dongles, and screwdrivers and wrenches to detach the data logger from the cookstove. This whole process took about 10 minutes per cookstove and certainly created quite a spectacle in the IDP camps. If we could have discretely and wirelessly transmitted the data, it is likely that our study could have been designed differently.

In addition to creating a custom case for securely holding and attaching the SUMS to the cookstoves, it was extremely important to identify a good placement location for the SUMS. Because these iButtons were a fully integrated miniature data logger in a small case, that meant that the sensitive microcontroller, memory, and battery would get just as hot as the temperature sensor. Want to measure a 100 °C temperature? Well, the whole data logger has to get that hot.

This posed a significant optimization challenge when measuring cookstove use in the hot Sudanese desert. The bounds of this design challenge were as follows:

1. In order to easily identify temperature spikes in temperature time series data, the temperature when cooking should be as high as possible.
2. In order to maximize battery life and minimize the risk of damaging the data logger and losing all the data, the temperature should be as low as possible.
3. The hot summer sun in Darfur could easily heat a cookstove's surface to 50 °C or more even when a user was not cooking.

This "get the data logger as hot as possible but not so hot that it destroys itself or drains the battery before the study is complete but also hot enough to unambiguously identify cooking apart from solar heating" design challenge required some significant engineering effort. This challenge was unique to the design of the iButton SUMS because of the quirk that the sensor was co-located with the battery, memory, and microcontroller. We used parameterized testing, thermal cameras, and some optimization calculations around battery lifetime vs. temperature to select a best-possible mounting location (Fig. 15.10).

Not only was this position-optimization process cumbersome and overwrought, it had very little potential for scalability or impact. How could any team of other than well-resourced engineering PhD students replicate this technique for future studies?

Double-Click vs. Single Tap

It's a muggy January 2013 day in Addis Ababa, Ethiopia. I am sitting on a loveseat in my hotel room with Abdel Rahman at my side. We're both looking at the screen of a laptop that Abdel has brought with him from his NGO office in Sudan. Together, we're working through how to use a particularly byzantine piece of data logger software, but we keep getting hung up on one particular step. Abdel is having a

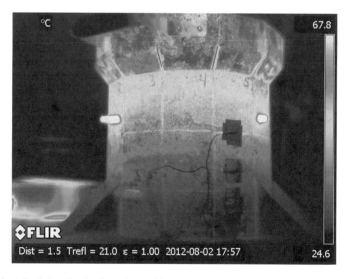

Fig. 15.10 A Berkeley-Darfur Stove imaged by an infrared camera to determine optimal iButton SUMS placement, August 2012

very difficult time double-clicking certain icons in the software's user interface. As in, he just cannot properly execute the double-clicking action. It's not clear what is leading to this confusion, because Abdel is clearly computer literate. But he can't quite seem to get the timing or positioning of the cursor quite right. Perhaps this is a new laptop, or maybe he is used to using a wired mouse and not fussing with the flimsy trackpad and buttons of this low-quality laptop.

We're spending long stretches of time, sometimes minutes on end, simply trying to open files, click menus, and generally get the cursor to behave the way Abdel wants. He's getting frustrated, and I can feel it. I'm starting to wonder to myself, "if Abdel, as the senior lead of this project, cannot succeed at this task from the quiet comfort of a hotel room, how are his dozens of less-educated field staff going to be successful in the middle of a Darfuri IDP camp in the blazing hot sun surrounded by countless challenges and distractions?" Abdel is fidgeting next to me in the loveseat and getting increasingly agitated. He asks for a break, pulls out his phone, and starts effortlessly navigating the apps and screens, checking email, messaging friends, and reading articles. I think to myself, "well, that's interesting."

This was a formative experience in my journey to build a better system for monitoring cookstove adoption. Whatever we did long term, we needed it to work on mobile phones and mobile phones alone.

My experience struggling to use the laptop in the hotel room in Ethiopia with Abdel Rahman caused me to become increasingly concerned that a laptop-based sensor data acquisition system would not be appropriate for the field staff in Darfur. In the spring of 2013, Javier Rosa, an electrical engineering and computer science PhD student at Berkeley, and I spent considerable effort trying to hack the iButton dongles to work with Android phones.

However, after much consternation, we were never able to read an iButton with a phone. But this desire for a system that would allow for a phone-only ecosystem to provision, deploy, and collect data from SUMS stayed with us (Kipf et al., 2015). For the time being, we resigned ourselves to the idea that one or two exceptionally well-trained field staff would need to run the laptop computer to download data from the iButtons, while the survey enumeration staff would administer the survey in another area of the Women's Center.

Given the ubiquity of literacy with mobile phones, we decided to design the survey-based data collection for our study around an open-source survey tool called Open Data Kit (ODK). We also designed a sidecar paper survey with identical questions just in case the phone survey failed. Also, in true mechanical engineering grad student overengineered fashion, we even designed a little jig to allow field staff to repeatably take high-quality photos of the survey since the SAG offices in Darfur lacked any kind of scanner (Fig. 15.11).

Part of the reason we were anxious about the ODK system not working correctly (and thus implemented the paper survey as a backup plan) was our discovery that global sanctions on the Sudanese government were blocking Internet traffic originating from Sudan to ODK's servers, which were hosted by Google at that time. We discovered this in late spring 2013, and the experiment was scheduled to

Fig. 15.11 The Open Data Kit (ODK) mobile phone and backup jig to take photos of paper surveys

start that summer. This incident led to a scramble to redeploy the open-source ODK backend system on Javier's own server. Pointing traffic originating in Sudan to a server on US soil was probably still a violation of some sort of international law, but our attempts to deploy a server in Sudan were not successful. This entire incident, which could have significantly delayed our experiment if it would have been discovered a few months later, helped me to understand the value of controlling the entirety of the data platform when performing field research in heavily sanctioned countries or countries with extremely restrictive Internet access. We also owe our success to the open-source feature of the ODK backend system, so it could be ported to Javier's server at UC Berkeley.

Meanwhile, around April of 2013, it was becoming clear to our team that Abdel Rahman was not going to be able to execute this study as we had planned. It had become increasingly difficult to maintain consistent lines of communication and continue momentum, progress, and meet schedules with Abdel and SAG. It seemed like the SUMS study was not a top-of-mind priority for SAG and was losing momentum. Additionally, it was unclear if Abdel was prepared to lead the complex administration and oversight of the study from Darfur. Abdel remained a critical member of our SUMS study team, especially as it related to cookstove assembly and distribution, but we needed to make an important decision about bringing in additional help.

On April 10, 2013, Jan Maes, Potential Energy's consultant, introduced me to Dr. Mohammed Idris Adam. Dr. Moh (as he liked to be called) was a professor at Al Fashir University in Darfur. He had been assisting Impact Carbon and Potential Energy with enumerators of surveys related to carbon credits, and he had become a trusted collaborator. Over the course of the coming months, Dr. Moh, Angeli, and I collaborated via email and planned to meet in Addis Ababa in July to coordinate final plans for the experiment.

Around this time, another important hire was made. Potential Energy hired a Sudanese woman then living in the Bay Area named Omnia Abbas. Omnia was an incredible resource at Potential Energy. She was fluent in Arabic and English, intimately familiar with Sudanese culture, and could travel with relative ease back and forth to Darfur.

I met with Dr. Moh, Omnia, and Potential Energy's associate director, Debra Stein, in Addis Ababa in July 2013. The content of this meeting was largely a repeat of the meeting just 6 months earlier. Unfortunately, the timing of this meeting was over Ramadan, and the fasting and prayer schedule made it difficult for Dr. Moh to participate past the midafternoon (this is a common rookie scheduling oversight made by researchers). But still, the effect of this second meeting was transformational for the project. In retrospect, the difficult decision to pivot project leadership just months before the study was slated to begin was one of the most important decisions we made for the overall success of this project. Before hiring Dr. Moh and Omnia, our team did not realize the incredible importance of having an experienced, motivated, and trusted champion for your research study stationed in the field. Today, I strongly believe that it is impossible for field research to be administered solely from outside the study site. To this day, one of the first questions I ask students who are planning field work is "Who is the champion who is located in the field?" (Fig. 15.12).

Fig. 15.12 Left: Dr. Mohammed "Moh" Idris Adam (top row, third from left) and Abdel Rahman (top row, fifth from left) with the SUMS field team they assembled in Darfur. Right: In the foreground, field staff administer surveys to Darfuri IDP women using paper and mobile phone-based techniques. In the background, Dr. Moh and an assistant interact with SUMS using a laptop computer

2.2.2   The Study

In August of 2013, the study in Darfur began in earnest. One of the first things we noticed was that far more of our sensors were failing due to overheating than expected. We had invested such careful attention in the placement of the sensors (e.g., the thermal imaging), but now, we noticed that about 20% of our data loggers were coming back with symptoms of overheating.

In our survey data, we found that participants who self-reported charcoal as a primary or secondary fuel source were much more likely to have burned-out sensors. After some sleuthing by SAG field staff, we discovered that some women like to burn charcoal in the Berkeley-Darfur Cookstoves even though the cookstove was only designed to burn wood. In order to adapt the cookstove to the charcoal fuel, women were flipping the cookstove upside down and packing the bottom of the stove (close to where the iButton was mounted) with charcoal. This behavior caused the iButtons to overheat and become unresponsive. It was very valuable to discover how common this charcoal-cooking behavior was, but it was at the heavy cost of losing almost 20% of our data. Additionally, this data was lost in a way that biased study results (participants who don't use their cookstove can't burn out their sensors and therefore were overrepresented in our surviving sensor data).

Throughout the fall, our team performed the hard work of administering a sensor-based data collection program from overseas. Dr. Moh and our team had regular weekly meetings late at night California time. At first, these meetings covered mundane issues such as how to buy new data plans for the SIM cards in the ODK cell phones, if we would be reimbursing the field staff for gasoline, and how to download and install Dropbox. However, as time went on, an increasing number of our conversations covered mounting issues related to data quality. "Where's the data from household 10?" "I looked through the data, and there are 3 baseline surveys that are all marked as being from the same respondent." "The sensor data from household 153 says that it was collected in the year 1970—what's going on there?" Our team was so busy administering the day-to-day activities of the study that we did not have the time or resources to preprocess any of the data coming back from Darfur.

While I was occupied with administration work, Angeli Kirk was looking through some of the early results from sensor data that had been downloaded and sent back to California via Dropbox. She noticed that a small subset of households had barely used their stoves yet. She wondered what impact, if any, returning to the Women's Center for the follow-up survey (and data download) would have on future adoption for this group. However, the plan had always been that the data collection would stop after the follow-up survey visit. Personally, I was struggling to hold the administration of the project together as it was, and I was not interested in mission creep to answer a new and unplanned research question. However, after some prodding from Angeli and an amendment to our institutional review board (IRB) protocol, I acquiesced. We asked Dr. Mohammed to send future tranches of follow-up survey takers home with SUMS-equipped stoves instead of taking the sensors off. A second and final follow-up was planned on the fly just to remove the

SUMS. This decision, which was enabled by an early peek at results, turned out to be vital for our research.

As the plan changes and late-night phone calls carried on, administering the Darfur SUMS program became a difficult contextual picture to maintain. The number of caveats and special cases and "oh yeah, that sensor is a different special case for ___ reason" issues began to mount. These gotchas were cataloged in ad hoc in emails, paper notebooks, spreadsheets, and mental notes. By the time the last of the sensor data was being collected in November 2013, we knew our team only had just a month or two of short-term memory acuity before it would become nearly impossible to stitch together all of the data into a coherent story.

2.2.3 After the Study

In the winter of 2013, we began assembling the data from the SUMS study. I personally had never analyzed data of this volume before. Up until that point, I had just "faked it until I made it" with data analysis. I was comfortable using Excel and had some really weak MATLAB skills. But the four million data point SUMS dataset was a huge step up from anything I, or anyone else on our team, had analyzed before. There were hundreds of files, each with about ten thousand rows of raw temperature and timestamp data, and our job was to create a coherent story about how women in Darfur adopted the Berkeley-Darfur Cookstove. We didn't know where to start.

A few months earlier, I had carpooled to Burning Man with another graduate student named Jeremy Coyle. Jeremy was a PhD student in biostatistics at Berkeley, and he lived across town from me in the student coops. Our team needed some help, so I rode over to his place on my bike to see if he could help me learn to use R, the statistical programming language, to analyze the Darfur SUMS dataset.

I had thought that Jeremy and I would just spend an afternoon getting me up speed with R, and I would be on my way. But eventually, to both of our surprises, Jeremy and I collaborated on the analysis of the Darfur dataset for the next 3 months. The amount of time that went into this analysis was far beyond my expectations. As an aside, Jeremy and I still collaborate on research, software development, and data analysis to this day.

The majority of this effort was spent creating a cooking-event-detection algorithm that could reliably find the start and end of individual cooking events from long stretches of temperature time series (Fig. 15.13). Also, there was significant data cleaning, organizing, and merging of survey and sensor data. Because the sensor and survey data acquisition systems were completely separate, all of the merging and comparison efforts needed to be hand-rolled in code. Our team estimated that we spent 400 person-hours on this analytics effort. In 2020, the going rate for quality data science consulting is about $300/hour. Although a professional senior data scientist would probably be able to finish this analysis faster than we did, an analysis like this could easily cost on the order of $100,000. However, more likely, a smaller NGO or development agency that wanted to deploy sensors to

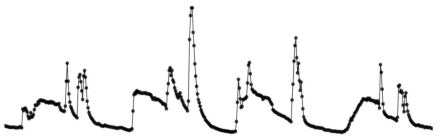

Fig. 15.13 An example of a temperature time series to illustrate the challenge of creating a deterministic algorithm to count and quantify "cooking." Where do the cooking events start and stop? Axis scales intentionally left blank, but vertical (Y) is temperature, and horizontal (X) is time

monitor cookstove adoption would find themselves stuck at the data analysis step and might resort to a much simpler analysis that didn't realize the full value of the data. This experience helped me to realize the massive investment in time and/or money to analyze sensor data for development research. This investment was likely a major barrier to ubiquitous deployment of sensors for research.

2.2.4 What We Learned

The Darfur SUMS study ended up teaching us many valuable lessons. We found that about 75% of women adopted the Berkeley-Darfur Stove (Wilson et al., 2016a). For the women who did not adopt it, we found out that about 80% of them could be converted into adopters simply through the act of conducting the first follow-up. Angeli was right. Without her early analysis and encouragement to change the research study, we would have never known this important insight that non-adopters just needed a small nudge to become adopters. In addition to this, we confirmed what we believed about the quality of survey data to assess adoption of technology; no matter how we asked the question, we were not able to assess adoption of the Berkeley-Darfur Stove through surveys. As shown in the figure below, when asked a question about how many times per day she uses her Berkeley-Darfur Stove, cooks almost always respond with the socially desirable response: three (every meal). The number of times someone self-reports cooking in surveys has essentially no correlation with their behavior measured by sensors (Fig. 15.14).

In addition to what we learned about how women in Darfur use cookstoves, we learned even more about what it takes to run an effect sensor-based impact evaluation. Some of the most important takeaways from Darfur were as follows:

1. Analytics is a major barrier to effective deployments of sensors for monitoring and evaluation.
2. Field staff are far more comfortable with mobile phones than with laptop computers, and mobile phones do not attract the attention that computers do.

Fig. 15.14 Cooking events measured by self-report vs. SUMS

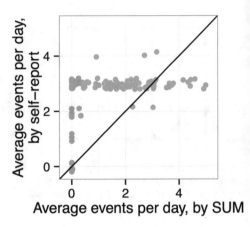

Average events per day, by SUM

3. Training field staff to deploy industrial data loggers using industrial tools is extremely difficult and error-prone.
4. Cookstove sensors that cannot survive cookstove temperatures are bound to break.
5. The higher a temperature a cookstove sensor can measure, the clearer its cooking signal will become, and the more easily analytics can be run on the data.
6. It's common for data loggers' batteries to die during a study. So, a dead battery should not erase all of the data from the logger (as happens with iButtons).
7. The ability to continuously and easily audit data in real-time as it is collected in the field is critical to maintaining data integrity and communicating feedback about data quality issues to your field team and for potentially asking interesting new questions during your study.
8. Every time a person needs to take an action in the sensor data chain, anything from naming a file to adding an attachment to an email, you are opening your project up to significant risks in terms of coordination, privacy, and data quality.
9. If it takes the undivided attention of a PhD student and 2 years of significant technical and administrative effort to execute a relatively small sensor-based impact evaluation, then sensor-based impact evaluations are too burdensome for all but the most well-funded academic research studies.

2.2.5 Where We Went Next

SUMSarizer One of the main takeaways from the seminal work in Darfur was that analyzing sensor data for impact evaluation was extremely difficult. Jeremy Coye, Ajay Pillarisetti from Public Health, and I were very interested in democratizing event detection from time series data. We wanted to make it easier for coding-naive users to summarize SUMS data, so we began work on a machine learning tool called SUMSarizer. SUMSarizer was a web-based "label and learn" tool that allowed users to import raw SUMS files from common SUMS data loggers. Once imported, users

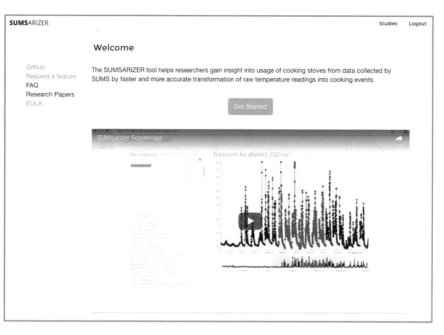

Fig. 15.15   The home screen of the SUMSarizer web application

could highlight which sections of the time series data they believed represented cooking. Over time, SUMSarizer would learn to identify cooking events, even in complex data. SUMSarizer would then summarize the cooking data and output results that could be more easily interpreted in a simpler tool like Excel. SUMSarizer allowed someone who did not know how to write code to repeat the Darfur analysis that took Jeremy and me 3 months in about 3 hours.

Unfortunately, we still had some hard lessons to learn. The development of the web application was funded by a single ~$30 K grant from Center for Effective Global Action (CEGA) at Berkeley. This money went toward research stipends for the creators, but we did not stop to think about the ongoing costs of maintaining a popular web application. As the user base of SUMSarizer grew over the years, the cost of maintaining SUMSarizer grew significantly. Tens of millions of data points to warehouse, cloud service subscriptions to maintain, and significant user technical support to maintain were costing Jeremy and I, personally, thousands of dollars per year. Without the ability to find ongoing support for the platform, in 2019, we made the hard choice to shut down SUMSarizer.com and open source its machine learning code[1] (Fig. 15.15).

ASUM   Some of the core challenges of using iButtons as SUMS inspired the creation of the Advanced Stove Usage Monitor (ASUM). The ASUM was designed

---

[1] https://github.com/geocene/sumsarizer

by Advait Kumar, Abhinav Saksena, Meenakshi Monga, and me at the Indian Institute of Technology (IIT) Delhi during a 2014 Fulbright Fellowship to India. Unlike the iButton, the ASUM was compatible with multiple sensor input channels, had nonvolatile memory that could survive a dead battery, had room for billions rather than thousands of samples, and used microSD card storage instead of a proprietary interface that required custom dongles and Windows-only software. The ASUM was powered by an Arduino-compatible Atmel microcontroller, and for our research program in India, we integrated its multichannel analog frontend with an advanced cookstove's internal thermoelectric generator, USB charging port, battery, fan, and a proximity switch (Fig. 15.16).

The flexibility of the ASUM allowed us to perform novel research about how advanced cookstoves with inbuilt thermoelectric generators are adopted, namely, how the ability of a cookstove to power a USB port for charging a cellphone and other small appliances influenced adoption of the cookstove (Wilson et al., 2018). Also, the ASUM allowed our team, for the first time, to be able to administrate an entire cookstove impact evaluation using only mobile phones. Most Android phones at the time had a slot for microSD card expansion storage, and we used this port to read the cards and then transmitted the data to the cloud with the phones. The data from ASUMs were analyzed using SUMSarizer.

However, ASUM was not without its problems. It was a boutique purpose-built device for our study. The device could not maintain charge on a reasonable-size battery for more than a couple weeks, it was not very rugged and had no integrated tool for metadata collection (e.g., which household the data logger was deployed in), and the flexibility of the microSD card was as much a liability as a feature; on a few of the ASUM microSD cards we retrieved from the field, we found all of our data was missing and had been replaced by MP3 files of Bollywood music.

Geocene The most recent iteration of a stove usage monitoring system is Geocene. Today, in summer 2020, I am the CEO of Geocene. Geocene makes a fully integrated SUMS product. The three pillars of this system are as follows:

1. The Geocene Temperature Logger (also known as the "Dot"). This data logger runs on replaceable AAA batteries that can be found in almost any small town in the world, uses rugged thermocouple probes that can withstand temperatures up to 500 °C, has over a year of storage and battery life, and communicates over Bluetooth Low Energy (BLE).
2. The Geocene Mobile App runs on Android or iOS and is the only tool field staff need to provision and collect data from Dots. The mobile application also has an inbuilt survey feature. This allows survey data to be collected alongside sensor data, making easy the process of merging, filtering, and analyzing data with survey questions as covariates. The mobile application is designed to run even without access to the Internet, but when it does have access, it syncs its data to the cloud.
3. Geocene Studies is a cloud-based web application that organizes users' sensor and survey data and leverages the SUMSarizer engine to analyze data from the

Fig. 15.16 Top: Evolution of the ASUM from breadboard to final manufactured device. Bottom: ASUMs installed on BioLite cookstoves undergoing functional testing at IIT Delhi

Dots. Today, Studies is analyzing data from about 7000 cookstoves every day, has hundreds of active users, and contains about 1 billion individual temperature samples (Fig. 15.17).

Geocene has supported several very large field trials including the Household Air Pollution Investigation Network (HAPIN) with about 3200 households and LEADERS Nepal with about 2000 households. In addition to cookstove projects, Geocene's platform also supports electricity-monitoring and GPS asset-tracking sensors. Using these tools, Geocene has also supported international development

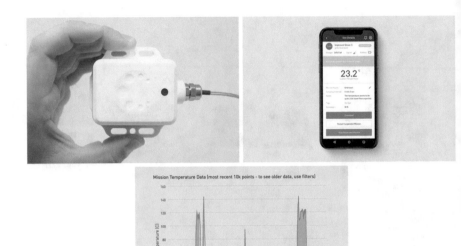

Fig. 15.17 Clockwise from top-left: a Geocene Dot-brand thermocouple data logger, the Geocene mobile application, and a screenshot of temperature time series data with cooking events detected and highlighted from Geocene's web application, Geocene Studies

projects monitoring electric grids and tracking the leakage of aid through supply chains and onto the black market.

However, despite these large projects, Geocene does not generate enough revenue from supporting cookstove programs to employ the talented professional engineering team it takes to support such a modern and highly integrated IoT product. Today, most of Geocene's revenue comes from consulting work for Silicon Valley companies building IoT products. Still, the founding and heart of Geocene are still sensor-based impact evaluation for the developing world.

3 Summary

As development engineers, we strive to develop technologies that will improve lives. However, measuring impact is a complex exercise that must consider product performance, adoption, and scale (IPAS). Historically, measuring adoption relied on self-reported surveys, but we have discussed and demonstrated in this chapter that sensor-based objective evaluation of product adoption is a more reliable and informative measure of adoption behaviors.

Still, deploying sensors came with significant challenges related to sensor provisioning, data collection, warehousing, and analytics. To solve these problems, we built three major iterations of cookstove sensor platforms from cookstoves and

ODK, to ASUM, to the Geocene Dot and Studies platform. In each subsequent iteration, we endeavored to move field work to intuitive mobile interfaces and analytics to cloud-based coding-naive point-and-click systems. In doing this, our goal was to make sensor-based monitoring and evaluation more accessible to broader groups of users.

This journey has been an incredibly enriching part of my life. I hope this chapter has helped the reader imagine some of the possibilities that a career in development engineering could afford. In your own work, my colleagues and I hope you take the lessons we learned into consideration. Aim for impact, measure reality, and keep on iterating toward a future with less poverty and more justice for all human beings.

Discussion Questions

1. When Internet-connected sensors are the dominant instruments used for data collection rather than surveys, how do we still engage local communities to participate in data collection? What happens to capacity building and job opportunities that used to be afforded to large teams of survey enumerators.

2. If you had $25 K to spend on an impact evaluation of 100 households in rural India but the sensors you needed to perform this evaluation cost $500 each, what would you do? Propose a rough budget that includes travel expenses, staff salaries, incidentals, and sensors.

3. Given what you have learned in this case study, if you were designing a sensor-based field study of technology adoption, what would be your top-three priorities in terms of designing the study?

4. Sensors can offer objective observations about the physical environment, but what are three ways you can imagine sensors fail to teach us what we need to know about technology adoption?

References

Burnett, R. T., Pope, C. A. I., Ezzati, M., Olives, C., Lim, S. S., Mehta, S., et al. (2014). An integrated risk function for estimating the global burden of disease attributable to ambient fine particulate matter exposure. *Environmental Health Perspectives, 122*, 397–403.

Burwen, J. (2011). *From technology to impact: Understanding and measuring behavior change with improved biomass cookstoves* (pp. 1–60). Energy and Resources Group, University of California.

Clasen, T., Fabini, D., Boisson, S., Taneja, J., Song, J., Aichinger, E., et al. (2012). Making sanitation count: Developing and testing a device for assessing latrine use in low-income settings – Environmental Science & Technology (ACS publications). *Environmental Science & Technology, 46*, 3295–3303. https://doi.org/10.1021/es2036702

Das, J., Hammer, J., & Sánchez-Paramo, C. (2012). The impact of recall periods on reported morbidity and health seeking behavior. *Journal of Development Economics, 98*(1), 76–88. https://doi.org/10.1016/j.jdeveco.2011.07.001

Edwards, A. L. (1957). *The social desirability variable in personality assessment and research.* Dryden Press.

Forouzanfar, M. H., Alexander, L., Anderson, H. R., Bachman, V. F., Biryukov, S., Brauer, M., et al. (2015). Global, regional, and national comparative risk assessment of 79 behavioural, environmental and occupational, and metabolic risks or clusters of risks in 188 countries, 1990-2013: A systematic analysis for the Global Burden of Disease Study 2013. *The Lancet, 386*, 2287–2323. https://doi.org/10.1016/S0140-6736(15)00128-2

Galitsky, C., Gadgil, A. J., Jacobs, M., & Lee, Y.-M. (2006). *Fuel efficient stoves for Darfur camps internally displaced persons report of field trip to north and South Darfur, Nov. 16 – Dec.17, 2005 – LBNL-59540* (pp. 1–43). Lawrence Berkeley National Laboratory.

Iribagiza, C., Sharpe, T., Wilson, D. L., & Thomas, E. A. (2020). User-centered design of an air quality feedback technology to promote adoption of clean cookstoves. *Journal of Exposure Science & Environmental Epidemiology*, 1–12. https://doi.org/10.1038/s41370-020-0250-2

Jetter, J. J., Zhao, Y., Smith, K. R., Khan, B., Yelverton, T., DeCarlo, P., & Hays, M. D. (2012). Pollutant emissions and energy efficiency under controlled conditions for household biomass Cookstoves and implications for metrics useful in setting international test standards. *Environmental Science & Technology, 46*, 10827–10834.

Kipf, A., Brunette, W., Kellerstrass, J., Podolsky, M., Rosa, J., Sundt, M., et al. (2015). A proposed integrated data collection, analysis and sharing platform for impact evaluation. *Development Engineering*, 1–9. https://doi.org/10.1016/j.deveng.2015.12.002

Landsberger, H. A. (1958). *Hawthorne revisited*. Ithaca.

Lewis, J. J., & Pattanayak, S. K. (2012). Who adopts improved fuels and cookstoves? A systematic review. *Environmental Health Perspectives, 120*(5), 637. https://doi.org/10.1289/ehp.1104194

Lim, S. S., Vos, T., Flaxman, A. D., Danaei, G., & Shibuya, K. (2013). A comparative risk assessment of burden of disease and injury attributable to 67 risk factors and risk factor clusters in 21 regions, 1990–2010: A systematic analysis for the Global Burden of Disease Study 2010. *The Lancet, 380*, 2224–2260.

McCarney, R., Warner, J., Iliffe, S., van Haselen, R., Griffin, M., & Fisher, P. (2007). The Hawthorne effect: A randomised, controlled trial. *BMC Medical Research Methodology, 7*(1), 30. https://doi.org/10.1186/1471-2288-7-30

Methodology, B. B. I. J. O. S. R., 2010. (n.d.). *The Hawthorne effect in community trials in developing countries*. Taylor & Francis. https://doi.org/10.1080/13645570903269096

Nunnally, J. C., & Bernstein, I. H. (1994). *Psychometric theory* (3rd ed.). McGraw Hill.

Pillarisetti, A., Vaswani, M., Jack, D., Balakrishnan, K., Bates, M. N., Arora, N. K., & Smith, K. R. (2014). Patterns of stove usage after introduction of an advanced cookstove: The long-term application of household sensors. *Environmental Science & Technology, 48*(24), 14525–14533. https://doi.org/10.1021/es504624c

Rapp, V. H., Caubel, J. J., Wilson, D. L., & Gadgil, A. J. (2016). Reducing ultrafine particle emissions using air injection in wood-burning cookstoves. *Environmental Science & Technology, 50*(15), 8368–8374.

Ruiz-Mercado, I., Lam, N. L., & Canuz, E. (2008). Low-cost temperature loggers as stove use monitors (SUMs). *Boiling Point*.

Ruiz-Mercado, I., Masera, O., Zamora, H., & Smith, K. R. (2011). Adoption and sustained use of improved cookstoves. *Energy Policy, 39*(12), 7557–7566. https://doi.org/10.1016/j.enpol.2011.03.028

Ruiz-Mercado, I., Canuz, E., & Smith, K. R. (2012). Temperature dataloggers as stove use monitors (SUMs): Field methods and signal analysis. *Biomass and Bioenergy, 47*, 459–468.

Ruiz-Mercado, I., Canuz, E., Walker, J. L., & Smith, K. R. (2013). Quantitative metrics of stove adoption using Stove Use Monitors (SUMs). *Biomass and Bioenergy, 57*, 136–148.

The World Bank. (2011). *Household cookstoves, environment, health, and climate change* (pp. 1–94). The World Bank.

Thomas, E. A., & Brown, J. (2020). Using feedback to improve accountability in global environmental health and engineering. *Environmental Science & Technology*, acs.est.0c04115–10. doi:https://doi.org/10.1021/acs.est.0c04115.

Thomas, E. A., Barstow, C. K., & Rosa, G. A. (2013). Use of remotely reporting electronic sensors for assessing use of water filters and cookstoves in Rwanda. *Environmental Science & Technology*.

Thomas, E. A., Tellez-Sanchez, S., Wick, C., Kirby, M., Zambrano, L., Abadie Rosa, G., et al. (2016). Behavioral reactivity associated with electronic monitoring of environmental health interventions—A cluster randomized trial with water filters and cookstoves. *Environmental Science & Technology, 50*(7), 3773–3780. https://doi.org/10.1021/acs.est.6b00161

Thomas, E. A., Kathuni, S., Wilson, D. L., Muragijimana, C., Sharpe, T., Kaberia, D., et al. (2020). The drought resilience impact platform (DRIP): Improving water security through actionable water management insights. *Frontiers in Climate, 2*, 258–211. https://doi.org/10.3389/fclim.2020.00006

Turman-Bryant, N., Sharpe, T., Nagel, C., Stover, L., & Thomas, E. A. (2020). Toilet alarms: A novel application of latrine sensors and machine learning for optimizing sanitation services in informal settlements. *Development Engineering, 5*, 100052. https://doi.org/10.1016/j.deveng.2020.100052

Wilson, D. L., Adam, M. I., Abbas, O., Coyle, J., Kirk, A., Rosa, J., & Gadgil, A. J. (2015). Comparing cookstove usage measured with sensors versus cell phone-based surveys in Darfur, Sudan. In S. Hostettler, E. Hazbourn, & J.-C. Bolay (Eds.), *Technologies for development* (pp. 211–221). Springer International Publishing. https://doi.org/10.1007/978-3-319-16247-8_20

Wilson, D. L., Coyle, J., Kirk, A., Rosa, J., Abbas, O., Adam, M. I., & Gadgil, A. J. (2016a). Measuring and increasing adoption rates of cookstoves in a humanitarian crisis. *Environmental Science & Technology, 50*(15), 8393–8399. https://doi.org/10.1021/acs.est.6b02899

Wilson, D. L., Talancon, D. R., Winslow, R. L., Linares, X., & Gadgil, A. J. (2016b, January). *Avoided emissions of a fuel-efficient biomass cookstove dwarf embodied emissions*. Retrieved August 9, 2017, from.

Wilson, D. L., Coyle, J. R., & Thomas, E. A. (2017). Ensemble machine learning and forecasting can achieve 99% uptime for rural handpumps. *PLoS One, 12*(11), e0188808. https://doi.org/10.1371/journal.pone.0188808

Wilson, D. L., Monga, M., Saksena, A., Kumar, A., & Gadgil, A. J. (2018). Effects of USB port access on advanced cookstove adoption. *Development Engineering, 3*, 209–217. https://doi.org/10.1016/j.deveng.2018.08.001

Wilson, D. L., Williams, K. N., & Pillarisetti, A. (2020). An integrated sensor data logging, survey, and analytics platform for field research and its application in HAPIN, a multi-center household energy intervention trial. *Sustainability, 12*(5), 1805–1815. https://doi.org/10.3390/su12051805

Zwane, A. P., Zinman, J., & Van Dusen Proceedings of the, E., 2011. (n.d.). Being surveyed can change later behavior and related parameter estimates. *Proceedings of the National Academy of Sciences*. https://doi.org/10.1073/pnas.1000776108/-/DCSupplemental

Chapter 16
Reimagining Excreta as a Resource: Recovering Nitrogen from Urine in Nairobi, Kenya

William A. Tarpeh, Brandon D. Clark, Kara L. Nelson, and Kevin D. Orner

1 Development Challenge (Problem Identification)

1.1 Sanitation as Development Challenge

Access to sustainable sanitation in the developing world is lacking, fundamental, and overlooked. Globally, 1 in 3 people still do not have access to improved sanitation, contributing to over 3.6 million water-related, preventable deaths each year (WHO/UNICEF Joint Water Supply et al., 2015). In Kenya, 75% of the population lives in rural areas with little access to sanitation infrastructure (World Health Organization, 2013). Nairobi, the capital of Kenya, only treats half of its wastewater and often uses untreated wastewater to irrigate crops (Scott et al., 2004). Irrigating food crops with untreated wastewater leads to a high risk of exposure to pathogens for consumers (United Nations, 2015, p. 6). Only 10–15% of Nairobi's informal settlements are sewered, and sewer pipes are often broken or clogged.

W. A. Tarpeh (✉)
Stanford University, Stanford, CA, USA
e-mail: wtarpeh@stanford.edu

B. D. Clark
Department of Chemical Engineering, Stanford University, Stanford, CA, USA
e-mail: bdclark4@stanford.edu

K. L. Nelson
Department of Civil & Environmental Engineering, University of California Berkeley, Davis Hall, Berkeley, CA, USA
e-mail: karanelson@berkeley.edu

K. D. Orner
Department of Civil & Environmental Engineering, West Virginia University, Morgantown, WV, USA
e-mail: kevinorner@mail.wvu.edu

© The Author(s) 2023
T. Madon et al. (eds.), *Introduction to Development Engineering*,
https://doi.org/10.1007/978-3-030-86065-3_16

In total, four million tons of human waste from Kenya's informal settlements are dumped untreated into waterways each year – polluting the environment, spreading disease, and threatening community health (Auerbach, 2016).

Sanitation is fundamental – it improves public health and environmental quality by separating excreta from water, food, and soil, reducing the prevalence of preventable diseases like cholera, dysentery, and typhoid; it also improves women's lives by increasing access to nearby toilets, saving time and decreasing the risk of sexual violence (World Health Organization, 2011). School sanitation facilities increase girls' attendance and literacy rates, thereby improving women's job opportunities (World Health Organization, 2019). When people are not sick at home, productivity at school and work increase, generating more income. The global cost of inadequate sanitation is $260 billion, including healthcare costs, premature deaths, and productivity losses (Ki-Moon, 2013). The loss of productivity due to sanitation-related illness costs Kenya's GDP a million dollars a day (O'Keefe et al., 2015; UN-Habitat 2006); instead, the average dollar invested in sanitation shows a fivefold return (Ki-Moon, 2013). In addition to improving public health, sanitation also addresses environmental quality. Untreated excreta cause nutrients in the form of nitrogen and phosphorus to be released in high concentrations, irreversibly altering aquatic environments and contaminating groundwater (Galloway et al., 2008; Kringel et al., 2016).

Sanitation is overlooked, with access expanding more slowly than for drinking water because of less investment, although both are needed to prevent water-related diseases (Ki-Moon, 2013). To inspire and facilitate a global response for sanitation progress, Millennium Development Goal 7(c) aimed to halve the number of people without sustainable access to sanitation by the end of 2015. Since this goal was not met (Fig. 16.1), the Sustainable Development Goals, approved by the UN in 2015, revised the MDG commitment to "achieve access to adequate and equitable sanitation and hygiene for all and end open defecation" by 2030 (United Nations, 2015). In Kenya, sanitation infrastructure has not kept up with population growth, leading to the net failure to meet Millennium Development Goal 7(c): 70% of the country in 2015 did not have access to improved sanitation (WHO/UNICEF Joint Water Supply et al., 2015). At current rates, reaching complete sanitation coverage in Kenya will take 150 years. As a result, low-income communities like Mukuru, an urban Nairobi informal settlement with over 185,000 residents, are left without adequate sanitation solutions. These implications extend beyond Nairobi as the majority of people in the developing world live with untreated feces and urine in their communities.

The developing world's sanitation needs will intensify in the next few decades, necessitating robust sanitation methods that can be rapidly and efficiently implemented. Providing access to sanitation will be particularly challenging in urban informal settlements, which will double in population over the next decade; this explosive population growth outpaces provision of essential services like water, sanitation, and electricity (United Nations Human Settlements Programme and United Nations Human Settlements Programme, 2003). At current installation rates, traditional, centralized wastewater treatment systems would still reach less than half

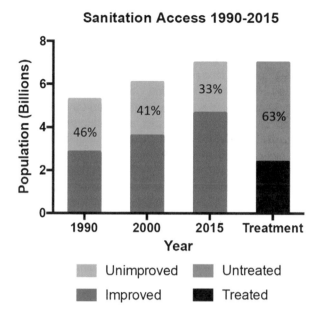

Fig. 16.1 Graphical representation of Millennium Development Goal 7(c): halving the population without access to improved sanitation. Here we compare the global population with access to sanitation as defined by excreta collection (improved vs. unimproved fixtures, gray/green bars) and as defined by excreta treatment in 2010 (untreated vs. treated, red/black bars). For excreta collection, the Millennium Development Goals were developed in 2000 and aimed to halve the 1990 population without sanitation access by 2015. Original data from (Baum et al., 2013; United Nations, 2015, p. 6)

of the populations of Asia and Africa by 2050 due to high infrastructure, staffing, and energy requirements (Larsen et al., 2016). On-site, decentralized options must be considered as an alternative to rapidly increase access to improved sanitation.

1.2 Partnership Between UC Berkeley and Sanergy

In this chapter, we describe a multi-year, ongoing partnership between researchers at the University of California, Berkeley, and Sanergy, a sanitation service provider in Kenya. The UC Berkeley team was led by Professor Kara Nelson in Civil and Environmental Engineering and continues to involve William Tarpeh, a former PhD student in Prof. Nelson's group and now an assistant professor of Chemical Engineering at Stanford. Sanergy is a social enterprise that provides comprehensive sanitation services, including manufacturing, installation, operation (via franchisees), maintenance, and resource recovery from source-separating toilets in informal settlements around Nairobi, Kenya. The Nelson group experience with pathogen inactivation and treatment processes was complementary, and both partners prioritized recovery of valuable products from excreta through their continual interactions over the past several years. After 2013, we refer to the team as ElectroSan, the name of the process engineering spinoff from the urine valorization technologies we developed during this project. Related, ongoing activities in the Tarpeh group at Stanford are also considered part of ElectroSan's efforts.

In 2012, Nelson led a Gates Foundation Grand Challenge team focused on in-toilet disinfection. Dubbed Safe Sludge, the innovation focused on using ammonia in urine to disinfect feces using an attractive, ergonomic toilet design. The project team attended the August 2012 Gates Reinvent the Toilet Exposition in Seattle, Washington, and met Sanergy, another Gates grantee. This conversation eventually led to several research exchanges, observation- and practice-driven innovations, and blended industrial and academic expertise that has catalyzed the development of options to recover nutrients from urine. Designing technologies for use in informal settlements around Nairobi (Sanergy's primary region) demands rapid scaling and constant adaptation to dynamic needs, such as the explosive population growth associated with informal settlements near sub-Saharan African cities (Jacobsen et al., 2012). Over the past decade, Sanergy has established itself as a pioneer in container-based sanitation by meeting the needs for rapid scaling of toilets, collaborating with local stakeholders, and designing diversified income streams. In terms of resource recovery, Sanergy has focused on the production of valuable end products from the collected feces and has successfully brought two products to market. The ElectroSan team has focused on novel processes for converting collected urine into valuable products. This narrowed scope was critical to problem definition and progress on a clearly defined challenge while staying connected to the broader development challenge of increasing access to sanitation.

1.3   Problem Iteration

The first iteration of our problem was addressing the pathogen risks from feces during toilet or pit emptying, especially for toilet users and sanitation workers. The Safe Sludge Toilet was developed as a self-disinfecting fixture that separated urine and feces, created disinfectant in urine through urea hydrolysis and lime addition, and then ensured sufficient contact time to disinfect the feces (Ogunyoku et al., 2016). We envisioned Sanergy adapting the Safe Sludge approach to its Fresh Life Toilets (FLTs) to reduce the risk of pathogen exposure for collection staff and community members.

While hydrolyzed urine could be used to disinfect feces, its high ammonia content might interfere with thermophilic composting, Sanergy's primary method of valorizing feces at the time. A field visit by the UC Berkeley team focused on the effect of ammonia on composting identified an additional problem to solve: how to separately valorize urine. Urine-diverting toilets were being increasingly installed in Nairobi and by other sanitation service providers; however, resource-oriented urine treatment technologies lagged behind feces valorization. We further specified this problem to what nitrogen products could be extracted from urine, because struvite precipitation, a mature phosphate recovery technique that generates a phosphate-rich fertilizer, recovers only 5% of nitrogen from urine (Etter et al., 2011). Nitrogen recovery from urine was less understood at the time (2012) and benefited from lessons learned during nitrogen fertilizer recovery from other wastewaters.

From 2013 onward, ElectroSan focused on developing treatment processes to extract nitrogen from urine. We made several choices early on to constrain the problem but remained flexible to revisit them as needed. First, we focused on nitrogen fertilizer because Sanergy was already focused on agricultural products with its feces-derived soil conditioner. Second, we focused on liquid fertilizers because they required minimal post-processing and could be combined with irrigation. After a preliminary market analysis, we identified ammonium sulfate as a promising product because it is among the highest used fertilizers worldwide and is typically sold in liquid form (Chien et al., 2011). We also chose to focus on treating urine once it was collected from several toilets at a collection depot, which provided relatively stable design constraints as the number of toilets and installation sites (e.g., schools, private, public, residential) grew rapidly. Based on this decision, we concentrated our efforts on hydrolyzed urine because urine was hydrolyzed by the time it reached the collection center and was added to containers of already hydrolyzed urine. The major difference between fresh and hydrolyzed urine is the form of nitrogen: uncharged urea in fresh urine and positively charged ammonium in hydrolyzed urine (see Sect. 3.1).

Thus, the large problem of crafting economically viable sanitation was reduced to the narrow challenge of designing technologies to recover ammonium sulfate fertilizer from urine. Establishing the problem in the context of Sanergy and economically viable sanitation facilitated translation of laboratory results to evaluation of factors critical for real-world implementation (see Sect. 3.2).

1.4   Lessons from Problem Definition

1.4.1   Problem Definition: Access to Sanitation and Fertilizer

Like many Development Engineering efforts, the process of defining our problem was iterative and time-consuming. Based on the team's expertise, we approached the problem primarily from the motivation of increasing access to sanitation. We hypothesized that recovering valuable products from excreta could generate revenue for sanitation service providers. While we started with the technical aspects of designing recovery processes, we also had to consider the market viability of the recovered products, which involved another value chain, agricultural inputs, including fertilizers. Because our innovations could affect two dire needs in Kenya (access to sanitation and access to fertilizer), we spent considerable time talking with partners and subject matter experts to situate our project among the many stakeholders involved. We also constructed business model canvases and theories of change for both sanitation and fertilizer (see Sect. 3). As can be seen from our project timeline, we focused first on sanitation, then added agriculture, and then refined our vision at the nexus of both problems among partners in both sectors.

1.4.2 Theory: Circular Economy and Resource Recovery

Our work on resource-oriented sanitation was motivated by two increasingly recognized theories: the circular economy and resource recovery from wastewater. For decades, fecal matter has been repurposed as compost, energy, and even irrigation water. Recently, efforts have intensified to recover additional valuable products from excreta, including microbial protein products (Calvert, 1979), building materials (Diener et al., 2014; Mohajerani et al., 2019), cooking charcoal (Ohm et al., 2013), and disinfectants (Huang et al., 2016). This resource recovery perspective aims to extract all possible value from waste before disposal. Most recently, a US National Academies of Engineering report identified one of the major twenty first-century environmental engineering challenges as "designing a future without pollution and waste" (Board and National Academies of Sciences 2019). Several of these extracted products intersect with other critical industries, including agriculture and energy storage. Another one of these five grand challenges is "sustainably supplying food, water, and energy"; resource recovery may play a major goal in realizing this vision over the next several decades.

More broadly, the circular economy re-envisions current linear, extract-and-emit chemical manufacturing as loops that use every molecule as many times as possible before disposal (Fig. 16.2). With this perspective, waste streams of all kinds, including excreta, should be maximally repurposed into useful products (Geissdoerfer et al., 2017). The circular economy perspective is a broader instantiation of resource recovery and can apply to gaseous carbon dioxide emissions (Amouroux et al., 2016), solid domestic waste (Tisserant et al., 2017), and consumer products (Singh & Ordoñez, 2016). The circular economy concept in sanitation is well-recognized and regularly discussed at conferences like the Fecal Sludge Management Conference and the International Water Association Conference on Resource-Oriented Sanitation. Recently, the concept has been synthesized as the Circular Sanitation Economy by the Toilet Board Coalition, a network of multinational industries in sanitation provision (Toilet Board Coalition, 2016).

1.4.3 Observation-Driven: Valorizing Urine

While our academic focus on converting excreta into valuable products embodies resource recovery and a circular economy, our approach was also radically informed by observations in resource-constrained communities. By the ElectroSan team's estimation in 2012–2013, resource recovery tended to focus on combined wastewater (urine and feces) most, followed by separately collected feces and urine as a distant third. From our informal review of resource-oriented toilet

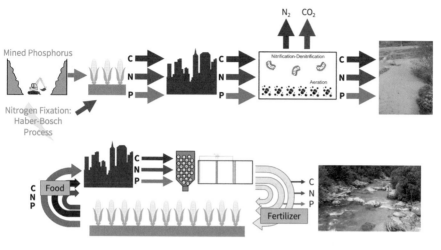

Fig. 16.2 Contrasting linear (top) and circular (bottom) economies for carbon, nitrogen, and phosphorus. Image adapted courtesy of Heather Goetsch

systems, the major motivation for separately collecting feces was to reduce water content by removing urine, which enhances composting and drying techniques. We observed that urine was underutilized, in part because of its large volume and high water content (Kvarnström et al., 2006), which would lead to high transport costs. Once we decided to concentrate our efforts on urine, we considered several possible recovered products. The most established product to recover from urine is struvite, or magnesium ammonium phosphate ($MgNH_4PO_4 \cdot 6H_2O$), a phosphate-rich fertilizer (NPK mass ratio 6:29:0). Urine contains the majority of macronutrients in excreta (80% N, 50% P, 60% K), making it an ideal stream from which fertilizer nutrients can be concentrated (Larsen & Gujer, 1996). At the time, we observed that urine efforts were more focused on phosphorus than nitrogen; thus, we concentrated our efforts on recovering nitrogen from urine. In addition to being understudied, nitrogen is the primary parameter by which fertilizer application rates are determined; furthermore, nitrogen recovery from excreta could benefit the global nitrogen cycle, which humans have severely imbalanced due to synthetic fertilizer production (Galloway et al., 2008). Ultimately, recovering nitrogen from urine was an untapped opportunity for recovery of valuable products from excreta.

2 Implementation Context

2.1 Partners and Implementation Setting

Sanergy operates over 3690 Fresh Life Toilets that are collectively used over 126,600 times daily in informal settlements around Nairobi. As of 2019, FLTs are

Fig. 16.3 (Left) Evergrow organic soil conditioner. Photo credit: William Tarpeh. (Right) The black soldier fly larvae are sold as an animal protein feed. Photo credit: Kevin Orner

primarily owned and operated by local entrepreneurs as pay-per-use, but others are open to the public, located at schools, or associated with multifamily dwellings. The toilet user interface is a squat plate with separate holes for urine and feces. Toilet operators pay for urine and feces containers (25 L urine, 30 L feces) to be collected and replaced twice per week by Sanergy employees (O'Keefe et al. 2015). The containers are taken to a central collection center where urine and feces are processed separately. Fecal waste is trucked 30 km to a processing facility and converted into the following value-added products: a soil additive branded as *Evergrow* through a co-composting process and a black soldier fly animal feed (Fig. 16.3). Urine from the toilets is stored in 1000 L tanks at the collection center and periodically disposed of in a nearby sewer or trucked 20 km to a wastewater treatment facility. Sanergy's New Technologies team identifies and scales novel methods to treat the separated feces and urine, establishing a sanitation value chain (Fig. 16.4).

2.2 Aligning Incentives

Both ElectroSan and Sanergy aimed to create a diversified portfolio of excreta-derived products, including valorizing urine as a critical business and innovation opportunity. Several practical questions posed by experiences in Nairobi inspired laboratory research at UC Berkeley; conversely, Sanergy provided an ideal testbed for several research questions (e.g., urine composition at different toilets) based

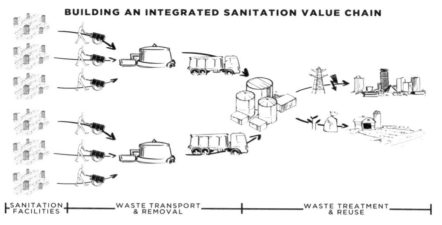

BUILDING AN INTEGRATED SANITATION VALUE CHAIN

SANITATION   WASTE TRANSPORT   WASTE TREATMENT
FACILITIES   & REMOVAL   & REUSE

Fig. 16.4 Conceptual sanitation value chain, including collection, treatment, and reuse. Image credit: Sanergy

on their large volumes of collected urine, which quickly outgrew urine-separating networks in the United States.

At times, our focus on fundamental research was not completely aligned with Sanergy's business objectives. After our pilot study of ion exchange (i.e., flowing urine through columns of negatively charged polymeric resin beads that adsorb cations; see Sect. 3.1.1 for more details) in Nairobi, we determined that urine-derived fertilizers cost 40% less than trucking urine to a wastewater treatment facility and cost 70% less than commercial fertilizers in Kenya (Tarpeh et al., 2018a, b). However, the supply chain for the resins and the customer based for urine-derived fertilizers were not well established. Thus, Sanergy pivoted more to scaling up feces valorization with anaerobic digesters and black soldier fly larvae, eventually favoring the latter. Technical challenges still remained, and ElectroSan focused on addressing those in the laboratory. Partners stayed in contact, but diverged efforts for a time period as the research team focused on reducing regeneration costs and combining nitrogen recovery with phosphate recovery. Once these practical questions were answered in the laboratory and urine valorization became a priority for Sanergy again, the partners planned another field visit and pilot-scale facility. Regular communication played a key role in identifying catalytic collaborative opportunities for ElectroSan and Sanergy when incentives aligned. Before each field visit, both partners iteratively agreed on a work plan and desired outcomes from the interaction.

2.3 Overcoming Constraints

One early challenge was establishing an analytical laboratory on-site. Our first field visits entailed battery-operated lab equipment brought from UC Berkeley. Within a few years, Sanergy purchased additional warehouse space near the operations office and collection depot in Mukuru and hired several laboratory staff. Berkeley researchers (both within and outside of ElectroSan) were involved as consultants and even on-site employees for the laboratory design, which facilitated future experiments and regular monitoring of Sanergy's excreta-derived products. Close collaborative preparations between ElectroSan and Sanergy facilitated successful equipment planning, and some equipment used during field visits was donated to the Sanergy lab (e.g., spectrophotometer for ammonia analysis).

To overcome the variability of data collected during field studies, field methods were benchmarked with more precise laboratory methods before and after deployment. Fieldwork in turn revealed issues to be resolved in the laboratory, including ion exchange column pressure management during long-term operation. Based on ion exchange resin regeneration consuming the majority of operating expenses, we also turned to alternative regeneration schemes in the laboratory with clear field motivation. Because the urine-derived fertilizer market was not as well established as its feces-derived counterpart, we also began considering other high-nitrogen waste streams and other products beyond fertilizers. In our preliminary field experiments, we demonstrated nitrogen recovery from anaerobic digester effluent to reduce ammonia inhibition and create ammonia fertilizers. These preliminary data catalyzed further laboratory experiments with adsorbent and electrochemistry-based approaches to recovering nitrogen from additional waste streams.

2.4 Lessons Learned from Implementation

One of our major lessons, which is common to many Development Engineering studies, is the value of close and continued partnership. Sanergy and ElectroSan have complementary and overlapping missions that allowed both entities to focus on their core business: increasing toilet access (Sanergy) and creating value from liquid waste streams (ElectroSan). Our shared interests in expanding the product portfolio for excreta-derived products and our willingness to experiment with bold ideas were fundamental ingredients of the successful development and implementation of fertilizer recovery from urine in Nairobi. Partnership also includes broader networks of academic researchers, practitioners, and funders from which Sanergy and ElectroSan have benefited. Conferences and alliances like the Sustainable Sanitation Alliance (SuSanA), the Container-Based Sanitation Alliance (CBSA), and the Rich Earth Summit have fostered catalytic conversations and created enabling environments within the sanitation community.

One of the largest challenges of Development Engineering is making decisions based on limited information. Throughout this experience, we became comfortable with making as informed a decision as possible, documenting our thought process, and remaining open to revisiting the decision later. We decided early on to focus on designing processes at the collection depot level for several reasons: we expected urine to be hydrolyzed (so that nitrogen was present as ammonia), variability would be less acute than at the toilet level, and samples would be easier to gather. This decision constrained some of our applications, but ultimately in a manner that led to meaningful progress on recovering ammonium sulfate fertilizer from hydrolyzed urine. We have since relaxed some of these assumptions to expand the applications of our technological innovations, as informed by our preliminary efforts. Expanded applications have included treating fresh urine and accelerating urea hydrolysis, producing alternative ammonia products from hydrolyzed urine, and recovering nitrogen from other waste streams besides urine.

During technology development, we used an iterative design process in the laboratory (Berkeley) and the field (Nairobi). We did not wait until the technology was "complete" before going into the field but instead aimed to learn as much as we could in the field to drive further laboratory research. For example, we learned early on that showing the robustness of the technology to deal with variable influent composition was crucial to implementation; thus, we focused on mechanistic laboratory studies to identify which components most influenced ammonia recovery performance metrics (e.g., recovery efficiency). Based on making the practical decision to focus on treatment of thousands of liters of urine, we were very interested in how the choice of scale would impact process design. After establishing performance in our lab-scale columns, we explored increasingly large columns capable of treating a larger volume of urine on-site with Sanergy. This scientific equivalent of rapid prototyping aligns with the lean business model canvas approach of failing fast and gathering rich information on the target customer. Focusing on one target application with one partner led to deep insights that informed and directed our laboratory development of nitrogen-selective extraction technologies.

3 Innovate, Implement, Evaluate, and Adapt

3.1 Innovation

Within the broader context of resource-oriented sanitation, our work addresses one overarching question: can valuable nitrogen products be made and profitably sold from separately collected urine? Within the potential urine-derived product portfolio, ElectroSan focused on nitrogen fertilizers because 90% of global ammonia is used to manufacture fertilizer (Nørskov et al., 2016); we focused on liquid fertilizers because they could be easily combined with irrigation (fertigation) and because they are simpler to produce from a process engineering perspective. Even more specific to Sanergy, we aimed to design unit processes that could treat thousands of liters of urine each day at the collection depot (Fig. 16.5). Based on this constraint, we designed for hydrolyzed urine treatment after phosphorus was recovered. We compared the aqueous chemical compositions of fresh and hydrolyzed urine to inform design of recovery processes focused on ammonia over urea (Table 16.1). We always imagined the ElectroSan processes within a treatment train that would

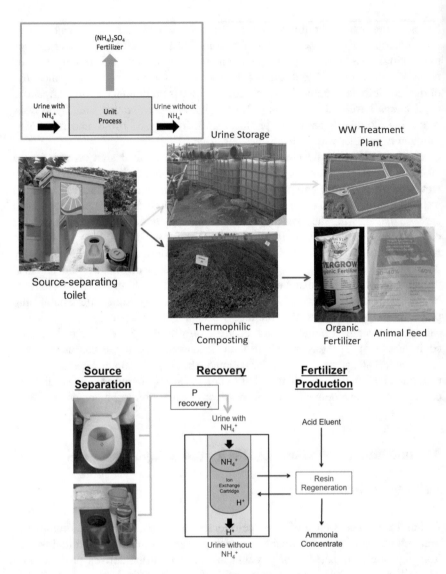

Fig. 16.5 (Top) Schematic of urine and feces treatment at Sanergy and placement of ElectroSan technologies. (Bottom) Schematic of ion exchange in urine treatment train

recover phosphorus precipitates as valuable products and mitigate fouling in the ElectroSan reactors. Designing for a unit process within a larger process scheme helped constrain the possible composition and informed repeatable laboratory influents (synthetic urine) for experiments.

Our core innovation is selective separation processes that extract ammonium from urine as high-purity ammonium sulfate, a common liquid fertilizer. Thus, we

Table 16.1 Composition of fresh and stored urine

	Fresh Urine	Stored Urine
pH	6	9
Species	Concentrations (mM)	
Urea	300	0
Cl⁻	100	100
Na⁺	100	100
K⁺	60	60
Total Ammonia	30	600
NH_4^+	30	400
Total Phosphate	25	20
SO_4^{2-}	15	15
Ca^{2+}	5	0
Mg^{2+}	5	0
NH_3	0	200
Total carbonate	0	300
Alkalinity	22	500

Rows highlighted in yellow describe parameters that change significantly during urine storage and urea hydrolysis (Larsen et al., 2021; Udert et al., 2003)

focused on both urine treatment and fertilizer production. For fertilizer production, we considered how liquid ammonium sulfate could be generated and used as a viable product. We considered three major value propositions: reducing urine treatment volume, supplementing Evergrow solid organic fertilizer from feces, and producing ammonium sulfate fertilizer. We evaluated these value propositions by considering several metrics, including volume of regenerant compared to volume of urine treated, purity of product, and ammonium concentration of the product.

3.1.1 Technology

After reviewing existing methods for nitrogen recovery from wastewater in practice and in academic literature, ElectroSan developed two technological solutions to achieve selective separations: ion exchange and electrochemical stripping. Ion exchange and electrochemistry were first identified as methods with untapped potential for selective nitrogen recovery; within those two areas, we iterated ideas over several months. Throughout this chapter, we focus primarily on ion exchange, as we chose to do during technology development in Kenya, because it was the more mature technology. The relative maturity of ion exchange eased communication with partners and integration with existing treatment processes.

Ion exchange leverages solid adsorbents in a fixed-bed column, where liquid flows over stationary adsorbent beads. For nitrogen recovery, positive ammonium ions are passed over a negatively charged ion exchange resin and trade places with protons (Fig. 16.6). After all adsorption sites are filled with ammonium, an acid

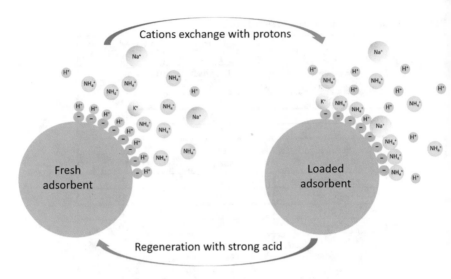

Fig. 16.6 Conceptual diagram of ammonium adsorption and regeneration via ion exchange adsorption

(proton-rich) solution is passed over the resin to regenerate the resin and produce concentrated ammonia solution. To produce ammonium sulfate fertilizer, we used sulfuric acid as an eluent. Previously, adsorbents were primarily used for nitrogen removal rather than recovery and rarely used in hydrolyzed urine with the aim of producing high-purity ammonia concentrate. In our first study, we compared the behavior of clinoptilolite, wood husk biochar, and two industrial ion exchange resins for ammonium adsorption and fertilizer production (Tarpeh et al., 2017). In addition to characterizing adsorption mechanisms in ideal solutions and real urine, we examined several performance criteria: maximum adsorption density (mg ammonia/g resin), regeneration efficiency (ratio of ammonia eluted to ammonia adsorbed), and cost per mass of nitrogen recovered.

Electrochemical nitrogen stripping was still in development as we concentrated our efforts on ion exchange in Nairobi. As electrons are removed from solution via the circuit, ammonium (NH_4^+) ions and protons cross the cation exchange membrane (CEM) to the cathode (Fig. 16.7). Both diffusion due to concentration gradients and migration due to applied current contribute to transmembrane flux. Protons are produced from the oxidation of water, which decreases the pH of the anode, making it strongly acidic. Once ammonium reaches the cathode, it is transformed into aqueous ammonia (NH_3) due to high pH resulting from the production of hydroxide ions. Aqueous ammonia is selectively transported to the trap chamber, which conserves the driving force for ammonia flux from cathode to trap by consuming the ammonia with dilute acid and producing ammonium concentrate, a key ingredient for fertilizer. The liquid can be further processed by distillation or freeze-drying to produce concentrated liquid and solid fertilizer. The

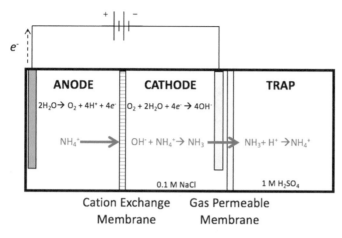

Fig. 16.7 Conceptual diagram of electrochemical stripping, where nitrogen as ammonia is selectively recovered based on positive charge (cation exchange membrane) and volatility (gas-permeable membrane)

cathodic reduction of oxygen produces hydroxide ions that raise pH, facilitating the production of ammonia from ammonium (Tarpeh et al. 2018a, b).

We regularly benchmarked the energy consumption, nitrogen recovery efficiency, and selectivity of electrochemical stripping with ion exchange and the other existing nitrogen treatment technologies. Even during our field study of ion exchange, we operated preliminary tests on electrochemical stripping on the same influent that accelerated development back in the laboratory. Ultimately, electrochemical stripping excels in its low chemical and energy input requirements, as well as its much higher selectivity than ion exchange. On the other hand, ion exchange requires less technical maintenance and exhibits potentially longer lifetime due to its relative simplicity compared to electrochemical stripping.

Designing with the user in mind, and even more specifically operators within Sanergy, drove design decisions like combining columns in series for easy operation. Working closely with Sanergy treatment operators helped the ElectroSan team understand their schedules, incentives, and challenges (e.g., intermittent electricity). These users were kept in mind when developing technologies and provided informal and formal feedback on the technologies during our field visits. Based on discussion with Sanergy, we have since aimed to reduce chemical inputs due to limited and potentially volatile supply chains to informal settlements.

3.1.2 Theories of Change: Fertilizer and Agriculture

There are two main direct beneficiaries of ElectroSan innovations: fertilizer consumers and sanitation collection services. Within fertilizer consumers, we characterized smallholder farmers who would benefit from access to locally produced

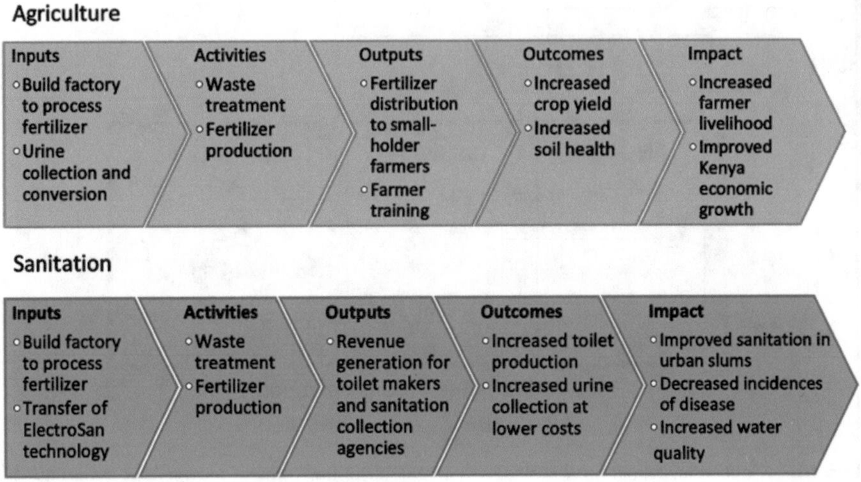

Agriculture

Inputs	Activities	Outputs	Outcomes	Impact
○ Build factory to process fertilizer ○ Urine collection and conversion	○ Waste treatment ○ Fertilizer production	○ Fertilizer distribution to small-holder farmers ○ Farmer training	○ Increased crop yield ○ Increased soil health	○ Increased farmer livelihood ○ Improved Kenya economic growth

Sanitation

Inputs	Activities	Outputs	Outcomes	Impact
○ Build factory to process fertilizer ○ Transfer of ElectroSan technology	○ Waste treatment ○ Fertilizer production	○ Revenue generation for toilet makers and sanitation collection agencies	○ Increased toilet production ○ Increased urine collection at lower costs	○ Improved sanitation in urban slums ○ Decreased incidences of disease ○ Increased water quality

Fig. 16.8 Agricultural and sanitation theories of change for ElectroSan. Image adapted courtesy of Rachel Dzombak, Ivan Buchanan, Woojin Jung, Martha Lesniewski, and William Tarpeh

affordable fertilizers to improve their crop yield and increase income. Currently, almost all fertilizer is produced outside of Kenya and distributed by either the national government or private wholesalers. Utilizing local distribution networks and incorporating education with product distribution, ElectroSan targeted farmers who need agricultural innovation the most. Sanitation collection services like Sanergy would benefit by avoiding transport of collected urine for disposal and by increasing revenue due to sales of urine-derived products. As such, we developed two theories of change: one for agriculture and one for sanitation (Fig. 16.8).

We characterized user needs by incorporating feedback from multiple stakeholders with experience in Kenya including agricultural experts, farmers, and sanitation and fertilizer distribution companies. Based on our team's expertise, we primarily focused on the sanitation theory of change but were informed by our agriculture theory of change. We engaged several potential agriculture partners, both with Sanergy and on our own. With Sanergy, we engaged potential organic fertilizer users, such as Lipton and horticulture farms. In parallel, the ElectroSan team consulted with businesses focused on farming inputs (e.g., One Acre Fund, Mavuuno). These did not lead to formal partnerships but informed our technology design and instilled confidence that these activities could be carried out once the fertilizer product was more firmly characterized.

3.1.3 Markets/Economic Sustainability

Several versions of both business model and social impact canvases have been iterated within ElectroSan (one example in Fig. 16.9). A key part of our business

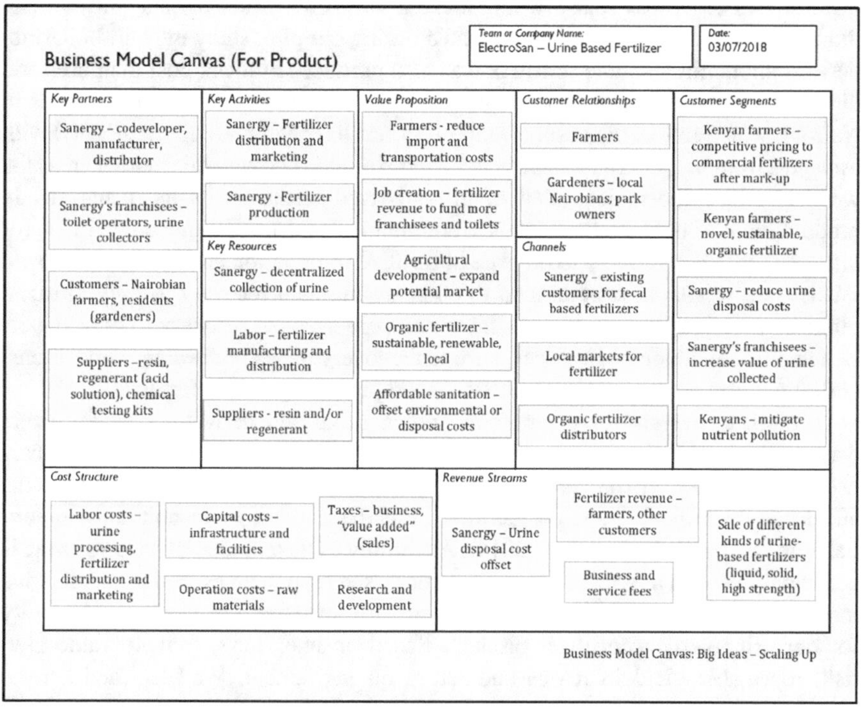

Fig. 16.9 Business model for urine-derived fertilizers. Image adapted courtesy of Hannah Greenwald, Luis Anaya, John Law, and William Tarpeh

model is partnering with fertilizer distributors so that ElectroSan can focus on fertilizer production. Ongoing work focuses on characterizing the market for liquid fertilizers in Nairobi, as well as the potential for producing solid fertilizer by freeze-drying or other similar methods. Anecdotally, most fertilizer is distributed in solid form, so we would need to identify submarkets open to liquid fertilizer (e.g., horticulture) or show that liquid fertilizer has marked advantages over solid fertilizer to elicit behavior change. Another key aspect related to behavior change is demonstrating the efficacy of our fertilizer product, which we have considered piloting during exhibitions, landscaping projects, or cooperative farms to double as research and demonstration sites.

In parallel with our laboratory development, we catalogued cost and forecasted costs at scale. Initially, we used experimentally determined adsorption densities to estimate operating costs of ion exchange adsorbents. Naturally derived adsorbents (clinoptilolite and biochar) were cheaper than conventional nitrogen removal via nitrification-denitrification but exhibited suboptimal recovery efficiencies. Of the synthetic resins tested, Dowex Mac 3, a weak cation exchange resin, exhibited the lowest cost because of its high nitrogen adsorption density and high regeneration efficiency (99%) (Tarpeh et al. 2017). Our preliminary cost estimates showed that just ten uses of Dowex Mac 3 would make it more cost-effective than conventional nitrogen removal, which we demonstrated during our pilot study in Nairobi. Urine-derived ammonium sulfate fertilizer was also produced at lower cost than urea and diammonium phosphate (Tarpeh et al., 2018a, b). We used local fertilizer prices in Nairobi, which reflect the costs of importing fertilizer and markups associated with transport from the port of Mombasa to inland Nairobi. Through this market analysis, we determined the major value of urine-derived fertilizer: its local and steady production. An ongoing challenge is to communicate the most fitting comparison for nitrogen fertilizer recovery to *the sum* of fertilizer production and nitrogen removal. Wastewater treatment experts tend to compare nitrogen recovery to only nitrogen removal, and fertilizer producers tend to compare nitrogen recovery to only Haber-Bosch nitrogen fixation. However, nitrogen recovery meets both value propositions: curtailing nitrogen wastewater discharges and producing usable fertilizer.

A market analysis for positioning human urine and derivatives as fertilizers was conducted by Sanergy and ElectroSan in 2014. Synthetic nitrogen fertilizers use the Haber-Bosch process, which requires million-dollar capital investments and large quantities of natural gas that limit the number of installations in sub-Saharan Africa (Buluswar et al., 2014). According to stakeholder interviews, urine is commonly viewed as a means to achieve increased crop yield as well as an effective insecticide (Sanergy, 2014). ElectroSan aims to leverage this market opportunity for our urine-derived fertilizer product. Based on interviews, farmers value low-risk, repeatable yield increases and return on investment. We identified several competitors and existing options for fertilizer, including synthetic fertilizers. Both solid synthetic fertilizers (e.g., calcium ammonium nitrate) and liquid synthetic fertilizers (e.g., solubilized diammonium phosphate) are commonly used across Kenya but are often cost prohibitive for low-income farmers. Import tariffs, a 30% markup due to transportation costs, and few importers contribute to high price points (Sanergy, 2014). Solid fertilizers tend to be easier to transport, but each formulation requires different production mechanisms. Liquid fertilizers may be more difficult to transport but can be easily combined with irrigation. ElectroSan exhibits several competitive advantages: local production, regular supply, customization to various chemical compositions, and application by fertigation.

3.1.4 Environmental Sustainability

Protecting environments for future generations is a major motivation of ElectroSan's vision of reimagining waste. Worldwide, 80% of wastewater is discharged without treatment (Larsen et al., 2016); re-envisioning toilets as collection centers for raw materials could incentivize collection and treatment. Reimagining waste as a resource is particularly beneficial for global nitrogen use, because the global nitrogen cycle is severely unbalanced due to anthropogenic interferences. Industrial ammonia production, large-scale agriculture, and fossil fuel combustion have added unprecedented amounts of reactive nitrogen (e.g., NO_3^-, NH_4^+, N_2O) to the biosphere from the atmosphere, outpacing denitrification that converts reactive nitrogen back to dinitrogen gas (N_2). Because of the widespread untreated wastewater discharge, much of the world's synthetic reactive nitrogen contributes to eutrophication and threatens aquatic ecosystems. Recovering nitrogen directly from wastewater could reduce the demand for atmospheric N_2 extraction and incentivizes restoration of polluted aqueous environments.

Even while we were still developing our ion exchange technology, we evaluated the environmental impacts of its potential implementation using life cycle assessment (LCA) and techno-economic assessment (TEA). LCA considers embedded energy and emissions associated with engineered processes from raw materials to disposal; TEA considers process modeling, engineering design, and cost estimation. These system-level assessments of new technologies can serve two major goals: (1) comparing to existing approaches and (2) prioritizing opportunities for optimization (Corominas et al., 2013a,b). We considered ion exchange cartridges or ammonia recovery from urine for both purposes and using both TEA and LCA, along with geospatial modeling for a citywide collection scheme (Kavvada et al., 2017). Our analysis demonstrated that ion exchange was not only less environmentally harmful than conventional nitrogen removal but net-negative for greenhouse gas emissions and energy input considering the avoided Haber-Bosch production. Within ion exchange-based ammonia recovery, we expected transport (collection of cartridges from households) to be the major emitter and energy consumer; however, the sulfuric acid for regenerating the resin represented 70% of greenhouse gas emissions and energy input (deriving from the manufacturing process). Although the city used for the environmental impact analysis (San Francisco, CA) differs in important ways from Nairobi, the finding that the sulfuric acid contributed such a large footprint would likely apply in both contexts. This insight inspired another laboratory iteration focused on alternative regenerants including acids (hydrochloric, nitric) and brine (sodium chloride). We found that the acids were similar in performance but the brine exhibited a trade-off: lower environmental impacts but also lower regeneration efficiency. Ongoing efforts focus on alternative regeneration schemes to further reduce the emissions and energy input required for nitrogen recovery from urine.

3.2 Iterative Implementation

The ElectroSan experience exemplifies a major aspect of Development Engineering: iterative implementation. Rapid and frequent iterations between laboratory research and field implementation, even with imperfect prototypes, accelerated the development of technologies to recover nitrogen from urine. Both laboratory and field research were supported by educational experiences (many through the Development Engineering program at UC Berkeley) and diverse funding sources.

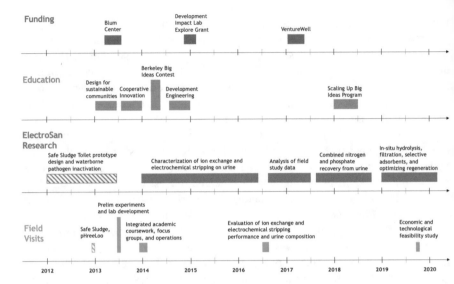

Fig. 16.10 Parallel timelines of funding, education, laboratory research, and field implementation. Safe Sludge activities are striped to identify them as preceding the formal start of ElectroSan but integral to the ElectroSan story

The interplay of funding, education, laboratory research, and field implementation is illustrated in the multicomponent timeline in Fig. 16.10.

3.2.1 Laboratory Research and Implementation

Early Visits

ElectroSan built on the foundation of the Safe Sludge project, which connected the Nelson group at UC Berkeley with Sanergy in 2012. Safe Sludge combined laboratory experiments with multidisciplinary design to create a conceptual prototype of a self-disinfecting toilet. In parallel, laboratory experiments documenting ideal ratios of urine to feces for inactivation of waterborne pathogens were determined (Ogunyoku et al., 2016). Once these results proved promising in Berkeley, Dr. Temi Ogunyoku planned a presentation at the second Fecal Sludge Management Conference in Durban, South Africa, followed by a field study of Safe Sludge for Sanergy excreta in December 2012.

The team returned to UC Berkeley to plan the next trip to Nairobi, with the major objective of identifying effects of the Safe Sludge process on compost in small-scale containers. We also explored the impact of lime addition as a cover material during toilet use on the thermophilic composting process at Sanergy's collection depot. Both the 2012 and 2013 field visits built a relationship with Sanergy and familiarized the UC Berkeley team with their operations by visiting toilets, meeting operations and collection team members, and working alongside treatment operators. In January 2014, two ElectroSan team members (Tarpeh and MBA student Ryan Jung) visited Sanergy to conduct focus groups with Fresh Life Toilet users. The major goals were to survey existing business resources in Nairobi, examine the price and cost needs for a successful business model to sell Fresh Life toilets to homeowners, and obtain feedback from users on prototyped treatment options.

These early trips also identified several challenges and opportunities for Safe Sludge. A major challenge was conducting experiments without a fully functional laboratory, which incentivized both partners to accelerate laboratory installation. Experiments during the 2012 trip were conducted some 15 km from Mukuru, the center of operations. During the 2013 trip, experiments were conducted at the composting treatment site; however, intermittent electricity required largely battery-operated experiments (e.g., pH meters, mass balances).

Starting ElectroSan

ElectroSan was formed out of the 2013–2015 visits and literature review on urine treatment technologies. We identified urine as underutilized during visits to Sanergy and conversations with other sanitation service providers visited in sub-Saharan Africa and Latin America. Working closely with treatment plant operators at Sanergy on a daily basis was a major precipitator, including regular visits from a truck to collect urine and transport it to a wastewater treatment plant. Based on these insights, Tarpeh and Nelson connected this opportunity with the ongoing experiments in Berkeley to selectively recover nitrogen from urine.

From January 2014 to December 2015, the ElectroSan team dedicated its efforts to developing ion exchange and electrochemical stripping in the laboratory. A major goal of laboratory efforts was to establish an experimental dataset and descriptive model that could be used to predict adsorption density for a given urine composition. We also successfully identified mechanisms of adsorption and the influence of composition by comparing simplified solution, synthetic urine, and real urine for both ion exchange and electrochemical stripping. During this focused time in the laboratory, the ElectroSan team still engaged with Sanergy through conference calls, meeting at conferences (e.g., Fecal Sludge Management Conference), and Sanergy visits to Berkeley. During this time, Sanergy grew rapidly in staff, number of toilets, and volume of excreta collected daily. We originally planned to visit Nairobi in Summer 2015 but decided to delay for a year to better characterize adsorption in realistic continuous operation and achieve publication-quality results during the field trial.

In 2016, the ElectroSan team had collected enough experimental data to plan for a field study of ion exchange. We met approximately monthly with Sanergy staff, particularly the laboratory manager and personnel. Pictures and supply lists were shared between the Berkeley and Sanergy laboratories to ensure compatibility. In 6 weeks during June to August 2016, Tarpeh, Nelson, and MS student Ileana Wald collected data on repeatable column performance using Dowex Mac 3 and Sanergy urine, surveyed urine composition across a representative set of Sanergy toilets, and monitored composition changes during collection. We also fabricated and operated columns at ten times the size of lab scale, which could then treat ten toilets' daily production of waste based on Sanergy's average across all toilets. Key performance metrics, including adsorption density and regeneration efficiency, were conserved over ten cycles at 6.5 L/d (one average Sanergy toilet) and for two cycles at 65 L/d (ten average Sanergy toilets). We continued analyzing the data collected in Nairobi over the 2016–2017 academic year that resulted in a publication in the new *Development Engineering Journal* in (Tarpeh et al., 2018b) (https://www. sciencedirect.com/science/article/pii/S235272851730074X).

Based on the encouraging results from our work in Nairobi, the ElectroSan team continued to develop larger-scale setups in Berkeley, as well as relax some of the assumptions that accelerated previous laboratory experiments. A master's student from ETH Zurich (Maja Wiprächtiger) explored urea hydrolysis and combining nitrogen recovery with phosphate recovery. We explored several options of combined nitrogen and phosphorus recovery: struvite precipitation followed by cation exchange, anion exchange followed by cation exchange, and simultaneous anion and cation exchange. We hypothesized that phosphate recovery could be conducted before nitrogen recovery with minimal effects on nitrogen recovery. In 2017 and 2018, we published several articles on our findings and presented our work at several conferences.

Dr. Kevin Orner, a postdoctoral researcher at UC Berkeley, conducted the most recent visit to Sanergy in September 2019. The objectives of the visit included (1) evaluating the economic feasibility of producing and selling urine-derived fertilizer, (2) exploring the technical feasibility of toilet-level nutrient recovery, and (3) determining if pit latrine waste could be used to produce liquid fertilizer via ion exchange. ElectroSan and Sanergy explored selling to Unilever, the owner of Lipton, which makes teas that require liquid fertilizer. Although Lipton could pay a premium for organic liquid fertilizer for their organic teas, we later learned that urine-derived fertilizer could not be classified as organic under current regulations. Another strategy for economically treating urine is to recover nutrients via ion exchange and locally dispose the treated urine effluent, which avoids transportation and disposal costs. Ion exchange can be used to recover ammonium and phosphate but will require additional treatment steps to meet effluent discharge requirements (e.g., pathogens, organic contaminants). This need motivates current research at UC Berkeley on urea hydrolysis, ion exchange, and media filtration that can take place underneath the Sanergy source-separating toilets. Liquid from fecal sludge could also be an influent to ElectroSan processes, which Orner explored with Sanergy and Sanivation, another sanitation company in Kenya.

3.2.2 Education and Funding

ElectroSan came of age with the Development Engineering program at Berkeley and benefited richly from interacting with the DevEng ecosystem of students, faculty, and collaborators. This enabling ecosystem also included the Blum Center for Developing Economies and the Development Impact Lab, supported by the US Agency of International Development through the Higher Education Solutions Network (HESN).

In Spring 2013, Tarpeh led a team of students in a Design for Sustainable Communities course to design an in-home toilet with Sanergy. This assessment validated the existing business model of shared public toilets being more profitable and user-friendly than in-home toilets. After pivoting away from in-home toilets, the ElectroSan team applied for class funds for the Summer 2013 field visit. The project continued during Fall 2013 as part of Cooperative Innovation, a new course focused on designing with low-income communities. The team designed and evaluated business models for excreta-derived products in Nairobi. Students traveled to Nairobi for several weeks in January 2014, supported by a Development Impact Lab Explore Grant ($5000) to seed new projects or pivots.

ElectroSan was first identified as an independent project through the Big Ideas at Berkeley grant competition, in which the team placed first in the Global Poverty Alleviation Category. As a Big Ideas recipient, the team benefited from mentoring from sanitation experts, funders, and impact investors. The work in progress was also presented at several conferences, which provided professional development to ElectroSan students, including Tarpeh.

In Fall 2014, the Development Engineering core course was offered for the first time at UC Berkeley. ElectroSan was pitched as a project and benefited from contributions from a multidisciplinary team (environmental/sustainability engineering, social work, business) that developed business models and surveyed potential users. This was a major step in articulating ElectroSan's relationship with Sanergy and role at the food-energy-water nexus: producing an agricultural input from wastewater at reduced energy. ElectroSan benefited from a free supply of urine to test its technologies, and Sanergy benefited from research and development that could expand the product portfolio of excreta-derived products by valorizing urine rather than paying for its disposal. A major output of this class was a Development Innovation Ventures Proposal, which is a USAID competition. ElectroSan did not actually submit the proposal, but articulating the ideas in a concise way guided both technology development and planning for future implementation.

ElectroSan continued to conduct research, funded in part by Big Ideas, NSF fellowship support, and the VentureWell business competition, which we won based on a presentation at the HESN annual meeting in 2016. VentureWell funds supported planning for installing a pilot source-separating toilet on the UC Berkeley campus to supply urine for research and attract visitors. ElectroSan successfully applied for the Scaling Up Big Ideas Program in 2018, which helped fund the next iteration of technology development. We also extended preliminary results on accelerating urea hydrolysis to facilitate toilet-level recovery. This iteration was again the subject of a Development Engineering course project that informed the most recent visit to Nairobi in Fall 2019. Several students have benefited from the ElectroSan project at UC Berkeley and Stanford, including undergraduates, MS students, PhD students, and postdoctoral researchers.

3.3   Adaptation

3.3.1   Reaching Scale: Scaling Down to Toilet-Level Recovery and Scaling Up to Pilot

Reaching meaningful scale for nitrogen recovery from urine could involve numerous small installations or a few large ones. We aim to identify optimal scale by combined experimental and modeling efforts. On the modeling side, we identified Pareto-efficient scales that minimize energy consumption and emissions in San Francisco. We will conduct similar analyses with Nairobi data to identify optimal scales for environmental impacts and costs. Experimentally, we have expanded our focus from larger-scale pilot installations to toilet-level treatment, which requires engineered urea hydrolysis.

Sanergy's rapidly scaled toilet collection inspired rapid scaling of feces and urine treatment. Several iterations of feces treatment have been considered, including thermophilic composting, anaerobic digestion, and, most recently, black soldier fly larvae. Sanergy now has a profitable full-scale plant for treating feces but transports and disposes urine without cost recovery due to no economically viable options. Thus, ElectroSan efforts were generally dedicated to developing pilot- and full-scale treatment options to recover nitrogen fertilizer from urine.

More recently, we have considered toilet-level urine treatment to recover nutrients and discharge treated effluent. Accordingly, ElectroSan researchers at UC Berkeley are currently investigating media filtration for removal of carbon and pathogens prior to local discharge (Fig. 16.11). This option would achieve scale by thousands of installations (one per Fresh Life Toilet). The real optimum may be a hybrid of toilet-level and centralized urine treatment, but further characterization of toilet-level nitrogen recovery is needed. Regardless, a practical, effective urine treatment solution could reach hundreds of thousands of users in Nairobi alone.

3.3.2   Current State of Technology

A challenge in recovering ammonium on-site from source-separated urine via ion exchange is accelerating the hydrolysis of urea. When urine exits the body, nitrogen is in the form of urea. The urea-N must be hydrolyzed into ammonium-N prior to nutrient recovery via ion exchange; however, the urea hydrolysis process can last multiple days. Accelerating urea hydrolysis by varying biofilm carriers or the bacterial inoculum could reduce reactor size to be feasible for individual source-separating toilets in informal settlements (Deleu 2020). Urine will be collected from a LAUFEN Save! urine-separating toilet installed in a bathroom on campus at UC Berkeley and used as an influent to the pilot treatment system that integrates urea hydrolysis, ion exchange, and media filtration.

In parallel with toilet-level efforts to recover urine, ElectroSan researchers at Stanford are investigating more selective adsorbents to improve the lifetime, cost-

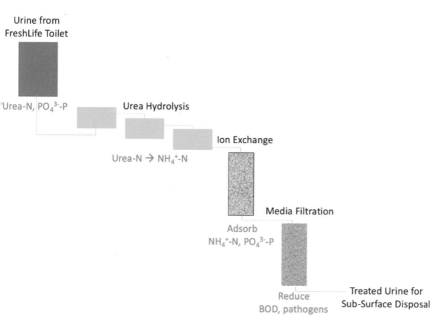

Fig. 16.11 Process flow diagram for treating urine from Fresh Life Toilet via urea hydrolysis, ion exchange, and media filtration prior to subsurface disposal

effectiveness, and purity of the process (Clark & Tarpeh, 2020; Clark et al., 2022). Electrochemical stripping has also been further developed with predictive models for removal and recovery and energy estimations that show that electrochemical stripping is on par with other nitrogen removal technologies (Liu et al., 2020). We have also demonstrated high recovery efficiencies in various wastewaters, which may expand the applications of electrochemical nitrogen stripping (Fig. 16.12). This line of research was first inspired by ElectroSan's 2016 visit to Sanergy, in which ion exchange was tested on anaerobic digester effluent because of potential scale-up of anaerobic digestion at the time.

3.3.3 Reaching Scale: Opportunities and Challenges

As is typical for Development Engineering, research questions have evolved over time from fundamental and proof-of-concept to applied, practical questions. Now that we have demonstrated that high-purity ammonium sulfate can be recovered from various wastewaters, including urine, several questions remain to reach scale. Establishing a regular supply chain for adsorbents is one such practical question. Making infrequent bulk purchases could be possible, but the overall lifetime of the adsorbent needs to be characterized to inform purchasing decisions. We have explored several alternative methods for regeneration in the laboratory but look

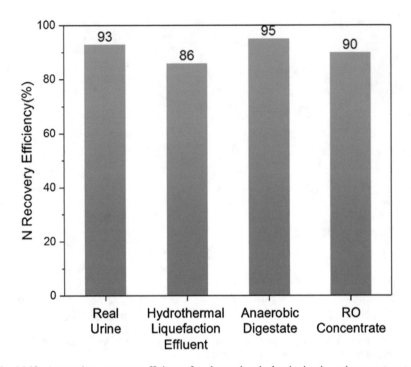

Fig. 16.12 Ammonium recovery efficiency for electrochemical stripping in various wastewaters. RO = reverse osmosis. Citations: real urine, Tarpeh et al. (2018a,b); hydrothermal liquefaction effluent, Li et al. (2018); anaerobic digestate, Liu et al. (2020); RO concentrate, unpublished

forward to contextualizing this regeneration decision based on cost and available supply chains for various acids and brine solutions. During a 2018 field trial in the United States, we encountered the practical challenge of how to apply an acidic ammonium sulfate fertilizer. Fertigation seems to be the best option, but dilution ratios must be devised and adapted from conventional ammonium sulfate application. Another question to reach scale is how to automate the continuous nitrogen recovery process, which we can adapt from other industrial ion exchange processes, such as water softening. Building on our cost analysis of ion exchange, we aim to identify marginal costs at scale. We expect economies of scale to exist, especially for resin purchasing and potential regenerant reuse, but aim to do a more long-term pilot study to further establish both technical and economic feasibilities. This was the aim of our pilot scale study with Lipton, but classifying urine-derived fertilizers as organic may be a prerequisite to running such a pilot study with fertilizer consumer interest.

Throughout its development, ElectroSan reported its results in scientific and gray literature, ranging from academic articles to reports and general media. Research collaborations have also played a critical role in disseminating our results, including with the University of Michigan where Tarpeh conducted postdoctoral research with Professors Nancy Love and Krista Wigginton. Ion exchange columns were constructed and operated at the University of Michigan to create ammonium sulfate fertilizer and applied to fields in Vermont. A growing community of urine-focused researchers now gathers annually at the Urine Summit, sponsored by the Rich Earth Institute in Vermont and most recently by the University of Michigan. Professor Tarpeh's research group at Stanford also led a recent short course in Senegal with master's students from Cheikh Anta Diop University in Dakar. Additional installations of urine-diverting dry toilets have led to informal communication among urine researchers, especially at conference symposia at several conferences, including the American Chemical Society and the Association of Environmental Engineering and Science Professors.

Another major challenge is reaching scale beyond Sanergy to other sanitation service providers (SSPs). To date, we have depended on building relationships with SSPs through our networks, including Sanivation and Delvic Sanitation Initiatives. The Container-Based Sanitation Alliance (CBSA) brings together several of these entities, some of whom have encountered ElectroSan at conferences and other meetings. To reach the needed scale of billions, we will need to leverage this increasingly robust network and develop regional models that cities can learn from and adapt.

3.3.4   Adapting to New Contexts

Since 2018, ElectroSan team members have explored electrochemical nitrogen stripping in partnership with Delvic Sanitation Initiatives (DSI), an SSP in Dakar, Senegal. This context differs considerably from Nairobi along several dimensions. In Dakar, the majority of the population is served by septic tanks that are exhausted by private trucks. DSI works with L'Office National de l'Assainissement de Senegal (ONAS), the national sanitation agency, to operate fecal sludge treatment plants in Dakar and recover valuable products. We have recently demonstrated high recovery efficiency from fecal sludge collected from anaerobic digesters at one such fecal sludge treatment plant and can produce both fertilizer and disinfectant from influents of varying composition. DSI benefits from strong collaborations with ONAS and Cheikh Anta Diop University (UCAD), from where its founders graduated. This enabling environment has accelerated field visits, funding, and student exchanges from 2018 to 2020. Most recently, a field visit in December 2019 included demonstration of electrochemical stripping, tours of fecal sludge treatment plants, meetings with ONAS and UCAD collaborators, and a short course on resource recovery for UCAD master's students. Diversifying ElectroSan's product portfolio will help identify additional markets and customers to incentivize urine collection and treatment.

4 Results/Lessons Learned

Our major lessons can be summarized as follows:

- Iterative development between laboratory and practical implementation accelerated innovation by facilitating rapid adaptation of laboratory findings to practical problems.
- Developing multiple technologies simultaneously helped establish relative advantages and value propositions.
- Flexible consideration of value propositions facilitated prioritization amidst changing partner priorities, constraints, and findings. Documenting rationale for our choices helped constrain our research questions and facilitated revisiting as needed.
- Strong in-country partnerships allowed ElectroSan to situate our expertise and focus on our core activity: converting urine into value-added products.
- Prospective assessment of costs and environmental impacts helped compare ElectroSan approaches to existing alternatives and prioritize future optimization opportunities.

4.1 Context and Vision for Our Results

Ultimately, our findings have advanced the case for a circular sanitation economy. Reimagining excreta as a resource has increasingly entered the public sphere through the collective efforts of the sanitation research community over decades, and ElectroSan is fortunate to contribute to this transformation and build on contributions from this dynamic community. Our vision remains to establish a diverse and profitable portfolio of urine-derived products, including fertilizers, disinfectants, fuel, water, and commodity chemicals. Recently, other researchers have extracted microbial proteins and brickmaking materials from urine (Randall et al., 2016; Christiaens et al., 2017). As the field matures, we expect to realize the vision of a suite of technologies and products that resource-oriented sanitation service providers can consider when they separately collect urine.

Within the field of resource-oriented sanitation, ElectroSan has focused primarily on a circular nitrogen economy. These efforts can help rebalance the nitrogen cycle by reusing reactive nitrogen, thus reducing discharges that foster eutrophication and harmful algal blooms, which damage aquatic ecosystems and threaten human health. Designing selective separation processes and materials will make this vision a reality. Just as we have integrated lab and field studies, we integrate investigations of novel materials that enable novel processes with process engineering that outlines criteria for groundbreaking materials.

While many problems are interdisciplinary, crafting profitable sanitation initiatives absolutely demands contributions from several fields (Hyun et al., 2019). The ElectroSan story demonstrates the multifaceted contributions required from engineering, business, policy, social sciences, public health, and humanities. Sanitation is often considered a "boundary object" to which many fields refer (Hyun et al., 2019), but will require communication and coordination between disciplines to meet the demanding challenge of providing billions of people with sanitation access.

4.2 Pivots

Like many Development Engineering projects, the story of ElectroSan is a series of pivots, design sprints, and iterative technology development. We have identified several pivot points throughout this chapter but highlight several significant changes in direction here:

1. We shifted from the Safe Sludge focus on using urine as a feces disinfectant to the ElectroSan focus of making separate valuable products from urine. This shift was motivated by separate collection of urine and feces and Sanergy already having thermophilic composting of feces. There was a major need to separately valorize urine, and ElectroSan developed to fulfill that need.
2. Although we developed two technologies in parallel, we chose to focus on ion exchange because it was more mature, logistically simple, easier to operate, and better proven at the time of our field trial. This decision was difficult at the time but in retrospect was a clear choice to accelerate the overall project and still apply learnings from ion exchange to electrochemical stripping that have facilitated transfer to other settings besides Nairobi.
3. Our brief expansion from urine to anaerobic digester effluent seeded the idea of using ion exchange and electrochemical stripping for other waste streams, but did not end up being installed at Sanergy. This decision allowed us to focus on urine at Sanergy, which we had most thoroughly characterized; at the same time, the decision opened up applications for ElectroSan technologies in the United States and other settings outside of urine.
4. Most recently, we pivoted from depot-level urine treatment to toilet-level recovery, which could obviate the need to transport urine. This was identified early on as a potential value proposition for urine treatment (during the 2016 field visit) but revisited with the scientific question of accelerating urea hydrolysis on site. The continued search for customer for urine-derived fertilizer also informed this choice, because scaling up will likely involve many installations rather than a few large installations.

We consider these pivots learning experiences more than failures or successes. Because hypotheses are constantly being generated in the laboratory and tested in the field or vice versa, we expect to continue to pivot as the idea of extracting value from urine matures.

4.3 Ongoing and Future Work

We have also identified several areas of ongoing work throughout this chapter. Significant highlights include:

1. Analyzing potential uses and buyers of ammonium sulfate fertilizer and other urine-derived products.
2. Engineering nutrient capture and urine treatment within the size constraints of a Fresh Life Toilet.
3. Adapting ion exchange and media filtration from treating urine to treating fecal sludge.
4. Developing more nitrogen-selective adsorbents to enhance adsorbent lifetime and product purity.

5 Summary and Interpretive Text Boxes

Guiding Framework
Throughout ElectroSan's activities, we have focused on several criteria: (1) maximizing nitrogen recovery efficiency, (2) creating profit by value of products exceeding costs of extraction, (3) operating at scale, (4) beating conventional methods in costs and/or environmental impacts, and (5) evaluating social acceptance from product users and/or treatment operators. Ideas have constantly been evaluated against this framework to inform our future research. For example, when we performed our life cycle assessment of ion exchange, we would have pursued other options if conventional nitrogen management was less costly and less environmentally harmful than ion exchange. We also conducted a preliminary estimate of the size of ion exchange cartridge needed for an average household (4 L); if it were prohibitively large, we would have pivoted to other adsorbents earlier on in the design process.

Pivots
One major pivot was our focus on ion exchange over electrochemical stripping, although electrochemical stripping was the newer and potentially more exciting technical option at the time. Another was our recent choice to pivot from selling liquid fertilizer at a premium because of its potentially organic source (urine) to instead reduce transport of urine, because the latter was easier to monetize. While we continue to work on establishing demand for urine-derived fertilizers, Sanergy's collection costs are a more sorely felt need that our innovations can address in the near term.

Responsible Research and Capacity
Many stories exist of organizations building a project far from home and then losing contact and thus failing to ensure that the project is sustainably operated and maintained. One best practice in international sustainable development efforts is partnering with a strong in-country partner who knows the social and economic

landscape. In this case, our in-country partner Sanergy employs 400 people, 60% of whom live in communities where Fresh Life Toilets are being utilized (Jobs, 2020). Maintaining close communication with Sanergy leadership and field staff promotes quick realistic feedback and iteration during laboratory and pilot research even while being located almost 10,000 miles away. The ElectroSan team has also engaged Sanergy laboratory staff to a high degree, including training laboratory staff on urine treatment technologies and having two New Technologies (Sanergy department) researchers as co-authors on our *Development Engineering Journal* publication (Tarpeh et al., 2018b). Similarly, training students and providing opportunities for research exchanges has been a major focus of our recent activities with Delvic in Senegal. UC Berkeley students also worked at Sanergy, helping establish the laboratory and build capacity for pathogen inactivation experiments and ammonia analysis.

Discussion Questions

1. How should "sustainability" be defined for sanitation service provision? How should trade-offs between human health, cost, and environmental protection be evaluated? For example, should Sanergy prioritize reducing nitrogen discharges to the environment at additional cost, given that the majority of human waste in Nairobi is discharged to the environment with minimal or no treatment? As another example, is it sustainable to treat urine using ion exchange resin materials that must be imported? Suggest other decisions facing Sanergy and ElectroSan, and evaluate them through your definition of sustainability.

2. What factors would most significantly affect the ElectroSan business model if it were implemented in Dakar, Senegal? Rural Kenya? India? Choose other settings and reflect on the same question. List the questions you would ask to ascertain the required changes. Describe the factors that companies like ElectroSan and Sanergy should consider when expanding their geographical scope and business verticals.

3. What is the role of beneficiary communities in the development of human waste collection services and waste-derived products? What is the importance of community input? What is the importance of culture?

4. How might fields other than economics/business and STEM provide value in Development Engineering contexts and specifically in this case study?

5. What are the benefits and potential drawbacks of franchising toilets instead of Sanergy maintaining complete ownership over toilets? How would changes in the toilet provision and excreta collection affect the ElectroSan business model?

6. How might businesses like Sanergy and ElectroSan be affected or adapt in the coming decades as developing areas change according to various forecasts (e.g., technological, environmental, political, economic)? Consider some best case, worst case, and moderate scenarios.

7. For researchers located far away from their Development Engineering enterprise or partners, what strategies are important for maximizing the effectiveness of trips to the enterprise location? Moreover, what are the most important aspects for maintaining effective productivity between both locations while distanced?

8. How might related enterprises be different or similar if sponsored or operated by a government instead of a private entity? What are the pros and cons of public vs. private sponsorship, for various countries?

9. When collaborating on a development venture, how might academic and business realms differ in the enterprise aspects that they emphasize or overlook? How might they overcome these differences to make integrated progress?

10. Comment on the timeline and workflow of this case study from inception to present day. Are there certain work strategies or order of events that you consider beneficial or would have modified?

11. How can mission-driven companies focused on social impact, like Sanergy and ElectroSan, balance ethical provision of services with profitability and financial sustainability, especially if business competition begins to arise?

References

Amouroux, J., & Auerbach, D. (2016). Sustainable sanitation provision in urban slums – The Sanergy case study. In E. A. Thomas (Ed.), *Broken pumps and promises: Incentivizing impact in environmental health* (pp. 211–216). Springer International Publishing.

Baum, R., Luh, J., & Bartram, J. (2013). Sanitation: A global estimate of sewerage connections without treatment and the resulting impact on MDG progress. *Environmental Science & Technology, 47,* 1994–2000.

Board, O. S., & National Academies of Sciences E and Medicine. (2019). *Environmental engineering for the 21st century: Addressing grand challenges.* National Academies Press.

Buluswar, S., Friedman, Z., Mehta, P., et al. (2014). *50 breakthroughs: Critical scientific and technological advances needed for sustainable global development.* LIGTT. Institute for Globally Transformative Technologies, Lawrence Berkeley National Lab.

Calvert, C. (1979). Use of animal excreta for microbial and insect protein synthesis. *Journal of Animal Science, 48,* 178–192.

Chien, S. H., Gearhart, M. M., & Villagarcía, S. (2011). Comparison of ammonium sulfate with other nitrogen and sulfur fertilizers in increasing crop production and minimizing environmental impact: A review. *Soil Science, 176,* 327–335.

Christiaens, M. E. R., Gildemyn, S., Matassa, S., et al. (2017). Electrochemical ammonia recovery from source-separated urine for microbial protein production. *Environmental Science & Technology, 51,* 13143–13150. https://doi.org/10.1021/acs.est.7b02819

Clark, B., & Tarpeh, W. (2020). Selective recovery of ammonia nitrogen from wastewaters with transition metal-loaded polymeric cation exchange adsorbents. *Chemistry – A European Journal n/a.* https://doi.org/10.1002/chem.202002170

Clark, B., Gilles, G., & Tarpeh, W. A. (2022). Resin-mediated PH control of metal-loaded ligand exchangers for selective nitrogen recovery from wastewaters. *ACS Applied Materials & Interfaces,* February 15. https://doi.org/10.1021/acsami.1c22316

Corominas, L., Larsen, H. F., Flores-Alsina, X., & Vanrolleghem, P. A. (2013a). Including life cycle assessment for decision-making in controlling wastewater nutrient removal systems. *Journal of Environmental Management, 128,* 759–767. https://doi.org/10.1016/j.jenvman.2013.06.002

Corominas, L., Foley, J., Guest, J. S., et al. (2013b). Life cycle assessment applied to wastewater treatment: State of the art. *Water Research, 47,* 5480–5492. https://doi.org/10.1016/j.watres.2013.06.049

Deleu, E. (2020). *Producing fertilizer from source-separated urine.* MS Thesis, Ghent University.

Diener, S., Semiyaga, S., Niwagaba, C. B., et al. (2014). A value proposition: Resource recovery from faecal sludge—Can it be the driver for improved sanitation? *Resources, Conservation and Recycling, 88,* 32–38. https://doi.org/10.1016/j.resconrec.2014.04.005

Etter, B., Tilley, E., Khadka, R., & Udert, K. (2011). Low-cost struvite production using source-separated urine in Nepal. *Water Research, 45*, 852–862.

Galloway, J. N., Townsend, A. R., Erisman, J. W., et al. (2008). Transformation of the nitrogen cycle: Recent trends, questions, and potential solutions. *Science, 320*, 889–892. https://doi.org/10.1126/science.1136674

Geissdoerfer, M., Savaget, P., Bocken, N. M., & Hultink, E. J. (2017). The circular economy–a new sustainability paradigm? *Journal of Cleaner Production, 143*, 757–768.

Huang, W., Zhao, Z., Yuan, T., et al. (2016). Effective ammonia recovery from swine excreta through dry anaerobic digestion followed by ammonia stripping at high total solids content. *Biomass and Bioenergy, 90*, 139–147.

Hyun, C., Burt, Z., Crider, Y., et al. (2019). Sanitation for low-income regions: A cross-disciplinary review. *Annual Review of Environment and Resources, 44*, 287–318. https://doi.org/10.1146/annurev-environ-101718-033327

Jacobsen, M., Webster, M., & Vairavamoorthy, K. (2012). *The future of water in African cities: Why waste water?* World Bank Publications.

Jobs. (2022). Sanergy. http://www.sanergy.com/jobs/

Kavvada, O., Tarpeh, W. A., Horvath, A., & Nelson, K. L. (2017). Life-cycle cost and environmental assessment of decentralized nitrogen recovery using ion exchange from source-separated urine through spatial modeling. *Environmental Science & Technology, 51*, 12061–12071. https://doi.org/10.1021/acs.est.7b02244

Ki-Moon B (2013) The millennium development goals report 2013.. United Nations Pubns.

Kringel, R., Rechenburg, A., Kuitcha, D., et al. (2016). Mass balance of nitrogen and potassium in urban groundwater in Central Africa, Yaounde/Cameroon. *Science of the Total Environment, 547*, 382–395. https://doi.org/10.1016/j.scitotenv.2015.12.090

Kvarnström E, Emilsson K, Stintzing AR, et al (2006) Urine diversion: One step towards sustainable sanitation.. EcoSanRes Programme.

Larsen, T. A., & Gujer, W. (1996). Separate management of anthropogenic nutrient solutions (human urine). *Water Science and Technology, 34*, 87–94.

Larsen, T. A., Hoffmann, S., Lüthi, C., et al. (2016). Emerging solutions to the water challenges of an urbanizing world. *Science, 352*, 928–933. https://doi.org/10.1126/science.aad8641

Larsen, T. A., Riechmann, M. E., Udert, K. M. (2021). State of the art of urine treatment technologies: A critical review. *Water Research X, 13*, 100114. https://doi.org/10.1016/j.wroa.2021.100114

Li, Y., Tarpeh, W. A., Nelson, K. L., & Strathmann, T. J. (2018). Quantitative evaluation of an integrated system for valorization of wastewater algae as bio-oil, fuel gas, and fertilizer products. *Environmental Science & Technology, 52*, 12717–12727. https://doi.org/10.1021/acs.est.8b04035

Liu, M. J., Neo, B. S., & Tarpeh, W. A. (2020). Building an operational framework for selective nitrogen recovery via electrochemical stripping. *Water Research, 169*, 115226. https://doi.org/10.1016/j.watres.2019.115226

Mohajerani, A., Ukwatta, A., Jeffrey-Bailey, T., et al. (2019). A proposal for recycling the world's unused stockpiles of treated wastewater sludge (biosolids) in fired-clay bricks. *Buildings, 9*. https://doi.org/10.3390/buildings9010014

Nørskov J, Chen J, Miranda R, et al (2016) Sustainable ammonia synthesis–exploring the scientific challenges associated with discovering alternative, sustainable processes for ammonia production.. US DOE Office of Science.

O'Keefe, M., Lüthi, C., Tumwebaze, I. K., & Tobias, R. (2015). Opportunities and limits to market-driven sanitation services: Evidence from urban informal settlements in East Africa. *Environment and Urbanization, 27*, 421–440. https://doi.org/10.1177/0956247815581758

Ogunyoku, T. A., Habebo, F., & Nelson, K. L. (2016). In-toilet disinfection of fresh fecal sludge with ammonia naturally present in excreta. *Journal of Water, Sanitation and Hygiene for Development, 6*, 104–114. https://doi.org/10.2166/washdev.2015.233

Ohm, T.-I., Chae, J.-S., Moon, S.-H., & Jung, B.-J. (2013). Experimental study of the characteristics of solid fuel from fry-dried swine excreta. *Process Safety and Environmental Protection, 91*, 227–234.

Randall, D. G., Krähenbühl, M., Köpping, I., et al. (2016). A novel approach for stabilizing fresh urine by calcium hydroxide addition. *Water Research, 95*, 361–369. https://doi.org/10.1016/j.watres.2016.03.007

Sanergy. (2014). Designing a product positioning strategy for human urine as fertilizer in Kenya.

Scott C, Faruqui N, Raschid-Sally L (2004) Wastewater use in irrigated agriculture: Confronting the livelihood and environmental realities.. International Water Management Institute.

Singh, J., & Ordoñez, I. (2016). Resource recovery from post-consumer waste: Important lessons for the upcoming circular economy. *Journal of Cleaner Production, 134*, 342–353.

Tarpeh, W. A., Udert, K. M., & Nelson, K. L. (2017). Comparing ion exchange adsorbents for nitrogen recovery from source-separated urine. *Environmental Science and Technology, 51*, 2373–2381. https://doi.org/10.1021/acs.est.6b05816

Tarpeh, W. A., Barazesh, J. M., Cath, T. Y., & Nelson, K. L. (2018a). Electrochemical stripping to recover nitrogen from source-separated urine. *Environmental Science & Technology, 52*, 1453–1460. https://doi.org/10.1021/acs.est.7b05488

Tarpeh, W. A., Wald, I., Omollo, M. O., et al. (2018b). Evaluating ion exchange for nitrogen recovery from source-separated urine in Nairobi, Kenya. *Development Engineering, 3*, 188–195.

Tisserant, A., Pauliuk, S., Merciai, S., et al. (2017). Solid waste and the circular economy: A global analysis of waste treatment and waste footprints. *Journal of Industrial Ecology, 21*, 628–640.

Toilet Board Coalition. (2016). *Sanitation in the circular economy. Transformation to a commercially valuable*. Self-Sustaining, Biological System.

Udert, K. M., et al. (2003). *Fate of major compounds in source-separated urine*. https://iwaponline-com.stanford.idm.oclc.org/wst/article-abstract/54/11-12/413/13916/Fate-of-major-compounds-in-source-separated-urine

United Nations. (2015). Goal 6: Ensure access to water and sanitation for all.

United Nations Human Settlements Programme. (2003). *The challenge of slums: Global report on human settlements, 2003*. UN-HABITAT.

WHO/UNICEF Joint Water Supply, Sanitation Monitoring Programme, World Health Organization. (2015). *Progress on sanitation and drinking water: 2015 update and MDG assessment*. World Health Organization.

World Health Organization. (2011). Women and girls and their right to sanitation.

World Health Organization. (2013). Health through safe drinking water and basic sanitation.

World Health Organization. (2019). Sanitation.

Chapter 17
Engineering Predictable Water Supply: The Humans Behind the Tech

Christopher Hyun, Tanu Kumar, Alison E. Post, and Isha Ray

1 Introduction

In the summer of 2007, graduate student Emily Kumpel, from the University of California, Berkeley (UC Berkeley), attempted to collect household water quality samples in Hubli-Dharwad, India. As she went from home to home, she found piped water service to be intermittent and unpredictable; even those who had lived in the city for years could not tell her when the water would turn on. She contacted the local water utility for more accurate information, but they could only give her a rough estimate. "They told me that water would arrive a certain number of times in a week, but not exactly when," Kumpel recalls. She realized then that significant numbers of households with piped network connections experienced uncertainty about when and if water would arrive.[1]

Around the globe, public services such as water and electricity are often not delivered continuously. Even though 89% of the world's population had access to piped water by 2012, Kaminsky and Kumpel (2018) estimate that at least one billion people around the world receive their piped water for fewer than 24 h a day. Service intermittency is often a consequence of inadequate supply for the entire service area.

Intermittency places many burdens upon households. First, it often leads to inadequate quantities of supply at the household level. A few hours of piped water

---

Tanu Kumar and Christopher Hyun are co-first authors of this chapter.

[1] E. Kumpel, personal communication, February 3, 2020

C. Hyun (✉) · A. E. Post · I. Ray
University of California, Berkeley, CA, USA
e-mail: chrish@berkeley.edu

T. Kumar
William & Mary, Williamsburg, VA, USA

© The Author(s) 2023
T. Madon et al. (eds.), *Introduction to Development Engineering*,
https://doi.org/10.1007/978-3-030-86065-3_17

supply, after all, may be insufficient for an urban household's cooking, cleaning, and bathing needs over the day or the week. Furthermore, intermittent water supply imposes burdens because it often arrives at unpredictable times. Water utilities are often unable to follow schedules for many reasons, including a lack of internal coordination, poor infrastructure, or problems with intermittent electricity supplies. Households are left not knowing exactly when their water supply will begin or end.

Water supply unpredictability can be particularly burdensome for poorer households. Many middle- and upper-class households can afford purchased water from tankers or automatically filling tanks to store the water whenever it arrives. For households that cannot regularly afford tanker water or do not have load-bearing roofs to support tanks, the only option is to wait for water and to fill household storage containers manually. For them, waiting for water could mean forgone wages or missed social events. These costs fall primarily upon women, who are most likely to be at home during the day.

Emily Kumpel, with a few colleagues, founded a social enterprise named NextDrop, which sent SMS (text message) notifications to urban households about water arrival times. The premise was that, since it is costly to address the underlying causes of water intermittency in piped systems, an information-based service could be a low-cost solution to ease the burden of water unpredictability. NextDrop rolled out and scaled up in the state of Karnataka in two urban areas of varying sizes, Hubli-Dharwad and Bangalore.

We studied NextDrop's services and found that while its pilot in the twin cities of Hubli-Dharwad was promising, the services rolled out on a larger scale in Bangalore did not have a measurable impact on many dimensions of household welfare. Why not?

NextDrop's system depended on water utility employees to provide timely information about water arrival to disseminate to the company's customers (see Fig. 17.1). These "valvemen" opened and closed water valves throughout the city, channeling water to particular segments of the piped network in rotation. NextDrop found it difficult to incentivize the valvemen to send the company regular and accurate data in metropolitan Bangalore. NextDrop's system was also not fully compatible with the existing social dynamics between households, valvemen, and utility supervisors. Moreover, even when information was accurately disseminated, customers often did not receive it; in many homes, for instance, the women who were responsible for waiting for and collecting water were not in possession of the household mobile phone during the day.

NextDrop's problematic rollout in Bangalore illustrates that the preferences and priorities of employees and customers are vital human components of technological innovations. They must be understood before and during the initial rollout of an innovation, and they can vary in unexpected ways across locations. It is essential to consider the possibility of this variation when scaling up an innovation from pilot locations.

In this chapter, we analyze NextDrop's evolution and develop this argument as follows: first, we provide an overview of the NextDrop system in its final form. We then go back in time and describe the system's development during a class

Fig. 17.1 NextDrop's phone-based water supply notification system depended on notifications from water valvemen. (Adapted from NextDrop)

on Information and Communications Technology for Development (ICTD) at UC Berkeley in 2009. We move on to discuss the company's initial, and apparently successful, pilot in Hubli-Dharwad, India. We follow the company's expansion to Bangalore with a pair of companion studies that we (the authors) conducted during 2014 and 2015. Drawing on insights from these studies, we show that the intervention had little impact on Bangalore's households, in part because key human elements in the system differed significantly and unexpectedly from the pilot location. We conclude with some takeaways for development engineers who wish to better understand and manage the human aspects of technological innovations.

2 The NextDrop System: An Overview

NextDrop's founders noticed that households faced major difficulties dealing with intermittent water supply and decided to develop a technological system that would provide them with advance warning of water arrival times. This was easier said than done, as water utilities delivering intermittent water supply tend not to have sensors throughout their piped network. In systems without such sensors, the water utility may not have any centralized information about where water is flowing throughout its pipes, especially at the household level. Though Bangalore's water utility released its schedule for piped water supply for residents to see in newspapers and local utility offices, responses from our 2014–2015 household survey revealed that this information was generally inaccurate in our study area (see Fig. 17.2). This shows that residents would have required a more accurate source of water supply information if they wanted to know when their water would turn on.

Scheduled Interval Between Supply Days

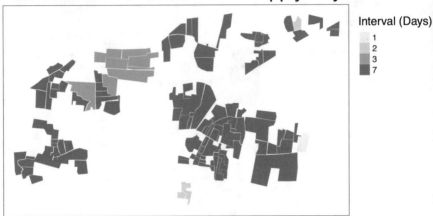

Reported Interval Between Supply Days

Fig. 17.2 Scheduled versus reported interval between supply days in Bangalore utility's E3 subdivision. (From Kumar et al., 2022)

After a fair amount of research, NextDrop's founders discovered that the valvemen who turned the water on and off were the best source for information on water arrival times. They developed a system through which they asked the valvemen to call a toll-free number whenever they opened a water valve, closed a valve, or did not open a valve when scheduled to do so (i.e., a supply cancellation). NextDrop developed proprietary software and then used it to send text or voice notifications with expected water arrival times to participating households that it had located in specific valve areas based on their GPS coordinates. Because it takes time for water to flow into a valve area and fully pressurize any specific portion of the network, notifications would arrive before water reached household taps, giving residents some time to get ready to receive water.

NextDrop relied on the valvemen's knowledge of the piped network in order to accurately map which valves served which households. It collected GPS coordinates for households and, with the valvemen's help, created maps of the valve areas, which Indian utilities typically do not possess. Each polygon in Fig. 17.2 is an example of one of Bangalore's thousands of valve areas; each valveman was responsible for several at a time. NextDrop could then identify which households fell in which valve areas based on household GPS coordinates.

Thus, we can see that the valvemen had unique knowledge of the day-to-day workings of the urban water system. They possessed information that was not available to even the chief engineers of the utility. As a result, their cooperation was essential to the success of NextDrop's innovation.

3 From Idea to Pilot to Scale-Up

3.1 Idea Formation at UC Berkeley, USA

After that summer in India, Kumpel, then a PhD student in the Civil and Environmental Engineering Department, took Tapan Parikh's project-based class at UC Berkeley, "Social Enterprise for International Development." Kumpel worked with a handful of other students—Ari Olmos, Thejovardhana Kote, Matt Gedigian, and Niranjan Krishnamurthi—to develop a technology that sent households text messages about their water supply status. The team settled on calling the system "NextDrop," inspired by an app called "NextBus." The team won a prize for the best project in the class. They eventually submitted an $8000 USD proposal to the Big Ideas Contest at UC Berkeley, an initiative providing funding along with training and contacts to help support student-led innovations. All this led Kumpel to do initial groundwork for a pilot in Hubli-Dharwad.[2]

3.2 Pilot in Hubli-Dharwad, India

The twin cities' area of Hubli-Dharwad is the second most populous urban center in the state of Karnataka, with a population of just under a million people by the time NextDrop started. Rapid population growth had put pressure on the municipal water

---

[2] Before this, Kumpel decided to conduct her dissertation research in Hubli-Dharwad. From 2007 onward, Kumpel and other UC Berkeley researchers established relationships with the municipality, so that when the NextDrop team put together its first business plan and won a grant in 2009, it could leverage this relationship to launch its system in Hubli-Dharwad.

distribution system operated by the Karnataka Urban Water Supply and Drainage Board. Most residents received piped municipal water intermittently, from every other day to once a week, with a median of every 5 days (Ray et al., 2018).[3]

By the time NextDrop began in Hubli-Dharwad, over 50 valvemen managed more than 800 valves, supplying water to various neighborhoods at different times across the twin cities. The utility announced water schedules in local newspapers, but they were frequently out of sync with actual water delivery times. Utility workers coordinated water delivery with one another mainly through ad hoc mobile phone calls and field visits. This, along with electricity outages and pipe breaks, led to not only intermittent but also unpredictable water supply timings. Households did not know when water would turn on and for how long the supply would last. For some (mostly higher-income) residents with large storage tanks, this may have not been a major inconvenience. However, water supply intermittency coupled with unpredictability can lead to high wait times, missed work or social events, and water-related stress for low-income residents with little water storage capacity (Kumpel et al., 2012).

In 2010, Kumpel spent almost a year in Hubli-Dharwad with the NextDrop team, setting up the first iteration of their innovation. Through her dissertation research, she had already established a solid working relationship with the chief engineer of the water board, who wanted to both address the problem of intermittent water supply and implement a novel mobile phone system. Ari Olmos and Anu Sridharan, a UC Berkeley undergraduate at the time, traveled to India to pilot the system. In 2012, after much discussion and with Sridharan as chief executive officer, NextDrop became a for-profit company, giving the organization greater opportunities for funding and tendering by the government. NextDrop won several awards, including the Clinton Global Initiative Award for Outstanding Commitment, the GSMA mWomen BOP App Challenge, and the Knight Foundation News Challenge.

NextDrop's first system was primarily manual and relied on households to report when they received their water supply. In this first iteration of the product, Kumpel worked with the Deshpande Foundation to collect mobile numbers from households, requesting customers to call in when they received water. Whenever NextDrop received a call from any household about water supply, they would use a spreadsheet to keep track and send messages to the households in the same valve areas. In later iterations, the valvemen would send notices via NextDrop's interactive voice response system when they were about to adjust a water valve, indicating (1) "advance notice" an hour or less before opening a valve, (2) "valve opening," (3) "valve closing," or (4) "last-minute changes" when water supply was interrupted. NextDrop would then send notifications to customers' mobile phones via voice or text messages based on household GPS coordinates. As the municipality did not supply funding or administrative capacity for their system, NextDrop recruited

---

[3] From 2007 to 2008, the Board installed new pipes with continuous water supply through a World Bank-funded project. However, only 10% of consumers received this 24/7 provision, and even then, there have been reports of service interruptions (Ray et al., 2018).

valvemen directly, who, for the most part, seemed content to contribute to an innovation that could potentially reduce the time they spent on fielding complaint calls from households.

NextDrop also implemented various incentives to encourage valvemen to use their system. The more often a valveman called NextDrop, the more they accumulated points for a list of items they could win. This worked so well that valvemen worked together to combine their points for large ticket items, such as televisions. NextDrop also recognized "Valvemen of the Quarter." During this pilot stage, NextDrop staff members worked closely with the valvemen, being considerate of their personal needs and providing a sense of camaraderie with NextDrop. Trust was built. According to NextDrop team members, this worked well for the valvemen and the company alike.[4]

In 2012, after a month of close monitoring, NextDrop found the valvemen's data to be relatively accurate for "opening" and "advance" notifications (Kumpel et al., 2012). They also found that advance notices reached customers 20 to 40 min before they received water supply, close to NextDrop's ideal 30-minute target. Customers reported being able to go home or arrange water collection by family or neighbors in time for the start of the water supply due to NextDrop's mobile messages. By 2012, NextDrop had over 7000 customers across 80 valve areas in Hubli-Dharwad. They offered new customers a free month of service and slowly rolled out a payment system with text messages for 10 INR ($0.20 USD) per month or voice messages for 15 INR ($0.30 USD) per month. Most of NextDrop's customers were middle- to high-income and opted for the text messages; however, the company hoped eventually to include more low-income customers who might have preferred voice messages due to lower literacy levels.

Though NextDrop collected fees from their customers in Hubli-Dharwad, these funds were not sufficient to cover costs. In order to find new funding streams and reap economies of scale, they decided to expand to a new city. They chose the state capital, Bangalore, a city over eight times the size of Hubli-Dharwad. As excitement and potential funding for the Bangalore rollout grew, NextDrop diverted most of its resources and personnel to the new site, eventually leaving operations in Hubli-Dharwad to run primarily on autopilot.

3.3 Scale-Up in Bangalore, India

Growing from 5.8 million people in 2001 to 8.6 million in 2011, Bangalore is considered the Silicon Valley of India. Though known in international water policy circles for providing extensive water coverage and achieving highly successful fee collection, piped water delivered by the Bangalore Water Supply and Sewerage Board (BWSSB) is still, like in Hubli-Dharwad, intermittent (McKenzie & Ray,

---

D. Miller, personal communication, January 30, 2020

2009). The utility's public website displays a water arrival timetable showing that neighborhoods can expect their water at varying intervals across a range of time—from once a week to everyday. For some locations, specific water arrival times are not given; for others, specific times are given but not specific days, leaving supply unpredictable for households.[5]

NextDrop initiated its Bangalore rollout by signing a memorandum of understanding (MOU) with the BWSSB in 2013. This agreement did not provide any funding from the utility, but it did provide NextDrop with some authority over the valvemen: they and their supervisors now had to comply with NextDrop's system. At the time, the BWSSB divided the city into 32 subdivisions. NextDrop initially rolled out in "Subdivision A," a pseudonym we use to protect the identities of the valvemen among whom we conducted our ethnographic study. At the time, Subdivision A had five service stations with two to four valvemen at each station. Each valveman was assigned 20 or more valve areas, and each valve area covered anywhere from 20 to 200 households.

From the start, NextDrop staff knew that the valvemen in Bangalore might be reluctant to comply with their notification regime. During its pilot in Hubli-Dharwad, NextDrop had garnered a workable level of cooperation from the valvemen. The company did not keep detailed data on incentive-specific performance but believed that their combination of individual and social incentives in Hubli-Dharwad had been effective. Scaling up this highly personalized incentive system to the megacity of Bangalore did not seem feasible, however, so NextDrop relied on the BWSSB's official hierarchy to encourage valvemen to submit the required data. In other words, the valvemen's supervisors were required to ensure that they sent notifications to NextDrop under the MOU. In effect, the Bangalore rollout substituted the reliance on individual incentives and personal goodwill for reliance on the utility's organizational structure—arguably a more scalable proposition.

At the same time, NextDrop continued to make changes to its technology, led by their chief engineer, Devin Miller. In Hubli-Dharwad, the messaging system was in Kannada, the primary language of the state. Bangalore is much more cosmopolitan, requiring multiple languages, such as Tamil and Telugu, in addition to Kannada, so these languages were added. NextDrop also created an online customer registration system, converted their spatial data to Google Maps, and added more specific information to customer notifications (e.g., power outages as a reason for supply delays).

---

[5] This information was retrieved on December 21, 2020, from the BWSSB's Water Supply Timing website (https://web.archive.org/web/20201221210514/https://www.bwssb.gov.in/watter_supply.php)

4 Research on the Scale-Up

In 2014, our team of researchers from UC Berkeley led by Professors Alison Post (Political Science) and Isha Ray (Energy and Resources Group) began an evaluation of NextDrop's innovation. Post researches urban water politics and service delivery in developing countries. Ray researches water and sanitation in developing countries and was one of the original authors for studies on the World Bank's continuous water supply demonstration in Hubli-Dharwad (Ray et al., 2018). Our evaluation consisted of two studies. The first was a randomized controlled trial to assess the effects of NextDrop's innovation on household welfare. The second was a mixed-method study aimed at understanding the NextDrop system from the water valvemen's perspectives. PhD students Tanu Kumar (Political Science) and Christopher Hyun (Energy and Resources Group) joined the team to work on the household evaluation and valvemen studies, respectively.[6]

4.1 Study 1: Impact Evaluation[7]

We conducted the household impact evaluation from June 2014 to November 2015. This evaluation focused on three main sets of outcomes at the household level: the costs of unpredictability, psychological effects, and political attitudes toward the government and the water utility. The costs of unpredictability included any costs associated with waiting for water, including the time spent, missed events (such as family gatherings) or missed work, missed or delayed religious observances (such as the need to wash in fresh water before Friday prayers for many Muslim families), and purchase of substitutes for piped water. Psychological effects included reported worry or stress about water arrival times or water running out. Political attitudes referred to citizens' perceptions of the BWSSB as a water provider.

We measured the effects of NextDrop's innovation on these outcomes for 3000 households using two waves of survey data on either end of the randomized allocation of NextDrop's services in the BWSSB's E3 subdivision. Overall, we detected null effects on all outcomes—except for stress related to household water management, for which we detected modest beneficial effects. We believe that these null-to-modest effects can largely be explained by the failure of accurate information about water arrivals to reach customers.

---

[5] Our research team was also advised by several others who had worked with NextDrop in Hubli-Dharwad, including Zachary Burt, Kara Nelson, and CS Sharada Prasad. This work was funded by a grant from UC Berkeley's Development Impact Lab (USAID Cooperative Agreement AID-OAA-A-13-00002), which aims to help students and researchers scale up and evaluate science and technology innovations for development.
[7] See Kumar et al. (2018) for full details on this study.

4.1.1 Site Selection

Our study site, E3, is a BWSSB subdivision near Bangalore's eastern periphery. To minimize interfering with NextDrop's operations and to keep subdivision-level characteristics constant across households, we decided to conduct the evaluation within a single subdivision. We selected E3, in consultation with NextDrop and the BWSSB, for two reasons. First, it was not among the subdivisions scheduled for immediate expansion of NextDrop's services, making possible the selection of control group households who would not receive NextDrop's services during the study period. Second, field visits and census-level data revealed that this was a subdivision with many low-income settlements and relatively intermittent water supply, suggesting this was an area where NextDrop's services might be particularly impactful.

4.1.2 Sample Selection and Randomization

Within areas receiving piped water in E3, we surveyed households and randomly enrolled them for NextDrop's notifications (the treatment condition) or not (the control condition). A concern common to randomized studies of information interventions such as this one is that treatment households might share information with control households. To reduce this threat to causal inference, we used cluster randomization and assigned households to treatment or control status in groups determined by their geographic location. These groups were separated by at least two streets to prevent individuals in our treatment and control groups from sharing information. Additionally, clusters were usually in different valve areas, meaning that it was unlikely for individuals across clusters to share information because water arrival times are relevant only to individuals within the same valve areas.

This cluster randomization was also stratified, or blocked, by neighborhood socioeconomic characteristics. Blocking increased the statistical precision of the study and allowed us to look for subgroup effects among the low-income households for which we predicted the effects of the information would be largest. Based on extensive site surveys, we designated blocks to be either low-income or mixed-income. Each block comprised four clusters (following Imbens, 2011) that we expected to be similar in terms of socioeconomic variables and the underlying water infrastructure. We randomly assigned two clusters in each block to receive treatment and two to the control condition.

Overall, the study included 3000 households, with the sample size calculated to detect a 30- to 45-min average reduction in the time spent waiting for water. The total sample had 10 low-income and 20 mixed-income blocks (Fig. 17.3), with 120 total clusters of 25 households each.

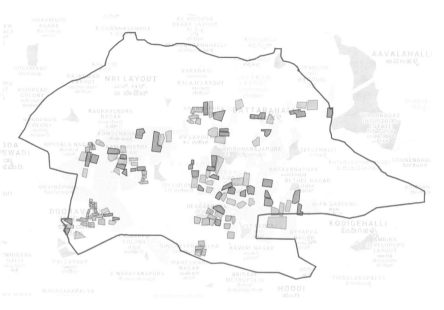

Fig. 17.3 The BWSSB E3 subdivision boundary is shown in blue. Areas receiving piped water supply are denoted in yellow. Green and purple polygons denote low-income clusters (treatment and control); black and red polygons denote mixed-income clusters (treatment and control). There are four clusters per block. (Adapted from Kumar et al. 2018)

4.1.3   Two-Wave Survey and Enrollment into NextDrop Services

In early 2015, we worked with a Bangalore-based survey firm to conduct a baseline survey of the 3000 study households. Enumerators contacted every third household in a cluster until a quota of 25 households was reached. They then administered a survey that we designed to obtain measures of our main outcomes and other variables of interest. Enumerators surveyed the individuals who reported managing and storing water for the household. In line with the expectation that women are generally responsible for managing water, 80% of our respondents were women.

The enumerators next offered all households the opportunity to enroll in NextDrop services by submitting their mobile phone numbers. They could sign up for text or voicemail notifications in English, Kannada, Telugu, or Tamil. Because surveys were conducted on tablets that collected GPS coordinates with 5-meter precision, NextDrop was able to correctly identify the valve areas in which households lived. NextDrop then enrolled the households in our treatment group clusters following the completion of the baseline survey. In October and November 2015, about 4 months after NextDrop had begun sending messages to customers in E3, we surveyed the study households once more. NextDrop and the research team estimated that 4 months was enough time for enrolled households in our treatment group to adapt to using the notification system and for potential impacts of the service to occur. The control group was enrolled after the completion of the study.

4.1.4   Results

Overall, we were unable to detect effects of NextDrop's intervention across our study population for any of our outcome variables, aside from those related to worry or stress. Even though our sample size would have enabled us to detect a decrease in wait times for water as small as 9 min, we observed no effects on time spent waiting for water, the likelihood of missing work or community events while waiting, or using substitutes for piped water. We similarly failed to detect effects for households living in low-income clusters or those we identified as NextDrop's target customers, namely, low-income households without automatically filling water tanks. We detected no effects on attitudes toward the utility, but this could be because of already positive attitudes at baseline.

Our baseline survey confirmed that E3 was the kind of site in which the notifications "should" have been impactful; our data indicated that over 85% of residents received water services just once or twice a week and 28% did not possess an automatically filling water tank. About 43% reported that they simply learned that water had arrived when it began to come out of their taps, rather than knowing when to expect it. Respondents spent roughly 1 h per supply day waiting for water, which they agreed led to missed work and community or family events. As we discuss below, we believe these null effects are a result of few customers receiving accurate and timely information about their water supply.

4.1.5   An "Information Pipeline" to Diagnose the Null Results

We created a framework consisting of each of the steps in the process of producing and disseminating information to understand our null results (Fig. 17.4). First, someone must collect the relevant information in full. Second, someone must analyze the information or compile it into a usable format. Third, someone must disseminate the information. There could be poorly aligned incentives or technical difficulties preventing any of these first three steps from happening. Organizations often cannot control those charged with executing their assigned tasks, and employees can exercise significant autonomy.

Fourth, this information must actually reach the intended recipients. Messages may go to the wrong person because phone numbers change or because the intended recipient does not have control over the household mobile phone. Fifth, these recipients must actually register that they received a message. Information may be sent in the wrong language, or go unnoticed if a recipient is inundated with messages, or be so useless that the recipient stops paying attention. Finally, the information must be accurate. Inaccuracies could reflect deliberate efforts to conceal information, carelessness, or some inability to measure the relevant information.

We identified several "leaks" in this information pipeline for NextDrop's model in Bangalore. First, valvemen had to submit information by calling NextDrop's automated interactive voice response system to log valve opening times. Yet logs

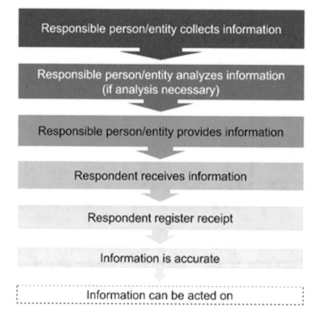

Fig. 17.4 Information pipeline: from collection to action. (Adapted from Kumar et al., 2018)

of valvemen notifications to NextDrop show that they sent reports to NextDrop only 70% of the time.

Furthermore, even while NextDrop passed on all of these messages to customers, only 38% of treatment group members reported receiving notifications. We argue that these low levels of information receipt can be explained by two factors. Even among the 854 households that were regularly sent notifications, 207 household "waiters" for water, usually women, did not have daytime access to the mobile phone registered with NextDrop; a male member of the family usually took the phone to work. Second, many respondents may simply have not noticed NextDrop's texts or voicemails, which may have become buried under the many solicitation and informational messages regularly received via SMS.

Finally, only 289 of our 1193 treatment group respondents reported that the information received in NextDrop's notifications was either always or usually accurate. Moreover, when we compared household survey responses about the last day they had received water to the valveman reports, we found that over one-third of households reported receiving water on a different day than that reported by the valveman. In fact, 62% of households received "advance" notifications after their water had already arrived. Figure 17.5 shows each of the steps preventing information from reaching our treatment groups. We can identify two "human" factors that are particularly important here. The first is intra-household dynamics that prevent the individual responsible for waiting for water from accessing the household mobile phone. The second is the valvemen's lack of cooperation in

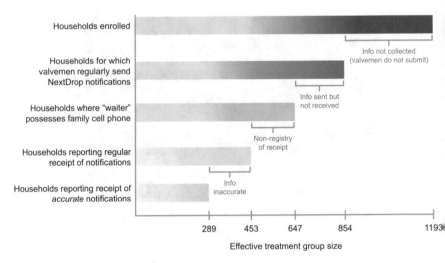

Fig. 17.5 NextDrop's leaky information pipeline and treatment group attrition. (Adapted from Kumar et al., 2018)

sending accurate information to NextDrop to be disseminated. Our second study explores the extent to which, and why, this latter dynamic was occurring.

4.2   Study 2: Valvemen's Role and "Compliance" in NextDrop's System[8]

Why were the valvemen not sending notifications when they were supposed to, and why were these notifications often inaccurate? In a parallel study in a different part of Bangalore, we examined factors that could contribute to the valvemens' willingness to comply with NextDrop's system. They were not sending notifications to NextDrop, as they were expected to do. So why did the valvemen not comply? We framed our second study in terms of three levels of factors that may have contributed to valvemen compliance: organization-, community-, and individual-level factors. In Fig. 17.6, we measured "compliance" by comparing actual to expected notifications for each valveman in Subdivision A, using NextDrop's database. Valvemen had low to moderate levels of compliance overall, but we also observed that the ratio of actual to expected valve opening notifications varied within the same subdivision (between 0.4 and 0.8) and even varied within the same service station (or team). This revealed that, when controlling for management, valvemen still complied at different rates, indicating that the characteristics of individual valvemen or of the

---

[8] See Hyun et al. (2018) for full details on this study.

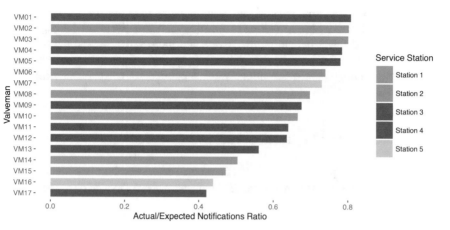

Fig. 17.6 Notification compliance: actual/expected valve opening reports per valvemen, August to December 2014, Subdivision A. (Adapted from Hyun et al., 2018)

different community neighborhoods within which they worked may explain why they did not comply with NextDrop.

4.2.1   Study Design to Understand Compliance Levels

Our field observations focused on Subdivision A because NextDrop's relationships with the valvemen were strong there, and it was relatively distant from our household evaluation study area in E3. We used an ethnographic approach to understand why valvemen complied with NextDrop's system at low rates. Our methods included interviews with valvemen, community residents, and staff members of the water utility and NextDrop. Out of the subdivision's 17 valvemen, we chose 9 for close observation, representing a variety of compliance levels. We followed the valvemen on their daily rounds, on breaks, and even into their homes, taking extensive notes and geolocated photographs of the neighborhoods that they serviced. Throughout our interviews and observations, we thought of the valvemen as "street-level bureaucrats" (following Maynard-Moody & Musheno, 2000, 2012) and encouraged them to share their personal narratives about their own jobs and their relationship with NextDrop.

We also created a dataset on the characteristics of the valvemen and the valve areas in which they worked. Through semi-structured interviews, we collected data on the number and gender of the valvemen's children, their wives' employment type, the vehicles they used for work, their ages, and the numbers of valves for which they were responsible. We also visited all 233 valve areas serviced by the 9 valvemen and collected data on the socioeconomic status (SES) of the neighborhoods, water infrastructure, and street activity, with most observations taking place around the same time during weekdays. We observed that "low" SES valve areas had narrow roadways, few trees, high noise levels, high numbers of people on the streets, and high domestic activity (e.g., cooking or washing dishes on the streets). "High" SES valve areas had fewer people on the streets and more water storage tanks, which implied less work for the valvemen. We ran linear regressions to establish whether or not our ethnographic observations were supported by correlations within our broader dataset.

4.2.2 Findings that Explained Overall Low Compliance

We found three explanations for low compliance with NextDrop's system: (1) the valvemen's views of their own jobs, (2) their overworked days, and (3) the insider knowledge valvemen had that gave them some level of job security.

First, in the water utility's view, the valvemen's job description was simple: listen to utility supervisors and adjust water valves as needed. However, valvemen perceived their job as responding to "the public" rather than merely to the utility's hierarchy. Their overriding sentiment was: "My main work is working with the public." The pressure from the public was so strong that another valveman claimed: "When I work I forget about my family and friends. These people are my family and friends."

In public administration literature, this phenomenon is discussed as the tendency of street-level workers to "cope toward" citizens, seeing themselves as working for ordinary citizens rather than just for the government (Tummers et al., 2015; Maynard-Moody & Musheno, 2000). We saw this when observing what valvemen do when water supply was not on time—the very situation in which NextDrop's text messages would be needed. If water arrived late, the conventional utility practice dictated that the valve area would be given less time, so as not to hold up water supply for subsequent areas. Valvemen, however, would not always follow this rule: "If I'm supposed to give them an hour of water, and due to power cuts they only get a half hour, then I will give them another half hour." Furthermore, residents would often complain to valvemen, and residents were not asking for NextDrop's text messages; they wanted water. We observed that valvemen routinely prioritized their time to meet residents' actual requests over sending notifications to NextDrop.

Second, though it took valvemen less than a minute to send a notification to NextDrop, even this small act was too much when they felt overworked. Many of them worked non-stop, constantly negotiating with residents, utility supervisors, and politicians on the phone and in person while moving from valve to valve. As a consequence of their low wages, they also made time for other income-generating activities, such as side plumbing jobs. If their wives had jobs outside of the home, the valvemen would also tend to domestic work. Since they knew the water system well, they would also be called in to assist with pipe breaks at night. Though some valvemen appreciated NextDrop's service, others found it annoying: it is "not helpful for valvemen"; "It's just an additional job." During these constant negotiations, moonlighting work, and moving from neighborhood to neighborhood, we rarely saw valvemen sending notifications to NextDrop—even though they knew this was why we were observing them. Many were just overwhelmed.

Finally, as mentioned earlier, valvemen had privileged knowledge of the water infrastructure and therefore did not take threats of dismissal from the utility seriously. NextDrop relied on their MOU with the water utility to get the valvemen to submit notifications, but the valvemen's tacit knowledge of the city's water system was highly valuable, and the utility would have been hard pressed to fire a valveman for not complying with NextDrop's system. On-the-ground understanding of the valve areas is passed on from valveman to valveman, bypassing the head engineers. When we questioned why one valveman spoke rudely to NextDrop's staff and even to his own utility supervisor, he was unconcerned: "I don't worry about being fired."

NextDrop had initially implemented positive incentives in Bangalore, as they had done in Hubli-Dharwad. They offered the valvemen prizes, such as phones and free mobile talk time. NextDrop could afford these prizes, but they were not motivating enough for valvemen who lived in a city with a higher cost of living than Hubli-Dharwad. For the Bangalore valvemen, the incentives did not match the extra work they had to do for the company. NextDrop also celebrated the "best" valveman, but this resulted in skepticism about the company's ranking system and led to social divisions. When NextDrop eventually secured an MOU with the water utility, the company was relieved; instead of expensive and ineffective incentives, they thought they could rely on the utility's hierarchy to keep their system going. However, as we saw, this was not enough to get valvemen to comply.

4.2.3 Findings that Explain Variation in Compliance

As explained above, overall compliance with NextDrop was low; however, we also uncovered explanations for why compliance varied across individual valvemen through our interviews and observations of the valvemen at work. We observed that valve area characteristics and a valveman's family circumstances could explain differences in compliance levels.

There were often stark community-level differences between middle-class and low-income valve areas, which influenced how the valvemen worked in each area. First, valve areas differed in terms of infrastructure. Middle-class areas had wide, paved roads, making it easier for valvemen to navigate. On the other hand, low-income areas had narrower and sometimes unpaved roads, with chickens and dogs that had to be avoided. Middle-class homes had more water storage tanks with maintenance holes outside of the home, while valveman often had to enter lower-income homes to check water pressure in their faucets, leading to greater inconvenience for both the households and the valvemen.

Second, valve areas differed in terms of face-to-face interactions with residents. In low-income areas with less household water storage capacity, residents stayed near their homes to wait for water. Sometimes residents stood outdoors with buckets, dirty laundry, and dishes ready as soon as they saw the valveman in their neighborhood. Such residents often confronted valvemen with complaints. According to the valvemen, this was a major difference across valve areas: "The higher class people call our superiors and the superiors tell the valvemen the problem. The lower class people come to me directly, and I have to explain to them directly....I lose a lot of time talking to people." This pushed NextDrop notifications to the bottom of the work list in low-income valve areas.

Along with these differences in community-level characteristics, the valvemen's individual characteristics affected levels of compliance, particularly in relation to their personal finance and family circumstances. More children indicated greater financial need, and this was especially apparent for girl children. Daughters come with the extra financial burden of (future) marriage dowry expenses, and this was on the minds of less compliant valvemen, who had to borrow money for dowries: "We ask our relatives for help—if you help us now, we'll help you when your daughters get married." These valvemen had side jobs, such as driving and plumbing, which diverted them from sending notifications to NextDrop. If their wives worked outside the home in inflexible jobs, such as babysitting and dishwashing, the valvemen sometimes went home in the middle of the work day to take care of a sick child or do other work (e.g., bring in the family's hanging laundry before it rained). A wife's occupation, then, indicated financial need and could mean more domestic responsibilities for the valvemen.

Our regression analysis supported our ethnographic observations of the valvemen's individual characteristics. We saw significant correlations between the number of children and levels of individual compliance, even in our small sample size. One additional child was associated with a 7% decrease in compliance. If that child was a girl, there was an 11% decrease. We also saw decreases in compliance related to the job type of a valvemen's wife. Though we created a dataset for 233 valve area communities, we could shadow only 9 valvemen, so we do not claim causality in this case. However, we see these results as hypotheses to pursue further explorations of frontline workers' compliance with new workplace expectations.[9]

5 Subsequent Status of NextDrop[10]

While we conducted our research in 2015, NextDrop made plans to roll out in different cities in order to expand their operations. During our study period, NextDrop visited other smaller cities in the state of Karnataka, as well as the cities of Varanasi and New Delhi. This was all done while operations in Hubli-Dharwad were slowing down and the rollout in Bangalore was still incomplete. Unilever decided to fund a new project in nearby Mysore, a city slightly smaller than Hubli-Dharwad.

---

[9] See Hyun et al. (2018) for more details on the analysis.

[10] This section is based on personal communication with A. Sridharan (February 3, 2020) and D. Miller (January 30, 2020).

To improve their data collection, NextDrop experimented with using smartphones in Mysore, creating a smartphone app that would collect data from the turning of a valve key—the large, metal T-shaped tool that valvemen use to turn water valves on. Valvemen would stick the smartphone to the valve key using Velcro, and, by employing the GPS and motion sensors already in the phone, algorithms would determine if a specific valve was opening or closing. This was an innovative idea but difficult to implement in practice. Threading on the valves was inconsistent and so were valve angles, smartphone placement on the key, and network connection.

Ultimately, it became an uphill battle for NextDrop to find venture capital for a valvemen-based call-in system. Raising investor funding on this idea became unsustainable, and our studies showed that households were not benefiting from the intervention, due in large part to dependence on the valvemen. NextDrop needed to reconsider their business model and technological approach. This, combined with struggles with personnel on the team, led NextDrop's leaders to downsize the company.

In 2016, the NextDrop team went from 20 to 5 employees. In Hubli-Dharwad, they let the system continue to run without oversight until 2019, when they officially shut it down. In Bangalore, NextDrop agreed to map out the rest of the city and register all of the remaining valvemen into the NextDrop system for the BWSSB. NextDrop shifted their maps to the BWSSB's servers even though valvemen were no longer sending in data. In 2017, NextDrop submitted a proposal to the BWSSB for a new system with sensors in major valve areas. Engineers at NextDrop had considered the use of a sensor-based system since day 1; however, it was cost-prohibitive. As of early 2020, the BWSSB had not floated a tender for the new project nor had any funding materialized from the utility. In the meantime, the heads of NextDrop, Sridharan and Miller, moved on to different water-related projects in India.

5 Key Takeaways for Development Engineers

In this chapter, we traced the evolution of a technological innovation designed to reduce the burden of unpredictable water supply, which affects millions of households around the globe. Climate change and the growth of cities will only exacerbate the problem, making it even more critical to study and mitigate such burdens. While NextDrop's water notification system proved unsuccessful at scale, it provides key lessons for development engineers, especially those working in the water or Information and Communications Technology (ICT) sectors.

Our main takeaways are best stated by NextDrop's own leaders: "We were asking the wrong question. We were asking, 'Can this [technology] work?', when we needed to ask, 'Is it worth it?'" says Sridharan, who, as of early 2020, continued as the CEO of NextDrop. "Worth" emphasizes the value of their innovation to people. In other words, even the most efficient technology must respond to felt needs in order to be successfully deployed at scale. There are needs across the human ecosystem of a technological intervention. Below we summarize three types of human needs to consider: (1) customer and end user needs, (2) needs of human intermediaries involved in implementation, and (3) changing needs across contexts during scale-up.

In our NextDrop study, we found that not everyone who actually received water supply messages on their phones acknowledged that they received them, suggesting that a number of customers felt no need to read NextDrop's messages. Devin Miller, head engineer at NextDrop, emphasizes the importance of market research, saying "Spend a whole year on market research if you need to! If you talk to a hundred customers to understand the problem, and they all say they want it, then you're good. But if they don't understand or don't say they want it, then you're going to have a hard time."

Technological interventions that aim to help women in particular will benefit from market research on gender roles or household bargaining structures. In many low- and middle-income countries, women may have different roles of domestic labor, access to resources, and less decision-making power than men. In our study, for example, we found that women tended to be responsible for waiting for water, but they did not have regular access to the phone to which water supply notifications were sent. Roessler et al. (2020) similarly find that, in a smartphone intervention for low-income women in Tanzania, over 40% of the intended beneficiaries reported that the main user of the phone was not themselves, but rather their husbands or sons.

Furthermore, it is important to assess how useful an intervention is to end users not only when delivered perfectly but in its actual, imperfect state. For NextDrop, this would have meant understanding the usefulness of providing information even when it was not completely accurate. Development engineers should not only continually refine their technology, but they also should invest in continual refinement of market research in order to evaluate the utility of their intervention as it evolves.

Incorporating the end users themselves, such as women residents, across the phases of a project may improve design and implementation overall. The Global Clean Cooking Alliance, an organization that delivers cleaner-burning cookstoves in low- and middle-income countries, aims to involve women from target countries in all phases of their projects, including design, production, distribution, promotion, investment, education, and product servicing. This practice increases the probability of creating a product that is actually useful and attractive to the intended end user.

Along with end users, it is also important to know what the innovation is worth to the people who help implement the innovation (e.g., the utility's employees in our study). Our research showed that human intermediaries, like the valvemen, are often ignored in business models or their needs are oversimplified. Often technological interventions, particularly ICT interventions, aim to eliminate the need for intermediary workers altogether. As we show, however, the intermediaries and the "last-mile" delivery for which they are responsible can be key to the success of an innovation.

For example, the valvemen in Bangalore focused mainly on providing water to the public, and sending data to NextDrop seemed unimportant to them in comparison. NextDrop could have worked more on convincing valvemen of the connection between their data input and helping the public. It could have been helpful for valvemen to meet satisfied NextDrop customers, either in person or through video, and get their feedback. The core problem here, however, is that neither the utility nor NextDrop understood the valvemen's jobs the way the valvemen understood them; what the company saw as non-compliance, the valvemen saw as compliance to their greater task—helping the public.

Relatedly, intermediaries are important sources of market information. Instead of attempting to simply align incentives, engineers and designers should understand that such intermediaries have insight into on-the-ground realities, which can hint at what is "worth" doing or not. In our study, we observed on the ground that the valvemen were "complying," i.e., actually helping customers, but in ways that the utility possibly did not sanction or would not approve of. Development engineers should account for their perspectives and needs along with those of the end users, especially when planning to scale up.

Lastly, it is crucial to consider differences in context when scaling up innovations, especially in terms of people's needs. When a pilot is deemed to be successful in one location, implementers and funders often decide to replicate it in another location. Yet, both the process for creating the innovation and its effects may vary from location to location. Indeed, even within areas of the same state and country (i.e., Bangalore and Hubli-Dharwad), differences in context were a key aspect of the difficulties NextDrop faced. In both cities, valvemen had similar job descriptions, but NextDrop found that incentives that seemed to work well for Hubli-Dharwad valvemen did not work well for those in Bangalore, reflecting differences in their needs. "Human behavior is really hard to change," Kumpel reflected, underscoring the key challenge of NextDrop's system in Bangalore.

The potential effects of the intervention also varied across the two locations. While customers in Hubli-Dharwad seemed to find the innovation useful, the relevant individuals in Bangalore were, for reasons of intra-household dynamics, prevented from even receiving the messages. Moreover, a 30-min advance water notification was actionable in Hubli-Dharwad, but in Bangalore—a larger city with traffic congestion—household members may have needed longer to reach home before their water was turned on. This longer interval between notification and water services was difficult to achieve with NextDrop's valveman-based system. It seems, therefore, that in this new context, the end users' needs could not be met.

All of these lessons ultimately point to the importance of not just understanding the technology that one designs but also the human ecosystem in which it is deployed and used. Kumpel believes that, in order to understand the human ecosystem, designers must know the context personally. She says, "You have to move there! There is no way NextDrop would have gotten off the ground without moving there." Reflecting on the NextDrop experience, Sridharan added another important point: "A lot of tech startup manuals and guides are based on companies that are based in Silicon Valley [USA], but the constraints faced by the customers and employees of places like PayPal and Dropbox are very different from those we saw." Understanding the diverse aspirations, constraints, and priorities of key stakeholders is fundamental to the success (or failure) of technology-based innovations designed to solve large-scale development challenges.

Discussion Questions

1. What types of information might have been helpful for NextDrop to collect prior to rolling out their services in Hubli-Dharwad? In the megacity of Bangalore? How would you design a process for collecting information over time as the company grew and developed?
2. How would you design a preliminary needs assessment to understand (a) whether women would have access to NextDrop notifications and (b) whether they really needed it? What information would you consider collecting (e.g., how women spend their time, their water use, how regularly they leave their home, and what you would want to know about these trips)? Consider why each type of information may or may not be useful.
3. How would you design an information collection process to learn about how women used (or might use) the innovation? Why might simply surveying women not be effective here?
4. What type of data collection and ongoing monitoring could NextDrop have performed to better understand the valvemen that were so critical to their system? What sort of information could they have collected on a continual basis about "compliance," and how could they have analyzed these data? What sorts of meetings with and/or programs for the valvemen might have improved valvemen's performance or led NextDrop to reconsider the way it worked with them?
5. How can we guide development engineering toward thinking about human systems first, allowing the technological interventions and/or designs to follow? How does NextDrop's experience illustrate the distinction between systems and interventions?

References

Hyun, C., Post, A. E., & Ray, I. (2018). Frontline worker compliance with transparency reforms: Barriers posed by family and financial responsibilities. *Governance, 31*(1), 65–83.
Imbens, G. W. (2011). Experimental design for unit and cluster randomized trials. *International Initiative for Impact Evaluation Paper.*
Kaminsky, J., & Kumpel, E. (2018). Dry pipes: Associations between utility performance and intermittent piped water supply in low and middle income countries. *Water, 10*(8), 1032.
Kumar, T., Post, A. E., & Ray, I. (2018). Flows, leaks and blockages in informational interventions: A field experimental study of Bangalore's water sector. *World Development, 106*, 149–160.
Kumar, T., Post, A. E., Otsuka, M., Pardo-Bosch, F., & Ray, I. (2022). *From public service access to service quality: The distributive politics of piped water in Bangalore.* World Development 151, 105736.

Kumpel, E., Sridharan, A., Kote, T., Olmos, A., & Parikh, T. S. (2012). NextDrop: Using human observations to track water distribution. In *NSDR*.

Maynard-Moody, S., & Musheno, M. (2000). State agent or citizen agent: Two narratives of discretion. *Journal of Public Administration Research and Theory, 10*(2), 329–358.

Maynard-Moody, S., & Musheno, M. (2012). Social equities and inequities in practice: Street-level workers as agents and pragmatists. *Public Administration Review, 72*(1), 16–23.

McKenzie, D., & Ray, I. (2009). Urban water supply in India: Status, reform options and possible lessons. *Water Policy, 11*(4), 442–460.

Ray, I., Billava, N., Burt, Z., Colford, J. M., Jr., Ercümen, A., Jayaramu, K. P., Kumpel, E., Nayak, N., Nelson, K. L., & Woelfle-Erskine, C. (2018). From intermittent to continuous water supply: A household-level evaluation of water system reforms in Hubli–Dharwad. *Economic and Political Weekly, 58*.

Roessler, P., Carroll, P., Myamba, F., Jahari, C., Kilama, B., Nielsen, D. L. (2020). *The economic impact of mobile phones on low-income households.* Working Paper.

Tummers, L. L., Bekkers, V., Vink, E., & Musheno, M. (2015). Coping during public service delivery: A conceptualization and systematic review of the literature. *Journal of Public Administration Research and Theory, 25*(4), 1099–1126.

Part V
Digital Governance

Arman Rezaee (iD)

Introduction

Governments are pivotal to economic development. But the capacity of governments to support economic development varies widely both across and within countries, especially in the case of the low- and middle-income countries (LMICs) that development engineers focus on. As an example, the World Bank's Worldwide Governance Indicators project ranks low-income countries as having an average percentile rank of 17.3 for government effectiveness as opposed to 87.9 for OECD countries (Dal Bo et al., 2017). It is no surprise, then, that some development engineers focus their attention specifically on improving government capacity.

While, at the end of the day, governments in developing countries face a set of problems to which the Development Engineering toolkit can be applied just like any other problem, there are some specific distinctions between governments and private or non-governmental actors. Dal Bo et al. (2017) summarize five key differences. First, governments operate at a longer time horizon than most others. Second, they are limited in what types of contracts they can write for their employees. Elections create very different incentives for politicians than those of a CEO, for example. Third, governments often heavily subsidize the goods and services that they provide, like education and healthcare, making it harder for market forces to ensure proper incentives for workers. Fourth, government careers differ from others in terms of the mission of the organization. Many go into government not to make money but to "make a difference." And fifth, governments often have more checks and balances built into their organization than private firms or NGOs.

The three cases in this chapter involve leveraging digital technology to improve government performance by tackling head-on these unique properties of governments. The first case is about attempts in Afghanistan, Uganda, Kenya, and South

A. Rezaee
Department of Economics, University of California, Davis, Davis, CA, USA
e-mail: abrezaee@ucdavis.edu

Africa to use technology, starting with a simple camera, to introduce greater checks and balances into elections themselves, or to ensure the incentives of those elected are to represent their populace. This case relies on working within the realities of electoral systems. The second case is about how ICT allowed for the collection of accurate and actionable data to improve the incentives for government healthcare workers to show up to work in Punjab, Pakistan. This case relies on understanding the existing post-colonial Pakistani bureaucracy and how it led to information bottlenecks. The third case is about two large-scale bio-authentication programs in India designed to increase the efficiency of massive public benefit schemes. Among many other things, this case relies on understanding the trade-offs between decreasing corruption by government officials and potentially stopping legitimate transactions from occurring.

Despite these differences, all three of these cases were selected because they embody the Development Engineering design process. In all three cases, authors and their collaborators developed an innovative, technology-based solution aimed at government capacity. In all three cases, iterative implementation was key to the solution's success. In all three cases, this success was measured through rigorous, researcher-led evaluation. And, in all three cases, the solution has been adapted to program or policies affecting hundreds of millions of the world's poorest citizens.

At the end of the day, all three of these cases can be thought of as being about increasing government transparency. In the first case, it is about private citizens holding their government accountable from the outside. In the second and third cases, it is about academics and NGOs working within government to increase accountability. Government accountability has become a centerpoint in the push for economic development in the last two decades, beginning most notably with the 2004 World Development Report, Making Services Work for Poor People (Ahmad et al., 2003). This report summarizes the impetus for this chapter:

> Learning from success and understanding the sources of failure, this year's World Development Report, argues that services can be improved by putting poor people at the center of service provision. How? By enabling the poor to monitor and discipline service providers, by amplifying their voice in policymaking, and by strengthening the incentives for providers to serve the poor.

References

Dal Bó, E., Finan, F., Folke, O., Persson, T., & Rickne, J. (2017). Who becomes a politician? *The Quarterly Journal of Economics, 132*(4), 1877–1914.

World Bank. (2003). *World development report 2004: Making services work for poor people* (No. 26895, pp. 1–356). The World Bank.

Chapter 18
Protecting Electoral Integrity in Emerging Democracies

James D. Long

1 Development Challenge

Inclusive political institutions are associated with improved service delivery (Bueno de Mesquita et al., 2002), reduced corruption (Kolstad & Wiig, 2016), and positive economic growth (Acemoglu et al., 2001). Two explanations for these patterns are that democratic governance reinforces political accountability by rewarding or punishing government performance (Barro, 1973), and by articulating the policy demands of the electorate (Manin et al., 1999). Accordingly, competitive elections represent a promising benchmark to guarantee government responsiveness and communicate clear mandates. A country's elections depend on robust participation and outcomes that reflect the will of the people to promote good governance and rule of law as a road to economic development.

But elections can only serve constructive purposes when conducted transparently with fair outcomes. While the holding of multiparty elections has become common across much of the developing world over the last 30 years, many emerging democracies still lack credible electoral institutions and processes. Problems include restrictions to party competition, voter suppression, and blatant fraud.[1] Stories

---

[1] From 1980 to 2010, upward of 70% of all developing country elections registered reports by independent observers of significant problems (Kelley, 2012). The inability of countries to manage elections properly has resulted in global "democratic backsliding" since 2010 (Hyde, 2020). According to Freedom House in 2020, while 115 of 190 (60%) countries qualify as "electoral democracies" (coded as having mostly free and fair elections) and elections technically occur in all but 2 sub-Saharan African countries, only 31% of African countries were rated as having electoral democracy.

J. D. Long (✉)
University of Washington, Seattle, WA, USA
e-mail: jdlong@uw.edu

© The Author(s) 2023
T. Madon et al. (eds.), *Introduction to Development Engineering*,
https://doi.org/10.1007/978-3-030-86065-3_18

abound documenting electoral manipulation in contests ranging from Afghanistan to Venezuela. There are many reasons such cheating persists. First, many election commissions – the managerial body tasked with electoral management – frequently lack the internal capacity and appropriate technology to oversee election day operations and obtain reliable ballot counts. Election infrastructure is also vulnerable to "hacking" of various sorts. Second, politicians often exploit weak legal safeguards to corrupt elections in illicit ways, from vote buying to rigging vote tallies. For their part, citizens often face political exclusion arising from numerous institutional and socio-demographic barriers that limit healthy electoral participation. The growing weaponization of election-related content on digital media further degrades voter mobilization, and a lack of quality elections increasingly undermines citizens' engagement (Norris, 2014). All told, while many developing countries have transitioned to democracy, problems with electoral integrity threaten democratic consolidation. The more governments reflect a corrupted vote process, the less likely its leaders are to pursue the reforms necessary for development.

This case study recounts attempts to confront threats to electoral integrity in emerging democracies. The set of actors critical to the story include electoral commissions, politicians, and citizens along with international organizations, academic researchers, and global publics. While the specific manner in which these groups work to protect or erode democracy forms a central narrative of a country's political cycles – including an array of public and hidden actions – elections are also shaped by a set of intuitive and fairly generalizable plot points derived from the political economy and behavioral dimensions of how some of these actors organize and respond to fraud. Not all electoral actors are corrupt, but enough of them are enough of the time that the quality of democratization over the past few decades falls short of democratic ideals and the public's aspirations in many countries. Within the governance sector, election administration is perhaps unique in requiring some of the most urgent, yet simple, fixes to existing systems. As I describe, these fixes do not require radical technological shifts as much as important shifts in theoretical orientation and reforms to programming.

My personal interest in the topic of election security arose from dissatisfaction with the status quo – of which I found myself a close observer and participant. In 2007, while collecting data on electoral integrity in Kenya for my PhD research, I uncovered fraud firsthand (Gibson & Long, 2009). Rigging claims in that election resulted in sustained post-election violence – upward of 2,000 deaths and 700,000 internally displaced people – and the collapse of a rapidly growing economy (Kanyinga & Long, 2012). Witnessing these events inspired me to study electoral corruption and work with organizations in the governance sector to combat it.

The international community has spent billions of dollars a year over the last three decades in democracy assistance to shore up elections in developing countries. Donor efforts target institutional strengthening through technical assistance to election administrative bodies to ensure accuracy in voting procedures and the deployment of non-partisan observers to oversee electoral processes and outcomes (Bjornlund, 2004). Third-party monitoring of elections should enhance election quality since independent oversight can provide important tools for improving the performance of election bodies (Hyde, 2011). Motivated to contribute positively to these activities, I joined a US Agency for International Development (USAID)-funded mission to support Democracy International's (DI) observation of Afghanistan's 2009 elections. Our team of advisors and monitors visited polling stations on election day, tracked the results, and lent other technical support. But in a manner eerily similar to what I saw in Kenya regarding problems in the tallying and certification of votes, failures of election management and producing accurate counts led to a contested outcome. Vote totals had been improperly aggregated, likely manipulated, and the election commission lacked the proper infrastructure to protect and certify results. Not from a lack of will and despite our presence, international observers could neither guarantee a fair process nor provide the requisite support to improve the election's management. Rather, like many observation missions, we were ill-equipped to anticipate and then action assistance necessary that would have better protected the vote count process.

I quickly learned that these experiences in Kenya and Afghanistan were not unique. Despite the near-universal independent observation of elections in developing countries over the past 30 years, electoral fraud abounds, and the presence of foreign democracy assistance is actually more likely to be associated either with a null or negative impact on election quality than a positive one (Kelley, 2012). I came to realize that the existing model of observation may not always work because the methods employed by these groups have not always taken into account changes in or threats to the electoral environment, evolving methods of hacking, or the strategic response by election workers and politicians to observation. Such missions also do not consistently monitor or audit the results transmission and lack reliable measures of fraudulent activity. These missions also tend to lack comprehensive coverage of polling stations, an important constraint we faced in Afghanistan where the security required for international observers becomes cost prohibitive beyond a handful of stations.[2] While the commitment to improve elections through donor aid remains, these limitations pointed to the need to re-examine aspects of democracy assistance. But after returning from Afghanistan in 2009, I saw that the lessons learned and technical fixes were not yet being implemented ahead of the country's upcoming parliamentary elections slated for 2010.

To address these gaps, I teamed up with a fellow graduate school colleague, Mike Callen, who had also been working in Afghanistan and examining fraudulent vote patterns from 2009. Looking at the evidence from Kenya and Afghanistan regarding changes to vote totals and the failure of other monitoring modalities, we did not see an obvious solution in existence to enhance the credibility of the count. Any improvement would need to confront core aspects of electoral management and the vote aggregation process, including the poor performance of

---

The largest international mission in Afghanistan in 2010 spent about $10 million USD and visited 35 stations due to security concerns (Callen & Long, 2015). While many missions do not report budgets, the European Union says it spends on average $4 million USD per observation mission (European Union, 2006). As calculated in Callen et al. (2016) and applied to the 643 stations the EU observed in Uganda in 2011, this costs on average $6,220 per station.

administrative bodies to conduct counts properly and outside political influence to corruptly change vote totals. No matter what solution we developed, we would be constrained by costs and security that would prevent anything like full monitoring of all activities at all polling stations. One way around this could be to leverage the evolving nature of advances in information and communications technology (ICT) in developing countries. ICT holds potential to better mobilize users – like election observers or citizens – to overcome barriers to participation in public acts and low-cost monitoring of bureaucratic performance, and to collect diffuse information regarding government service delivery. Curiously, prior election observation had often ignored these positive uses of ICT and instead focused technologies in expensive and complex ways without measurable improvements. Our aim then was to harness the reality of Afghanistan's institutional context, understand the (often hidden) threats to election integrity, and incorporate ICT adoption of low-cost forms of monitoring to inspire new ways to overcome the programmatic and technical challenges of previous approaches.

The innovation we developed – "photo quick count" – is a low-cost, ICT-capable, independently managed platform of audited election results performed by citizen monitors that provides polling station level photographic records of tally sheets to compare alongside scanned and certified results from a country's election commission. By obtaining original records of vote results on tallies, the audit detects procedural failures by election officials and aggregation fraud (rigging that occurs in results transmission as observed in Kenya and Afghanistan), including from officials directly changing tally results forms and computer hacking into results transmission systems. Photo quick count also has the ability to deter these administrative problems and corruption by announcing the audit to officials at polling stations. The inception and innovation for photo quick count was supported by DI and USAID's Development Innovation Venture (DIV) grant of phased funding.

Thus far, photo quick count has been conducted by a group of in-country deputized professional citizen monitors that have been activated, managed, and trained by our research team and deployed for project purposes in Afghanistan, Uganda, and Kenya. Recently, we pivoted to broadening adoption and functionality using crowdsourced election monitoring in South Africa with "VIP:Voice," a bespoke ICT platform that recruits dispersed users entirely through ICT channels with no pre-existing infrastructure or direct engagement by our team (other than via the platform). Available across user devices, VIP:Voice allows any citizen to access election-related content and obtains volunteers who agree to report on polling station activities and perform photo quick count.

Through rigorous evaluations, photo quick count and VIP:Voice have documented reductions in aggregation fraud and improvements to a variety of methods of citizen participation to improve electoral integrity. They further scientific knowledge and the evidence base on instruments for policy guidance on the mechanisms and cost-effective tools to bolster institutional performance and elections at scale. But many of the problems recounted here also plague the election security of many industrialized democracies, including the United States, who are increasingly incapable of managing and securing their own vote counts and lack consistent and credible audits. The knowledge transfer often promoted by research and policy communities from developed countries to developing countries has a different directionality in this instance, with vital contributions that developing countries are providing to protect electoral integrity on a global scale.

2 Implementation Context

Combating election fraud must confront manifold contextual, technical, and implementation challenges. Although every country presents its own unique constraints to fair elections, our innovation grew out of both how we understood these problems comparatively and attempts to overcome them within each case.

A first challenge is perhaps the most obvious: politicians lacking enough virtue to uphold clean races regardless of the outcome confront the temptation to cheat if doing so helps them win and they are unlikely to get caught or face sanction. Because developing countries are more likely to face fewer safeguards to protect rule of law compared to industrialized democracies, fraud is more likely in emerging democracies. Yet the possibility of corruption must be weighed against the activities of citizens, civil society, and donors working to fight it. Governance is therefore one of the most sensitive development sectors to work in and requires balancing a number of conflicting political interests of those running for office and voting in elections with the administrative duties of delivering a free and fair process.

A second obstacle involves the countries themselves, where levels of state capacity and the nature of political competition hold the potential to shape election quality in fundamental ways. Holding an election is never easy, and in fragile states with low bureaucratic capacity like Afghanistan, it is administratively difficult, financially burdensome, and dangerous.[3] Taliban insurgents frequently disrupted Afghan elections (Condra et al., 2019); even higher capacity states like Uganda and Kenya still often suffer perennial swells in violence during campaigns. These fragile and middling states contrast with countries like South Africa, which enjoy strong bureaucratic capabilities and less fraud but still have room to improve election management. The nature of political competition is also an important contextual variable; party systems and electoral rules influence how candidates gain office and subsequently patterns of likely corruption.[4]

Condra et al. (2018) document that Afghan elections cost the government about $210 million USD per round (not including security costs), most of the funds coming from donors.

As discussed in Callen and Long (2015), Afghanistan's 2010 parliamentary election had thousands of candidates running for hundreds of seats with a single non-transferable vote (SNTV) across 34 provinces; fraud is therefore relatively decentralized without party linkages to predict ex ante and large returns to fraud for numerous candidates. But in dominant party systems like Mexico under the PRI (Magaloni, 2006) or competitive party systems where parties enjoy strong "vote bank" areas like Ghana (Ferree & Long, 2016), corruption is likely to be either centralized nationally and controlled by one party, or localized to areas of strong party support.

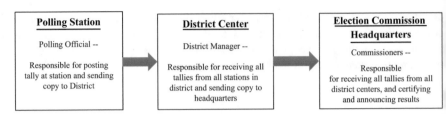

Fig. 18.1 Bureaucratic organization and process of aggregation of vote totals within an election commissions. (Adapted from Callen & Long, 2015)

A third set of challenges recognizes that elections are just unavoidably difficult to manage. To understand why, it is helpful to embed the political economy and behavioral dimensions of election fraud within an institutional analysis of election management to define the set of relevant actors and their roles managing elections in developing countries. This helps delineate and focus on dimensions for corruption, and from that, improvement. One helpful aspect of implementation context is that many countries in the developing world run their elections in very similar ways that follow analogous, if not the same, procedures. For demonstration purposes, I will adopt a uniform nomenclature to describe relevant electoral institutions and processes here, even though many words and concepts differ slightly depending on the case.

Election commissions are bureaucratic agencies mandated with delivering a credible election. They are responsible for registering voters, hiring and overseeing poll workers, managing voting processes, tabulating results, and certifying winners. Refined over the last three decades, developing countries often share the same or similar procedures for ensuring ballot secrecy, protecting ballot boxes, and transmitting results tallies. To run an election, election administrators are embedded within a standard bureaucratic chain of delegation. Starting from the most local unit in Fig. 18.1, at each polling station, what I term a *polling official* oversees voting and compliance with procedures opening the polling station, during the vote, and tabulating results. That official has a *district manager* in charge of running the election process at administrative units higher than the polling station (e.g., constituency or province), who oversees and monitors all of the officials and polling stations in their area. This includes receiving the election results that are sent by officials from individual polling stations to be aggregated and organized at the district level by the manager. At the national level, district managers are overseen by *election commissioners*, who run the nationwide electoral process from the agency's headquarters and oversee all managers and officials. Commissioners declare a winner after a final vote tabulation, but election workers at every stage are important to bring the process to conclusion.

An election commission is a necessary body for running an election, but its mere existence is not sufficient for protecting electoral integrity because commissions face many of the institutional pathologies of public sector governance in developing countries (Olken & Pande, 2011). Within the chain of delegation, three problems arise that threaten the ability of the commission to secure proper administrative procedures and without undue political interference. First, commissioners and managers face difficulty in the hiring process of polling station officials. These officials are not permanent workers with long-term contracts or incentives for performance pay; rather they are temporary laborers who are hired for a very short period (the election). Officials receive only minimum training, are provided wages that typically do not offset any wages they obtain from permanent employment, and while officials must demonstrate literacy and numeracy, such requirements are hard to enforce and heterogeneous across contexts.[5] Performance pay could theoretically help improve the quality of the labor pool but is prohibitively costly, does not address underlying problems of training, and would need to outweigh the amount an official could receive in a bribe.

Second, managers and commissioners lack reliable procedures and technology to oversee their agents. With a national exercise like an election occurring in a truncated period, managers are often not able to properly monitor the conduct of officials at individual polling stations or keep close watch on the vote tally and aggregation process.[6] Commissioners are unlikely to be in a position to action every reported problem from managers across districts on a single day. While standardized chain-of-custody practices of protecting the transmission of tallies in the results process exist on paper, they are not properly enforced in many countries, particularly at levels of the aggregation process beyond the polling station (e.g., district and national levels in Fig. 18.1). This occurs for a number of reasons. Officials frequently find it difficult to adjudicate discrepancies and may not be incentivized to report irregularities. Unfortunately, this structure provides opportunities to cheat, including administrators artificially and illegally changing vote totals for candidates directly on tally forms at district centers.[7] As a result, managers may overturn results from polling stations, change them, or fail to enact investigations into differences. The national intake center at commission headquarters scans a carbon copy of the tallies they receive, which should be the same as what is posted at the polling station to publicize and certify results on their website. But these totals often fail to reflect original totals, and candidates often seek to directly destroy evidence of the polling center count (including its tally) to then manufacture an entirely new tally that they insert into the chain of custody (Fig. 18.1).

South Africa requires at least 250,000 individual polling officials and managers to run the election and tally votes over a week's period every election. Kenya hires about 45,000 individual officials and the equivalent of 230 district managers. In Afghanistan, where men and women voters cast ballots in separate polling stations, the commission often finds it difficult to recruit female presiding officers who meet numeracy and literacy requirements.

While South Africa has one of the best performing commissions in Africa, our observers still noted tallies missing in nearly 60% of polling stations in clear violation of administrative procedure.

For example, illegal changes to tally sheets in Kenya's 2007 election are documented in a comprehensive investigation by Kanyinga et al. (2010).

Third, election commissions – which should remain non-partisan and independent of political influence – are weak and vulnerable to outside influence from collusion by candidates with officials to rig in their favor, either on the transmission of paper tallies or computer hacking into results systems from outside computer networks.[8] Collusion can involve direct bribe paying at the level of official, manager, or commissioner; but the returns to bribing a manager or commissioner are much higher given their control over the aggregation process.

A final set of challenges potentially involves actors who operate outside of the election management delegation chain but nonetheless have and could work to support electoral integrity. Civil society and the international community desire fair results and lend diplomatic, financial, and technical assistance to elections. But curiously, much of the technology advocated by these organizations has actually made it harder to secure results transmission by overburdening officials with unnecessary tasks that slow them down and are prone to error, while at the same time failing to protect or integrating original ballots or tallies into stream-lined, easy-to-audit, and difficult-to-hack ways.[9] Moreover, to stop corruption at scale, an election needs polling station monitoring beyond areas typically visited by these organizations and in ways that actually catch cheating. Methods to recruit citizen volunteers over ICT to report on polling stations and vote returns is one avenue to increase coverage, but citizens face many institutional and personal constraints to free and fair participation in elections. Even if people have technology readily available, it does not mean that they are motivated or equipped to monitor actions by bureaucratic actors embedded within election administrative bodies.

3 Innovate, Iterate, Evaluate, and Adapt

To combat threats to electoral integrity, our theory of change had to recognize the industrial organization of fraud in emerging democracies' elections as arising from institutional and behavioral elements of the actors described above. Specifically, we innovated a new way to detect and deter cheating with "photo quick count," which overcomes many of the bureaucratic pathologies of election commission by

---

[8] In Kenya's 2017 presidential election, forensic evidence revealed tampering of tally forms and computer hacking into the results transmission system.

[9] Afghanistan's biometric voter identification system (supported by $20 million USD from the international community) failed on election day in 2018 and did not protect the integrity of results transmission. Kenya's Integrated Electoral Management System (KIEMS), which reportedly costs the government upward of $1 billion USD ($24 million USD donated by the United States) was meant to upgrade and improve results sent over ICT for the 2017 election. The inability of thousands of officials to properly transmit results due to user, network, and administrative error over the system was apparent in real time as the commission was attempting to aggregate and certify results electronically. They would eventually have to revert to original hard-copy tallies. While the Supreme Court eventually nullified this election, the opposition boycotted the revote in part given a lack of any upgrades to KIEMS and the results transmission system.

rotecting against illicit actions that occur in the transmission of results along the chain of bureaucratic delegation. Photo quick count also seeks to improve the role that outside actors (including civil society, donors, and citizens) can play supporting elections. The cycle of innovation, iteration, evaluation, and adaption followed from pilot phasing in Afghanistan, to nationwide scaling in Uganda (2011) and Kenya 2013), pivoting to ICT-recruited and widespread citizen adoption of photo quick count with "VIP:Voice" in South Africa (2014).

3.1 Innovation

Photo quick count occurs in two phases, described here temporally. Phase 1 consists of a randomized announcement of an audit of polling station tallies. The announcement occurs by the delivery of a letter from one of our deputized election monitors given to the polling station official on election day during voting.[10] The letter states that the official's station has been randomly selected to have its results audited, which will occur the next day when monitors return to the station to photograph the tally that officials are required by law to post publicly at the conclusion of the station's vote count (typically the evening of the election). In effect, the letter "reminds" officials that they are responsible for publicly posting the tally and indicates that our monitor will photograph the tally to then compare to the certified result published at the conclusion of the election. The letter explains that this procedure helps to verify compliance with procedures at the polling station e.g., the official's posting of the tally) and against the certified result (e.g., by the commission), recording any discrepancies and differences. The letter asks officials to sign acknowledging having received it. The monitor takes the signed copy and leaves a copy with the official. Phase 2 occurs the next day when, as indicated in the letter, monitors return to the same polling station to record whether the tallies have been posted and photograph the ones that are present. This is the technological component of the intervention, photo quick count, which records whether procedures were followed in that tallies were posted, undamaged, and not removed after posting and, if properly posted, whether a polling center's tally matches the final and certified count.

The announcement of monitoring via the letter delivery and verification with photo quick count functions similar to other audits with measurement tools to detect irregularities, and in this context, to encourage compliance with electoral procedures as proscribed by countries' laws to guard against aggregation cheating (conducted by an outside actor to the delegation chain). This is because of how the tally itself functions in the results transmission process (Fig. 18.1). Importantly, all of the

---

[*] These monitors are in-country citizens recruited and managed by our team. They received the equivalent of 2 days of paid work for Phases 1 and 2 and attended one training session. They were not previously part of an election monitoring organization.

Fig. 18.2 Afghan voters examining a tally form posted at a polling station. (Berman et al., 2019)

countries in our studies require posting of tallies by officials – failure to do so is an abrogation of responsibilities, and not posting is a bureaucratic failure of managers to not properly oversee officials. If the tally is posted, photographed, and the same as the scanned and certified tally published by the commission (made available on their website for public viewing), the audit verifies that no changes were made in the aggregation process by managers, commissioners, or other political actors. As designed, these original tallies are supposed to provide checks on certified results since the latter are carbon copies of the former that were originally filled at each station with identical copies sent along the delegation chain to managers at the district level and then commissioners at the national level (Fig. 18.1). Electoral laws mandate that the results of the original tally be posted in a public and conspicuous place since they are the only official means by which an individual can see how their precinct votes (and citizens are legally allowed to view and photograph them (Fig. 18.2)). It is typical that many citizens and political agents from local communities examine tallies since they plausibly send a signal to those communities about the fairness of the election at that station. However, failure to post tallies, stealing or damaging of tallies, or inconsistencies in results between the original tallies and copies scanned by the commission reflect at best administrative failures and at worst the possible intention to manipulate the vote count process.[11]

---

[11] Berman et al. (2019) report that thousands of complaints received by the Afghan Electoral Complaints Commission demonstrate that tallies were often stolen by political agents in order to take tallies to a secret location or another part of the aggregation chain (e.g., provincial election centers) to falsify results and reinsert tallies back into the count.

Counterfactuals

From our theory of change and photo quick count's tally fraud detection, we hypothesized that the announcement of monitoring of stations receiving the letter in Phase 1 would be more likely to improve their procedures and result in fewer tally discrepancies, compared to a station that did not receive a letter. To evaluate this possibility against a counterfactual, we selected a sample of audit-eligible stations. In Phase 1, we randomized the delivery of the letter in treatment stations with control stations receiving no letter. In Phase 2, for both treatment and control stations, our researchers followed the same protocols to conduct photo quick counts. The evaluation compares these results to estimate the effect of the announcement via letter delivery.

To understand how the two phases work concretely, I describe the first deployment in Afghanistan's 2010 parliamentary elections (Callen & Long, 2015). We created an experimental sample of 471 polling centers (about 7.8% of the total) that we determined were audit-eligible and could be visited by our team.[12] Of the 471 centers, we randomly selected 238 to receive the letter on election day (Phase 1) delivered by a team of Afghan monitors that we hired and trained; the remaining stations received no letter. The next day (Phase 2), our monitors returned to the stations that had received a letter on election day and the stations in the sample without a letter to perform photo quick count the morning after the election at which point officials should have posted tallies. They photographed tallies with digital cameras purchased by us for the project, which were not capable to transmit data in real time but could save images on removable memory chips. (While our initial hope was to crowd-seed smartphone devices to our monitors, we lacked the budget, time, and programming capability in the project period.) If tallies were not posted or there was evidence they had been torn down, monitors investigated as to why and by whom (without interacting with any polling officials) and recorded that information.

Photo quick count documents numerous suspicious activities regarding tally postings by our monitors. Figure 18.3 shows a comparison taken by our monitors at a polling station (left-hand panel) and the scan of that polling station's tally received and published by the election commission in Kabul (right-hand panel) scraped from their website. These should be carbon copies and thus otherwise identical, but they differ in obvious ways. Someone has converted an original Dari script from the polling station into Arabic numerals scanned at the commission, the polling station official's name has changed, and the station tally records vote totals for several candidates that are entirely deleted from the commissions' tally. To see how these differences operated over the whole sample, Table 18.1 records levels and types of differences between tallies. Importantly, tallies were damaged or removed at 62 of

---

[2] The sample was limited to precincts within provincial capital cities due to safety, administrative, and cost limitations.

Fig. 18.3 Tally forms from the same polling station in Afghanistan (Callen & Long, 2015); left-hand panel shows photograph posted at station and right-hand panel shows scan at the election commission

Table 18.1 Patterns of discrepancies in tally aggregation process between photographed tallies at polling stations via photo quick count (sample of 346 stations) and commission tallies in Afghanistan

Pattern of discrepancies	Number of polling stations	Share of sample
None	74	21.4%
Adding votes (only)	70	20.2%
Subtracting votes (only)	15	4.3%
Adding and subtracting (equally)	15	4.3%
Adding more than subtracting	127	36.7%
Subtracting more than adding	45	13%

Adapted from Callen and Long (2015)

the sampled polling stations (13%), preventing any direct comparisons. But for the tallies observed in the remaining 346 stations, Table 18.1 shows that while there are no discrepancies recorded in 21%, the remaining stations had differences of addition, subtraction, or both.

Table 18.1 indicates that one limitation we confronted with photo quick count is that many times the tallies were not posted or torn down at polling stations. This generated a lack of consistent measurement across the entire sample regarding candidate's originally posted vote totals to compare against their certified vote totals.[13] Our intervention and evaluation in Afghanistan and subsequent countries

---

[13] We address treatment-related attrition for experimental estimates in Callen and Long (2015) and Callen et al. (2016).

therefore make use of other administrative data that further reveal measurement of likely administrative violations and fraud. Because commissions still publicize both scanned copies of tallies at the national center and separately certified data on polling station level results for each candidate on their website (even if tallies are not posted at stations), we obtain copies of commissions' scanned tallies and published results. We then employ election forensics, which uses statistical analyses of observed vote distributions to detect dubious or systematic irregularities indicative of fraud that deviate from theoretical distributions that accord with a fair vote (Mebane, 2008; Myagkov et al., 2009). Because we believed the audit announcement might differentially impact candidates based on their linkages to election administrators and therefore their ability to collude to change vote totals, we also used qualitative data on candidates' backgrounds that identifies likely patterns of political linkages between politicians and election administrators.[14] The combination of these data and tests with photo quick count helps measure the full impact of the intervention.

In a first set of results from Afghanistan, comparing the treatment sample relative to the control, the letter delivery positively improved measures directly observed by photo quick count and other administrative data sources. The letter delivery decreased the theft of tallies by 60 percent (from 18.9% to 8.1%) and reduced discrepancies in vote totals between photographed tallies and those certified by the electoral commission.[15] The letter delivery had the largest reduction on votes for candidates who gained the most votes at individual polling stations (from about 21–15 ballots, or a 25% drop).

3.2 Iterative Implementation

Aspects of the operational, political, and technical environment of Afghanistan influenced iterative implementation of photo quick count in Uganda and Kenya. First, we wanted to deploy photo quick count but to do so at scale in a nationwide and nationally representative sample of polling stations, which we had lacked in Afghanistan. Laterally, this scaling had to be cost-effective, so we considered ways to leverage photo capture with real-time data transmission over ICT using relatively cheap smartphones. We also wanted to test the effectiveness of the intervention in

---

[14] For Afghanistan, the main parliamentary candidates' known connections with government officials were recorded from ethnographic profiles produced by DI since there are no party affiliations; in Uganda and Kenya, we relied on more straightforward relevant party, ethnic, and regional linkages.

[15] Given sample attrition per Table 18.1, we use a bounding exercise to estimate that this effect was between about 9 and 17 fewer votes changed during aggregation for candidates with known connections to the equivalent of the district manager.

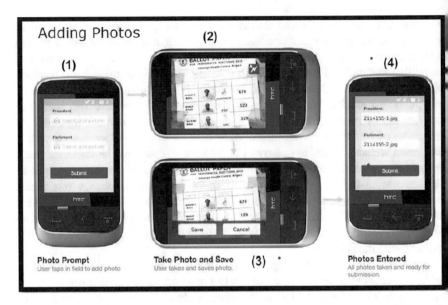

Fig. 18.4 Photo quick count smartphone application on Android for Uganda. (From Callen et al., 2016)

different institutional environments with variation in political dynamics that likely affect how political agents might influence bureaucratic functioning of commissions. While Afghanistan's commission was decentralized and the parliamentary elections featured no formal political parties, the institutional and political dynamics in Uganda and Kenya were different (and more tractable to measure).

For Uganda's 2011 election, we expanded polling station coverage to 1,002 in a nationwide sample (representative by region; see Callen et al., 2016). While this sample was more comprehensive of Uganda's electoral and institutional landscape than what we achieved in Afghanistan, it was also harder to access and manage, particularly in remote areas. We tried to offset this by having our monitors conduct photo quick count with an Android-enabled smartphone app developed in partnership with Qualcomm that cost less than half the price of the digital cameras in Afghanistan (Figure 18.4 displays the app interface). We had to crowd-seed phones since many of our team did not yet have personal devices capable of photo quick count. The political dynamics that affect institutional performance in Uganda are different than Afghanistan. Rather than multiple sites of contact and influence by dispersed candidates with no formal political linkages, the Ugandan election commission is staffed from a single dominant political force, the National Resistance Movement (NRM) and incumbent President Yoweri Museveni, who gained office after winning a civil war in 1986 and has won multiparty elections since 2006. Uganda therefore has a high degree of centralization of political influence in the commission (at all levels) which was easier to predict *ex ante* but could also differentially influence the effects of monitoring. With these dynamics in mind, we worried that the simple announcement of monitoring would potentially not operate as well in Uganda, so we developed a second version of the letter that announced monitoring and included a message that reminded officials of the legally proscribed penalties for violations of the country's electoral law (Fig. 18.5).

The third deployment in Kenya in 2013 built on the features of Uganda, with a truly nationally representative sample of 1,200 polling stations, an improved open-sourced photo quick count app, and novel administrative data sources to look at measurement strategies built into political context as detected by photo quick count, including conducting a parallel exit poll of voters. Patterns of corruption in Kenya are less centralized overall than Uganda given the lack of a dominant party. Instead, two large and polarized political forces have dominated recent elections and are backed in part by ethnic and regional coalitions that exert control in their "vote bank" areas (Long, 2020), allowing localized and regional avenues of collusion with managers; one of these coalitions reflecting the incumbent government is also strong nationally with influence in the commission.

3.3 Evaluation

Similar to Afghanistan, we evaluated the randomized announcement of the tally audit of the vote results via photo quick count. In Uganda, where we observe a lack of posted tallies in about 78% of the entire sample, letter delivery decreased the practice of not publicly posting tallies between about 6 and 11% points (depending on specification and letter version) and also reduced discrepancies in vote totals between photographed tallies and certified results for Museveni's votes per station from 8 to 16% (or 26–49 votes from an average of 307 per station). While Uganda introduced a second version of the letter reminding of penalties, we found no additional consistent effect of this letter compared to the announcement of auditing itself. We also observed that the letter reduced the propensity for adjacent digits (a forensic measure of fraud) by between 6 and 10% points in Museveni's votes. Although preliminary, early analysis from Kenya shows similar effects of the announcement increasing the propensity of posted tallies and decreasing the likelihood of suspicious vote totals and differences between tallies and commission results.

In a second set of results, we document that the audit intervention seems to affect political actors differently given how political dynamics map on to institutions and likely rigging strategies. In general, the intervention appears to have the largest impact on candidates most likely to rig, followed in different ways by evidence of strategic re-adjustment (analogous to negative or positive spillovers). In Afghanistan, the treatment effects were largest for candidates with known linkages to election administrators (specifically, the equivalent of the district manager), but we also found evidence of a chilling effect from letter delivery. Using geolocations of polling stations in our sampled areas, polling stations that had a neighboring

Urgent   February 18, 2011

ATTENTION: The Presiding Officer, ONIGO CENTRE Polling Station

Re: Election Monitoring at ONIGO CENTRE Polling Station

   Greetings! I am working with the University of California, an accredited election observation organization. We are providing this letter to tell you about some important information about your polling station.

   As part of our effort to help Uganda have free and fair elections, we would like to take this opportunity to remind you of an important part of Uganda's electoral law. As you know, the Presidential Election Act of 2010 stipulates a punishment of up to a 2.4 million UGS fine and/or imprisonment of up to five years for any election officer who knowingly gives inaccurate information about the vote returns.

   As another part of the observation effort, I will return to this polling station tomorrow in order to take pictures of the "Declaration of Results" forms that you are required by law to post publicly at this polling station. We will compare the results from the photos with the certified final count published by the EC in Kampala. AFTER the official EC certification, we will report these results on the internet (at www.uganda2011.org) and to newspapers. By doing this, the people of Uganda will be able to see if any changes have been made to the vote at ONIGO CENTRE Polling Station after the recording and posting of the "Declaration of Results" form. All Ugandans will be able to tell whether there have been any changes to the vote total, and they will know which candidate any change benefits. The following example shows how we will report this.

   Please note that we are only doing this in a small number of randomly selected polling stations, yours included, but not every polling station. As an accredited observer, we are legally authorized to complete this activity.

   In recognition that you have read and understood this letter, please sign here:

_____

Thank you kindly for your help and cooperation!

The following is an example of how we will report results:

Candidate:	Polling station: ONIGO CENTRE		
	Certified Vote from the EC in Kampala	Total Votes from Photographs at Polling Station	DIFFERENCE
Candidate A	100	600	+500
Candidate B	600	100	-500
Candidate C	14	14	0
Candidate D	0	0	0

Figure A1

Treatment Letter Example 1. An example of the treatment letter including both Monitoring and Punishment messages.

Fig. 18.5 Election day letter announcing photo quick count in Uganda. (From Callen et al., 2016)

station within 1 km treated with a letter also saw an additional loss of about seven votes for the most politically connected candidates. In Uganda, letter delivery had an impact on the polling station level results for Museveni, reducing his votes by about three percentage points. However, we see evidence that in our sample, the central election commission appears to have slightly added votes back in his favor in monitored stations. In Kenya, rigging appears to occur at the behest of both incumbent and opposition coalitions in their local and regional vote banks, but the more powerful coalition closely connected with commissioners at the national level had additional avenues for cheating and re-adjustment.

3.4 Adaption

Photo quick count produced scientific evidence on the use of randomized audits to bolster institutional performance and improve electoral integrity by an independent team of deputized citizen monitors that we trained and deployed. At the same time, scaling was still linearly expensive and limited by the availability of ICT that required crowd-seeding devices. Beyond organizational adoption, we had yet to investigate the opportunities and constraints to widespread citizens' participation and their capacity for ICT-enabled monitoring. If we wanted adaption at scale, we needed to pivot and re-orient our programmatic thinking and technological ambitions.

> **Pivot**
> Pivoting to citizen-based crowdsourced monitoring required thinking through some core components of photo quick count. Broad coverage of polling station observations by voters obviates the need for a team of researchers announcing monitoring via a letter, but many factors shape voters' political participation and technology use in positive and negative ways. Many crowdsourced platforms have encountered problems of uptake and usage in development applications; barriers to action and ICT usage must therefore inform the design and functionality of monitoring platforms.

To do so, I worked with colleagues to build a new multi-channel ICT and digital media system for an entirely citizen-based election monitoring platform financed through USAID Phase 2 funding. We designed the system, branded "VIP:Voice," and launched it in South Africa before its 2014 election in partnership with developers at the Praekelt Foundation (Ferree et al., 2020). VIP:Voice allowed citizens who registered with the system to engage with the electoral process by reporting their opinions on politics, campaign activities in their area, and other election-related matters in the lead up to polling day, including user reports on their election day experiences and monitoring of tallies. The platform was available on a variety of ICT channels accessible by users of basic phones, feature and smartphones, and web users. VIP:Voice differed from other interventions that leverage predefined organizations or lists of users obtained from registration drives, organizational memberships, or household surveys. Instead, it advertised on social media channels and "please call me" text messages (Fig. 18.6).[16] This design and functionality therefore did not limit usage to any segment of society, and multi-channel development eliminates the need to seed devices (they are able to rely on the devices participants already possess). Through our recruitment over SMS and social media, VIP:Voice was able to reach 50 million South African citizens, engaged with 250,000, and registered more than 90,000. Half of registered users came in through SMS "please call me" pushes on standard phone channels and half through social media channels (particularly "Mxit," which was South Africa's most popular social media channel at the time). From this, we also registered citizen volunteers in 37% of the countries' electoral wards from a set of identified highly engaged users, hundreds of whom provided reports of polling station activities and hundreds of photographed tallies.

Gender
Digital access is not equal across demographic groups in developing countries. In South Africa, older, female, rural, and Black citizens are all likely to lack the web-enabled smartphones that other groups employ, but they are still likely to own basic mobile phones. Because a goal of VIP:Voice was to obtain a nationwide yield of users that included people from all backgrounds, we had to explicitly consider how gender dynamics vary across ICT channels.

VIP:Voice featured mechanisms to evaluate the efficacy of different types of engagement across the different channels and a series of randomized experimental interventions on incentives and cost in the standard phone channel. Channels that were easier to use (with social media apps) had higher rates of engagement compared to text messages, but these user-friendly channels were more likely to see more engagement attrition over time. Many users registered with VIP:Voice with no external incentive offered, but small offers of incentives (free airtime or lotteries advertised within "please call me" alerts) improved levels of engagement. In a proof-of-concept experiment, from a list of about 42,000 highly engaged users, 17% volunteered to observe elections (which was further improved by incentivization) Actual photo quick count monitoring documents numerous problems of tallies and differences in the commissions result, but these differences did not appear

---

[16] Please call me messages are free-to-user alerts that people can send to a contact requesting a (paid) call back – they are popular when individuals lack airtime credit to make direct calls and are paid by advertising space embedded in the message.

Fig. 18.6   Advertising for VIP:Voice platform in South Africa. (From Ferree et al., 2020)

to systematically benefit any particular party and instead may have resulted from administrative failures.[17] Although the salience of one important design consideration was not apparent at the time, VIP:Voice had the added benefit of providing citizens with real news about the campaign and election results under the control of researchers to guard against misinformation.

Responsible Research
Allowing citizens to serve the function of election monitor does nothing more than allowing them to perform the same democracy-enhancing function as voting by providing reports on polling stations. But does that guarantee that all citizens are safe to monitor? This is a question that researchers must answer depending on context and that individual users must answer for themselves. We chose one of the safest election environments, South Africa, to build VIP:Voice but any crowdsourced platform must rely on trusting its users to know if and when it is too dangerous to report on government activity. And boosting participation, even if sometimes risky, should be weighed against the realities of corruption and inadequate service delivery that citizens in developing countries face and must collectively organized to overcome.

VIP:Voice shows that citizens can be mobilized to take an active role in protecting institutions in the public realm, even where they do not receive immediate private benefits. But engaging and registering them over a digital platform to engage in real-world activities is also beset with many recruitment, technical, and programmatic challenges (Erlich et al., 2018).

4 Results/Lessons Learned

Photo quick count provides a competitive advantage to alternatives, such as expensive international observation missions, given its effectiveness at detection and deterrence of aggregation fraud and potential for citizen adoption. It also works in most countries at less than 1% of the per polling station cost of international missions. The lessons learned have provided important insights informing ongoing work to translate previous project phases into improving the technical aspects of VIP:Voice. We aim to further address participation barriers, recruitment, and engagement at the same time as maintaining and growing the capacity for a professionally managed photo quick count of tallies with enough coverage to provide the minimum risk-limiting coverage.

---

[17] We also conducted a parallel study on locally enabled ICT professional monitors using the traditional photo quick count.

Results tie to numerous literatures and generalize to other contexts. They contribute to studies on the political economy of public sector corruption, evaluations of democracy and governance programs in emerging democracies, studies of election fraud, and connection to citizen-based ICT platforms in other development sectors.[18] While election commissions form only one critical institution within the governance sector, there are common institutional challenges that apply broadly across bureaucratic functioning and rule of law related to oversight and monitoring of frontline officials and managers, how political linkages and connections outside of bureaucracies often undermine the performance of those bureaucracies, and whether audits work to improve agency capacity and citizen monitoring of government performance. Results offer guidance to organizations, activists, and developers regarding the usability of platforms like VIP:Voice for elections in other contexts where problems of participation, data quality and reporting, and electoral integrity can be improved with ICT.

Capacity

An independently managed audit via photo quick count can help election commissions improve their capacity and output where those commissions lack strength. For example, if a commission had real-time exposure to the information coming from audits, it could respond immediately or preserve records to include in adjudication of disputes later. This possibility is likelier in cases where commissioners continue to be blamed for disputed results or where the commission is high performing already but wishes to use our tools to manage the process better.

These positive aspects should be considered alongside numerous lessons learned for policy, some of which reveal a new set of challenges and fruitful avenues for research, technical development, and programming. First, adaption of photo quick count by election commissions, civil society organizations, and donors can help improve electoral integrity; but photo quick count only provides a check on integrity when it itself is non-partisan and independent, or done in conjunction with credible partners. If used in partnership with or by commissions directly, it also requires institutional actors that are fair and forthright. Even though many commissions lack capacity or are subject to influence, the growing precedent to prosecute commissioners and managers (as in Afghanistan) or nullify elections from poor management (as in Kenya in 2017 and Malawi in 2020) could change the thinking on the need for audit, the modalities required to conduct one, and

---

[8] Because many features of VIP:Voice were agnostic as to the particular use of those features, we provide guidance on scaling to other sectors in international development where citizens' interaction with government agencies could be improved with monitoring and data capture to receive services, like health, education, agriculture, and financial inclusion.

how to preserve evidence to its effect. Given the alarming rate at which electronic systems are being hacked, photograph- and paper-based audit methods are growing increasingly important to provide original copies of results.

International organizations and civil society groups are also plausibly well placed to serve as third-party "monitors-of-monitors" to facilitate the use of these tools given the other important advocacy work they are already doing to support democracy programming and technical assistance to commissions. In our experience, beyond DI, it has been hard to find dependable international and civil society organizations to work with, which has revealed a number of second-order challenges that involve diplomatic and political considerations for donors and non-political interference for civil society. The fidelity of the design of photo quick count and VIP:Voice would need to ensure that the technology itself is deployed robustly and cannot itself be hacked, misused, or suppressed. Our fruitful partnership with USAID provides hope that these organizations could still play an important role funding or supporting independent applications of VIP:Voice as a component to democracy assistance.

5 Summary

Our research contributes important evidence regarding Afghanistan, Uganda, Kenya, and South Africa's electoral processes, the political economy of governance in election commissions in developing countries, and the viability of using photo quick count and VIP:Voice to improve electoral integrity. Although institutional pathologies of election administration persist, audits can work to improve accountability and guard against threats to results transmission. VIP:Voice provides important benefits to citizens both directly by engaging them in the election process and monitoring of their elections, and indirectly by building confidence in the credibility of the process. Because the posting and comparison of results via tallies has important resonance for certifying the election, citizen monitoring plausibly could occur in a large enough sample to provide something like a minimum threshold, particularly alongside an independent manual audit. But recruitment problems and rates of attrition from registration to monitoring with citizen platforms also suggest that if a large number of users are desired, it takes more development time and continued pushes across ICT channels. These technical challenges must be considered in the design phase, and the time is ripe to action these issues. Citizens have now spontaneously begun to employ version of photo quick count in recent elections. In 2017, thousands of dispersed and unorganized Kenyans photographed tallies and posted results to social media with a series of dedicated hashtags where comparisons often did not comport with the election commission's certified results. The Supreme Court then conducted its own audit of original paper tallies and a review of the electronic results transmission, a process revealing so many discrepancies that they nullified the election and declared a revote. The importance of independent tally audits and citizen-based monitoring is only growing more salient as the integrity of elections around the world faces new and evolving threats.

Discussion Questions

1. Under what conditions should government officials and bureaucratic agents be involved in performance audits?
2. What are the most productive ways to galvanize donors, civil society, and citizens to adopt new technologies to address needs in the governance sector?
3. How best might ICT improve citizens' political participation and ability to monitor government performance, at the same time as protecting respondent privacy and data security?

References

Acemoglu, D., Johnson, S., & Robinson, J. A. (2001). The colonial origins of comparative development: An empirical investigation. *American Economic Review, 91*(5), 1369–1401.

Barro, R. (1973). The control of politicians: An economic model. *Public Choice,* 19–42.

Berman, E., Callen, M., Gibson, C., Long, J. D., & Rezaee, A. (2019). Elections and government legitimacy in Afghanistan. *Journal of Economic Behavior & Organization, 168,* 292–317.

Bjornlund, E. (2004). *Beyond free and fair: Monitoring elections and building democracy.* Woodrow Wilson Center Press.

Bueno de Mesquita, B., Morrow, J., Siverson, R., & Smith, A. (2002). Political institutions, policy choice and the survival of leaders. *British Journal of Political Science, 32*(4), 559–590.

Callen, M., & Long, J. D. (2015). Institutional corruption and election fraud: Evidence from a field experiment in Afghanistan. *American Economic Review, 105*(1), 354–381.

Callen, M., Gibson, C. C., Jung, D. F., & Long, J. D. (2016). Improving electoral integrity with information and communications technology. *Journal of Experimental Political Science, 3*(01), 4–17.

Condra, L. N., Long, J. D., Shaver, A. C., & Wright, A. L. (2018). The logic of insurgent electoral violence. *American Economic Review, 108*(11), 3199–3231.

Condra, L. N., Callen, M., Iyengar, R. K., Long, J. D., & Shapiro, J. N. (2019). Damaging democracy? Security provision and turnout in afghan elections. *Economics and Politics, 31*(2), 163–193.

Erlich, A., Jung, D. F., Long, J. D., & McIntosh, C. (2018). The double-edged sword of mobilizing citizens via Mobile phone in developing countries. *Journal of Development Engineering, 3,* 34–46.

European Union. (2006). *European Union: External action FAQ.* www.eeas.europa.eu/. Accessed 4 Nov 2013.

Ferree, K., & Long, J. D. (2016). Gifts, threats, and perceptions of ballot secrecy in African elections. *African Affairs, 115*(461), 621–645.

Ferree, K., Gibson, C., Jung, D., Long, J. D., & McIntosh, C. (2020). *How technology shapes the crowd: The 2014 south African election.* Center for Effective Action Working Paper.

Gibson, C., & Long, J. D. (2009). The presidential and parliamentary elections in Kenya, December 2007. *Electoral Studies, 28*(3), 497–502.

Hyde, S. D. (2011). *The pseudo-democrat's dilemma: Why election observation became an international norm.* Cornell University Press.

Hyde, S. D. (2020). Democracy's backsliding in the international environment. *Science, 369*(6508), 1191–1196.

Kanyinga, K., & Long, J. D. (2012). The political economy of reforms in Kenya: The Post-2007 election violence and a new constitution. *African Studies Review, 55*(1), 31–51.

Kanyinga, K., Long, J. D., & Ndii, D. (2010). Was it rigged? A forensic analysis of vote returns in Kenya's 2007 election. In K. Kanyinga & D. Okello (Eds.), *Tensions and reversals in democratic transitions: The Kenya 2007 general elections* (pp. 373–414). Society for International Development and Institute for Development Studies.

Kelley, J. G. (2012). *Monitoring democracy: When international election observation works, and why it often fails*. Princeton University Press.

Kolstad, I., & Wiig, A. (2016). Does democracy reduce corruption? *Democratization, 23*(7), 1198–1215.

Long, J. D. (2020). Civil conflict, power-sharing, truth and reconciliation (2005–13). In N. Cheeseman, K. Kanyinga, & G. Lynch (Eds.), *The Oxford handbook of Kenyan politics* (pp. 82–95). Oxford University Press.

Magaloni, B. (2006). *Voting for autocracy: Hegemonic party survival and its demise in Mexico*. Cambridge University Press.

Manin, B., Przeworski, A., & Stokes, S. (1999). Elections and representation. In A. Przeworski, S. C. Stokes, & B. Manin (Eds.), *Democracy, accountability, and representation* (pp. 29–54). Cambridge University Press.

Mebane, W. R., Jr. (2008). Election forensics: The second-digit Benford's law test and recent American presidential elections. In R. M. Alvarez, T. E. Hall, & S. D. Hyde (Eds.), *Election fraud: Detecting and deterring electoral manipulation* (pp. 162–181). Brookings.

Myagkov, M., Ordeshook, P. C., & Shakin, D. (2009). *The forensics of election fraud: Russia and Ukraine*. Cambridge University Press.

Norris, P. (2014). *Why electoral integrity matters*. Cambridge University Press.

Olken, B., & Pande, R. (2011). *Governance review paper*, J-PAL Governance Initiative. MIT.

Chapter 19
Monitoring the Monitors in Punjab, Pakistan

Arman Rezaee (iD)

1 Development Problem

It is well documented that government frontline service providers such as teachers and health workers are often absent from their posts in developing countries. Chaudhury et al. (2006) conducted unannounced visits to primary schools and primary health centers across six developing countries spanning three continents. The authors found widespread absence, with rates averaging 19% for teachers and 25% for health workers. The authors also found that lower-income areas had higher absence rates, suggesting absence may exacerbate socioeconomic divides. Many studies have since documented similar patterns (e.g., Banerjee & Duo, 2006; Banerjee et al., 2008; Olken & Pande, 2012; Dhaliwal & Hanna, 2017; Finan et al., 2017).

While the specific impacts of absenteeism on outcomes of the world's poor are less well documented, evidence suggests the potential for large negative impacts from missing workers. For example, Duflo et al. (2012) leveraged cameras coupled with financial incentives to boost teacher attendance in India. The authors found that lower teacher absenteeism translated directly to increased student achievement. Using a nationwide representative sample of public schools in rural India, Muralid-aran et al. (2017) found that "investing in reducing teacher absence through better monitoring could be over ten times more cost effective at reducing the effective student teacher ratio (net of teacher absence) than investing in hiring more teachers."

Based on research conducted by Michael Callen, Saad Gulzar, Ali Hasanain, Yasir Khan, and Arman Rezaee

A. Rezaee (✉)
Department of Economics, University of California, Davis, CA, USA
e-mail: abrezaee@ucdavis.edu

C. Madon et al. (eds.), *Introduction to Development Engineering*,
https://doi.org/10.1007/978-3-030-86065-3_19

These seminal studies, as well as anecdotes about absent frontline health workers in Punjab, Pakistan, motivated this case study. Indeed, in November 2011, my co-authors and I conducted surprise, independent inspections of a representative sample of 850 of Punjab's near 2500 provincial rural health clinics (dubbed Basic Health Units (BHUs)). We found that government doctors posted to BHUs were absent from their facility during open hours two-thirds of the time (Callen et al., forthcoming). This is particularly alarming as BHUs provide primary care for nearly all of Punjab's rural population, including outpatient care, pre- and postnatal care and deliveries, and vaccinations.

Myriad factors contribute to the problem of absenteeism among frontline workers in developing countries, including poor incentives to show up to work (Callen et al., 2015; Dal Bó et al., 2013); political-economic factors such as patronage (Callen et al., 2016), political alignment (Callen et al., 2020), and/or political turnover (Akhtari et al., 2017); and poor management (Dal Bó et al., 2018; Rasul et al., 2018).

Another factor contributing to absenteeism is a lack of accurate and actionable data. Before the ICT-based program that is the subject of this case study took shape in late 2011, there was an elaborate, colonial-era, paper-based system for collecting operational data on public health facilities in Punjab. This system was similar to those still in existence across many developing country bureaucracies today. The centerpoint of this system is the physical register—log books at each facility that are used to keep diligent records on staff attendance, medicine stock, patients seen, vaccinations given, etc. During their required monthly inspections of each BHU, government health inspectors record aggregate information from these registers on a standard paper form. Once collected, forms are brought to a central district facility, manually entered into a spreadsheet, and aggregated into a monthly report for senior health officials.

While the pre-existing system led to the collection of vast amounts of data, it ultimately was ineffective in the case of Punjab's Health Department. Registers upon registers were filled. But then they collected dust. According to our independent inspections, only 23% of facilities received their required monthly health inspection prior to November 2011, essentially halting the flow of information from registers to health inspectors and their managers. And what little information did make its way to a District Headquarter was often inaccurate, for the same political economy reasons that allowed health worker absence in the first place (Callen et al., 2015). This could be seen as a self-fulfilling prophecy—data is collected but isn't used, so there is no incentive to collect good data in the first place, so data isn't good enough to be used, and so on.

Enter ICT. In late 2011, 2G networks and smartphones had just made their way to Punjab. Zubair Bhatti at the World Bank in Pakistan recognized this as an opportunity to make Punjab's health data more accurate and more actionable.[1] At the same time, Punjab's provincial government had just set up the Punjab Health

---

[1] We thank Asim Fayaz in addition to Zubair Bhatti for envisioning and designing the smartphone monitoring program.

Sector Reform Program (PHSRP), which had the mandate to push reform within the Department of Health but also had substantial autonomy. PHSRP was run by Farasat Iqbal, a well-respected career bureaucrat in Punjab and someone who was quick to recognize the potential of ICT. He was one of the first champions for the use of technology within the bureaucracy and remained a champion throughout. We partnered with Mr. Bhatti's World Bank unit and Mr. Iqbal's PHSRP to take advantage of this opportunity. Together, we designed, implemented, and evaluated a smartphone monitoring program, officially termed "Monitoring the Monitors," across Punjab's 36 districts. The program simply replaces the paper-based system of old with an app-based system, otherwise collecting the same data and involving the same health workers. Instead of using paper forms, government health inspectors are equipped with a smartphone and an app to collect data and feed it to an online dashboard system. Cutting out the aggregation step assures real-time information on BHU performance, aggregated into simple charts and tables, for the review of senior health officials.[2] Importantly, the app also includes several fail-safes to ensure accurate reporting and thus to break the self-fulfilling prophecy described above: reports are geo-stamped and time-stamped and all staff reported present must be photographed with the inspector (Fig. 19.1).

We evaluated the Monitoring the Monitors system through a large-scale randomized controlled trial. Beyond improving the flow of data, the system changed the behavior of inspectors, at least temporarily. In independent audits conducted 5 months after the system launched, the inspection rate increased from 25.5% to 51.9%. After a year of operation, inspection rates were 33.8% in the treatment districts and 23.5% in control districts.

It was our hope that this increase in inspections would increase doctor attendance. On average, this was not the case. But interestingly, we find that the Monitoring the Monitors system did lead to increased doctor attendance exactly in the cases when the system's data was made actionable for senior health officials. To show this, we built into the system an A/B test of sorts. Specifically, if more than three of the seven health workers that are supposed to staff a rural clinic were absent during a health inspection, we "flagged" a facility as underperforming by highlighting it in red on the dashboard.[3] We find that flagging increased subsequent doctor attendance from 33.6% to 41.3%. We interpret this as evidence that policymakers do indeed use data when making decisions, if it is made actionable.

This result brings us back to the development problem here—a lack of accurate and actionable data for public sector managers to use to ensure frontline workers show up to work. In this case, ICT ensured more health inspections feeding more accurate data on performance to senior health officials and subsequently much

---

The data include staff attendance, availability of medicine, patient visits, vaccines provided, cleanliness, and so on.

See Fig. 19.1, Panel B, for the red flagging. Note this arbitrary threshold was not known to anyone outside of the research team.

Health Department, Government of Punjab

| Compliance Status | Facility Status | Recent Visits | Indicators | Time Trend Charts | Photo Verification | Map | Change Password | Logout |

You are currently viewing [PUNJAB]   (Please click to change view)   🖨 Print

Officer Compliance Report

Officers are required to make the assigned number of visits to facilities in each calendar month. If the number of facilities is less than the assigned number of visits, the officer should repeat visits to some facilities to complete the quota of visits. View Detailed Report

Compliance - Last Month (by facility type)

- BHU
- RHC
- THQ
- DHQ

(bar chart with y-axis 0, 25, 50, 75, 100 and x-axis: Attock, Bahawalpur, Bhakkar, Chiniot, D.G. Khan, Faisalabad, Gujrat, Hafizabad, Jhang, Kasur, Khanewal, Lahore, Mianwali, Multan, Muzaffargarh, Narowal, Pakpattan, Rawalpindi, Vehari)

| Compliance Status | Facility Status | Recent Visits | Indicators | Time Trend Charts | Photo Verification | Map | Change Password | Logout |

You are currently viewing [District Attock]   (Please click to change view)   🖨 Print

Recent Facility Visits

⬛ Visits highlighted indicate significant staff absence.

| BHU | RHC | THQ | DHQ |

(Filter by Period) [   ] [   ]   (Clear Filter)

Showing all entries

Displaying 1-30 of 734 result(s).
Go to page: < Previous **1** 2 3 4 5 6 7 8 9 10 Next >

Facility	Tehsil	Visiting Officer	Date	MO	Other Absent Staff	Report Summary
		⬍		⬍		
BHU KANI	JAND	DDO Jand	2012-07-11	Absent	LHV, SHNS,	📋
BHU BHANGAI	HAZRO	DDO Hazro	2012-07-11	Present	Computer operator,	📋
BHU HAJI SHAH	ATTOCK	DDO Attock/Hassanabdal	2012-07-11	Present		📋
BHU TRAP	JAND	DDO Jand	2012-07-11	Present	Dispenser, LHV, SHNS,	📋
BHU DHURNAL	FATEH JANG	DDO Fateh Jang	2012-07-11	Present	Computer operator,	📋
BHU DAKHNAIR	ATTOCK	DDO Attock/Hassanabdal	2012-07-11	Present		📋
BHU SOJANDA	ATTOCK	DDO Attock/Hassanabdal	2012-07-11	Position Not Filled	Dispenser,	📋
BHU SHAMSABAD	HAZRO	DDO Hazro	2012-07-11	Present	Computer operator,	📋

Fig. 19.1 Online dashboard screenshots

higher rates of government doctors showing up to work in rural health clinics, but only when data on those doctors' (lack of) performance was actionable.

2 Implementation Context

Pakistan is the world's fifth most populous country, equivalent in population to all of Western Europe combined. Punjab is Pakistan's largest province, with over 110 million people. Most of these people rely on free government health services. These are provided by Punjab's Department of Health, which is headed by the Secretary of Health. The Monitoring the Monitors system facilitates flows of information through the Department's existing chain of command, so describing that chain is fundamental to characterizing the program.

This provincial Department comprises 36 District Health Departments, each headed by a "senior health official." Senior health officials are thus in charge of health service delivery for several million people, and they report directly to the Secretary.

Senior health officials are, in turn, each supported by several "health inspectors," typically one for each sub-district (there are, on average, 3.4 sub-districts per district). Health inspectors are charged with inspecting all of the health facilities in their sub-district at least once every month. Health inspectors are themselves trained doctors who likely began their government service lower down in the bureaucracy.

There are 5 classifications of health facilities, all inspected by health inspectors; we focus on the frontline tier, called Basic Health Units, of which there are 2496 spread throughout Punjab (on average 17 per sub-district). Each BHU is headed by a doctor, who is generally supported by six additional staff. Doctors are general practitioners who have completed 5 years of medical school and are therefore the most trained health professionals in rural areas.

Figure 19.2 depicts this administrative hierarchy.

While we have already discussed general reasons that frontline workers might be absent from their post, in this particular context, it is important to note that doctors also face particularly weak career advancement prospects. Very few doctors rise through the ranks to become health inspectors: compared to the 2496 doctors posted to BHUs, there are only 123 health inspectors. This is not the case with health officials higher up in the bureaucracy. Strong performance by senior health officials is commonly rewarded with appointment to a higher office. Health inspectors are also more likely to move up the ranks.

Another important institutional feature in this context is the fact that it is very uncommon for health workers to be fired from their jobs at any level. However, there are still tools for managers to ensure the performance of their subordinates. In the case of senior health officials ensuring the performance of health inspectors, interviews reveal that they typically begin simply by having a conversation with a problematic health inspector. The next step is to refer the matter to a senior provincial-level official in charge of general administration. This can result in a

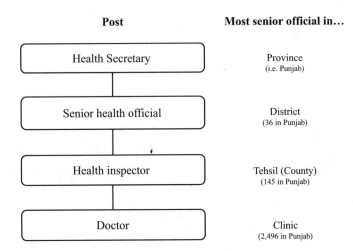

Fig. 19.2 Organization chart for the Punjab Department of Health

formal inquiry and ultimately in pay cuts. In the case of health inspectors ensuring the performance of doctors, health inspectors may issue a "show-cause notice," which requires doctors to explain their absence to senior health officials. In severe cases of persistent absence, health inspectors can transfer doctors to less desirable locations.

3 Innovate, Iterate, Evaluate, and Adapt

3.1 Innovation

While an app that collects data and sends it to an online dashboard may seem trivial today, in 2011 Pakistan, it was novel. The app itself was built using the Open Data Kit open-source platform (https://getodk.org/). As described above, other than the geo-stamps, time-stamps, and pictures, the data collected by the app was not innovative. In fact, it intentionally collected identical data to that collected by the original paper-based forms, to ensure buy-in from the Health Department.

The smartphones required to run the app were the first smartphones that any of Punjab's 123 health inspectors had ever used. Health inspectors had to be trained to use the phones and app in person, district by district, over the course of more than a month. And there were persistent challenges ensuring smartphones remained charged, working properly, and able to regularly sync data with the online dashboard via 2G networks. This required a full-time staff member to help health inspectors to troubleshoot technical issues and to work directly with phone manufacturers and a major telecom operator in Pakistan.

There was a more subtle innovation with this app in the context of 2011 Pakistan, and this is perhaps the reason this case is in this *Development Engineering* textbook. That was the way in which the app and dashboard were developed through a partnership between academics, the World Bank, and the Health Department. The idea of an app and online dashboard was not on the Health Department's radar prior to the World Bank offering the system. And the World Bank-based technology team that developed the system had no experience fulfilling the needs of an entrenched government bureaucracy, such as not changing what data was being collected by the app to allow for sufficient political buy-in. Finally, neither the World Bank nor the Health Department had experience implementing what was essentially a policy reform in a way that could allow for a careful evaluation of its impacts before any scale-up. For example, given the flow of information to senior health officials, it was important to do a clustered randomized controlled trial at the district level. The academics filled this need.

While such a process again may seem trivial today, it was quite innovative at the time, to the point where the system's technology lead, Asim Fayaz, subsequently started a non-profit center based at Lahore University of Management Sciences with support from Google to repeat the same needs assessment and app design process for other departments of Punjab's provincial government. This center, the Technology for People Initiative (TPI), went on to design apps for Punjab's schools, police, courts, and more. Eventually, Punjab's provincial government's in-house app development team, based at the Punjab IT Board (PITB), extended the use of mobile apps in many of those departments. PITB developed more than 270 apps for the provincial government between 2012 and 2017.

As stated in Chap. 2, "DevEng integrates the theory and methods of development economics with the principles and practice of engineering, resulting in technical innovations that are tightly coupled with the social or economic interventions required for success in the 'real world'." In this case, it was crucial that development economists and engineers at the World Bank, later TPI, came together. There was an added group that was crucial, however—the government—that had to be integrated in the process from the beginning.

One caveat is worth mentioning related to the development of the Monitoring the Monitors app and dashboard—the design approach was very top-down. Few doctors were consulted in the design process, as well as few health inspectors. As a result, once the novelty of the smartphones wore off, many health inspectors were unhappy with the system, and they left their phones at home. Of course, the dashboard allowed senior health officials to pick up on missing data and sanction delinquent health inspectors, ensuring the system continued to operate at a level deemed sufficient. We understand that researchers in fields such as Human-Computer Interaction (HCI) have conducted extensive research to suggest bottom-up app development may often be more effective than top-down (Hartson & Pyla, 2019).

3.2 Iteration

While in the end the Monitoring the Monitors system was relatively straightforward in terms of the data it collected and how it displayed that data (excluding the flagging A/B test built into the dashboard), the design, implementation, and evaluation process required several rounds of iteration.

The app and dashboard design process itself required several rounds of iterations between the World Bank technology team and the Health Department. As mentioned above, this type of direct interaction was new for those involved but ultimately successful. This success can be most attributed to the fact that Zubair Bhatti had served in the civil service in Pakistan prior to joining the World Bank. The understanding that this brought, coupled with sufficient patience, ensured the project did not stall at this stage.

Once a beta version of the system had been built, our research team piloted it in one district of Punjab—Khanewal. Khanewal was selected specifically because the District Coordination Officer (DCO), its highest ranked civil servant, was willing to serve as a champion for the new system within a bureaucracy that was resistant to change, especially if it might involve breaking down existing patronage networks that allowed doctors to shirk from their work. This dovetailed with Farasat Iqbal serving as the system's champion within the Health Department. That Rashid Langrial had become Khanewal's DCO shortly before the project launched was luck from the project's perspective. With his help, we were able to pilot the system in a real-life setting with several health inspectors and one senior health official. This proved instrumental in working out several kinks, especially in the process of training health inspectors and troubleshooting issues with data collection and aggregation. The pilot was also instrumental for final Health Department buy-in. The Department needed to see that the system could work.

The pilot also allowed for us to get DCO Langrial's perspective on the likelihood of success of the system. He was the one who recommended the dashboard flagging that ultimately proved successful, arguing that the system would be ineffectual unless we made sure those that wanted things to improve had what they needed.

In the last box, we stressed the importance of adding government bureaucrats themselves as key members of the team when turning DevEng toward governance. It is important that these champions have a deep understanding of political realities. It is also important that you identify champions at each step of the project, from design to pilot to the intervention. In our case, we had the head of a reform unit in the Health Department as well as the head of a district as champions. These individuals were critical.

The large-scale randomized controlled trial that followed can itself be thought of as another round of iteration, as we only rolled the system out across half of Punjab for approximately 1 year. We will now discuss this evaluation.

3.3 Evaluation/Results

The randomized controlled trial (RCT) of the Monitoring the Monitors was conducted between November 2011 and October 2012. To collect data for the RCT, independent survey enumerators made 3 unannounced visits to 850 health clinics randomly sampled from the province's 2496: one before smartphone monitoring began, in November 2011, and two after smartphone monitoring began (in treatment districts), in June and October 2012.

During these unannounced visits, enumerators collected the same information that health inspectors record—information on health clinic utilization, resource availability, and worker absence—as well as information on the occurrence of health clinic inspections themselves. Enumerators also physically verified health clinic staff presence. We also conducted face-to-face surveys with all doctors, health inspectors, and senior health officials in Punjab during the RCT.

We began the experiment after baseline surveys concluded. We randomly selected 18 of Punjab's 36 districts to receive the Monitoring the Monitors system ("treatment" districts), while 17 districts maintained the status quo, paper-based system ("control" districts).[4] We randomized at the district level for two reasons. First, the intervention channels information about health inspections to district-level senior health officials. Second, all inspectors in a district are required to attend monthly meetings and so interact frequently, while these relations are much weaker across districts. District-level randomization therefore makes sense in terms of the design of the program and also reduces concerns about contamination.

At the end of the day, the fact that our evaluation leveraged an RCT allowed us to measure the impacts of Monitoring the Monitors by simply comparing inspection and attendance rates in treatment districts with control districts in the post period (our second and third survey waves). Of course, many small decisions were made in the process of conducting the RCT and measuring the impacts of treatment, such as how to select districts for treatment or control (we chose to "stratify" on several variables to ensure the treatment and control groups were as comparable as possible with each other apart from receiving treatment, for example), which empirical specifications to use, how to conduct hypothesis tests carefully, and so on. Details can be found in Callen et al. (forthcoming).

Our results are summarized in Figs. 19.3 through 19.6.

Note 35 of Punjab's 36 districts were in our experimental sample. Khanewal was left out of our sample for the main RCT because it had already received the system during our pilot.

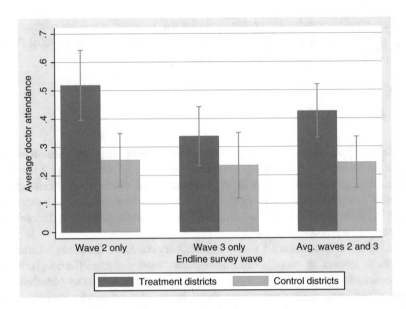

Fig. 19.3 The effect of smartphone monitoring on the rate of inspections. Notes: This figure reports unconditional average treatment effects of the "Monitoring the Monitors" program on the rate of health clinic inspections. The data come from primary unannounced surveys after the treatment was launched (survey waves 2 and 3). The dependent variable is the average inspection rate across clinics within a month prior to the survey. Whiskers depict 95% confidence intervals, calculated using standard errors clustered at the district level. See Table 1 in Callen et al (forthcoming) for additional details

Figure 19.3 shows the impact of Monitoring the Monitors on inspection rates across our sample facilities. We find that health clinics in treatment districts were 18.1% points more likely to be inspected in the previous month during the treatment period. This represents a 74% increase in inspection rates in treatment districts relative to control districts. Breaking this up into the two waves of post-treatment data collection, we find comparable effects, though there is evidence that the effect of treatment had attenuated by October 2012, a year after the introduction of the program. It is interesting to point out that Monitoring the Monitors still did not achieve anywhere near-perfect health inspector performance. This represents the reality that there are many factors, not just a lack of accurate and actionable data affecting health worker performance.

We should keep in mind that apps and dashboards are agnostic to political economy realities and that they cannot solve every problem. They might even exacerbate some problems. They are at best thought of as one part of a larger

(continued)

solution toward ensuring public sector absenteeism, or other governance failures, is not a problem in a particular setting. Development economists who choose to tackle issues one app at a time despite larger realities might think of themselves as a plumber engaged in fixing one leak at a time (Duflo, 2017).

Figure 19.4 shows the subsequent impact of Monitoring the Monitors on doctor attendance at BHUs (an explicit goal of the program). We find mixed results here that are highly dependent on our decisions about empirical specification. Conservatively, it seems the program did not have an impact on doctor attendance, on average. That is not to say that it didn't have an impact in some areas and not others. And, in fact, our flagging results do suggest large impacts when data was made actionable.

Figure 19.5 shows these heterogeneous impacts between BHUs that were flagged as underperforming on the dashboard (those that were likely to draw the attention of very busy senior health officials) and those that were not flagged, comparing

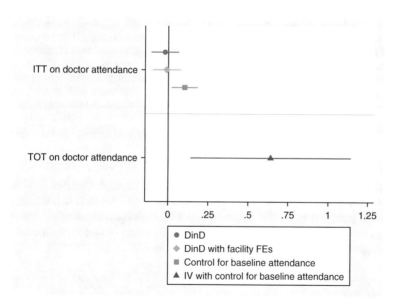

Fig. 19.4 The effect of smartphone monitoring on doctor attendance. Notes: This figure reports estimated intention-to-treat and treatment-on-the-treated (TOT) effects of the "Monitoring the Monitors" program on the doctor attendance. Effects are estimated from OLS regressions in which the unit of observation is the clinic, and data come from primary unannounced surveys after the treatment was launched. The dependent variable is an indicator for whether a doctor was present at the clinic during an announced visit. All models include randomization block fixed effects. DinD refers to a difference-in-difference specification. IV refers to a TOT specification that estimates the effect of inspections on doctor attendance, instrumenting for inspections with treatment. Whiskers depict 95% confidence intervals, calculated using standard errors clustered at the district level. See Table 2 in Callen et al. (forthcoming) for additional details

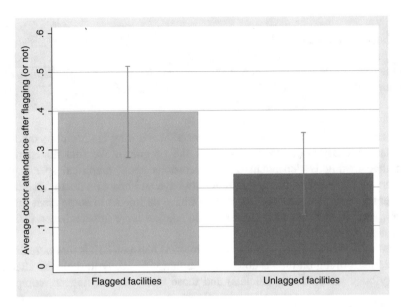

Fig. 19.5 The effect of flagging underperformance on the dashboard. Notes: This figure reports on the effect on subsequent doctor attendance of flagging on an online dashboard the fact that a clinic had three or more staff absent to a senior policymaker. Clinics were flagged in red on an online dashboard if three or more of the seven staff were absent in one or more health inspections of the clinic 11 to 25 days prior to an unannounced visit by our survey enumerators. The data sample limits to facility reports in which either two or three staff were absent (the threshold to trigger the underreporting red flag). In addition, the sample is limited to "Monitoring the Monitors" treatment districts due to the necessity of the web dashboard for flagging clinics. All underlying regressions include survey wave fixed effects. Whiskers depict 95% confidence intervals, calculated using standard errors clustered at clinic level. See Table 3 in Callen et al. (forthcoming) for additional details

facilities just above and below the flagging threshold to ensure the BHUs were otherwise very similar. We find that doctor attendance subsequent to flagging on the dashboard increases by 17.7% points or about 75%. Conversations with government partners suggest that the most likely driver of this effect is verbal reprimands from senior health officials to doctors in charge of clinics.

For our Monitoring the Monitors system to have an impact on doctor attendance, data needed to be made both accurate and actionable. In this case, making data actionable meant making our online dashboard skimmable by high-level, quite busy senior health officials. We did this by drawing particular attention with red highlights to the lowest performing health inspectors and doctors for senior health officials to quickly sanction.

Finally, Fig. 19.6 breaks down the impact of Monitoring the Monitors by two different dimensions than whether a facility was flagged on the dashboard. First, a growing literature documents the role of individuals (as opposed to institutions) in ensuring (or not) the level of public service delivery across the developing world, summarized in Finan et al. (2017). We examine whether personality characteristics of the health workers in our study predict their response to monitoring and to information. To do this, we measured personality characteristics—the Big Five Personality Index and the Perry Public Service Motivation Index[5]—of all of the doctors in our sample clinics and the universe of health inspectors and senior health officials in Punjab. In Callen et al. (2015), we explore this dimension of heterogeneity in great detail. As an example of the role of personality characteristics in this setting, we present heterogeneity of our flagging results in Fig. 19.6, Panel A. We find that personality characteristics systematically predict responses by senior health officials to our dashboard experiment, as measured by future doctor attendance in flagged facilities: a one standard deviation increase in the Big Five Personality Index of senior bureaucrats increases the effect of flagging on the likelihood that doctor is present at the clinic during a subsequent unannounced visit by 28% points. We see a similar though smaller differential effects of flagging in the case of Public Service Motivation. Of course, these results are merely predictive as we did not randomize personality traits.

When seeking to evaluate a program such as Monitoring the Monitors, it is crucial to consider the fact that impacts of the program will likely differ across time, across space, and across individuals. If you stop to develop hypotheses about likely dimensions of heterogeneity from the start (by talking to those

(continued)

Developed by psychologists in the 1980s, the Five-Factor Model, or Big Five personality traits, is now one of the most widely used personality taxonomies in the field. We measure the Big Five traits using a 60-question survey developed specifically in Urdu and validated for use in Pakistan by the National Institute of Psychology at Quaid-i-Azam University, Islamabad. Each of the 60 questions offers the respondent a statement such as "I see myself as someone who does a thorough job" and asks them to agree or disagree with the statement on a five-point Likert scale (disagree strongly, disagree a little, neutral, agree a little, or agree strongly). In addition to measuring Big Five traits separately as the mean response to 12 questions (where disagree strongly is assigned a 1, disagree a little a 2, etc.), all traits are normalized into z-scores and averaged to form a single Big Five index. The Big Five personality traits are agreeableness, emotional stability, extroversion, conscientiousness, and openness. The Perry Public Service Motivation (PSM) battery is designed to measure intrinsic motivation for public service. Also developed in the 1980s, it comprises a total of 40 questions measuring 6 traits—attraction to policymaking, commitment to policymaking, social justice, civic duty, compassion, and self-sacrifice. We similarly combine these traits into one index. See John et al. (2008) for the Big Five measure and Perry and Wise (1990) for the Perry Public Service Motivation measure. See Callen et al. (2015) for more information on these surveys and indices.

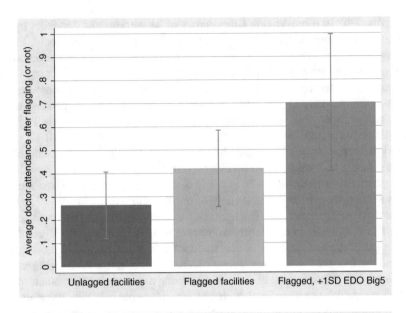

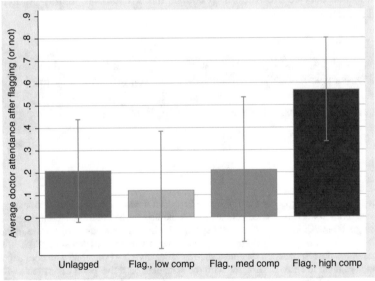

Fig. 19.6 Flagging effects by senior policymaker personality and by political competition. Notes Panel A reports average doctor attendance in clinics 11 to 25 days after the clinic was flagged or not on the dashboard as well as for flagged facilities in districts with a senior health official (EDO with a Big Five index one standard deviation or higher above the mean. The Big Five index is a z-score average across five Big Five personality traits, elicited through surveys of senior health officials. Panel B also reports average subsequent doctor attendance for flagged facilities in districts with low, medium, and high political competition based on the most recent election. Notes for Fig 19.5 provide additional details about flagging. Appendix Tables A12 and A13 in Callen et al (forthcoming) provide additional details about the personality and political competition measures

who know the context), you will have a chance to test for differential impacts and potentially adapt your program to them before long.

Second, we explore whether measures of political competition predict flagging effects. The bureaucrats we worked with to create the program felt strongly that the program would break down when politicians interfered with senior health officials' attempts to sanction their subordinates. Indeed, in our surveys, senior health officials report that politicians routinely interfere in this way.[6] In Callen et al. (2016), we match each clinic in our sample to a provincial assembly constituency and examine in detail the extent to which these political moderators affect the efficacy of the Monitoring the Monitors program. As an example, we find that the treatment effect of flagging on doctor attendance varies by the degree of competitiveness in the previous election. Fig. 19.6, Panel B, presents the results. We find that while flagging increases subsequent attendance by 35.9% points in the most politically competitive third of constituencies, flagging has no apparent effect in the least competitive third. In addition, flagging works better on doctors who do not report a direct connection with a local politician. Indeed, the point estimates, though noisy, suggest the program may have negatively affected attendance of connected doctors.

Taken together, these results drive home that the effectiveness of ICT-based programs such as Monitoring the Monitors very much depends on the individuals making use of the program as well as the political environment in which the program is implemented. Similar findings have been documented in many settings, including by Kentaro Toyama (see Chap. 4). These results also drive home the importance of the Development Engineering framework—the fact that despite the technology is more or less universal, local context is paramount to the effectiveness of potential interventions.

3.4 Adaptation

Perhaps the primary goal of the evaluation of Monitoring the Monitors was to determine the extent to which it should be adapted into a permanent, province-wide policy. The results above suggest the program could be quite effective, but only in certain circumstances. To determine whether the effects were worth the cost, we would have liked to conduct a thorough cost-effectiveness analysis as part of our evaluation. This was not easy, however, since increased inspections or doctor attendance was not the benefits we were ultimately interested in. The ultimate goal was to increase health outcomes for Punjab's rural poor, outcomes that were too

---

[*] Based on interviews with all senior health officials in Punjab, we find that 44% report a politician interfering in their decision to sanction an underperforming employee during the previous year.

costly for us to measure well in this setting. As a result, we had to rely on one fact about our ICT-based intervention to make a recommendation about continuing the policy or not—it was very cheap to implement. The Monitoring the Monitors program cost 17,800 USD to set up and 510 USD per month to operate.[7] While the results of the program are mixed, given this low cost, we would expect it to pass a cost-effectiveness test.

Thanks to the excitement of our Health Department partners about this project, we had the opportunity to present the results from our experiment and our conclusions at one of Punjab's Chief Minister's (the province's highest level bureaucrat) monthly health stock-taking meetings. One slide and 1 minute of presenting led to a nod by the Chief Minister. The government of Punjab then scaled the program up to cover the entire province. The hand-off from our team to the government included several adaptations. First, the government rebranded our system to be called HealthWatch, both for design reasons and also so that the government could take credit for developing a new system. Second, the Health Department had to learn how to manage the system. As an example, when we first handed off the system, it crashed for several months, while the Department worked out a process to pay server, airtime, and phone repair bills. Our full-time staff member also spent a great amount of time meeting with those in the Health Department put in charge of HealthWatch to ensure they had the ability to train new health inspectors and to troubleshoot problems with the system.

It is important to understand the goals of politicians and bureaucrats and how they may differ from those of development economists and/or engineers. While academics often care about academic journal publications as a way to impact their discipline (and gain credit toward promotion), bureaucrats may place a much higher weight on photo opportunities as a way to ensure they receive credit toward their own career advancement.

HealthWatch remains in place today, and it is largely the same as first designed. It is unusual for policies and programs to preserve for so long in Punjab as political tides have regularly shifted. We see HealthWatch's persistence as testament to the strength of the initial Development Engineering collaboration. If any one of the groups involved had attempted to design, implement, and evaluate the program by themselves, it would likely have been unsuccessful. By coming together, we were able to innovate, iterate, evaluate, and adapt our way to a sustainable, impactful program.

The hundreds of similar apps and dashboards that have followed HealthWatch in Punjab can also be considered further adaptations. Some of these adaptations have literally been cut-and-paste jobs. While the photo ops and political payoffs

---

[7] The setup costs included 4470 USD to develop the app and 13,330 USD for smartphones.

to such apps were high, our observation is that such quick adaptations were much less successful than cases when the TPI or PITB teams repeated the entire process that we followed, including the initial design stage. This is in keeping with best practices in Development Engineering (Chap. 3). We have also observed many online dashboards that failed to take the lesson from our flagging results in making data expedient. Bureaucrats have told us when it takes them too long to browse an online dashboard, they tend to delegate to assistants to create paper reports of what is found on the dashboard. This is a clunky and inefficient process at best.

One final anecdote about adaptation—in a recent trip to Punjab, I met with a government vaccinator. When I asked him about the app he was charged with using to keep track of his performance, he pulled out his phone and showed me not one, not two, but three different apps he was required to open, fill out, and monitor for the same set of activities (on top of registers that are still filled out and kept collecting dust in a closet at the BHU). Because it is so easy to create new apps, PITB and the Health Department have taken to doing so rather than fixing and/or adapting old apps. But the bureaucracy is not as nimble and often still relies on old apps for some time, creating redundancies.

4 Lessons Learned

An investment in developing ICT-based solutions to public sector management problems by Punjab and others like it has driven a revolution in the amount of data that can quickly and cheaply be accessed for policy decisions. This trend is only likely to accelerate with the rise of remote sensing, digital trace (e.g., cell phone call and mobile money transaction records), smartphones, and other research innovations. A key lesson from this exercise is that, appropriately channeled, these data streams can improve policy outcomes.

Other lessons are highlighted in the text boxes above:

- The Development Engineering focus of bringing together development economists and engineers holds when working to make changes within a government bureaucracy, but there is another group that must also be involved from the beginning—government bureaucrats themselves. It is particularly important to build this political buy-in from the start of a new project. In our case, we ensured buy-in through partnering with individuals with a deep understanding of political realities and through identifying champions within the government at each step.
- Relatedly, it is important to understand the goals of politicians and bureaucrats and how they may differ from those of the rest of the team. While everyone may agree that health outcomes are paramount, bureaucrats may place a much higher weight on photo opportunities as a way to ensure they receive credit toward their own career advancement. As an academic, it is important to be reminded that many don't care one bit about an economics journal publication.

- Relatedly, it is important to understand that apps and dashboards are agnostic to political economy realities and that they cannot solve every problem. They might even exacerbate some problems. They are at best thought of as one part of a larger solution toward ensuring public sector absenteeism is not a problem in a particular setting.
- For our Monitoring the Monitors system to have an impact on doctor attendance, data needed to be made both accurate and actionable. In this case, making data actionable meant making our online dashboard skimmable and drawing particular attention to the lowest performing health inspectors and doctors for senior health officials to quickly sanction.
- Finally, when seeking to evaluate a program such as Monitoring the Monitors, it is crucial to consider the fact that impacts of the program will likely differ across time, across space, and across individuals. If you stop to develop hypotheses about likely dimensions of heterogeneity from the start (by talking to those who know the context), you will have a chance to test for differential impacts and potentially adapt your program to them before long.

Discussion Questions

1. The success of Monitoring the Monitors was in part due to building off an existing system of data collection. What if there had been no data collected in the first place as is often the case in these settings?
2. We make the point that it may be important to allow specific politicians and bureaucrats to claim credit for a new app. But we have also seen in Pakistan that apps associated too much with one particular individual or party are quickly removed with political turnover. How can a project strike the balance between political expediency and long-term sustainability?
3. Why bother partnering with governments to create monitoring systems in the first place? We could have created a crowdsourced data collection system outside of the government such as a yelp.com or angieslist.com. Indeed, in follow-up work to this project, co-authors and I did this for government veterinarians to great success (Hasanain et al., 2020). The case study in this book on Protecting Electoral Integrity is another example of a political intervention from the outside. When might it be better to work within the bureaucracy rather than from outside?
4. Relatedly, this case has been about improving the flow of information on health service delivery within the bureaucracy and ultimately the performance of frontline health workers in Punjab. Health services could improve dramatically but still not translate to improved health outcomes for poor Punjabis for many reasons, such as a lack of trust in the government. How could you imagine improving the Monitoring the Monitors system, or coupling it with other interventions, to ensure end-goal improvements in health?
5. As more and more systems such as Monitoring the Monitors are maintained across the developing world, the amount of data available to understand public service delivery outcomes, shortcomings, and the political economy reasons behind such has been growing rapidly. If you suddenly had access to all of PITB's data, how could you use it to an effect?

6. Imagine you were the Secretary of Health for Punjab and you learned that personality characteristics of bureaucrats and local political competition are pivotal to whether data is used to improve public sector performance. How could you make use of this information to improve health outcomes in your province?

References

Akhtari, M., Moreira, D., & Trucco, L. (2017). *Political turnover, bureaucratic turnover, and the quality of public services*. Working paper.

Banerjee, A., & Duo, E. (2006). Addressing absence. *The Journal of Economic Perspectives, 20*(1), 117–132.

Banerjee, Abhijit V., Esther Duo, and Rachel Glennerster, "Putting a band-aid on a corpse: Incentives for nurses in the Indian public health care system," Journal of the European Economic Association, 04-05 2008, 6 (2–3), 487–500.

Callen, M., Gulzar, S., Hasanain, A., Khan, M. Y., & Rezaee, A. (2015). *Personalities and public sector performance: Evidence from a health experiment in Pakistan* (No. w21180). National Bureau of Economic Research.

Callen, M., Gulzar, S., Hasanain, S. A., & Khan, Y. (2016). *The political economy of public sector absence: Experimental evidence from Pakistan* (No. w22340). National Bureau of Economic Research.

Callen, M., Gulzar, S., & Rezaee, A. (2020). Can political alignment be costly? *The Journal of Politics, 82*(2), 612–626.

Callen, M., Gulzar, S., Hasanain, A., Khan, M. Y., & Rezaee, A. (forthcoming). Data and policy decisions: Experimental evidence from Pakistan. *Journal of Development Economics.*

Chaudhury, N., Hammer, J., Kremer, M., Muralidharan, K., & Rogers, F. H. (2006). Missing in action: Teacher and health worker absence in developing countries. *Journal of Economic Perspectives, 20*(1), 91–116.

Dal Bó, E., Finan, F., & Rossi, M. A. (2013). Strengthening state capabilities: The role of financial incentives in the call to public service. *The Quarterly Journal of Economics, 128*(3), 1169–1218.

Dal Bó, E., Finan, F., Li, N. Y., & Schechter, L. (2018). *Government decentralization under changing state capacity: Experimental evidence from Paraguay (No. w24879)*. National Bureau of Economic Research.

Dhaliwal, I., & Hanna, R. (2017). Deal with the devil: The successes and limitations of bureaucratic reform in India. *Journal of Development Economics.*

Duflo, E. (2017). The economist as plumber. *American Economic Review, 107*(5), 1–26.

Duflo, E., Hanna, R., & Ryan, S. P. (2012). Incentives work: Getting teachers to come to school. *American Economic Review, 102*(4), 1241–1278.

Finan, F., Olken, B. A., & Pande, R. (2017). The personnel economics of the developing state. In *Handbook of economic field experiments* (Vol. 2, pp. 467–514). North-Holland.

Hartson, R., & Pyla, P. (2019). Chapter 13 – Bottom-up versus top-down design. In *The UX book* (2nd ed., pp. 279–291). Morgan Kaufmann.

Hasanain, A., Khan, Y., & Rezaee, A., (2020). *No bulls: Asymmetric information in the market for artificial insemination in Pakistan*. Working paper.

John, O. P., Naumann, L. P., & Soto, C. J. (2008). Paradigm shift to the integrative big five trait taxonomy: History, measurement, and conceptual issues. In *Handbook of personality: Theory and research*. The Guilford Press. chapter 4.

Muralidharan, K., Das, J., Holla, A., & Mohpal, A. (2017). The fiscal cost of weak governance: Evidence from teacher absence in India. *Journal of Public Economics, 145*, 116–135.

Olken, B. A., & Pande, R. (2012). Corruption in developing countries. *Annu. Rev. Econ., 4*(1), 479–509.
Perry, J. L., & Wise, L. R. (1990). The motivational bases of public service. *Public Administration Review*, 367–373.
Rasul, I., Rogger, D., & Williams, M. J. (2018). *Management and bureaucratic effectiveness: Evidence from the ghanaian civil service*. The World Bank.

Chapter 20
Digital Public Services: The Development of Biometric Authentication in India

Ashwin Nair and Burak Eskici

1 Development Challenge

Throughout the world, government welfare programs face two intransigent challenges: (1) how to deliver public assistance to the neediest, the "targeting" problem, and (2) how to minimize the "leakage" of resources from public coffers (World Bank, 2003). Targeting assistance to those in need is particularly difficult in developing countries because government agencies often lack reliable data on household income and government systems for welfare monitoring are weak (Niehaus et al., 2013). There is also widespread corruption in the delivery of services, a problem documented by many researchers over several decades (Bardhan, 1997; Olken & Pande, 2012). Corruption in public services can come in many forms, from skimming off beneficiary payments to creating "ghost" beneficiaries and extracting bribes from eligible beneficiaries. In 2006, Olken found that 18% of subsidized rice distributed to poor families in Indonesia was unaccounted for, when comparing household reports with administrative data. Even more stark, a study in Uganda

The authors were affiliated with the Payments and Governance Research Program in the capacity of Research Associate and Project Director, respectively. We thank Karthik Muralidharan, Paul Niehaus, Sandip Sukhtankar, and the Payments and Governance Research Program for their support. In addition, we would like to thank Anustubh Agnihotri, Temina Madon, Karthik Muralidharan, Arman Rezaee, and Sneha for their comments and suggestions.

A. Nair (✉)
University of Virginia, Charlottesville, VA, USA
e-mail: an2cg@virginia.edu

B. Eskici
University of California at San Diego, San Diego, CA, USA

Table 20.1 Fiscal cost of leakages in India

Commodity	Government allocation	Public consumption	Leakage	Fiscal cost of leakage
Wheat	18.77 M tons	8.59 M tons	54%	USD 2.1 billion
Kerosene	9.03 M kiloliters	5.35 M kiloliters	41%	USD 1.6 billion
Rice	24.33 M tons	20.69 M tons	15%	USD 1.0 billion

Source: Indian Economic Survey 2014/2015 (Volume 1)
Note: The fiscal cost of leakage has been calculated by multiplying the subsidy cost of a unit of the commodity by the government allocation in excess of household consumption. The total consumption by households has been estimated using survey data from a random sample of households

estimated that on average just 13% of reported government expenditure on primary schools was actually delivered to schools (Reinikka & Svensson, 2001).

The diversion of public resources away from intended beneficiaries is a development challenge on multiple fronts. First, the social and economic value of public welfare programs is drained when those dependent on assistance are prevented from accessing their entitlements (Ferraz et al., 2012; Niehaus & Sukhtankar, 2013). Second, leakage of public funds substantially reduces the efficiency of governments. This is particularly detrimental for developing countries that have limited fiscal capacity, and it undermines citizens' confidence in the public sector (Muralidharan et al., 2017; World Bank, 2003).

Our case study focuses on India, where it is estimated that a significant proportion of spending on key forms of public assistance does not reach its intended beneficiaries (see Table 20.1, from the 2014 to 2015 Indian Economic Survey and Dreze & Khera, 2015). It is estimated that these leakages cost the Indian government an estimated Rs. 28,534 crores annually – translating to roughly 4.7 billion dollars, or 0.2% of India's GDP.

Digital identification (ID) technology offers a promising solution to the challenges of public welfare distribution. In this chapter, we examine the application of two different biometric identification systems in India: a "Smartcards" pilot in the state of Andhra Pradesh and the use of Aadhaar, the national digital ID, in the state of Jharkhand. The applications combined advances in biometric authentication technology with a suite of policy reforms, public-private partnerships, and service delivery innovations. Taken together, these innovations represent a profound investment in India's state capacity.

To grasp the novelty and significance of India's *Aadhaar*, it is helpful to first understand how most governments operate welfare programs today. Welfare services are typically "targeted" to those most in need, with eligible beneficiaries identified through proxy means testing (PMT). These tests establish eligibility for assistance based on readily observable household characteristics that closely proxy income – such as household size, land holdings, quality of housing, and consumer durables. PMT models or algorithms are designed and validated using data from representative surveys (which include more detailed measures of household income, assets, and consumption). A household's score is usually assigned by government workers, who directly observe the outcomes that determine eligibility.

However, targeting is prone to two types of errors: (1) inclusion errors, giving benefits to households who are not eligible for the program thereby increasing leakage and fiscal burden on the government, and (2) exclusion errors, denying benefits to eligible households.

In India, household scores are assigned at the state level, and those falling below the government's poverty line are issued a below poverty line (BPL) ration card (Alkire & Seth, 2013; Niehaus et al., 2013). To access public benefits, registered individuals or households have traditionally presented their BPL card alongside documents that authenticate their identity. However, many people lack this documentation, particularly households and individuals that are socioeconomically marginalized or disadvantaged. It is estimated that over a billion people in the world lack legal identities (World Bank, 2019a).

Assuming eligible citizens or households are correctly identified, the next challenge is reaching and authenticating the targeted beneficiary. Recent advances in biometric technology present a solution to the challenge of rapidly, accurately, and uniquely identifying welfare beneficiaries (Gelb & Metz, 2018). Biometric technologies capture and digitize the physical traits of individuals including fingerprints, iris scans, and speech. As these traits are unique and attached to a person, they cannot be lost or forgotten and are difficult to replicate (unlike paper identification cards; see Goldberg et al., 2010). In addition, ID programs that use biometric technology are more suitable for populations with low literacy and numeracy, particularly compared to alternatives like passwords and pin numbers (Muralidharan et al., 2016).

In the past two decades, the number of national ID programs that use biometric technology has grown rapidly (Gelb & Metz, 2018). India's ambitious *Aadhaar* initiative was one of the first, providing unique biometric IDs to over a billion residents in under a decade. Similar large-scale initiatives are currently underway in Kenya (Huduma Namba), Indonesia (Kartu Tanda Penduduk Elektronik), Nigeria National Identification Number), and the Philippines (PhilSys).

In addition to authenticating citizens' identities, biometric technology can also be linked to the banking ecosystem, enabling governments to use existing payment systems to deliver welfare transfers. Money can be deposited directly into the bank accounts of beneficiaries, which creates more credible records for government accountability and transparency compared to cash-based systems. Even better are digital payments, which can be continuously monitored and audited to expose fraud.

With the proliferation of mobile phones, biometric IDs and bank accounts are also being interlinked with beneficiaries' mobile numbers (i.e., SIM cards), making it possible for banks to remotely authenticate an individual's identity, complete digital payments, and provide customer support without brick-and-mortar infrastructure. Digital payments can also be delivered to the doorstep of the beneficiary, by deploying mobile banking correspondents equipped with point-of-service (PoS) devices to authenticate users and then disburse cash on the spot.

While there are many potential benefits of digital ID technology, these systems also raise important concerns about citizens' privacy and the exclusion of eligible beneficiaries. Many developing countries lack the legal safeguards, legal capacity, and civil society organizations needed to protect individuals from privacy violations and other abuses by the state. For example, there are concerns that biometric IDs can be used by governments to profile citizens and carry out mass surveillance activities, under the guise of national security (Dreze, 2016). There are also fears that governments can monetize interlinked databases, by selling data to private entities (Bhatia, 2018) without citizens' permission.[1] In some countries, biometric ID enrollment efforts have failed to reach the most marginalized households, and it is unclear whether this is politically motivated or a failure of the bureaucracy. Those lacking official identification are made more vulnerable when biometric IDs are made mandatory for accessing welfare schemes and other government services like education and health care (Dahir, 2020) .

The Aadhaar program in India is the largest application of biometric technology in a national ID program; and given its scale, it is a unique opportunity to study how the design of a new technology influences economic development. The effort has exposed a number of intertwined technological, bureaucratic, social, and ethical challenges. Technical problems like hardware failures and lack of Internet connectivity have led to the exclusion of genuine beneficiaries and the starvation deaths of a few. Government collection of biometric data en masse has raised important privacy and security concerns among members of India's civil society (Dreze et al., 2017; Khera, 2019a, b). The opportunities for learning – both what works and what has failed – are unparalleled (Box 20.1).

Box 20.1: Evolution of Biometric IDs in India

Calls for a biometric identity card in India can be traced back to national security concerns related to illegal immigration across India's borders (Government of India, 2001). In 2003, the government launched a biometric Multi-Purpose National Identity Card (MNIC) for all citizens. The MNIC project aimed to provide a "credible identification system" for speedy delivery of public and private services and which could also serve as "deterrent for illegal migration" (Census of India, 2003).

With these aims in mind, a pilot of the MNIC was initiated in 2006 in 13 districts. The pilot ran for a period of 3 years but failed to establish

(continued)

---

[1] Taking cognizance of these issues, the Kenyan High Court had in fact temporarily suspended the implementation of the Huduma Namba until adequate legal safeguards were put in place. On the other hand, the Supreme Court in India in a majority opinion (with one judge dissenting) allowed the use of the Aadhaar biometric ID program for welfare schemes, as it was of the opinion that it collected minimal data for a larger purpose, i.e., to provide services to the poor.

Box 20.1 (continued)

citizenship for more than half of the residents of participating districts. Rural residents, with a weak document base, found it particularly difficult to prove their citizenship (Singh, 2020).

In parallel, the Indian government was developing a biometric identification project focused on identifying families whose incomes were below the poverty line (BPL), with the aim of improving targeting of public assistance to poor households. Challenges in establishing citizenship for these households – and the weak results of the MNIC pilot – led the government to shift its investment in biometrics away from citizenship and toward welfare targeting (Misra, 2019).

In the following years, different states in India introduced their own biometric ID programs to improve last-mile delivery of welfare schemes. Following the success of some state-level biometric ID programs including the AP Smartcards project, the union government created the Unique Identification Authority of India (UIDAI) in 2009, charged with implementing a nationwide biometric identity project. This came to be known as *Aadhaar*. The first Aadhaar card was issued in September 2010; and as of 2020, over one billion Aadhaar cards have been issued to Indian residents – making it the world's largest digital identity program (Abraham et al., 2018).

Although the need for a biometric ID card initially emerged out of national security concerns, Aadhaar is now viewed by the government as a tool to reduce leakages in social programs and ensure better targeting of welfare schemes (Government of India, 2015).

1   Overview of the Case Study

In this case study, we will examine two different biometric identification systems in India: a "Smartcards" pilot in the state of Andhra Pradesh (AP) and the full-scale launch of the *Aadhaar* system in the state of Jharkhand. In each case, the implementation of the new technology was achieved through partnerships between state governments and private entities such as banks. Here we will describe the policy innovations and digital technologies used to form a new digital welfare state across India. We will also outline results from two large-scale randomized controlled trials (RCTs) conducted by the Payments and Governance Research Program[2] evaluating the social and economic impacts of these systems (Muralidharan et al.,

---

[2] The Payments and Governance Research Program is a team of researchers led by Karthik Muralidharan, Paul Niehaus, and Sandip Sukhtankar based at the Jameel Poverty Action Lab (J-PAL), South Asia (see https://sites.google.com/ucsd.edu/pgrp/home).

2016, 2020b). The RCTs were designed to credibly estimate the impact of programs implemented by governments at scale. As a result, these studies also document the complex administrative, logistical, and social factors that can influence the implementation of large-scale technology initiatives.

The chapter is organized as follows: Section 2 provides contextual details of welfare schemes in India, including earlier methods of targeting and delivering benefits. Section 3 describes how biometric identification systems facilitate reforms in beneficiary authentication and service delivery. Section 4 outlines the main challenges in implementing biometric ID programs in India. Section 5 presents results from the large-scale evaluations in two states, and conclusions are provided in Sect. 6.

2 Implementation Context

Despite rapid economic growth since liberalizing the economy in the early 1990s, India's development indicators remain poor. As of 2020, India currently languishes at 131st (out of 189 countries) in the United Nations' Human Development Index (UNDP, 2020). The 2020 Global Hunger Index placed India at 94th out of 107 countries, far behind South Asian peers with slower growth rates. The country also continues to have one of the highest infant mortality rates in South Asia (second only to Pakistan) (UNICEF et al., 2020).

It is argued that India's sub-par development indicators are partly a result of the government's weak implementation of welfare schemes (Pritchett, 2009). Last mile delivery of many government programs has been undermined by low state capacity (Muralidharan et al., 2016; Sukhtankar & Vaishnav, 2015). In this section we motivate the introduction of biometric IDs in India. We describe common modes of failure in the nation's service delivery and how these failures undermine India's investment in poverty reduction and economic development.

2.1 Public Distribution System

The Government of India's Public Distribution System (**PDS**), established at independence as a way to regulate food prices and prevent famines, provide subsidized food and fuel to over 800 million beneficiaries in India through a network of fair price shops (FPS). The Indian government spends approximately 1% of its GDP and 12% of its social services expenditure on the PDS (Government of India 2020).

Beneficiaries of the PDS are assigned a ration card which lists household members and displays a photograph of the household head. They are then assigned to a geographically close FPS. They visit the FPS in person (ration cards in hand) to purchase rations each month at a highly subsidized price.

Prior to the introduction of biometric authentication, shopkeepers at the FPS (known as FPS dealers) were expected to record transactions on both ration cards and their own paper ledgers, neither of which were regularly audited by the government. The informal nature of authentication, alongside the dual price system created by public subsidies, created opportunities for leakage. This resulted in the diversion of PDS commodities to the open market, both directly from government warehouses and from trucking networks established to deliver goods to FPS dealers. In addition, beneficiaries have reported the adulteration of goods, overcharging for rations, and reduction in their entitlements at FPS (Muralidharan et al., 2020b).

2.2 National Rural Employment Guarantee Scheme

The National Rural Employment Guarantee Scheme (**NREGS**) was implemented across India after the enactment of the National Rural Employment Guarantee Act in 2005. The Act mandated that state governments set up employment programs guaranteeing 100 days of paid employment per year to any rural household. The Indian government spends approximately 0.3% of its GDP and 4% of its social services expenditure on the NREGS (Government of India, 2020).

Work provided through the NREGS can vary across villages but typically involves minor irrigation work or improvement of marginal lands. There is no eligibility criterion for the scheme, and those in need of wage labor can access the scheme (although the poorest households are more likely to self-select into the program). Households that participate in the scheme receive a job card which lists household members and has empty spaces for recording employment and payment. The job cards are issued by the local village or sub-district government offices. Workers with job cards can apply for work at will, and officials are legally obligated to provide either work or unemployment benefits (although, in practice, the latter are rarely provided; see Muralidharan et al., 2016).

Worksites are managed by officials called Field Assistants, who record worker attendance and output on "muster rolls." These are then sent up to the sub-district office for digitization, after which the work records are sent up to the state level. This triggers the release of funds to pay workers. The state government transfers money to district offices, which then pass the funds to sub-district offices, which in turn transfer it to the beneficiary's post office savings accounts at the village level (see Fig. 20.1).

Prior to the roll-out of biometric authentication, workers would withdraw funds by traveling to their village's post office, where they would establish their identity using job cards and bank passbooks. In practice, it was common for workers, especially those who were illiterate, to give their documents to the Field Assistant who would then control and operate their bank accounts. This would entail taking sets of bank passbooks to the post office, withdrawing cash in bulk, and returning to distribute it in villages.

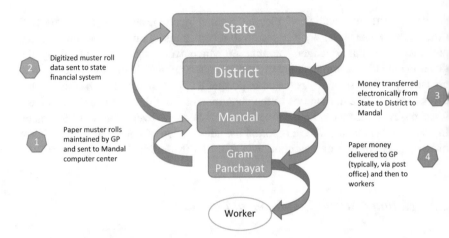

Fig. 20.1 Payment mechanism of the NREGS prior to the roll-out of biometric authentication
Note: The black arrows denote the flow of information on beneficiaries of the NREGS. The arrows
in blue denote the flow of funds to the beneficiary. Source: Muralidharan et al. (2016)

The payment mechanism was susceptible to leakage of two forms – over-
reporting of hours worked and underpayment – with the former being more
prevalent (Niehaus & Sukhtankar, 2013).[3] Two extreme forms of over-reporting are
"ghost" workers who do not exist, but against whose names work is reported by
Field Assistants, and "quasi-ghost" workers who exist, but who have not received
any work or payments, even though their work is reported and payments are made.
In both cases, the payments are typically siphoned off by government officials
(Muralidharan et al., 2016). Estimates of leakages in the NREGS have varied
across districts in India, ranging from 5 to 40% (Mookherjee, 2014). Additionally
payments made to beneficiaries through the NREGS have also been slow and
unreliable. In some extreme cases, delayed payments have led to worker suicides
(Pai, 2013).

2.3   Social Security Pensions

The Social Security Pensions (**SSP**) scheme, part of the Indian government's
National Social Assistance Programme, offers monthly payments to vulnerable
populations. Eligible households must live below the poverty line (BPL), meet
an additional vulnerability criterion,[4] and not be covered by any other pension

---

[3] To illustrate how over-reporting and underpayment took place in practice, consider a worker who
has genuinely earned Rs. 100; the Field Assistant might report in the muster rolls that he is actually
owed Rs. 150 but pay the worker only Rs. 90 thereby pocketing Rs. 50 through over-reporting and
Rs. 10 through underpayment, resulting in a leakage rate of 40% (Muralidharan et al., 2016).

[4] Old age (>65), widow, disabled, or certain displaced traditional occupations.

scheme. Lists of eligible households are prepared by local village assemblies and sanctioned by the administration at the sub-district level. Pensions pay Rs. 200 per month except for disability pensions, which pay Rs. 500. A fund flow similar to that illustrated in Fig. 20.1 existed for the SSP (Muralidharan et al., 2016).

As the scheme mostly has a fixed list of beneficiaries who received a fixed payment, at a fixed time every month, for every month of the year, the SSP has been less prone to leakages than the NREGS and the PDS (ibid.). Leakages in the SSP have mainly taken the form of postmen disbursing less cash than the beneficiary's entitlement, the existence of false beneficiaries (including deceased persons), demand for bribes, or theft. A study in the state of Karnataka found leakages in the SSP to be relatively lower, at about 17% (Dutta, 2008; Dutta et al., 2010).

3 Innovation

Biometric authentication technology has the potential to transform the delivery of public services in resource-poor settings, where state capacity is limited. It promises to reduce leakages resulting from ghost beneficiaries and improve the quality of last-mile service delivery. In the case of India, the technology's adoption took place over a decade, beginning with initial state-level pilots and eventually scaling up across the country. This case study covers the pilot phase, in the form of an offline smartcard biometric system in the state of Andhra Pradesh (referred to as "AP" or 'AP Smartcards") between 2010 and 2012, as well as the central government's fully scaled program, *Aadhaar*, as implemented in the state of Jharkhand in 2016–2017.

Both state governments introduced biometric ID systems alongside other technology advances and policy reforms. For example, they linked biometric IDs to bank accounts and to the existing databases of large welfare schemes, complementing the central government's push to expand financial inclusion and improve service delivery (Government of India, 2015). However, the two states differed in the application of technology: AP used biometric IDs for rural employment (NREGS) and social security pensions (SSP), while Jharkhand used biometric IDs for food and fuel subsidies (PDS). The resulting solutions are described here. We first explain how the solutions modified government identification protocols and then describe how payment systems and service delivery were reformed.

3.1 Reforms in Identification

As discussed earlier, failures in paper-based government identity systems have allowed for several forms of leakage from public programs – from the creation of "ghost" welfare beneficiaries and the exclusion of intended beneficiaries to the extraction of bribes. Biometric authentication enables more accurate, auditable, and secure identification and targeting of beneficiaries.

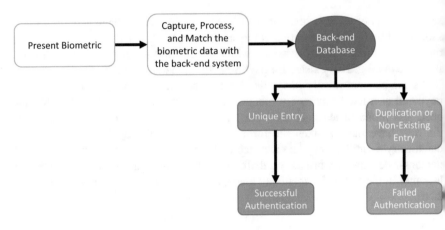

Fig. 20.2 Process flow of a biometric authentication system
Note: A backend database may be a central server that stores biometric data (such as the Central Identities Data Repository in the case of Aadhaar) or can also be biometric data that is stored on the card itself

To establish a biometric ID system, individuals must first be registered or enrolled. Their physical traits are recorded using devices that contain cameras, scanners, or other hardware. A backend software system then extracts, encodes, and stores this information in a database. Any request for authentication of a registered user compares an "input" provided in person (like a fingerprint) against the database of encoded biometric data (see Fig. 20.2). As the entire process is automated, authentication takes only a few seconds in most cases, including for Internet-connected systems (Xavier et al., 2012).

An important design choice is around the storage location for an individual's biometric data. It can be stored centrally on a server or stored locally on a smartcard.[5] If biometric data is stored on the card, authentication can occur offline. A card reader is used to confirm that the data stored on chip matches the input presented in person. However, this approach has a number of limitations. First, each smartcard requires a chip for storing data, making large-scale programs quite expensive (World Bank, 2019b). Second, without a central database that assigns a unique ID to each individual, the biometric data stored on a smartcard is prone to duplication, which can be difficult to discover and remediate. Therefore, stand-alone biometric smartcards are prone to some of the same issues that affect traditional paper-based identification processes (Banerjee & Sharma, 2018). Lastly, smartcards place limits on usability, as the physical smartcard must be presented for access to

---

[5] Another (more costly) option is combining both designs with the use of a chip-based smartcard with a unique ID linked to a central server. This allows for both online and offline authentication. National biometric ID cards in Indonesia and the Philippines utilize this design (Asian Development Bank, 2016).

services. If a beneficiary loses or damages their smartcard, there is no backup: they are now unable to authenticate themselves.

At the time the AP government initiated its biometric ID project in 2007, there was no universal ID database, and high-speed mobile networks were not universally available. As a result, the government chose to issue smartcards with local storage of data. This meant using biometric data to authenticate welfare beneficiaries only at the point of benefit disbursement (Mukhopadhyay et al., 2013).

In the case of Jharkhand, the state was able to use a newly created central government database, called the Central Identities Data Repository (CIDR). Each person's biometric data, once stored in the CIDR, is linked to a unique ID assigned to the individual. Biometric data of over a billion Indian residents is now stored on the CIDR, and the central database is used to match beneficiary biometrics across the country, at the point of service.

The central government made a conscious design choice to not use smartcards. The Unique Identification Authority of India (UIDAI), a unit established to implement *Aadhaar*, did not want to restrict the use of biometric authentication to applications requiring card readers (Dholabhai, 2012). A disadvantage to this choice is that the system requires Internet connectivity to match a biometric input with data stored in the CIDR. Technical errors due to poor Internet connectivity are therefore a potential source of exclusion. Despite concerns raised by other government agencies, the UIDAI maintained the view that with the expansion of high-speed mobile networks, Internet connectivity would grow more reliable with time, particularly in rural areas (Zelazny, 2012).

Another design choice for both smartcards and *Aadhaar* is where to set the target accuracy for biometric authentication. Accuracy is dependent on the quality of the biometric data that is captured when beneficiaries are enrolled. However, it is also possible for an individual's biometrics to change over time. For example, fingerprints tend to wear off, and eye surgeries can alter iris texture patterns. These can result in technical errors, due to a mismatch between the original biometric records (captured during registration) and those being captured in real time (Ramanathan, 2018; Nigam et al., 2019). Populations such as senior citizens and casual laborers are particularly vulnerable to such authentication errors (Khera, 2019a, b).

One way to improve the match accuracy is to base authentication on multiple biometrics rather on a single biometric. Accuracy levels with a combination of fingerprints and iris scans are an order of magnitude better than using only one or the other (UIDAI, 2012). With multiple biometrics, vulnerable populations whose fingerprints have worn off can still authenticate themselves with iris scans. With each additional biometric, the chances of duplication reduce, increasing the accuracy of the biometric match.

The AP government based its smartcards project on fingerprints (of eight to ten fingers). This may have been due to limitations in the storage capacity of biometric data on smartcards. *Aadhaar*, on the other hand, is based on a multimodal system of biometric authentication; in addition to fingerprints, iris scans can also be used. However, authentication through iris scans does depend on an additional hardware component that can read the iris. Without this, *Aadhaar*-based authentication remains dependent on a single biometric.

3.2 Reforms in Payment Systems and Benefits Delivery

Prior to the introduction of biometric IDs, payments and benefits delivery systems for welfare schemes in India were susceptible to delays and inefficiencies, imposing significant costs for both households and the government. Beneficiaries incurred significant time costs in travelling to bank branches and post offices to access benefits. Governments lacked the ability to track and reconcile transactions in the paper-based system, exacerbating the problem of leakages.

In contrast, biometric IDs linked to bank accounts allow for remote authentication of beneficiaries and the issuance of digital payments. Banks can also hire agents (called business correspondents) to make last-mile payments to beneficiaries using PoS devices that support biometric authentication. This eliminates the need for beneficiaries of the NREGS or the SSP programs to travel to a bank branch or a post office. Instead, payments can be made by business correspondents, at the beneficiary's doorstep. Figure 20.3 illustrates a sample process flow for a biometrically enabled payment system.

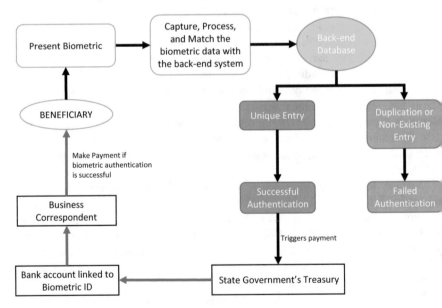

Fig. 20.3 Process flow of a biometrically enabled payment system
Note: The blue arrows show the flow of funds from the state government's treasury to the beneficiary

Biometric authentication on PoS devices also allows for digital record-keeping, as digital receipts are generated for each transaction (Muralidharan et al., 2020b). Governments can therefore reconcile each transaction made to beneficiaries. This is useful particularly in the context of the PDS. The digital transaction records created at an FPS allow the government to reconcile household purchases with the subsidized food rations allocated to each shop. If an FPS dealer records transactions of an amount lesser than the entitlements allocated, the government can reduce future allocations, to adjust for the previous month's balance. This penalty disincentivizes the FPS dealer from over-reporting the volume of goods disbursed to beneficiaries each month. Regular reconciliation can also, in principle, increase government savings by restricting expenditures to only the quantity of commodities required by beneficiaries.

4 Implementation Challenges

As this textbook argues, in order to achieve development impact, basic technologies need to be modified and adapted for local context. The process of building an appropriate biometric ID technology, and then implementing it on a large scale, is not without challenges. As highlighted in the prior section, certain design choices for biometric technology can lead to the exclusion of significant numbers of genuine beneficiaries. In this section, we show how these choices resulted in exclusion errors and how the two state governments in Andhra Pradesh and Jharkhand dealt with challenges in each of the following steps involved in implementing a biometric authentication system.

4.1 Enrollment of Beneficiaries

The most resource-intensive step in any biometric ID program is the enrollment of individuals. In both the AP Smartcards and Jharkhand *Aadhaar* programs, the government entered into partnerships with private service providers to execute enrollment.

In the case of the AP government, it was difficult to incentivize private players (mainly banks) to participate in the biometric program. Partnership with the AP government required private service providers to bear the upfront cost of infrastructure required for biometric smartcards, in return for a 2% commission on each payment transaction. The government hoped that banks would view this as a business opportunity, enabling them to establish a foothold in rural areas. However, banks were uncertain about the profitability of the program and were therefore reluctant to participate. The banks that did participate most likely did so under pressure from the state government (Mukhopadhyay et al., 2013).

These service providers faced a number of issues while enrolling beneficiaries. There was a lack of coordination with local government officials, which resulted in weak mobilization of beneficiaries at enrollment camps. Technical problems also cropped up, including hardware and software glitches and errors in the lists of eligible beneficiaries supplied by the AP government (ibid.).

Because a substantial share of beneficiaries remained unenrolled, the government was concerned that the high volume of payments to non-enrolled beneficiaries would re-open the door to corruption and leakages. They deployed two strategies in response. First, to incentivize beneficiaries to enroll for smartcards, the government ordered the stoppage of payments to non-enrolled beneficiaries. NREGS and SSP funds were placed in suspended accounts until beneficiaries completed their enrollment. But due to the slow pace of enrollment, the government finally had to relax its stance, allowing manual payments as long as beneficiaries were simultaneously enrolling for biometric IDs. Second, the government tried eliminating commissions for payments that were not biometrically authenticated. The enforcement of this rule required government officials to regularly monitor transaction-level data. However, the government was unable to enforce this rule, as banks were slow to respond to the government's requests for detailed transaction-level data (ibid.).

Unlike in AP, enrollment in Jharkhand took place at a rapid pace. The government of Jharkhand's efforts of introducing biometric authentication in the PDS had the advantage of being built on top of the national biometric ID program – Aadhaar which was already being rolled out by the central agency, the UIDAI.

The UIDAI worked through state governments to hire private entities known as enrollment agencies. These had to be certified by the UIDAI to carry out enrollment (Sathe, 2014; Sen, 2019). Contracted agencies set up mobile *Aadhaar* enrollment centers. At these centers, fingerprints for all ten fingers and iris scans were obtained. Individuals also were required to submit existing identity documents as proof of their demographic details (including name, address, date of birth, and gender). The information collected by the enrollment agency was then encrypted and sent to the UIDAI, which would compare the incoming data with data already stored on the CIDR to check for duplicates. If there were no duplicates, the UIDAI would generate a 12-digit unique ID number (called the *Aadhaar* number) and issue a card by post with the individual's *Aadhaar* number and demographic details. If there were any discrepancies identified in the deduplication step, the individual would be informed of next steps for remediation (Misra, 2019). The enrollment agencies were paid a commission for every successful enrollment (Zelazny, 2012).

Over the course of a decade, the UIDAI issued over a billion cards to Indian residents. This was aided by the incentive structure the UIDAI had put in place (i.e., a fixed incentive paid for each successful enrollment). The Government of India also made *Aadhaar* enrollment mandatory for a number of welfare schemes (including the PDS and the NREGS) and made the *Aadhaar* card a legitimate identity document for a variety of other purposes (Perrigo, 2018).

The *Aadhaar* incentive structure and rapid pace for enrollments did, however have an impact on the quality of data collected by enrollment agencies and made enrollment vulnerable to malpractice. Demographic data collected by enrollmen agencies were prone to error, since they were dependent on the quality of paper documents submitted. Misspelt names and incorrect dates of birth have been widely reported (Khera, 2019a, b). In addition, mix-ups in biometric data were reported, with the biometrics of one family member mixed together with other family members' records (Mohammed, 2018).

Ultimately, the Indian government had to blacklist 34,000 enrollment agencies in the period 2010–2017 (Hebbar, 2017). Reasons for dismissal from the program included charging individuals a fee for enrollment (even though enrollment for *Aadhaar* was supposed to be free) and using weak security protocols that led to leaks in the demographic data of a large number of applicants (Khaira, 2018; Surabhi, 2017). As with AP Smartcards, technical issues in enrollment were also common, especially in remote areas. This made it difficult for individuals in some regions to apply for *Aadhaar* cards (Matthew, 2019). Those lacking high-quality biometrics and those unable to physically travel to an enrollment agency (owing to disability or old age) found it difficult to enroll (Drolia, 2018).

4.2   Deduplication of Records

To ensure that the same individual is not able to obtain more than one biometric ID, it is important to run regular deduplication checks using algorithms that can accurately identify duplicates. In AP, the government made banks responsible for carrying out deduplication checks. With deduplication, the state hoped to establish a clean beneficiary database of eligible and legitimate beneficiaries for its welfare schemes. However, banks made limited progress in implementing robust deduplication checks. As the development of advanced deduplication software was still in its early stages, banks were at best able to carry out rudimentary checks, such as a village-level text deduplication of beneficiaries (Mukhopadhyay et al., 2013).

In comparison, *Aadhaar*'s software and hardware architecture enabled superior detection of duplicates. The system uses multiple biometrics (fingerprints and iris scans) for authentication, and the underlying data are stored in a central server called the CIDR. While a biometric database of ten fingerprints can ensure a deduplication accuracy of more than 95%,[6] with iris scans, the accuracy can increase to more than 99%. However, as noted earlier, the accuracy rates are highly dependent on the quality of data collected during the enrollment stage (UIDAI, 2009).

This is referred to as the true acceptance rate (TAR) of a biometric system and is a measure of the number of times a biometric system can correctly verify an individual's claim of identity. For example, a TAR of 95% implies that if 10% of individuals in a population of 1000 try to register for a biometric ID twice, the system would incorrectly generate duplicate biometric IDs for 5 individuals. Similarly, for a population of 1.2 billion, this would imply duplicate biometric IDs for 6 million individuals. An increase in the TAR to 99% brings this number down to 1.2 million individuals.

As deduplication algorithms had never been tested at this scale, the UIDAI decided to select three vendors rather than one. It hoped that competition among the three vendors would yield greater accuracy. The vendors were paid according to the number of deduplication checks they were able to carry out, and the accuracy of the checks determined the allocation of payments across the three vendors (Sharma, 2016). It is not technically feasible for any deduplication algorithm to achieve 100% accuracy, so there is always a possibility of duplicate *Aadhaar* cards being generated for the same individual (Abraham et al., 2018; Agarwal, 2013). Given the scale at which the program has been rolled out, and challenges with the quality of the enrollment process, accuracy rates of more than 99% for a population of over one billion can still create a substantial number of duplicates[7] (Matthews, 2016; Zelazny, 2012).

4.3   Authentication of Identity

Errors in authentication occur when genuine beneficiaries are unable to authenticate themselves using biometric IDs. This can occur for many reasons, and the consequences can be severe: authentication errors without a manual override option (which is discussed in the next section) can strip beneficiaries of their legal entitlements.

In AP, biometric data were stored on the smartcard itself, so authentication could occur offline. However, the system was still prone to authentication errors, mainly due to technical problems with the PoS machines used. In some cases, the PoS reader was unable to recognize fingerprints, despite multiple attempts; in other cases, intermittent lack of power rendered the POS devices useless. Another source of authentication errors was the poor quality of biometric and demographic data collected during enrollment (Mukhopadhyay et al., 2013).

Aadhaar has been vulnerable to some of the same authentication errors experienced in the AP Smartcards program. In addition, it has suffered from poor Internet connectivity in some regions. With biometric data stored on a central server (the CIDR), all real-time authentication requests require connectivity to the Internet. In rural areas with poor Internet connectivity, online authentication proved a major challenge and a cause for significant exclusion of beneficiaries from welfare schemes (Dreze et al., 2017; Sneha, 2017).

The UIDAI did create certain alternatives to online authentication, although their use has been limited. One such alternative is the use of a one-time password (OTP) sent to the mobile number linked to an individual's UID. However, the use of an OTP naturally requires an active mobile phone number with good mobile connectivity. Early on, mobile phone ownership and connectivity were still lacking

---

[7] The results of an intervention implemented by an NGO at a small scale, for example, may be very different if the same intervention is rolled out by a state/provincial government at a much larger scale.

in rural areas in India, especially in states with low teledensity such as Jharkhand. Administrative errors in the linkage of mobile numbers to *Aadhaar* IDs – including the entry of incorrect mobile numbers during enrollment, and the use of "head of household" mobile numbers rather than individuals' numbers – reduced the reliability of the OTP alternative (Dreze et al., 2017; Muralidharan et al., 2020b). In addition, awareness of this fallback mechanism has been quite low (Abraham et al., 2018).

A second alternative for offline *Aadhaar* authentication is the use of offline PoS devices. Offline PoS devices capture the beneficiary's fingerprints but cannot perform real-time authentication. Beneficiaries can still collect their entitlements after scanning their fingerprints. The operators of the offline PoS devices are required to regularly synchronize their transaction logs with the central government server by accessing the Internet. To ensure that logs are regularly synchronized, new transactions are not authorized if the operator has not synchronized earlier logs. To minimize inclusion errors (wherein non-eligible beneficiaries access welfare schemes), the use of PoS devices for offline authentication had been limited to areas with extremely poor connectivity (Muralidharan et al., 2020b).

Another source of authentication errors in Jharkhand was the incomplete or incorrect linking of Aadhaar IDs with existing government beneficiary records, such as that of the PDS. Once beneficiaries of the PDS had obtained an *Aadhaar* ID, they were required to link their *Aadhaar* numbers with their existing PDS ration card numbers, a process known as "seeding." This was a non-trivial process and imposed significant costs on beneficiaries, as they often had to make repeated trips to their local government offices to ensure that ration cards had been seeded. Families with even a single member lacking the seeded ration card were unable to access their PDS benefits. The seeding process was also prone to data entry errors (Dreze et al., 2017; Sneha, 2017).

4.4   Manual Overrides

Given the many challenges described so far, there have been legitimate concerns that making biometric IDs mandatory for public benefits can lead to exclusion of genuine beneficiaries. To reduce exclusion errors, one approach is to create manual overrides. This entails the use of non-biometric authentication, for use when it is impossible to biometrically authenticate an individual. However, manual overrides bring the risk of increased inclusion errors. In the absence of sound protocols for manual overrides, non-eligible beneficiaries can easily access welfare schemes, undermining one of the primary benefits of biometric ID programs.

As discussed earlier, a significant proportion of beneficiaries in AP were never enrolled for biometric smartcards, because the government was unable to incentivize banks and beneficiaries to increase their participation in the smartcard system. This resulted in heavy reliance on manual overrides. Business correspondents continued to make payments (in cash) to unenrolled beneficiaries and to individuals unable to biometrically authenticate themselves. To minimize inclusion errors, the override protocol required business correspondents to authenticate using paper-based ID documents, and for record-keeping, beneficiaries were made to stamp their fingerprints on a pre-filled beneficiary roster (Mukhopadhyay et al., 2013).

In Jharkhand, *Aadhaar* was in effect a mandatory requirement for accessing welfare schemes such as the PDS; and compared to AP, the use of manual overrides was limited. Offline PoS machines and OTPs were the only fallback options available to those with an *Aadhaar* ID who could not authenticate biometrically (Dreze et al., 2017). Those without an *Aadhaar* card (and those whose entire family had failed to "seed" their Aadhaar cards with the ration database) were, in effect, excluded from the PDS. Although the central government did issue notifications to maintain exemption mechanisms, these mechanisms were rarely used (Dutta, 2019).

4.5   Political Buy-In

The successful implementation of any large-scale reform hinges on strong political support. This is often hard to achieve, as government officials and frontline service delivery workers who have benefited under the status quo have little incentive to ensure that reforms are successfully implemented – even when these reforms improve welfare outcomes for society as a whole (Hoff & Stiglitz, 2008; Olson, 1965). Fortunately, AP's smartcard program had strong political support among senior state government officials. Senior bureaucrats invested significant time and effort to ensure that the smartcards were successfully rolled out. The AP government was also strongly committed to making welfare schemes such as the NREGS and the SSP easily accessible for beneficiaries. This is reflected in the fact that the state had one of the highest fund utilization rates for welfare schemes in the country (Mukhopadhyay et al., 2013).

However, support for the program from local government officials (at the district and sub-district levels) was weak, and this hampered progress in implementing the reform. In addition, business correspondents (BCs) hired by banks to deliver payments to beneficiaries became a sought-after position, which politicized the process of selecting BCs. Politicians would lobby for their own candidates, for political gains. This stalled the selection process in many villages, affecting the delivery of payments. There were also instances of active resistance to the smartcard program from local government officials, and local officials were more likely to convey negative anecdotes about the reform. In fact, opposition from local officials was strong enough for the government to consider abandoning the reform in 201? (ibid.).

Aadhaar, on the other hand, received strong support early on from important decision-makers in the central government. This helped the program gain political mileage, despite opposition from prominent politicians and external actors related to exclusion risks and privacy concerns (Misra, 2019). Importantly, *Aadhaar* aligned with the government's objective of reducing the fiscal burden from leakage in welfare schemes (Government of India, 2015). Indeed, fiscal savings from biometric authentication was the primary benefit highlighted in government report of *Aadhaar* pilots in various welfare schemes (Dreze & Khera, 2018; Government of India, 2016). A change in the central government's ruling party midway through the program's implementation had little impact on the pace at which the *Aadhaar* cards were issued to Indian residents and linked with welfare schemes (Sathe, 2014).

State governments that implemented Aadhaar-Based Biometric Authentication (ABBA) in their welfare schemes devoted significant resources to the reform, following the central government's policy priority of reducing leakages. At the time, state-level reforms were politically popular thanks to strong voter support for reforms addressing corruption and fraud (Sukhtankar & Vaishnav, 2015). As such, the Jharkhand government was willing to institute ABBA in its welfare schemes even if it came at the cost of excluding some genuine beneficiaries (Muralidharan et al., 2020b).

The Jharkhand government's main policy priority was to increase fiscal savings through ABBA, and this is reflected in the limited use of manual overrides. The state also increased the rate of beneficiary deletion from the PDS rolls shortly before ABBA was introduced. The government claimed that beneficiary names lacking an *Aadhaar* number were primarily "ghost" beneficiaries. The government also decided to reconcile grain stocks at PDS shops, with the aim of reducing its expenses on the PDS. Reconciliation entailed adjusting the delivery of commodity grains to shops based on earlier transaction records. Reconciliation proved quite unpopular, as FPS dealers generally passed a significant proportion of the government's reduction in benefits along to beneficiaries. In fact the government had to temporarily suspend and relax its reconciliation protocols in the face of opposition from FPS dealers and beneficiaries (ibid.).

4.6 Innovation and Iteration at Scale

The implementation challenges highlighted here illustrate how biometric ID technology, like any other technological innovation, must be iteratively modified to the context in which it is being deployed. The digital identity solutions deployed in AP and Jharkhand were influenced not only by technology design decisions but also by the broader financial ecosystem and local policy environment. As solutions were scaled across India by UIDAI, the database architecture and hardware became somewhat fixed; but variation in delivery was facilitated through policy and protocol design. Banks needed the right incentives, and state-level agencies needed tools to monitor implementation to ensure that the gains from innovation were reaching beneficiaries. In line with the framework for Development Engineering outlined in this book, the features of the solution were continuously revisited and modified as implementation expanded.

One lesson from this is the importance of researchers engaging with governments and not shying away from the politics of implementation. As this case study demonstrates, the AP and Jharkhand governments weighed the benefits from reduced leakages against the accompanying increase in exclusion errors resulting from biometric authentication. Their decisions were ultimately a function of policy priorities. The AP government's focus on improving beneficiary experience explains some of the design details of AP Smartcards, including their commitment to offline authentication, manual overrides, and payments to non-enrolled beneficiaries. The Jharkhand government's focus on fiscal savings explains why *Aadhaar* was in effect made mandatory for access to welfare services and why limited override mechanisms were made available for beneficiaries.

5 Results from Large-Scale Evaluations

Whether a technological innovation can be viewed as contributing to economic development depends on the rigor with which it is evaluated. A priori, it was unclear whether welfare beneficiaries in India would be better off with biometric IDs since these systems can exclude real beneficiaries. They can be very costly, and faulty implementation can compromise performance. It is important to measure rates of inclusion and exclusion, and other user experiences, to assess the overall welfare impacts of such a large-scale program. In addition, the massive scale of deployment of digital IDs across multiple states of India made evaluation crucial for understanding impacts on the broader economy. Implementation within large geographies can create unique dynamics that are never captured by small-scale evaluations.

Here, we will describe the results of two large-scale randomized evaluations implemented by the Payments and Governance Research Program (Muralidharan et al., 2016, 2020b). What makes these evaluations unique is the scale at which they were conducted, across two varied contexts.

5.1 Experimentation at Scale

The use of RCTs to evaluate the impact of welfare programs has become a common tool in social science research. RCTs involve randomly assigning a treatment (or policy intervention) across a population and then comparing outcomes of those who receive the intervention with those who do not. As treatment and control groups are randomly assigned, RCTs generate credible estimates of the causal impact of an intervention, within the context of the population under study.

There have been concerns about the representativeness of results from RCTs beyond the immediate context of the study sample. If the study's sample is not representative be of the population a government is interested in, which findings can be generalized? How should the evidence from a small trial inform a larger-scale policy intervention? (see footnote 7) Spillover effects – in particular, how the effects of the intervention interact with those who do not receive the intervention – can become pronounced at large scale. An evaluation that does not measure how changes in welfare programs affect the wider economy[8] may underestimate the impact of an intervention like biometric IDs (Muralidharan & Niehaus, 2017).

The researchers on the team dealt with these issues by carrying out RCTs at scale, which offers the following benefits:

1. The studies consist of samples that are representative of larger populations.
2. The unit of randomization[9] is selected such that it is large enough to capture spillover effects.[10]
3. The evaluations can credibly estimate the impact of policies that are implemented by governments at a large scale.

Given the nature of these evaluations, the results can provide valuable input on the impacts of biometric ID systems for policymakers in other developing countries.

5.2 Results from the Evaluation of AP Smartcards

The evaluation of AP's smartcards initiative was made possible by an unexpected glitch in the program's roll-out. Because of the challenges in beneficiary enrollment described earlier, there were still pockets in AP where smartcard coverage was limited circa 2009. In eight districts, the banks responsible for enrollment had made little progress, so the project needed to be re-launched in these areas. The re-launch in 2010 provided a unique opportunity to randomize the roll-out of the intervention, enabling the first randomized evaluation of a large-scale biometric ID program (Muralidharan et al., 2016).

The study leveraged the phased introduction of biometric smartcards across 20 million beneficiaries in the eight districts; the results would therefore be representative of a large population. For the study, 112 sub-districts were randomly assigned to receive smartcards immediately (treatment sub-districts), while 45 sub-districts were randomly selected to receive the smartcards 2 years later (control sub-districts). In control sub-districts, the status quo mode of non-biometric authentication continued. By selecting a large unit of randomization (i.e., the sub-district with an average population of 70,000), the study team was able to measure spillover effects of the Smartcards program on broader economic outcomes.

The main outcomes of interest for the study were (1) beneficiary experience and satisfaction with NREGS and SSP and (2) leakage from the programs. Beneficiary experience was captured through household surveys. Leakage was measured as the difference between the dollar value of welfare benefits disbursed by the government

---

For example, as we show later, an improvement in access to the NREGS forced landowners in Andhra Pradesh to increase the wages for laborers as the laborers now had better outside options.

The unit of randomization essentially refers to the level at which the treatment/policy is introduced. For example, the policy may be introduced randomly across individuals in which case the unit of randomization would be at the individual level, or it may be introduced randomly across sub-districts in which case the unit of randomization is at the sub-district level.

For example, if the unit of randomization is the village or the sub-district, then researchers can capture spillovers at the village or sub-district level.

(obtained from administrative records) and the value received by beneficiaries, net of transaction costs incurred by beneficiaries in accessing the various welfare schemes (as reported in surveys).

By July 2012, 2 years after the roll-out of the intervention, only 50% of payments were being made via biometric smartcards in treatment sub-districts. However, despite low enrollment, the evaluation found that payments were faster and more predictable – especially for beneficiaries of the NREGS in treatment sub-districts. Beneficiaries in treatment sub-districts spent significantly less time collecting payments, and they received their payments earlier. On the leakage front, the evaluation found an increase in earnings of NREGS and SSP beneficiaries in treated areas. This increase in earnings did not coincide with any major increases in government outlays, suggesting a significant reduction in the leakage of funds. Indeed, leakage in treatment areas was estimated to have reduced by 12.7 percentage points for NREGS beneficiaries and 2.7 percentage points for SSP beneficiaries. These results may be explained by a reduction in government officials pocketing money for beneficiaries who neither worked nor claimed payments. Biometric IDs had made it harder for officials to over-report, since biometric authentication requires beneficiaries to be physically present to verify and receive payment.

Decomposing the effects in treatment areas further, the evaluation found that improvements in payment timeliness were concentrated entirely in villages that had shifted to the new smartcard-linked payment system. In these villages, timeliness improved regardless of whether or not beneficiaries had received biometric smartcards although the reduction in leakages was concentrated entirely among beneficiaries who had received biometric smartcards. These results suggest that the organizational reforms associated with the new payment system – which shifted the responsibility for payments from local government officials over to banks – were a key factor in improving the quality of service delivery and the possession of a biometric ID was key to reducing leakages.

Interestingly, the AP Smartcards program also had a positive impact on the wider economy. In treatment sub-districts, the introduction of smartcards led to broad increases in the earnings of the poor. Most of this came from sources other than the NREGS, suggesting increases in market wages and employment. This helped increase the bargaining power of workers in the labor market, who now had better outside options (Muralidharan et al., 2020a). The evaluation was instrumental in showing that biometric ID systems can have a positive impact both directly – by improving beneficiary experience and reducing leakages – and indirectly, via positive spillover effects on market wages.

5.3   Results from the Evaluation of Aadhaar in Jharkhand

As *Aadhaar* was rolled out across India, Jharkhand (and its state PDS program) became one of the frontrunners in completing the prerequisites for participation. After being satisfied with the results from small-scale pilot deployments, the Jhark hand government scaled up biometric authentication to ten districts. To evaluate the impacts of the reform, Jharkhand's government agreed to randomize the ordering of the roll-out across 15 million beneficiaries in the 10 districts. This was after the results of the AP experiment had been shared, building interest of the government in collaborating with the research team (Muralidharan et al., 2020b).

Similar to AP, the roll-out was randomized at the level of a large administrative unit, to capture spillover effects. In the experiment, 132 sub-districts were randomly selected to receive ePoS devices supporting ABBA, of which 87 sub-districts (treatment) would receive the ePoS devices first and the remaining 45 sub-districts (control) would receive the ePoS devices at a later stage.

The first phase of the study consisted of introducing ABBA into the PDS treatment group, without any reconciliation based on FPS transaction data. Despite Jharkhand being a relatively low capacity state, by the time of the follow-up surveys, 96% of the shops in treatment areas had introduced ePoS devices, and 91% reported regularly using the devices to biometrically authenticate and record transactions.[11]

The evaluation results demonstrated a limited impact of ABBA on leakage: the value of food entitlements received by beneficiaries did not vary significantly between treatment and control groups. In addition, the reform did not decrease government spend on food entitlements in treatment areas, suggesting the reform did not reduce diversion of goods to the black market. However, the use of ABBA came at a cost to beneficiaries. Many beneficiaries made multiple unsuccessful trips to the FPS before they were able to authenticate and access food entitlements, resulting in a 17% increase in transaction costs for those in treatment sub-districts. The reform also had a significant negative impact on the value of food entitlements for certain subgroups. For households with a single member failing to seed their *Aadhaar* card with a ration card, the reform led to a significant drop in the value of rice and wheat received and increased the probability of receiving no commodities at all.

Shortly after ABBA was introduced in the control districts, the state government introduced stricter reconciliation protocols. This "second phase" of intervention involved the government adjusting downward the amount of grain disbursed to each FPS, reflecting stocks that the FPS should have maintained (based on digitized transaction records from prior months). Reconciliation was introduced at the same time in both control and treatment sub-districts, but treatment areas had ePoS devices for a longer period of time (11 months) than control sub-districts (2 months). So in treatment areas, 11 months' worth of commodities diverted away from beneficiaries would have been adjusted for at the introduction of reconciliation. The analysis suggests that stricter reconciliation of commodities did reduce leakage.

---

[11] This difference was mainly because enrollment rates for Aadhaar cards were already quite high prior to it being introduced in Jharkhand's PDS. The central government had made high Aadhaar coverage and ABBA for welfare schemes a nationwide priority. In AP, there was no supporting infrastructure such as Aadhaar that was available to the state government (Muralidharan et al., 2020b).

In treatment sub-districts, reconciliation led to a sharper decrease in the value of food grains disbursed by the government, relative to control sub-districts. It also coincided with a sharp reduction in the value of goods received by households in treatment sub-districts. According to the reconciliation protocols used by the government, since FPS dealers in treatment areas had ePoS devices for longer, they were more likely to have undisbursed food grains captured in transaction records. But in practice, it was more likely that FPS dealers had already diverted the undisbursed grains to the open market. This explains why there was a sharper decrease in grains disbursed to households in the treatment sub-districts after reconciliation.

Had the government decided not to hold FPS dealers accountable for past diversions – and had the government given all dealers a "fresh start" – a more favorable outcome may have been obtained. Starting reconciliation on a "clean slate" basis (rather than basing disbursements on past transaction records) would have achieved a more modest reduction in leakage, with minimal impact on the value received by beneficiaries. These results show that the introduction of biometric IDs in Jharkhand, on its own, did not reduce leakage. But combined with reconciliation, there was a reduction in leakage that coincided with a sharp reduction in the value of benefits received by beneficiaries of the PDS.

5.4 Summary

In both states, the introduction of biometric IDs coupled with other reforms (such as reconciliation in Jharkhand) did lead to a substantial reduction in leakage of benefits. However, in AP, the benefits from the reduction of leakage were passed onto beneficiaries, whereas in Jharkhand, the main policy priority was to increase fiscal savings, which ultimately resulted in beneficiary loss. The policy implication is that while biometric IDs can bring down leakages, differences in program and technology design can have an important bearing on whether beneficiaries gain or lose.

6 Conclusions

The studies discussed in this chapter demonstrate that technological interventions like biometric IDs have the potential to increase state capacity and improve last-mile service delivery in developing countries. However, differences in policy priorities and the details of solution design influence the extent to which households benefit from the digitization of welfare schemes.

Results from two large-scale evaluations find that more accurate biometric ID systems, coupled with payments and policy reforms, reduced leakages in welfare schemes in two different states in India. However, there were varying results. In Jharkhand, reduced fiscal leakage came at the expense of excluding genuine beneficiaries who were unable to meet new standards for identification. Exclusion of beneficiaries was low in Andhra Pradesh, where the government was more focused on improving beneficiary experience with welfare programs.

Differences in policy priorities of the two states are reflected in differences in the design of the two ID programs. Manual overrides were available for those unable to biometrically authenticate themselves in AP, but limited overrides were made available to beneficiaries of the PDS in Jharkhand. The use of past transaction records to reconcile grain stocks, combined with the deletion of ration cards from old beneficiary rolls, underlines the Jharkhand government's priority of fiscal savings. These policies are arguably legitimate political choices. ABBA was introduced into Jharkhand at a time when there was a strong public movement in India against corruption and there was a push from the central government to roll out ABBA in social service delivery. Despite reports of exclusion and preventable starvation deaths, surveys of beneficiaries indicated that while views were polarized,[12] there was still strong support for *Aadhaar* reforms to the PDS (Muralidharan et al., 2020b). This highlights that reforms that are harmful to a significant minority can still be politically viable, if they are perceived as being in the larger public interest.

The evaluations also highlight the importance of incorporating regular data collection on beneficiary experience as part of any large-scale technology reform. For example, in AP with access to only administrative data, one may have mistakenly concluded that there was no reduction in leakage, as there was no change in government expenditure. Households' surveys showed otherwise. Conversely, in Jharkhand, one might conclude that there was a sharp reduction in leakage due to the reduction in government expenditure post-reconciliation. As we now know, Jharkhand's reduction in government expenditure post-reconciliation came at the cost of excluding eligible beneficiaries and reducing their overall benefits.

Ultimately, the design of technologies for public service delivery is highly political. There are many non-technical failures in the performance of digital identity solutions, and these expose the governance challenges underlying any public sector technology solution.

Discussion Questions

. Is there a trade-off between inclusion (which leads to leakage) and exclusion errors? If yes, how should governments weigh the trade-off between inclusion and exclusion errors?

---

[2] The ruling government in Jharkhand at the time ABBA was introduced in fact lost its majority in the next elections.

2. In addition to the manual overrides discussed in this chapter, are there any other technical/non-technical solutions to the problem of exclusion errors?
3. What role can civil society play in protecting citizens from potential misuse of biometric data?
4. Should the Government of Jharkhand have introduced a "clean slate" reconciliation of PDS stocks, to protect beneficiaries from loss of benefits? Could they have known, ex ante, that it would have disadvantaged beneficiaries?
5. The AP Smartcards program has since been replaced with *Aadhaar*. Was the state government wise to experiment with a local, offline smartcards solution, given the high costs of transitioning to the central government's *Aadhaar* technology?

References

Abraham, R., Bennett, E. S., Bhushal, R., Dubey, S., Li, Q., Pattanayak, A., & Shah, N. B. (2018). *State of Aadhaar report 2017–18*. Technical report, IDinsight.

Agarwal, S. (2013, March 5). *Duplicate Aadhaar numbers within estimates: UIDAI*. Mint.

Alkire, S., & Seth, S. (2013). Selecting a targeting method to identify BPL households in India. *Social Indicators Research, 112*(2), 417–446. https://doi.org/10.1007/s11205-013-0254-6

Asian Development Bank. (2016). *Identity for development in Asia and the Pacific*. Asian Development Bank.

Banerjee, S., & Sharma, S. (2018). *An offline alternative for Aadhaar-based biometric authentication*. Ideas for India. https://www.ideasforindia.in/topics/productivity-innovation/an-offline-alternative-for-aadhaar-based-biometric-authentication.html

Bardhan, P. (1997). Corruption and development: A review of issues. *Journal of Economic Literature, 35*(3), 1320–1346.

Bhatia, R. (2018, October 5). The Indian Government's astonishing hunger for citizen data. *The New York Times*.

Census of India. (2003). *Project review: Multi-purpose national identity card*. eCensus India. Issue 17.

Dahir, A. L. (2020, January 28). Kenya's new digital IDs may exclude millions of minorities. *The New York Times*.

Dholabhai, N. (2012, January 8). Unique ID better than PC smart card: Montek. *The Telegraph*.

Dreze, J. (2016, March 15). The Aadhaar Coup. *The Hindu*.

Dreze, J., Khalid, N., Khera, R., & Somanchi, A. (2017). Aadhaar and food security in Jharkhand pain without gain? *Economic & Political Weekly, 52*(50), 50–59.

Dreze, J., & Khera, R. (2015). Understanding leakages in the public distribution system. *Economic & Political Weekly, 50*(7), 39–42.

Dreze, J., & Khera, R. (2018, February 8). Aadhaar's $11-bn question. *The Economic Times*.

Drolia, R. (2018, February 15). 81 year old seeks exemption for super elderly. *The Times of India*.

Dutta, P. (2008). *The performance of social pensions in India: The case of Rajasthan*. SP Discussion Paper No. 0834. World Bank.

Dutta, P., Murgai, R., & Howes, S. (2010). Small but effective: India's targeted unconditional cash transfers. *Economic & Political Weekly, 45*(52), 63–70.

Dutta, S. (2019, February 8). Refusing to learn from tragedy, Jharkhand Demands Aadhaar for rations. *The Wire*.

Ferraz, C., Finan, F., & Moreira, D. B. (2012). Corrupting learning. *Journal of Public Economics 96*(9–10), 712–726. https://doi.org/10.1016/j.jpubeco.2012.05.012

Gelb, A., & Metz, D. A. (2018). *Identification revolution: Can digital ID be harnessed for development?* Center for Global Development.

Global Hunger Index Scores by 2020 GHI Rank. https://www.globalhungerindex.org/ranking.html. Accessed 28 Dec 2020.

Goldberg, J., Gine, X., & Yang, D. (2010). Identification strategy: A field experiment on dynamic incentives in rural credit markets. *Policy Research Working Papers*. https://doi.org/10.1596/1813-9450-5438

Government of India. (2001). *Report of the Group of Ministers on National Security*.

Government of India, Ministry of Finance. (2015). *Economic Survey 2014–15*.

Government of India, Ministry of Finance. (2016). *Economic Survey 2015–16*.

Government of India, Ministry of Finance. (2020). *Economic Survey 2019–20*.

Hebbar, N. (2017, April 12). Aadhaar robust; the poor have no complaints about it: Ravi Shankar Prasad. *The Hindu*.

Hoff, K., & Stiglitz, J. E. (2008). Exiting a lawless state. *The Economic Journal, 118*(531), 1474–1497. https://doi.org/10.1111/j.1468-0297.2008.02177.x

Khaira, R. (2018, January 3). Rs 500, 10 minutes, and you have access to billion Aadhaar details. *The Tribune*.

Khera, R. (2019a, March 10). The many pitfalls of Aadhaar. *The Hindu Business Line*.

Khera, R. (2019b). Aadhaar failures: A tragedy of errors. *Economic and Political Weekly*. https://www.epw.in/engage/article/aadhaar-failures-food-services-welfare

Matthew, A. (2019, October 21). Aadhaar failure: Several thousands in three districts in Odisha denied ration for two months. *National Herald*.

Matthews, H. V. (2016). Flaws in the UIDAI process. *Economic & Political Weekly, 51*(9), 74–78.

Misra, P. (2019). *Lessons from Aadhaar: Analog aspects of digital governance shouldn't be overlooked*. Pathways for Prosperity Commission.

Mohammed, A. (2018, July 26). Biometric mix-up affects nearly 2 crore Aadhaar holders. *The New Indian Express*.

Mookherjee, D. (2014). *MNREGA: Populist leaky bucket or successful anti-poverty Programme?* Ideas for India. https://www.ideasforindia.in/topics/poverty-inequality/mnrega-populist-leaky-bucket-or-successful-anti-poverty-programme.html

Mukhopadhyay, P., Muralidharan, K., Niehaus, P., & Sukhtankar, S. (2013). *Implementing a biometric payment system: The Andhra Pradesh experience*. Technical Report. University of California, San Diego.

Muralidharan, K., Das, J., Holla, A., & Mohpal, A. (2017). The fiscal cost of weak governance: Evidence from teacher absence in India. *Journal of Public Economics, 145*, 116–135. https://doi.org/10.1016/j.jpubeco.2016.11.005

Muralidharan, K., & Niehaus, P. (2017). Experimentation at scale. *Journal of Economic Perspectives, 31*(4), 103–124. https://doi.org/10.1257/jep.31.4.103

Muralidharan, K., Niehaus, P., & Sukhtankar, S. (2016). Building state capacity: Evidence from biometric smartcards in India. *American Economic Review, 106*(10), 2895–2929. https://doi.org/10.1257/aer.20141346

Muralidharan, K., Niehaus, P., & Sukhtankar, S. (2020a, March). *General equilibrium effects of (improving) public employment programs: Experimental evidence from India*. Working paper 23838, National Bureau of Economic Research.

Muralidharan, K., Niehaus, P., & Sukhtankar, S. (2020b). *Identity verification standards in welfare programs: Experimental evidence from India*. https://doi.org/10.3386/w26744.

Niehaus, P., Atanassova, A., Bertrand, M., & Mullainathan, S. (2013). Targeting with agents. *American Economic Journal: Economic Policy, 5*(1), 206–238. https://doi.org/10.1257/pol.5.1.206

Niehaus, P., & Sukhtankar, S. (2013). The marginal rate of corruption in public programs: Evidence from India. *Journal of Public Economics, 104*, 52–64. https://doi.org/10.1016/j.jpubeco.2013.05.001

Nigam, I., Keshari, R., Vatsa, M., Singh, R., & Bowyer, K. (2019). Phacoemulsification cataract surgery affects the discriminative capacity of iris pattern recognition. *Scientific Reports, 9*(1). https://doi.org/10.1038/s41598-019-47222-4

Olken, B. A. (2006). Corruption and the costs of redistribution: Micro evidence from Indonesia. *Journal of Public Economics, 90*(4–5), 853–870. https://doi.org/10.1016/j.jpubeco.2005.05.004

Olken, B. A., & Pande, R. (2012). Corruption in developing countries. *Annual Review of Economics, 4*(1), 479–509. https://doi.org/10.1146/annurev-economics-080511-110917

Olson, M. (1965). *Logic of collective action: Public goods and the theory of groups* (Harvard economic studies. v. 124). Harvard University Press.

Pai, S. (2013, December 29). Delayed NREGA payments drive workers to suicide. *Hindustan Times.*

Perrigo, B. (2018, September 28). India has been collecting eye scans and fingerprint records from every citizen. Here's what to know. *Time.*

Pritchett, L. (2009). *Is India a flailing state?: Detours on the four lane highway to modernization.* HKS Faculty Research Working Paper Series RWP09-013.

Ramanathan, U. (2018, September 29). More equal than others. *The Indian Express.*

Reinikka, R., & Svensson, J. (2001). *Explaining leakage of public funds.* Policy Research Working Papers. https://doi.org/10.1596/1813-9450-2709.

Sathe, V. (2014). Managing massive change: India's Aadhaar, the world's most ambitious ID project (innovations case narrative: Project Aadhaar). *Innovations: Technology, Governance, Globalization, 9*(1–2), 85–111. https://doi.org/10.1162/inov_a_00204

Sen, S. (2019). *A decade of Aadhaar: Lessons in implementing a foundational ID system.* ORF Issue Brief No. 292. Observer Research Foundation.

Sharma, R. S. (2016). *UIDAI's public policy innovations.* NIPFP Working Paper Series No. 176. National Institute of Public Finance and Policy, New Delhi.

Singh, S. (2020, February 9). Once upon a time, a national ID project found: Complicated to prove citizenship. *The Indian Express.*

Sneha. (2017). Aadhaar-based biometric authentication for PDS and food security: Observations on implementation in Jharkhand's Ranchi District. *Indian Journal of Human Development, 11*(3), 387–401. https://doi.org/10.1177/0973703017748384

Sukhtankar, S., & Vaishnav, M. (2015). Corruption in India: Bridging research evidence and policy options. India policy forum. *National Council of Applied Economic Research, 11*(1), 193–276.

Surabhi. (2017, January 9). Touts cashing in on Aadhaar card rush. *The Hindu Business Line.*

UNICEF, et al. (2020). *Levels and trends in child mortality.* UNICEF.

Unique Identification Authority of India. (2009). *Biometrics design standards for UID applications.* UIAI.

Unique Identification Authority of India. (2012). *Role of biometric technology in Aadhaar enrollment.* UIAI.

United Nations Development Programme. (2020). *Human development report 2020.* UNDP.

World Bank Group. (2003). *World development report 2004: Making services work for poor people.* World Bank.

World Bank Group. (2019a). *Identification for development (ID4D) 2019 annual report* (English). http://documents.worldbank.org/curated/en/566431581578116247/Identification-for-Development-ID4D-2019-Annual-Report

World Bank Group. (2019b). *Identification for development (ID4D) practitioner's guide* http://documents1.worldbank.org/curated/en/248371559325561562/pdf/ID4D-Practitioner-s-Guide.pdf

Xavier, G., Goldberg, J., Sankaranarayanan, S., Sheerin, P., & Yang, D. (2012). Use of biometric technology in developing countries. In R. Cull, A. DemirgucKunt, & J. Morduch (Eds.), *Banking the world: Empirical foundations of financial inclusion.* MIT Press.

Zelazny, F. (2012). *The evolution of India's UID program: Lessons learned and implications for other developing countries* (CGD policy paper 008). Center for Global Development. http://www.cgdev.org/content/publications/detail/1426371

Part VI
Connectivity: Digital Communication Technology

Richard Anderson

Introduction

One of the most important technological revolutions of the early twenty-first century is digital communication technology. While this is, in fact, multiple technologies maturing simultaneously—mobile broadband, smartphones, social networks, and cloud infrastructure—the combined impact has transformed the way that people live. The impact of digital communication technologies is arguably greater in developing countries than developed ones. This is because technologies brought new services to people, as opposed to upgrading existing services. For example, basic mobile phones have been the first telephone connection for the majority of people in the world, with phone access (measured as subscriptions per 100 people) growing from 12 to 106 between 2000 and 2018.

Smartphone penetration has also become a global phenomenon which started with the iPhone launch in 2007 and the release of Android in 2008. With the development of low-cost smartphones from countries such as China, smartphones have become widely used in emerging countries, with smartphones being 37% of the market in India and 44% of the market in Africa. The combined effect of these technologies has changed how people interact with information and communicate. The reader is directed to many good references on the subject including the book *After Access* by Jonathan Donner.

With these advances, digital communication technologies have become important to development engineering projects. These technologies allow different types of things to be included in interventions; for example, it is now possible to provide remote monitoring of systems and to get real-time reporting of system failures. The ability to contact people in the field by mobile phone had a profound impact on reducing the need for expensive field visits and to support processes such as

R. Anderson
Department of Computer Science and Engineering, University of Washington, Seattle, WA, USA
-mail: anderson@cs.washington.edu

troubleshooting. There are also a range of development engineering interventions that can be based directly on communication technologies, such as citizen reporting on services or crowdsourcing of information. As multiple chapters in this text demonstrate, digital communication technologies are now core to development engineering.

This section focuses on how development engineering can be applied to communication technologies. The emphasis is on how to extend these new communication technologies to make them appropriate for the development context. In particular, the chapters identify innovations that can complement and extend the activities of mobile network operators (MNOs) and the global technology companies.

Kurtis Heimerl and co-authors look at the problem of expanding the reach of mobile cellular networks to populations that are not currently served by the global MNOs. While a very large population is reached by cellular communication, even in the poorest countries, there are still areas with no cellular service. Economics requires an "economic density" for a company to profitably deploy infrastructure. Standard estimates for the cost of a cell tower are about $200,000 USD when set up by a mobile network operator. The technology idea is to develop low-cost cellular systems which can be locally run, replacing the tower with a lower-cost cellular radio and establishing a backhaul connection to support connections with global networks. There are technology questions related to implementation, and there are also very important economic questions on sustainability and political questions on administration. The chapter reports on deployments in the Philippines and discusses the full range of considerations for implementation—ranging from spectrum negotiations to managing customer payments.

Aditya Vashistha and Agha Ali Raza present work on voice-based social media. Social media has enormous impact, both for personal access to entertainment and information and society-wide implication on politics and economy. However, there is a divide in access to the global social media platforms, with a large number of people in developing countries excluded. There are many barriers, including technological barriers such as access to devices and connectivity as well as others such as literacy and language. The idea pursued by Vashistha and Raza is to develop social media technology that can be accessed using a basic mobile phone through voice and the keypad, with users leaving voice messages and navigating messages with menus. They present multiple examples and field experiences and address technological and social challenges. A key question is how to scale these services and manage audio content as the number of users increases. They describe how community moderation can be used to prioritize and edit user submission. The financial sustainability of voice-based social networks is a big question, with advertising models being difficult to maintain, but other creative models such as microwork are considered. A final challenge is managing behavior of users and the risk of creating hostile environment to different groups of users.

Finally, Waylon Brunette and Carl Hartung discuss mobile data management tools for development organizations. They focus on the Open Data Kit (ODK) project which has developed a collection of widely used open-source tools for mobile data collection and management. The original challenge was to provide data collection tools to organizations that worked in remote areas with limited network connectivity and had limited technical capacity for managing technology. When the ODK project started in 2008, the team made the technology bet that smartphones would be a viable technology for use in development settings. At the time, this was a controversial notion—but turned out to be correct. The chapter traces the history of Open Data Kit and discusses the different types of data management that are conducted with mobile devices, explaining why multiple iterations of the system have been deployed to handle dramatically different use cases. The system has been deployed globally, and applications to various disaster management and immunization logistics scenarios are discussed. An important meta-issue discussed in the chapter is the process of developing, deploying, and maintaining global goods software with a complicated network of stakeholder organizations.

Reference

Donner, J. (2015). *After access: Inclusion, development, and a more mobile internet.* MIT Press.

Chapter 21
Connecting Communities Through Mobile Networks: A Case Study of Implementing Community Cellular Networks in the Philippines

Mary Claire Barela, Maria Theresa Cunanan, Cedric Angelo Festin, Kurtis Heimerl, Esther Jang, Matthew William Johnson, Philip Martinez, and Ronel Vincent Vistal

1 Introduction

No recent technology has had a greater impact on economic development than mobile networks, which comprise the largest networks on Earth and cover over five billion subscribers. Cellular networks have allowed easier access to communication and information crucial in making informed decisions about education, employment, social and entrepreneurial activities, and other aspects of daily life. Mobile coverage gaps often exist in remote, rural areas, sometimes called the "last mile" (or "first mile") of connectivity, where the typical economics of providing network service do not hold (Cherry, 2003). For-profit companies often find it difficult to justify expanding coverage to areas where profits are low due to the large investments required for traditional cellular infrastructure, for example, places where the potential subscriber base is not large or wealthy enough to recover the total capital and operational expenditures (CapEx and OpEx) involved in ongoing operation. In addition, many remote rural areas lack the supporting infrastructure needed for cell networks, such as stable grid electricity and a network connection back to the global Internet, called "backhaul." Furthermore, technical support and maintenance for these networks is often slow and difficult due to the long distances that network support engineers, typically based in city centers, must travel to provide service.

M. C. Barela (✉) · M. T. Cunanan · C. A. Festin · P. Martinez · R. V. Vistal
University of the Philippines Diliman, Quezon City, Philippines
e-mail: mabarela@up.edu.ph; theresa.cunanan@upd.edu.ph; cmfestin@up.edu.ph; philip.martinez@up.edu.ph; ronelvincent.vistal@eee.upd.edu.ph

K. Heimerl · E. Jang · M. W. Johnson
Paul G. Allen School of Computer Science and Engineering, Seattle, WA, USA
e-mail: kheimerl@cs.washington.edu; infrared@cs.washington.edu; matt9j@cs.washington.edu

© The Author(s) 2023
B. Madon et al. (eds.), *Introduction to Development Engineering*,
https://doi.org/10.1007/978-3-030-86065-3_21

To address these coverage gaps, the cellular infrastructure and operational model must be redesigned, taking the aforementioned issues into account to make deployment in remote areas more economically viable. The CoCoMoNets project (Connecting Communities through Mobile Networks), implemented jointly by researchers at the University of the Philippines Diliman (UPD) and international research partners, tackled the challenge through **community cellular networks** (CCNs). CCNs fall under the broader domain of community networking – an alternative approach to standard telecom service in which communication infrastructure is built by or with local people who will use the network (Song, 2017). The researchers designed a low-cost GSM base station that allows voice calls and text messaging at a fraction of the capital and operational costs of traditional cell networks. To address the challenges around maintenance and support, local community members and community structures were involved in the deployment, operations, management, and ownership of the network (Bidwell & Jensen, 2019).

This chapter documents the research team's experiences and challenges as they attempted to pilot community cellular networks in the Philippines from 2015 to 2019. Despite initial successes, they encountered barriers to continued operation which led them to ultimately shut down sites or pivot to Wi-Fi Internet access networks.

This chapter is organized as follows. First, Sect. 2 presents an overview of the project and its context. Section 3 narrates the challenges and implementation experiences, and a discussion of open challenges for CCNs follows in Sect. 4. Section 5 closes with a summary and conclusion.

2 Innovation

2.1 Technology

Designing cellular network infrastructure for rural areas requires careful consideration of its context. First, many remote areas in the Philippines do not have access to reliable grid electricity, so the network equipment's power consumption must be low enough to be supported by off-grid renewable energy systems. Second, backhaul Internet connectivity is expensive and intermittent in remote areas, especially those prone to bad weather such as typhoons. Ideally, the network must be able to provide service locally even with an unstable or non-existent Internet connection. These requirements led the researchers to develop a low-cost 2G GSM cellular base station for community cellular networks (CCNs).

To reduce the capital expenditures for the equipment, the base station was built using off-the-shelf hardware and open-source software. It is able to run on renewable energy sources such as solar power to reduce the reliance of the service on grid electricity. Each CCN has a satellite (very small aperture terminal or VSAT

Fig. 21.1 A typical installation of a community cellular network cell site piloted in the Philippines

backhaul via which it eventually connects to the broader public telephone network. A typical installation is shown in Fig. 21.1.

Unlike other community network deployments using Wi-Fi as the access technology, 2G GSM cellular technology was chosen because there is still a large 2G subscriber base in the Philippines, especially among low-income and rural residents. GSM is supported by most handsets, even basic feature or "candybar" phones, so most people already owned devices compatible with the network at the time of deployment.

Given the absence of available radio frequency spectrum and regulatory frameworks for last-mile initiatives such as CCNs, legal network operation in GSM frequencies relied on a public-private partnership with the Philippines' leading mobile network operator (MNO), Globe Telecom, for sharing permission to broadcast in their nationally licensed cellular spectrum. Globe also allowed the project to use their SIM cards, phone number allocations, and cloud services which provided interconnection services between the CoCoMoNets network and other telephone networks.

To allow the co-existence of the community cellular networks alongside the mainstream Globe network, the researchers used the CommunityCellularManager (CCM) software stack, a novel IP-based cellular core network that allows multiple separate community networks to be managed (i.e., configured, monitored, and provided with interconnection and billing services) under one technical domain.

The CCM system is divided into two modules: the Client and the Cloud. In this deployment, Globe managed a CCM Cloud instance and provided the researchers at UPD with an account for the project's Client networks. The CCM Cloud handled integration with the Globe network, which provided routing, interconnect, and phone numbers. The CCM Client was installed on the remote base stations. Network installations were then handled independently of Globe by the researchers and community partners.

2.2 Operational Model

The project team partnered with several local institutions and organizations in the deployment sites for network operation. At the grassroots level, local business cooperatives were recruited for day-to-day technical and business operations. Local government units (LGUs) at higher municipal and barangay levels were also engaged, as they had administrative jurisdiction over the deployment areas. Finally, faculty and students at the nearby Aurora State College of Technology (ASCOT) were onboarded for more advanced technical support of the community cellular networks.

The community-based operational model had two aspects: technical and commercial operations. Each aspect employed a three-tiered structure for scoping the roles and responsibilities of the actors. The three tiers, simply referred to as Level 1 (L1), Level 2 (L2), and Level 3 (L3), are detailed in Table 21.1. L1 comprised community partners, L2 comprised cooperative and LGU partners, and L3 comprised the researchers, who acted as the liaison with Globe Telecom (see Fig. 21.4).

For the technical operations, the tiers were defined according to foreseen maintenance, troubleshooting, and repair activities. The simplest and most frequent issues (e.g., power failure) were designed to be resolvable at the community level (L1). This eliminated the need for the research team to travel to the sites all the way from Manila (a full day's journey) for simple checks or repairs, ensuring quicker service restoration.

For the commercial operations, the tiers were defined based on the reach and resources of the relevant actors to carry out the needed business operations. The L1 tier was assigned to local store owners, leveraging their existing infrastructure, capability, and capital to perform retail transactions. A local business cooperative filled the role of L2, using its capacity to coordinate with multiple L1 retailers within its base municipality. The MNO (Globe) filled the role of L3, tasked with provision of SIM cards and wholesale sale of prepaid airtime (also called electronic load or e-load). A revenue-sharing agreement was set up between Globe and the cooperatives so that revenues could be funneled back to the community to finance the network's operating expenses.

Table 21.1 Technical and commercial operational structure

	Technical		Commercial	
Tier level	Actors	Responsibilities	Actors	Responsibilities
L1	On-ground partners (maintenance officer, security personnel, site leader)	Basic day-to-day upkeep of system Site security Reporting to L2 and L3	Retailers	Sell e-load to subscribers Buy e-load from cooperative Distribute remaining SIM cards to identified subscribers Receive subscribers' concerns
L2	SUC, cooperative, and representatives from LGU	Intermediate troubleshooting and repair tasks beyond L1 capability and resources Coordination between L1 and L3	Cooperative	Sell e-load to retailers Buy wholesale e-load from MNO Pay off maintenance and security personnel
L3	Research team and MNO	Remote monitoring of the network Interfacing between local actors and Manila-based stakeholders	Research team and MNO	Sell wholesale e-load to coop Process revenue share Provide SIM cards

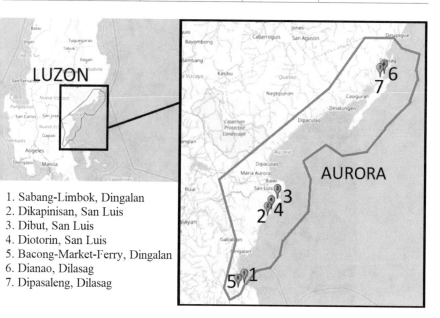

1. Sabang-Limbok, Dingalan
2. Dikapinisan, San Luis
3. Dibut, San Luis
4. Diotorin, San Luis
5. Bacong-Market-Ferry, Dingalan
6. Dianao, Dilasag
7. Dipasaleng, Dilasag

Fig. 21.2 Map of the deployment sites

3 Implementation

3.1 Deployment Context

The CCN deployments, shown in Fig. 21.2, are located in the province of Aurora, in barangays scattered across three municipalities along the Pacific coastline and Sierra Madre mountain range. From Manila, it takes 8 to 10 hours of land travel to reach Baler, the provincial capital. From Baler, several hours' travel by sea or unpaved roads is required to reach these remote and isolated coastal communities, a trip sometimes made hazardous or impossible by seasonal typhoons. Community members mainly depend on fishing and farming for income and sustenance. A few locals earn their living by reselling retail goods brought from the town centers to their respective communities. Owing to their geographic location, these sites do not have access to terrestrial radio and television broadcasts. Prior to the commencement

Fig. 21.3 SIM card distribution during community network launch event

Table 21.2 Final list of sites and their status as of March 2020

Site no.	Treatment site	Municipality	Population estimate (as of 2016)	Date launched	Status as of Feb 2020
1	Sabang-Limbok	Dingalan	450	Sept. 13, 2017	Inactive
2	Dikapinisan	San Luis	2177	Oct. 25, 2017	Active (Wi-Fi)
3	Dibut	San Luis	1032	Feb. 1, 2018	Active (Wi-Fi)
4	Diotorin	San Luis	578	May 30, 2018	Active (Wi-Fi)
5	Bacong-Market-Ferry	Dingalan	500	Aug. 29, 2018	Inactive
6	Dianao	Dilasag	300	Oct. 17, 2018	Inactive
7	Dipasaleng	Dilasag	500	Jan. 25, 2019	Inactive

of the CCN deployments, they were beyond the reach of cellular coverage; locals would need to travel several hours to use cellular services like calls, SMS, and data.

The project installed seven community cellular networks in Aurora from September 2017 to January 2019. These sites were randomly selected from a pool of 14 candidate sites as part of a randomized controlled trial (RCT) impact assessment study.

Prior to the networks' installation, all candidate sites underwent a household-level baseline survey in December 2016. The baseline survey involved three parts: (1) a household survey, (2) a listing of all adults 15 years or older, and (3) a one-on-one adult survey. The household survey queried household demographic composition, asset ownership, and economic activity. The core adult survey asked about social networks and included a travel diary. Across the 7 CCN sites, 1131 households were interviewed, and we listed 3057 adults (Keleher et al., 2020).

The survey data was used as input to a pairwise matching procedure as described in Bruhn and McKenzie (2009). First, potential cell tower locations were sorted into pairs that were as similar as possible along observable characteristics, and then one location within each pair was randomly selected to receive a cell tower in the first wave (treatment) of deployments. The other would receive a cell tower in the second wave (control). Specifically, pairs were formed so as to minimize the Mahalanobis distance between the values of all selected characteristics within pairs, using an optimal greedy algorithm. The research design required the CCNs to be deployed in a specific order.[1] Table 21.2 shows the final list of pilot sites. Several months after a site's network was launched, a household-level randomized experiment was performed on promotional pricing such as free credits and discounted rates.

Finally, from May and September 2019, surveyors returned to all 14 candidate sites to conduct an end line survey to measure changes after the network installations. Surveyors administered the same household and adult surveys, prioritizing interviews with the same individuals as in the baseline.

---

[1] Throughout the chapter, sites may be referred to by either their site number or name.

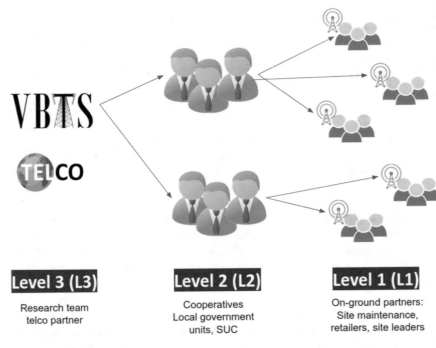

Level 3 (L3)

Research team
telco partner

Level 2 (L2)

Cooperatives
Local government
units, SUC

Level 1 (L1)

On-ground partners:
Site maintenance,
retailers, site leaders

Fig. 21.4 An illustration of the operational tiers

As part of the RCT study, SIM card distribution was tightly controlled as shown in Fig. 21.3. While SIM cards were given for free, only eligible individuals could receive them. Eligible individuals were defined to be 15 years old and above, residing in the community for at least 6 months at the time of site launch. The impact evaluation team had initially wanted to limit SIM cards only to residents of the barangay. However, upon request, exemptions were given to civil workers such as public school teachers and soldiers, as they were often assigned from other municipalities.

The network was dubbed "VBTS (Village Base Transceiver Station) Konekt Barangay" and branded separately to differentiate it from the mainstream Globe network. Pricing for the network services was already set by Globe Telecom and the national regulator. As the CCN was categorized as an experimental network, the per-minute or per-SMS service rates were lower than those of the mainstream network. Any price changes (such as the time-limited promotional pricing or "promos") were to be approved by Globe first and then by the national regulator. Service rates are shown in Table 21.3.

Table 21.3 Table of service rates

Traffic stream	Tariff (in PHP)	Unit
Calls from a VBTS barangay number to another VBTS barangay number	1.00	Per min
Calls from a VBTS barangay number to a regular Globe number	3.00	Per min
Calls from a VBTS barangay number to a non-Globe/VBTS barangay number	5.50	Per min
Call from a VBTS barangay number to an NDD number	5.50	Per min
Text/SMS from a VBTS barangay number to another VBTS barangay number	0.25	Per SMS
Text/SMS from a VBTS barangay number to a regular Globe number	0.50	Per SMS
Text/SMS from a VBTS barangay number to a non-Globe/VBTS barangay number	1.00	Per SMS

3.2 Implementation Challenges

Aside from developing the technical intervention, the project required coordination and work with a wide variety of stakeholders. These included a national mobile network operator (Globe), the national regulatory agency (National Telecommunications Commission – NTC), local government units, research collaborators in the academe, local cooperatives, and network end users within the remote communities. A large amount of time and resources were invested to establish the necessary partnerships and agreements with these stakeholders.

3.3 Spectrum Negotiations

With the core technical pieces in place, the researchers quickly discovered that starting these community cellular networks was not as straightforward as they had originally envisioned. The Philippines does not have a dedicated spectrum policy for last-mile service delivery, and the current regulatory framework forces small operators to adapt to the model used for national telecoms and other large organizations. MNOs are given licenses that span the whole country, even in areas where they are not providing service. Furthermore, current regulations on radio equipment use, SIM card production, and interconnect limit these activities to only MNOs.

There was no path for community networks to apply for their own licenses. In response, the project team first attempted to acquire a license exemption from the national regulator, the NTC. However, the regulatory officers advised the researchers to instead reach out to any of the current license holders and ask if they would allow co-use of frequencies under their respective licenses. The NTC would allow the project, provided an official agreement was acquired with a current licensee.

The researchers sent out proposals to the major Philippine MNOs, to which Globe Telecom responded.

The long negotiation process with Globe took more than 2 years to close.[2] The project worked its way internally through various departments, explaining how different components of the project would pan out upon deployment. The research team also had to reassure Globe that the project would shoulder the CCN's capital and operational expenses. Eventually, the project was taken under Globe's corporate social responsibility arm and was granted approval to use Globe's spectrum for an initial 1-year pilot period. One of the risks the project took on due to this spectrum arrangement was that the CCNs' operation was critically dependent on Globe's continued support.

3.4 Stakeholder Needs and Interests

Over the course of discussions and consultations between the researchers, Globe, and LGUs, we determined that local community partners would be needed to handle the day-to-day operations, management, and first-level maintenance for the CCN installations. The researchers initially wanted to recruit only local organizations based in the same communities as the installations, but were not always able to find organizations that fit these criteria. We enlisted the help of the LGU to nominate cooperatives and organizations and ended up with cooperatives not based in any of the target communities, but local to the municipality. Table 21.4 describes the partnership setup per municipality.

Moreover, the researchers wanted to maximize community involvement to create an operational model for the network that was as close as possible to the ideals of community networking yet that would also satisfy the other partners' requirements. While the team had past deployment experience on which to base the operational model, Globe's stipulations had to be accommodated. Globe preferred to have a single point of contact, rather than dealing with many independent communities. Its existing trade and distribution processes also relied heavily on partners having easy access to financial institutions such as banks and means of electronic communications such as email.

The cooperatives were primarily worried about the capital and potential financial liability that this venture could bring should the project not become sustainable. They also recognized that they would need to visit the retailers at the sites often, which might become inconvenient or infeasible due to the danger of travel in bad weather. They were skeptical about the business viability, as some of the treatment sites were very small in population. These were all valid concerns, especially as these deployments would be a test of the CCN model. However, they were convinced to participate primarily by the 80% revenue share they would eventually

---

[2] https://www.up.edu.ph/index.php/up-globe-sign-moa-for-village-base-station-project/

Table 21.4 Cooperative setup per municipality

Municipality and sites covered	Partnership setup
Dingalan, Aurora Site 1: Sabang-Limbok Site 5: Bacong-Market-Ferry	The nominated cooperative is the Paltic Mangingisdang Nagkakaisa Producers Cooperative (PAMANA). However, PAMANA is not based in the same barangay as the site installations, so another local group named the Samahan ng Mangingisda ng Sitio Limbok at Sabang (SAMAHAN) was engaged. These two groups are already acquainted with each other and have already worked together for a previous government project. SAMAHAN performed the day-to-day operations and maintenance duties for the sites, while PAMANA is in charge of the distribution of load to the on-ground retailers
San Luis, Aurora Site 2: Dikapinisan Site 3: Dibut Site 4: Diotorin	The nominated cooperative is the Dibayabay Primary Multipurpose Cooperative (DPMC). The cooperative is based in Barangay Dibayabay, which currently has no VBTS installation, but has extended membership in Sitio Diotorin, one of the current VBTS sites. The DPMC performs the distribution of load to the on-ground retailers, as well as other functions
Dilasag, Aurora Site 6: Dianao Site 7: Dipasaleng	The nominated cooperative is the Dilasag Municipal Employees Credit Cooperative (DMECC), which is based in the town proper. The sites are 20–30 min away. DMECC has members residing in sites 6 and 7

receive from the network's gross revenues. As L2 actors, the cooperatives would also receive a discount on the wholesale e-load purchases from Globe as well as their 80% share. Finally, the project team assured the cooperative that all initial investments for the infrastructure were funded by the project.

3.5 Unexpected Changes in the Field

While the spectrum negotiations were taking place, the deployment team from UPD had been visiting and surveying potential sites where little to no cellular coverage existed. The team had set criteria for selecting pilot sites for CCN deployments. A primary consideration was for a village to be outside existing cellular coverage. Initially, five isolated coastal barangays in San Luis, Aurora, were selected as deployment sites. However, the introduction of the RCT study required additional sites to meet the minimum requirement for the matched pairs design (at least 14 sites). Hence, the search was extended to include nearby coastal communities in the town of Dingalan.

The initial site listing was vetted by Globe to ensure that they were indeed excluded on their existing coverage map. At the time, the Dingalan sites were on the fringes of existing cellular coverage. This meant that coverage was not present in the village, but residents had identified spots several kilometers outside the village where they could acquire a signal by walking.

The data gathered from the site surveys were then passed to the impact assessment study team for evaluation. The researchers then generated a final listing of seven treatment sites, with the research design necessitating that the CCNs had to be installed and deployed in a specific order. As a result of this process, the sites were primarily selected according to RCT requirements and not for business viability. Some of the target sites were very small in population and had been identified from the onset as having potential difficulty with generating enough revenue to cover recurring costs without external support.

Over the course of site preparations (mid-2017), two of the treatment sites in Dingalan needed to be relocated due to unforeseen security threats from military insurgent groups in the area. This forced the research team to abort some of the initially selected sites, and the deployment team had to scout other nearby areas for candidate locations, prompting a re-evaluation of the control-treatment pairs. The project team then had to forge new partnerships with local stakeholders in these locations.

Once the treatment sites were identified, the next step was to secure a small lot (7 m × 7 m) in each site where the CCN tower and equipment shelter would be built. If possible, the project team preferred to have the CCNs erected on government land, as the project already had partnerships with the LGUs that had assisted in expediting site clearances and other permits.

However, the project team had a hard time acquiring land for the sites in Dingalan, as a private corporation, Green Square Properties, claimed ownership over almost the whole municipality. The team initially attempted to seek permission with this private entity, but the company wanted lease payments in exchange. As we could not guarantee that income would be generated, the negotiations with them failed. The team tried by all possible means, including leveraging the land-use agreement that the Dingalan LGU had with Green Square. The negotiations for land use took a considerable amount of time and was further complicated when sites had to be re-randomized. In the end, the team ended up making informal arrangements with private homeowners in the area.

In subsequent visits to Dingalan, the researchers observed that network coverage from mainstream networks had improved since these locations were first assessed. By mid-2018, mainstream network coverage had expanded so that residents could now utilize them in the comfort of their homes. With this change, residents in Dingalan preferred to use the mainstream networks over the community network since residents perceived them as more reliable and affordable, and they offered additional services such as mobile data and promos for "unlimited" usage. The community network could not compete with the incumbent MNOs. While the coverage expansion was detrimental to the survival of the community network, we acknowledge that it ultimately brought benefits to the community.

4 Evaluation

4.1 Subscription and Usage

At their peak, the CCNs had about 2000 subscribers, equivalent to about 90% of the total eligible population across all sites. We believe the high adoption rate can be attributed to the RCT study giving away SIM cards for free. About 40% of subscribers topped up monthly, spending $1.20 per month. Monthly ARPU (average revenue per user) across all sites was around $0.60. Voice calls dominated the overall traffic, with 15 times more inbound call minutes to subscribers on the network than outbound call minutes. On the other hand, subscribers made three times more outbound text messages compared to inbound text messages. This was indicative of a "call-me" behavior, since subscribers were only charged for user-initiated calls or SMS.

4.2 Technical Operations

Stable network performance is crucial to smooth network operations. While the technical team took measures to make the systems robust to rural conditions, the system is still not fault-proof and occasionally encounters technical issues that disrupt service. During the first few months of operation, subscribers considered such service disruptions and downtime acceptable, from the viewpoint of their prior condition where they had no network coverage at all. However, their service expectations and attitudes changed throughout the course of the project. Subscribers started to expect continuous and reliable operation similar to their experience with mainstream cellular networks in Baler.

The deployed systems had occasional hardware issues, which were most frequently power-related. Most power failures were due to low battery charge during the rainy season (May–November). The system's battery bank was designed to be sufficient for 3 days of autonomous operation, but limited sun hours prolonged the duration to fully charge the battery bank. The technical team initially tried to operate the network 24/7, but this became very challenging once the rainy season arrived. In response, the researchers and community maintenance staff agreed to limit the operational hours from 5 AM to 10 PM. Initially, the staff had to manually turn off the system, but it turned out that sometimes they forgot to perform the shutdown routine or found it too inconvenient. To resolve this, a digital timer switch was installed that turned the system off and on at the specified times. The switch could be manually overridden by the local staff, in case the community needed extended hours or for emergencies.

Another common cause of network failure was broken inverters, which required replacement hardware to be shipped from L3 or L2. For worst-case scenarios, the researchers were pleased to discover instances where the local maintenance staff had taken the initiative and coordinated with the barangay council to allow powering the CCN hardware using the barangay's electric grid connection or emergency power systems.

In the event that a network hardware component fails, it can take weeks to get replaced or repaired. Specialized equipment is almost impossible to procure in Baler and most often needs to be supplied from Manila. Anticipating this, the project team practiced providing backup inventory for commonly failing components. To expedite replacement of commonly available off-the-shelf hardware, an arrangement was made with the partner LGUs so that they would supply the needed hardware and ship it off to the sites. The on-ground personnel at the site would then perform the hardware installation.

4.3 Commercial Operations

The local business cooperatives were responsible for e-load replenishment, a two-part process that required them to (a) remit payments to the MNO through banks or remittance centers and (b) travel to the sites to provide load to the retailers. Replenishment was challenging primarily because these tasks required personnel travel, which varied in feasibility (e.g., favorable weather conditions and available transportation). Banks and remittance centers were only located in municipal town centers, and the CCN sites were also separated from the cooperative's home base. While e-load transactions can be done electronically, payments and remittances were particularly difficult for cooperatives to receive due to the geographical distances of the sites from the town centers. Payments needed to be remitted regularly for the prescribed commercial model to run smoothly, as cooperatives had a very limited amount of capital to circulate for commercial operations.

As a result of the challenges of travel, some hard-to-reach sites experienced e-load shortages lasting 1–2 weeks, with the worst one lasting for about a month. A straightforward solution was to increase inventory capacity per retailer, also preferred by the cooperative partners as it would require fewer visits for payment collection. However, some retailers expressed an inability to spend larger amounts of capital at a time for more e-load. The retailers and the cooperative were able to work out possible schemes to optimize their process and ease capital requirements. Schemes included setting larger load orders to reduce the frequency of cooperative visits while using a consignment-based structure to reduce capital requirements on retailers. In the "consignment" scheme, the cooperative agreed to transfer the retailer a large amount of e-load for a 50% down payment and 50% balance to be paid on the next collection date.

4.4 Community Response

In general, community buy-in was not that difficult to gain since locals immediately saw the benefits that the project would bring to their respective communities. In informal interviews during the site survey phase, locals highlighted that the network would be useful for emergencies and for contacting loved ones far away. There were a few concerns about radiation and potential negative health effects, but residents were assured that the community base stations transmit at a fraction of the power that typical base stations use. Eventually, the positive benefits of being able to communicate long distance, i.e., make a call without needing to leave their villages, outweighed these concerns.

Though the CCNs were made possible by the partnership with Globe, the researchers wanted to emphasize to the community that this was a separate initiative whose primary purpose was research and not income generation. Specifically, the project team wanted to avoid the risk of insurgents misidentifying the community network as a commercial enterprise and possibly extort "revolutionary taxes" from the project.[3] As a response to frequent queries about the project and to avoid misinformation, the research team held town hall meetings prior to and during the CCN's inauguration and formal launch.

The network launch was conducted as a whole community event where the team had a chance to formally introduce the project and explain the details of the research and network operation. The team explained the capabilities and limitations of the system, tariff rates, and information about future promotions. As an experimental network, the VBTS-CoCoMoNets network could be expected to suffer downtime or outages and would not guarantee the same grade of service as mainstream cellular networks. This and other capacity limitations were communicated clearly to all stakeholders, especially to subscribers prior to their sign-up. The event was also an opportunity to address questions and concerns from the community, which included the privacy of their communications, potential health effects of radiation from CCN towers, value-added services such as mobile Internet, and other comparisons of the CCN to mainstream networks.

Subscribers' appreciation of the project depended highly on their distance to existing coverage and the effort required to reach it. The communities of San Luis, Aurora (including Dikapinisan, Dibut, and Diotorin), who are extremely geographically isolated from the rest of the world, were highly appreciative of the community network even though the service offerings were not perfect. Network adoption was also highest in San Luis.

Resistance to the project was most salient in the Dingalan sites. Some community members outspokenly mentioned that VBTS-CoCoMoNets did not have a positive impact on them at all, as it allegedly did not keep its promise of improving mobile reception, and actually weakened the scant signal of the networks they

Philippine Institute for Development Studies. 2018. "The telco duopoly has become the CPP-CPA's biggest funder" https://pids.gov.ph/pids-in-the-news/2247

had been using prior to the CCN's installation. To make matters worse, the delegated maintenance personnel took a job elsewhere, leaving no one to maintain the equipment. Maintenance and repair had to originate from Manila, making the upkeep of the site much more difficult. The team made several attempts to explain the situation to the community, but the subscribers had lost interest. Due to the dissatisfaction, the team received requests from community members to pull out the equipment, as they did not want liability in the case of theft or damage. Unfortunately, they could not act on the pullout request until the end line survey was completed.

4.5   Personnel Retention

Personnel retention has remained a challenge for CCNs, as delegated personnel find better opportunities beyond their village, limiting their long-term participation. In the designation of tasks, the project team had initially hoped that the required effort and frequency of maintenance (e.g., checking battery voltage daily) would be of little consequence to the personnel everyday routines. However, keeping the L1 and even L2 presence failed in several sites, severely affecting operations and sustainability. In sites with low network traffic and adoption, L1 personnel understandably did not exhibit enthusiasm in fulfilling their duties, as perhaps the subscribers' behavior indicated the CCN's lack of value to them. L1's failure to perform maintenance may have led to further degradation of the service, so that disappointed subscribers returned to their old usage of the mainstream network if available.

The researchers looked deeper into the reasons behind L1 absence and found the following: (a) insufficient compensation and (b) unappreciation of the network's value. Although L1 personnel received an allowance from the LGU or cooperative, they often opted to venture to more rewarding livelihoods for higher compensation. Some allocated more of their time to their fishing or farming activities, while some accepted better jobs outside their community. This tendency to relocate to "greener pastures" for work echoes the overseas Filipino worker (OFW) phenomenon wherein Filipinos resort to opportunities in more developed countries due to the unavailability of local high-paying jobs. For the context of the project's CCN sites, this translates to migration to more urban areas. The researchers accept the L1 personnel standpoint, as they need to provide for their families. While these migrations are often not permanent, with workers shuffling intermittently between their hometowns and Baler, unavoidable gaps are created when those left behind lack the knowledge to maintain the network. While there is certainly promise and value in providing training to produce local expertise, this training must be consistently available and repeated to be effective. Issues related to personnel turnover can be expected to continue until the community as a whole becomes more technically savvy.

Fig. 21.5 System diagram created by L1 personnel

In one case of L1 absence, the project team was fortunate enough to have a local resident step up to understand the system and take over operations. The formerly untrained resident, with a high school/early college level background in electrical principles, received cursory instructions from the former personnel when he left for work in Baler and took up his functions. He was in charge of the system's upkeep for several weeks before receiving official training from the project team, performing well on his own through self-study of the reference operational manuals and deployed equipment layout. A diagram from the manuals with his own drawing and labels, produced before training, is shown in Fig. 21.5.

4.6 Trust and Community Relations

Forming partnerships and trust with the communities required a tremendous amount of effort and time. The project encountered instances where negotiations took much longer to close than expected, causing delays in the installation and deployment timelines. The team also wanted to ensure participation of local partners and give them the opportunity to exercise their decision-making rights. Moreover, local politics and leadership changes interrupted the continuity of community relationships that had taken a long time to build. In some cases, the project had to re-introduce itself to new local leaders after a new administration was elected. In Philippine politics, it is common to replace or reshuffle all staff when a new administration takes office, so the team had to re-establish connections and re-identify liaisons or L2 personnel for the project.

Despite the difficulties, gaining and maintaining trust with local stakeholders is highly important, as it will influence an intervention's ultimate value and impact for the community. A sense of distrust, as arose in Dingalan, meant that the community would never use and appreciate the network. Building trust requires researchers to spend significant time with the communities and try to understand their way of life. While most of the researchers were Filipinos and understood the local context, they were still living in Manila. Unfortunately, the difficulty of travel and the RCT requirement to minimize survey/response bias limited the researchers' ability to maximize their presence in the communities.

5 Adaptation

By September 2019, four CCN sites had been terminated due to non-performance. They were Site 1 (Sabang-Limbok), Site 5 (Bacong-Market-Ferry), Site 6 (Dianao), and Site 7 (Dipasaleng). "Non-performance" in these cases referred to a lack of subscriber activity, lack of L1 maintenance support, and lack of cooperative business support. While the research team strove to keep these sites running, for example, by performing L1 repairs, replacing any defunct equipment, or initiating pricing promos, the lack of interest from stakeholders signaled that the intervention was not useful for these sites.

In November 2019, a contact at Globe notified the researchers that Globe would be terminating their own parallel set of small-scale cellular deployments based on the same CCM software stack. This was due to various technical and business reasons including hardware vendor problems, target revenues not being met, and lack of support from Facebook, another peripherally involved partner adopting the CCM software. Corporate supports for the existing VBTS-CoCoMoNets sites were bundled with this initiative, and thus they would also be terminated.

With this news, the project team decided to install community Wi-Fi networks alongside the existing 2G cellular networks. Although this intervention was beyond the scope of the current project, the researchers felt the need to provide an alternative for communities that had become reliant on CCN services. Both networks operated in parallel until March 2020, when the 2G networks were taken offline.

Several community consultations were held in the remaining sites of Dikap inisan, Diotorin, and Dibut to discuss the status and future of the project. Inputs from the communities were gathered regarding operation beyond the project period. The main concern for these sites' sustainability was the high cost of the satellite backhaul. Two main proposals surfaced from these discussions. The first was to let the network be free and open for anyone to use, the catch being that the service would eventually end once the VSAT contract paid for by the project ended, if no subsidies were acquired from the local government. The second was to accumulate funds for the monthly recurring costs through sales of Wi-Fi voucher codes to community members wishing to access the Internet. Two out of the three communities (Dikapinisan and Dibut) decided right away to adopt the management of the Wi-Fi network. They would sell voucher codes for Wi-Fi Internet access for PHP 10/h. Sales would be accumulated to pay for the VSAT backhaul starting in August 2020.

Community reception of the Wi-Fi service has been mixed. Some are happy about the introduction of the Internet to their communities, as they had been longing to use the Internet for Facebook, research, or other purposes. Dikapinisan, being the most populous and urban of the three communities, has received the Wi-Fi service very well; residents largely have a prior understanding about what Wi-Fi is and how it can be used. However, some feature phone users were dismayed because their phones do not have Wi-Fi capability, or they still find voice and text easier to use over video calls and chat. Wi-Fi also has shorter propagation characteristics compared to cellular 2G, meaning the coverage area of the community network is now smaller. Community members who were not covered by the Wi-Fi signal coverage would have to walk or congregate near the community access point. Finally, the termination of 2G service has disenfranchised some community members who do not have the capacity to upgrade to Wi-Fi-enabled devices. The situation has been difficult for the research team to accept, but we have had to acknowledge our own limited capacity as well.

In spite of this outcome, efforts initiated by the CoCoMoNets project are being continued through other research initiatives. Researchers from UPD and the Department of Science and Technology Advanced Science and Technology – Institute (DOST-ASTI) saw an opportunity with the planned digital TV migration and proposed the opening of the 600 MHz spectrum for use of community cellular networks, and thereby addressing the challenges encountered by the CoCoMoNets project revolving around spectrum usage. To support this proposition, researchers from DOST-ASTI have developed an LTE base station prototype operating at 600 MHz (Hilario et al., 2020). Recently, the ongoing COVID-19 pandemic has brought again to spotlight the long-standing problem of last-mile connectivity access in the Philippines. As part of the government response to address the connectivity gap, the Philippine's Department of Science and Technology announced that it is planning to build community cellular networks in remote areas across the country to provide Internet connectivity that could aid in distance learning.[4] This pronouncement is a positive step forward as community networks are being

---

Jaehwa Bernardo. September 2020. ABS-CBN News. "DOST plans to build community-based cellular networks in remote areas" https://news.abs-cbn.com/news/09/29/20/dost-plans-to-build-community-based-cellular-networks-in-remote-areas

acknowledged as a solution to last-mile access. Moreover, the government's adoption of community networks is expected to bring the necessary support for crafting responsive regulatory policies and additional spectrum reform that will help narrow the digital gap in the country and resolve some of the issues surfaced through this study.

6 Discussion: Open Challenges

The geographical remoteness of the sites, while important for the networks' business success, resulted in a number of setbacks in creating a feasible trade and distribution process. The sites are far away from formal financial institutions like banks and remittance centers; travel time and difficulty reaching the sites also pose ongoing dangers to local intermediaries.

MNOs operating at massive scale are not well suited to using iterative or rapid design processes, which may be crucial when working with new technologies or marginalized populations to make interventions appropriate for adoption. For example, during negotiations with Globe, it was agreed that co-ops would receive 80% of the revenue share, which would be used to cover operating costs of the site beyond the project duration. However, the co-ops' receipt of their share was delayed, as Globe later required them to enter into separate contracts directly with Globe. These contracts took almost another 2 years to get approved. Moreover, executing the revenue share in practice was a longer process than initially expected. Globe's revenue share disbursement was not completed until July 2020.

While our counterparts at Globe were committed and worked hard to complete the disbursement, most of the delays were due to procedural precedents in a massive organization attuned to working with other similarly large enterprises. The project team recognized the administrative challenges and potential concerns that arose from this delay, so the team constantly kept lines of communications open for coordination between Globe and the cooperative. Unfortunately, the delays broke the projected operational model, as it was assumed that cooperatives would be able to tap into this revenue for their operational expenses or for additional capital for e-load distribution. To help alleviate their concerns during this unfortunate development, persistent and continuous communication with both the cooperatives and Globe was important and necessary. The team also extended assistance to both parties to expedite completion of the requirements (e.g., getting required documentation or assisting cooperatives in digital account setup).

7 Summary and Conclusion

Despite all the operational challenges, the researchers still believe that community networks will play an important role in rural communications, especially when market forces fail to deliver service in last-mile areas. However, CCNs will continue to face challenges without supportive and enabling environments, requiring coordination and commitment from government, academia, industry, and community. A systematic review of community network strategy by policy makers and stakeholders will be required to allow CCNs to flourish. Finally, meaningful and sustainable impact requires more than a 2-year pilot project. To achieve larger-scale, more holistic development impacts in rural areas, we will need long-term roadmaps for rural development with respect to both physical infrastructure and human resources.

References

Bidwell, N., & Jensen, M. (2019). *Bottom-up connectivity strategies: Community-led small-scale telecommunication infrastructure in the global South.* Association for Progressive Communications, Technical Report.

Bruhn, M., & McKenzie, D. (2009). In pursuit of balance: Randomization in practice in development field experiments. *American Economic Journal: Applied Economics, 1*(4), 200–232.

Cherry, S. M. (2003). The wireless last mile. *IEEE Spectrum, 40*(9), 18–22.

Hilario, C. A. G., Barela, M. C., De Guzman, M. F. D., Loquias, R. T., Raro, R. V. C. B., Quitayen, J. J. J., & Marciano, J. J. S., Jr. (2020). LokaLTE: 600 MHz community LTE networks for rural areas in the Philippines. *IEEE global humanitarian technology conference (GHTC 2020).*

Keleher, N., Barela, M. C., Blumenstock, J., Festin, C., Podolsky, M., Troland, E., Rezaee, A., & Heimerl, K. (2020, June). Connecting isolated communities: Quantitative evidence on the adoption of community cellular networks in the Philippines. In *Proceedings of the 2020 international conference on information and communication technologies and development* (pp. 1–16).

Song, S. (2017, October). *Policy brief: Spectrum approaches for community networks.* Internet Society, Technical Report. [Online]. Available: https://www.internetsociety.org/wp-content/uploads/2017/10/Spectrum-Approaches-for-Community-Networks_20171010.pdf

Chapter 22
Voice Interfaces for Underserved Communities

Aditya Vashistha and Agha Ali Raza

1 Evolution of Voice Forums

Information and Communications Technologies (ICTs) have the potential to enable socioeconomic development where information and connectivity are the missing components. ICT-based interventions can lead to better management of available resources, improved monitoring and reporting of corruption, and more awareness and connectivity among people. To achieve impact at such a large scale, such solutions need to be robust enough to reach the target populations using available means with minimum resource expenditures. The Internet is a phenomenon that has enabled this enormous metamorphosis, and services like social media and online discussion forums have transformed how people participate in the information ecology and digital economy.

Unfortunately, most Internet services currently only empower urban, affluent, and literate people and exclude billions of "othered" people who are too poor to afford Internet-enabled devices, too remote to access the Internet, or too low-literate to navigate the mostly text-driven Internet. Up to 81% of the people living in developing countries are offline (compared to the 13% in the developed world).[1] This disproportionate access is even worse for marginalized populations

---

[1] "Measuring Digital Development – Offline population." https://itu.foleon.com/itu/measuring-digital-development/offline-population/ (accessed Aug. 31, 2020).

A. Vashistha (✉)
Cornell University, Ithaca, NY, USA
e-mail: adityav@cornell.edu

A. A. Raza
Lahore University of Management Sciences (LUMS), Lahore, Punjab, Pakistan
e-mail: agha.ali.raza@lums.edu.pk

© The Author(s) 2023
S. Madon et al. (eds.), *Introduction to Development Engineering*,
https://doi.org/10.1007/978-3-030-86065-3_22

within these countries. The Internet access gender gap is as high as 43% in developing countries (compared to 2.3% in developed countries).[2] Also among these offline communities are low-literates and native speakers of languages that do not have written forms (46% of all languages[3]). Hence, lack of access is strongly associated with poverty, low literacy, tech naivety, and being marginalized. This makes mainstream Internet services inadequate to meet the needs of billions of people living in underserved settings. Sadly, many of these communities are unable to find alternate means of communication as television and radio are non-interactive, print media assumes literacy, and computers require stable electricity and Internet connectivity.

Clearly, textual interfaces and Internet access cannot be relied upon to provide information access to the populations identified above. Speech as an alternate communication medium is more suitable especially because speech-based information can even be provided over simple phone calls without any Internet connectivity. Given the rapid proliferation of mobile phones in developing countries, speech-over-simple mobile phones is a viable way to connect underserved populations. In light of these considerations, global development researchers and practitioners have used interactive voice response (IVR) technology to create voice-based services that overcome connectivity barriers by using ordinary phone calls, literacy barriers by using local language speaking and listening skills, and socioeconomic barriers by using toll-free (1–800) lines. These services assume no more than the capability to make and receive a voice call from users and let them call a phone number to record and listen to voice messages in their local languages. Over the last two decades, such services have included efficient marketplaces; common interest groups; message boards, blogs, mailing lists, and social networks that facilitate social and political activism; information campaigns on themes of education, training, employment opportunities, health, agriculture, and emergency response; citizen journalism; and automated surveys and polls to gather information. Because of their accessible and usable design, these services have found applications in diverse domains and have profoundly impacted underserved communities in low-resource environments. This section follows the evolution of these services over the last decade and their big challenges and new frontiers.

Voice forums can be designed as (1) top-down information push services where the content is developed by experts and the interface simply allows individuals to consume the content, (2) peer-to-peer services where individuals can communicate with other individuals via audio messages, and (3) social services where users can record and broadcast content to the community. The first wave of voice-based services focused on improving information access for people in low-resource communities. The target audience of Project HealthLine – one of the first such

---

[2] "Measuring Digital Development – Gender gap." https://itu.foleon.com/itu/measuring-digital development/gender-gap/ (accessed Aug. 31, 2020).

[3] "How many languages in the world are unwritten? | Ethnologue" https://www.ethnologue.com enterprise-faq/how-many-languages-world-are-unwritten-0 (accessed Aug. 31, 2020).

endeavors – was low-literate community health workers in rural Sindh Province, Pakistan (Sherwani et al., 2007). It enabled them to retrieve relevant information by speaking out predefined commands. The goal was to provide telephone-based access to reliable and up-to-date health information, and the speech interface performed well once the health workers were trained to use it via human-guided tutorials. This project also highlighted the challenges in eliciting informative feedback from low-literate users. While services like HealthLine allowed users to only consume information, subsequent services took the form of voice forums and enabled marginalized communities to also produce and share information.

The impact of the community involvement in voice forums has been thoroughly studied and has been shown to promote social inclusion among underserved communities. Notable research in this direction includes Avaaj Otalo (an agriculture discussion forum in India), CGNet Swara (a citizen journalism service in India), Mobile Vaani (a social media service in India), Ila Dhageyso (a civic engagement portal in Somaliland), and IBM's Spoken Web (a user-generated information directory in India) (Patel et al., 2010; Mudliar et al., 2012; Moitra et al., 2016; Gulaid & Vashistha, 2013). For example, Avaaj Otalo was designed to connect farmers in Gujarat, India, and offered three features: an open forum where users could post and answer questions, a top-down announcement board, and a radio archive that allowed users to listen to previously broadcast radio program episodes. The most popular service turned out to be the open forum, constituting 60% of the total traffic, and users found interesting unintended uses for it like business consulting and advertisement. Mudliar et al. examined participation of rural communities in India on CGNet Swara, an interactive voice forum to record and listen to local news, grievances, and cultural content, and found that it serves as a vehicle for digital inclusion for indigenous communities. Koradia et al. (2012) integrated voice forums with community radio technology to amplify its reach and involve radio listeners in content creation, feedback, and station management. The influence of peer-generated content available on social platforms on the target communities has been compared against expert-generated content by Patel et al. (2012). In a 2-week trial, 7 agricultural tips were disseminated to 305 farmers in Gujarat, India. Each tip was recorded in the voices of university scientists and farmers. The study showed that farmers preferred to hear agricultural tips in the voice of their peers, even though in interviews they maintained their more socially acceptable inclination toward scientists. Table 22.1 provides a summary of the comparison of popular voice forums that have been deployed in various developing regions (Fig. 22.1).

An important question in developing voice forums has been that of the input modality: speech vs. key press. Project HealthLine found that speech input performed better than key press in terms of task completion, for both literate and low-literate users (Sherwani et al., 2007). However, it provided no clear answer in terms of subjective user preference. Lee et al. reported key press input to be more efficient for linear and simple tasks and speech input to be better for nonlinear tasks. Grover et al. found task completion rates for speech and key press input modes to be similar; and tech literacy is a more important factor than overall literacy for task completion. Patel et al. (2009) found key press input to be more intuitive and

Table 22.1 Summarized comparison of major voice forums

	Avaaj Otalo	CGNet Swara	Ila Dhageyso	Mobile Vaani	Gurgaon Idol	VoiKiosk	Polly	Sangeet Swara	Baang
Domain	Agriculture	Citizen journalism	Civic engagement	Grievance redressal	Social media	Social media	Social media	Social media	Social media
Subsidized airtime	Yes	Yes	Yes	Yes	No	N/A	Yes	Yes	Yes
Voting	No	No	No	Yes	Yes	No	No	Yes	Yes
Audio comments	Yes	No	Yes	Yes	No	No	No	No	Yes
Sharing posts	No	No	No	Yes	No	No	Yes	Yes	Yes
Deployment length	7 months	2009–now	5 months	2012–now	1 month	4 months	1 year	4 months	8 months
Calls	6975	137,000	N/A	10,000/day	306	20,499	636,000	25,000	269,468
Users	45	100,000	N/A	1.5 million	252	976	165,000	13,500	10,721
Posts	896	13,595	4300	300/day	31	2532	387,301	5000+	44,178
Female users	0%	21%	15%	N/A	Low	N/A	11%	6%	10%
Users with visual impairment	N/A	N/A	N/A	N/A	N/A	N/A	<1%	26%+	69%

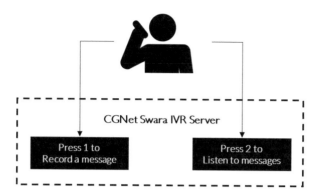

Fig. 22.1 High-level user interface design of CGNet Swara

reliable than speech. Over the subsequent years, key press input became the de facto standard for IVR services targeting developing regions.

The success of these initial services demonstrated their great potential to enable information access and connectivity among underserved populations in diverse contexts. However, voice forums also presented a wide array of implementation, scalability, and sustainability challenges, including how to (1) train the users to use speech interfaces, (2) spread and advertise these services to tech-novice and poorly connected masses, and (3) engage users in potentially involved voice-based interactions while keeping them oriented, motivated, and grounded. These traits of training, spread, engagement, and retention were found to be necessary for effective transmission of knowledge but were very hard to achieve simultaneously.

2 Scaling Voice-Based Services

In 2010, the biggest roadblocks to scaling voice forums among underserved populations were usability, motivation, and spread. Target populations faced difficulties in using even the simplest of voice forums, they did not exhibit interest or trust in using such services, and it was hard to advertise and spread these services to under-connected people. Researchers tried to overcome these barriers by conducting lab trainings, demo sessions, and door-to-door field campaigns, but it was quickly realized that these approaches too are not scalable. The identified challenge was to inform and train the users without the benefit of explicit, in-person sessions. To reach people at a large scale, a need to advertise and promote these voice services was also identified that was challenging using traditional advertising mechanisms as these populations are low-literate and not very well connected with technology. The way forward, unexpectedly, was found to be entertainment!

In 2010, Smyth et al. described the remarkable ingenuity exhibited by low-literate users when they are motivated by the desire to be entertained and concluded

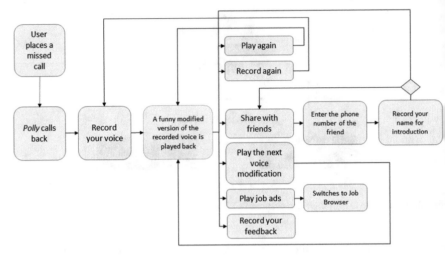

Fig. 22.2 High-level simplified user interface design of Polly

that such powerful motivation "turns UI (user interface) barriers into mere speed bumps" (Smyth et al., 2010). Inspired by this powerful demonstration, Raza et al. set out to systematically develop practices for entertainment-driven mass familiarization and training of low-literate users in the use of telephone-based services and created Polly (Raza et al., 2013a). Polly engaged low-literate and non-tech-savvy users in light entertainment to spread useful development-related information to them as they became more comfortable with the voice interface. Polly allowed users to record their voice, modify it using funny voice modifications, and send the modified recording to their friends (as shown in Fig. 22.2). The entertainment appealed to the target users and allowed Polly to spread virally. It acted as a soft incentive for users to train themselves and overcomes the scalability hurdle of explicit user training and motivation. It also allowed Polly to organically spread among the population through the word of mouth as well as scheduled message deliveries from one user to another. As users became more comfortable with the interface, Polly introduced them to development-related services like job search and health information.

Polly was first pilot-tested in 2011 with 32 low-literate users who were handed out Polly's phone number and were asked to explore its functionality. Within 3 weeks, Polly had organically reached over 2000 users and logged more than 10,000 calls before it was shut down due to insufficient telephone capacity and unsustainable cellular airtime cost. In 2012, Polly was relaunched in Pakistan with an increased telephone capacity. The system was seeded via automated phone calls to five users from the pilot launch. Within 1 week, the 30 phone lines were saturated, and usage quotas had to be imposed. A job audio browser had also been introduced in Polly that allowed users to listen to entry-level job ads from local newspapers in their local language. Callers could browse job opportunities and could even forward promising ones to friends. Polly remained online for a year and amassed 165,000 users who participated in 636,000 interactions, including over 200,000 forwarded voice messages and 22,000 forwarded job ads. At its peak, it was spreading to a thousand new users every day. The 728 job ads were listened to 386,000 times by 34,000 users. Polly was used primarily by low-educated young men for entertainment and other creative uses like voicemail, group messaging, and telemarketing. Its viral spread crossed gender and age boundaries and also attracted a large number of visually impaired users but remained primarily in the similar socioeconomic strata.

Raza et al. showed that Polly's users improved at using speech interfaces with time. A behavior study of the 165,000 users of Polly revealed that with more experience, users responded faster to menus (using more intentional menu interruptions aka barge-ins) and made fewer mistakes and abortive attempts (Raza et al., 2013a, b). Users' choice of activity also evolved over time, and with experience, they showed an increasing interest in message sending, became more explorative of the system's capabilities, and better adapted themselves to its constraints. Long-term users engaged in lengthier calls from the start and took a more active interest in voice modification and forwarding features. The forwarding feature that allowed users to send audio messages to friends brought 85% of all new users to the service. An analysis of the geographical spread of Polly showed that there were calls from all over Pakistan (Raza et al., 2013a, b).

Since 2010, Polly has been successfully used in three countries to rapidly spread useful information to underserved populations at a large scale. In 2014, at the peak of the Ebola crisis in West Africa, Polly-Santé (Polly-Health) was deployed as an emergency disaster response service in Guinea to spread reliable information about prevention, symptoms, and cure of Ebola (Wolfe et al., 2015). The information originated from the US Centers for Disease Control and Prevention (CDC), and the service was funded by the US Embassy in Conakry. A hurdle to scale information dissemination in the Guinean context was great linguistic diversity and the lack of a widely understood common language. This did not turn out to be a major impediment for voice forums, and Polly-Santé was launched in 11 local languages and reached more than 7000 users within a few months. In India, Polly was used by Babajob.com to advertise a voice directory of available jobs to thousands of low-literate job seekers and by Jharkhand Mobile Vaani, a popular citizen-radio-over-phone platform, to spread awareness about their platform using a "cross-selling" model of advertisement (Raza et al., 2016). These deployments highlighted the significance of committed local partners and showed that seeding via promos and advertisements has the potential to induce viral spread and that the content, mood, and tone of the promos play a vital role in influencing a user's understanding of the service and its capabilities.

Despite their demonstrated impact on user training, trust, and service spread, voice forums like Polly face three major challenges that significantly impede their scalability and sustainability: (1) how to moderate, filter, and manage the massive user-generated local language audio content in near real time; (2) how to retain the users of the service for long-term interactions, as most users of Polly just stop using the platform within days as the novelty of the simple entertainment wears off; and (3) how to manage the high call costs required to subsidize these voice forums for the target users who are mostly not able to afford such expenses on their own. These challenges were further amplified as voice forums evolved from being peer-to-peer (individual) to broadcast (social) platforms.

3 Managing Local Language Audio Content

Voice forums quickly evolved to be a de facto social media platform for basic phone users where they record, listen to, and vote on audio messages generated in local languages. However, since such voice forums generate a large amount of audio content in low-resource languages and accents that are unsupported by advancements in natural language processing, it is very difficult to automate categorization and moderation of posts and responses, which are needed for respectful use of the service. Lack of categorization makes it very hard for users to browse data and providers to regulate these services. This is a major hurdle for voice modality where users must listen to audio content in a sequential manner and cannot skim it like textual content. Also, voice forums are often deployed in low-resource environments where most people lack technical know-how of social media, making them particularly vulnerable to disinformation and fake news. For example, recent acts of mob violence in India have been attributed to fake WhatsApp[4] messages,[5] which can also be easily recorded and shared on voice forums. These reasons make content moderation critical to present users with respectful, accurate, and high-quality recordings.

Large voice forums typically employ a dedicated team of moderators, who screen posts, offer feedback to contributors, and perform tagging, categorization, and moderation of audio posts. For example, both CGNet Swara and Mobile Vaani currently employ 10–15 full-time moderators who carefully review each submitted post. Although manual moderation is highly accurate, it becomes difficult to scale as these services grow, due to high cost, delayed response, and challenges in hiring moderators who are familiar with local context.

Several news and social networking sites like Reddit and Stack Overflow draw on collaborative filtering and community moderation algorithms to manage user-generated content. They use community votes and recency to determine high-quality and contextually relevant user-generated content. Since most of the content on these platforms is textual, a number of natural language processing techniques have been employed to categorize, annotate, and moderate content and even predict emotions. However, no prior work has focused on using community moderation on a voice

---

[4] S. Biswas, "How WhatsApp helped turn a village into a mob," *BBC News*, July 19, 2018.

[5] V. Goel, S. Raj, and P. Ravichandran, "How WhatsApp Leads Mobs to Murder in India," *The New York Times*, July 18, 2018.

forum, which is different in several ways from community moderation on a text-based forum. For example, audio content is more difficult to skim than textual content, meaning that users may lose patience in hearing and ranking lower-ranked posts. An IVR system can track exactly what a user listened to and what content they skipped, which is difficult to do on a webpage. Finally, the limited affordances of an IVR interface and the limited digital skills of voice forum users add more constraints to the design of community moderation algorithms.

To address the content moderation challenge, Vashistha et al. harnessed collaborative filtering and community sourcing and showed that the users of voice forums, although socioeconomically marginalized and technologically inexperienced, can themselves be entrusted with the tasks of audio content moderation and categorization (Vashistha et al., 2015a). Recognizing that entertainment drives technology adoption by low-income people, they built a new voice forum called Sangeet Swara, a community-moderated social media service that enabled users to record, listen to, and vote on songs, poems, and other cultural content. Figure 22.3 shows the high-level user interface design of the voice forum. As users listened to messages, Sangeet Swara requested them to annotate the quality and category of the content by pressing phone keys (e.g., press 1 to upvote or 2 to downvote the message) and used collaborative filtering techniques to rank, order, and categorize audio posts based on users' votes. A key aspect of the system was that it *ranked* the posts based on feedback from the community. The ranking aimed to order the posts according to what was most likely to be enjoyed and appreciated by listeners. There was a single, global ranking computed across all posts and all listeners in the system. In addition to the rank order, the system calculated a separate playback order that determined which post a listener heard at a given time. The playback order balanced the interests of listeners (who desired to hear high-quality posts) with the interests of content contributors (who desired to have as large of an audience as possible). Both the rank order and playback order were dynamically updated based on listeners' ratings of the content.

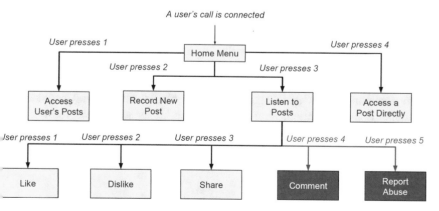

Fig. 22.3 High-level simplified user interface design of Swara and Baang. Additional components in Baang's UI design are represented in red color

To create awareness about Sangeet Swara in rural and small-town India, Vashistha et al. seeded Sangeet Swara with 15 songs and poems that appeared previously on CGNet Swara and posted a message on CGNet Swara to invite participation from its users. The post was accessible to CGNet Swara callers for 2 days, during which time it was heard by 393 unique callers. Out of those, 73 people placed a call to Sangeet Swara. In an 8-month deployment in India, Sangeet Swara received 53,000 phone calls from 13,000 users who submitted 6000 voice messages in 11 languages as well as 150,000 votes. Nearly 80% of users had never used any social media platform before, 50% lived in low-income environments in rural India, and 25% were people with visual impairments (Vashistha et al., 2015a, b). Many users attributed great value to their interactions on Sangeet Swara. They recorded strong positive sentiments about the service and shared interesting anecdotes about how the service was impacting their lives. They considered it to be a platform where people show their creativity, voice their opinions, and record interesting content. This sentiment was often strongest among visually impaired users. The findings indicate that users took an active role in policing the system, for example, by urging others to record cultural content, to follow community guidelines, and to avoid posting abusive comments. Community moderation was 98% accurate in content categorization, made meaningful distinctions between high- and low-quality posts, and performed judgments that were in 90% agreement with expert moderators. The ability to vote, comment, and share led to viral spread, deeper engagement, and the emergence of true dialog among participants. Beyond connectivity, Sangeet Swara provided its users with a voice and a social identity as well as a means to share information and get community support. Moreover, it demonstrated that a community of low-income, low-literate people can moderate themselves without any outside support, thereby addressing the content management challenge of these voice forums.

Although Sangeet Swara demonstrated the feasibility, acceptability, usability and efficacy of community moderation, several aspects of community moderation remain untested. For example, Sangeet Swara focused on the domain of entertainment, where the content is relatively uncontroversial. Extending to domains such as politics and citizen journalism will require sensitivity to stronger disagreements between callers, which could impact their ratings as well as their flagging of posts for deletion. Similarly, the community moderation algorithm can be improved by making it more sensitive to who is voting and when they are voting. For example, discounting votes of users who acted too soon or those who deviated from community standards (e.g., people who abused other users) could reduce randomness in moderation. Similarly, assigning higher weights to votes of users who call consistently and record high-quality posts could improve the quality of moderation. Future work should identify features to predict casual voting by community members.

4 Increasing User Retention

Unlike Polly users that churn at a high rate as the novelty wears off, Sangeet Swara saw much lower churn rate and high user engagement throughout its deployment. To investigate these effects more systematically, Raza et al. created Baang, a social media voice forum with in-built mechanisms to achieve greater spread and uptake as well as deeper and long-term engagement (Raza et al., 2018). The interface of Baang was inspired by both Polly and Sangeet Swara and included two novel features. Like Sangeet Swara, Baang allowed users to create and consume audio content and express their feedback via likes and dislikes (see Fig. 22.3). Similar to Polly, users could share audio content with friends, and Baang actively called up the recipients to deliver the messages. In addition to these features, Baang also allowed users to record audio comments on posted messages and also allowed them multiple options to control the order of playback of messages (popularity, recency, and trending).

Deployed in Pakistan in 2015, Baang organically reached 10,000 users (69% of them were blind) within 8 months who participated in nearly 270,000 calls and contributed more than 44,000 voice messages that were played more than 2.8 million times and received 340,000 votes, 124,000 audio comments, and about 45,000 shares. The user retention of Baang was significantly higher compared to Polly. The differences between the two services became more pronounced after a week when more than 20% of users returned to Baang, while Polly only retained less than 5% of its users beyond the initial week of exposure. Up to 20% of Baang's users (compared to 1% to 3% in case of Polly) kept returning after 4 weeks. Quantitative analysis of usage patterns revealed that the user retention was highly associated with the act of posting audio comments. Interestingly, posting of comments was found to be a better predictor of continued use compared to any other single action of the platform including posting of messages. The viral spread of Baang was found to be largely through the message sharing feature where 60% of all new users were introduced to the platform through forwarded messages.

The analysis found that Baang created a community of users from diverse socioeconomic and linguistic backgrounds including 69% blind people, 10% females, and mostly low-educated, unemployed, young men from all over Pakistan. Baang's open community included people from remote areas and linguistic minorities. Social network features like voting, content sharing, and voice comments led to viral and enthusiastic uptake of the service, high user engagement and retention, and true dialog among the community. Baang provided a window into the collective values of a community as they raised their voice against disability abuse, female harassment, foul language, hatred, and terrorism and united for their rights and in support of the oppressed. Like Sangeet Swara, Baang showed that orality-driven social platforms have the potential to provide under-connected and tech-naive individuals with a voice and social identity.

5 Managing Cost

Although the success of services like Polly, Sangeet Swara, and Baang is very encouraging, there is a growing concern among global development researchers and practitioners about the high operating costs of voice forums especially when they reach a large scale. Providers of these services often pay for the cost of acquiring toll-free lines so that low-income people can access these services for free. However, this cost becomes prohibitive as the usage increases, often putting these services at risk of being shut down. For example, Polly was discontinued several times because of the lack of resources to meet growing call volumes. Many voice forums rely on external funding in the form of grants and awards to subsidize the cost of voice calls; however, the non-deterministic nature of such opportunities makes this approach unsustainable. For example, the founder of CGNet Swara expressed frustrations about how limited funding to subsidize phone calls may cause them to "shut down completely".[6] Some voice forums, including Sangeet Swara, have conducted experiments to examine users' willingness to bear the cost of voice calls; however, the outcome of such experiments have consistently shown that low-income users are unable to afford the cost of voice calls regardless of the benefits offered by the service (Vashistha et al., 2015a, b).

A few voice-based services such as Kan Khajura Tesan[7] and Mobile Vaani have used advertising revenues to subsidize the cost of voice calls. These services advertise products and services that cater to low-income consumers in rural and peri-urban areas (e.g., small sachets of washing powder, toothpaste, and soap). Although these services are existential proof of advertising as a viable approach to financially sustain large-scale voice forums, the initial investment required to gain critical mass for advertising is often beyond the reach of most bottom-up, development-focused voice forums.

Some voice forums such as Ila Dhageyso and 3-2-1 Service[8] have partnered with government agencies and mobile network operators to subsidize the cost of voice calls. Although such partnerships greatly reduce the burden of voice call costs, building and maintaining such partnerships is seldom possible due to mismatch in goals, expectations, and values. Given these limitations in existing approaches to financially sustain voice forums, there is an urgent need to find alternatives to reduce the burden of phone calls on voice forum providers.

Recently, Vashistha et al. examined an alternative approach to address the financial sustainability challenge. They investigated whether low-income users of

---

[6] A. S. Writer, "Amid fund crunch, CGNet Swara eyes shift to Bluetooth radio tech," https://www.livemint.com/, Sep. 25, 2016. https://www.livemint.com/Politics/UcrYsrB8fIAGTDiIoC452N/Amid-fund-crunch-CGNet-Swara-eyes-shift-to-Bluetooth-radio.html (accessed Sep. 04, 2018).

[7] "Kan Khajura Tesan." http://www.kankhajuratesan.com/ (accessed Feb. 23, 2017).

[8] "3-2-1 – On-Demand Messaging for Development." http://hni.org/what-we-do/3-2-1-service (accessed Sep. 04, 2018).

these services could complete useful work on their mobile phones to offset their participation costs on services like Sangeet Swara. To do so, they designed and built Respeak, a new crowdsourcing marketplace that works on basic mobile phones and offers tasks that do not require familiarity with English language and advanced skills like typing, unlike mainstream crowdsourcing marketplaces like Mechanical Turk and CrowdFlower. Respeak is a voice-based, crowd-powered speech transcription system that pays users to transcribe audio files vocally (Vashistha et al., 2017, 2018, 2019a, b). To obtain transcription of an audio file, Respeak partitions the file into small audio segments and sends these segments to multiple users. Instead of typing the transcript on a phone's keyboard with constrained physical space, users re-speak (i.e., repeat) audio content into an off-the-shelf speech recognition engine and submit the speech recognition output as a transcript. Once multiple users submit transcripts for a particular segment, Respeak combines the transcripts using sequence alignment algorithms to reduce random speech recognition errors. Based on the overlap between aligned transcript and individual transcripts, Respeak sends rewards to users via mobile airtime or mobile payment, thereby incentivizing them to complete more tasks accurately. Figure 22.4 shows the high-level illustration of Respeak's design.

Before launching the Respeak system widely, Vashistha et al. conducted a range of cognitive experiments, usability studies, and experimental evaluations to assess its feasibility, usability, and acceptability. For example, they investigated how audio segment length and presentation order affect content retention and cognitive load on Respeak users, whether speaking is indeed a more efficient and usable output medium for transcription than typing, and how different phone types, channel types, and modes to review transcripts affect task accuracy and completion time. After iteratively incorporating insights from these evaluations into Respeak's design, they deployed it to 73 low-income students, blind people, and rural residents in India for nearly 2 months by partnering with Indian Institute of Technology Bombay

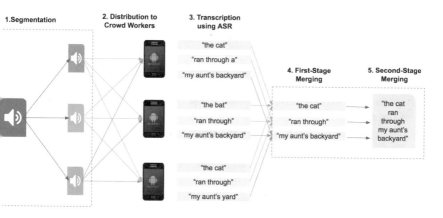

Fig. 22.4 A high-level illustration of Respeak's design. Areas inside dotted lines represent the processes of the engine

(IIT Bombay), Enable India, and Nehru Yuva Sangathan Tisi (NYST), respectively. Collectively, users transcribed 70 hours of audio data by completing 50,000 micro tasks with an average accuracy of 70% and earned INR 31,000 at an hourly rate that exceeds the average hourly wage in India. Respeak then merged transcripts from multiple users to produce the transcript with over 90% accuracy at one-fourth of the market rate, generating sufficient profit to subsidize participation costs of other voice-based services. The analysis indicated that 1 min of crowd work on Respeak could subsidize 8 min of airtime on services like Sangeet Swara. User evaluations also indicated that voice forum users were willing to do tasks on Respeak to get subsidized airtime to use other voice-based services like Sangeet Swara. Also, switching between these two services – Respeak to complete crowd work and Sangeet Swara to use free credits – did not affect the usability and experience of users on both services. Although Vashistha et al. demonstrated the promise of using crowd work to financially sustain voice forums, future work should look at conducting long-term deployments, large-scale field evaluations, and real-time integration with existing voice forums like CGNet Swara and Mobile Vaani to uncover issues that may arise at scale.

6 Replication and Scale

Despite the enthusiasm surrounding voice forums, the unfortunate reality is that it remains quite complex to install and configure them. Many services like CGNet Swara, Avaaj Otalo, and PhonePeti utilize open-source platforms like Asterisk[9] or FreeSWITCH[10] for the telephony interface and require hosting a web server to connect with moderators (Koradia & Seth, 2012). Although tractable for technology researchers, using these platforms requires Linux expertise that is usually beyond the reach of many grassroots and non-governmental organizations that have no in house developers.

Another issue in scaling voice forums is the implementation of the underlying architecture. Most voice forums have a centralized architecture that provides a single access point (or calling number) for users, making it difficult to scale the service to new geographic locations. For example, if an organization would like to scale a voice forum operating in region A to another region B, then either users living in B need to make an expensive long-distance phone call to the access point in region A or the organization needs to set up a local service in B, thereby disconnecting people in the two locations. Also, most voice forums currently operate in silos and are disconnected from mainstream social media platforms, impairing information exchange between different local communities as well as global audience, which might be desirable in cases like political activism, human rights violation reporting etc.

---

[9] "Asterisk.org," Asterisk.org. http://www.asterisk.org/ (accessed Sep. 15, 2016).

[10] "FreeSWITCH.org | Communication Consolidation." https://freeswitch.org/ (accessed Sep. 15, 2016).

Current toolkits to build voice forums like Asterisk and FreeSWITCH do not offer such features to support distributed, scalable, and connected operations. Although cloud telephony systems – like Twilio, Tropo, Exotel, KooKoo, and engageSPARK – make it easy for organizations that lack technical expertise to build and maintain voice forums, these services are *very* expensive to use, are disconnected from social media platforms, and do not synchronize content across distributed call centers, making them less scalable and disconnected.

To address these bottlenecks in building, replicating, and maintaining voice forums at scale, Vashistha et al. built IVR Junction: a free and open-source toolkit that enables organizations to build and replicate voice forums (Vashistha & Thies, 2012). IVR Junction has three main advantages over existing IVR toolkits like Asterisk, Freedom Fone,[11] and FreeSWITCH:

1. Easy to build and set up: IVR Junction makes it easier for organizations with limited technical skills to build, set up, and maintain voice forums. Using IVR Junction, anyone with basic computer literacy can use templates and configure simple options to set up a robust voice forum as an ordinary program on a Window-based commodity machine.
2. Distributed architecture: IVR Junction enables distributed access points, thereby connecting multiple geographically distributed communities via inexpensive local calls as well as enabling robustness to regional power outages or crackdowns by repressive regimes.
3. Global reach: IVR Junction integrates voice forums with free Internet services and social media platforms – recordings contributed over the phone are immediately broadcast on YouTube and Facebook, and posts made on the Internet can also be listened to over the phone. Thus, IVR Junction enables anyone with a basic mobile phone to participate in global social media; low-income populations can record and listen to posts via mobile phone, while the global community can access and contribute recordings via the Internet. This capability enables remote communities to create their own repositories of highly relevant information while also sharing them with audiences worldwide.

In the last few years, IVR Junction has been used by many organizations to connect people in low-resource environments and provide them access to information, news, and governance. For example, in Somaliland, IVR Junction was used to build a voice forum that established a direct communication channel between the rural tribal population and government officials to bring transparency and trust in the political processes (Gulaid & Vashistha, 2013). Somaliland – an autonomous region of Somalia – has fragile political institutions, fragmented and polarized media, and unstable government. Parliamentarians in the capital city were unable to convey their policies and receive feedback from low-literate constituents living in remote, rural, disconnected regions due to misinformation by partisan media. The solution to this intractable problem appeared simple: connect parliamentarians

---

[1] "Freedom Fone." http://www.freedomfone.org/

directly with their communities. However, Facebook and Twitter were infeasible solutions in a region with extremely low Internet penetration and adult literacy rate. To overcome these challenges, IVR Junction was used to build Ila Dhageyso, a voice forum that enabled parliamentarians and constituents to call a phone number and record and listen to asynchronous audio posts in a discussion forum format. Ila Dhageyso also automatically posted these audio messages to a YouTube channel to engage with Somaliland's diaspora. The voice forum was supported by the Office of the Communication of the President and Telesom (the largest telecommunication company in Somaliland) and was launched as a toll-free line so that people living in poverty could participate. The deployment received an enthusiastic response both from the constituents and parliamentarians who recorded over 4300 audio messages in just 5 months.

In war-torn Mali, the Broadcasting Board of Governors and Voice of America used IVR Junction to provide on-demand, reliable, and up-to-date news in the local language. People in Mali called the service to listen to a 3-minute news broadcast by Voice of America, thereby getting access to breaking news and health information as well as sharing their feedback.

In India, IVR Junction was used by women's rights activists in response to a gang rape incident in New Delhi that sparked international outrage. They built a voice petition forum where supporters from all economic backgrounds and varied literacy levels raised their voice for women's safety and empowerment. The recordings, which spanned from support for the victim to plans for sensitizing local communities, were available not only on the voice forum but also on a YouTube channel and Facebook page, thereby dramatically amplifying the local voices.

7   Mechanisms of User Acquisition

Until now we have discussed the unique advantages of voice forums in isolation without establishing a comparison between such forums and other available means in developing regions to reach various development goals. As discussed, technology-based interventions typically require significant resources to achieve scale beyond the pilot stage. Spreading awareness, acquiring users, and retaining them over time are all significant barriers to scale. While we have discussed that voice-based entertainment services can be used as vehicles for spreading development services, it is not clear how well such entertainment-driven proliferation performs in comparison with traditional advertising channels in terms of cost, extent of spread, and quality of user engagement.

Raza et al. compared various advertisement channels side by side as they attempt to scale the same development service – a maternal health hotline, Super Abbu connecting expectant parents anonymously to trained gynecologists (Naseem et al. 2020). They also show how users acquired through various channels fare in terms of overall activity, engagement, seriousness of purpose, and IVR sophistication. An 11-week campaign was conducted where users were acquired for Super Abbu through seven different advertising channels: (1) paper flyers, (2) banners displayed at the back of auto rickshaws, (3) cable TV ads, (4) radio ads, (5) robocalls, (6) sponsored Facebook ads, and (7) an IVR-based entertainment service. Across these channels, they reached 21,770 users who engaged in over 32,500 interactions on Super Abbu. To assess the efficacy of the channels, three main user acquisition metrics were considered: conversion rate, cost of user acquisition, and retention rate. Furthermore, to understand whether the IVR users interact with Super Abbu differently from users acquired through other channels, users were compared in terms of their activity, engagement, and IVR use sophistication.

The results show that IVR platforms (robocalls and Polly) performed better than other platforms (Facebook, flyers, radio, cable TV, and rickshaw ads) in terms of user acquisition. Users acquired through the IVR entertainment service (Polly) performed better than other channels in all interface-related measures (activity, feature engagement, use of sophisticated interface features, and retention). Only robocalls, Facebook, and the entertainment service were able to acquire users at relatively low cost per user (around $1 or less). In contrast, most users acquired from outside of the entertainment service did not end up becoming long-term users of the development service. Their findings also show the comparatively lower performance of increasingly popular social media advertising platforms (Facebook) to recruit low-income users (91 recruits out of over 100,000 people reached).

This study established that voice forums perform better compared to other means as advertisement channels that recruit users for development intervention. It also revealed the surprising lack of success of mainstream social media platforms for user acquisition in developing regions.

Measuring Impact

Generally, impact assessment of information campaigns is carried out via follow-up surveys to measure knowledge retention. However, measuring impact presents a unique set of challenges when information is mass disseminated using voice forums. To encourage spread, inclusiveness, and anonymity, such services do not require users to go through any formal recruitment or registration, and a significant fraction of users only engage with a handful of calls. As a result, manual or telephonic baseline and end line surveys are not feasible. In addition, such surveys are hard to scale and prone to delays. Consequently, despite a growing number of telephonic campaigns, there has been little focus on the measurement of retention of the delivered information. Due to a lack of rapid, scalable, and reliable mechanisms to quantify and measure knowledge retention, campaigns mostly resort to measuring only the extent of delivered information (for instance, number of calls and number of users who listened to the information content).

The services described in this section make use of voice-based quizzes to simultaneously spread information and measure its impact on the knowledge of the target users. These services use various incentives to engage the users in answering multiple-choice questions over IVR interactions. As the users provide their responses, they are informed about the correct answers to these questions. Such interactions allow the users to actively participate in learning about topics that are instrumental for their development. Their score as they keep engaging with the quizzes acts as an indicator of how well the information dissemination influences their knowledge and beliefs.

Raza et al. merge knowledge gap discovery, information dissemination, and knowledge retention measurement into one service, using voice-based quizzes. They show that voice-based quizzes over simple mobile phones, consisting of multiple-choice questions, can be used to simultaneously measure the existing knowledge gaps as well as to disseminate information. Rephrased versions of quiz questions are repeated at regular intervals to measure retention of conveyed information. Long-term user engagement is encouraged by allowing users to contribute their own questions and with social connectivity, gamification, and spirit of competition that make the service engaging and fun for the target audience. The resulting service, Sawaal, allowed its open community of users to post and attempt multiple-choice questions and to vote and comment on them (Raza et al., 2019). Sawaal was designed to spread virally as users challenge friends via shared quizzes and compete for high scores. Administrator-posted questions allowed discovering knowledge gaps, spreading correct information, and measuring knowledge retention via rephrased, repeated questions. Community-contributed quiz content and an ability to play against friends for high scores led to inclusion and ownership, active collaborative learning, and a spirit of competition among the users. Sawaal spread organically among the target audience, received an enthusiastic response and successfully retained a significant fraction of the users for several weeks. In 14 weeks and with no advertisement, Sawaal reached 3400 users (120,000 calls) in Pakistan, who contributed 13,000 questions that were attempted over 450,000 times by 2000 users. Knowledge retention remained significant for up to 2 weeks. Surveys revealed that 71% of the mostly low-literate, young, male users were blind.

Along a similar theme, Swaminathan et al. created Learn2Earn to spread awareness in rural areas of developing regions about critical issues in health governance, and other instrumental topics (Swaminathan et al., 2019). While Sawaal incentivized platform usage through gamification, competition, and social media soft incentives of likes and votes, Learn2Earn leveraged mobile payments to bolster public awareness campaigns in rural India. Users who interacted with the service listened to a brief tutorial followed by a multiple-choice quiz. Users who passed the quiz received a mobile top-up (approximately $0.14) and got the opportunity to earn additional credits by referring others to the service. The pilot deployment of Learn2Earn reached 15,000 people within 7 weeks via word of mouth. Most of the users were young men including a large fraction of students. Learn2Earn was shown as a viable way of building awareness among target users about important topics.

While these techniques of remotely contacting low-income, low-literate populations, engaging them in interactive quizzes, and gauging their awareness around important topics were created originally in the development context, the lesson learned from these services have a broader appeal during crises like COVID-19 where it is difficult to conduct door-to-door surveys. Services like Sawaal and Learn2Earn may be employed to remotely examine knowledge gaps and gather user feedback in such crisis scenarios.

9 Open Challenges

Most of the work focusing on voice forums demonstrates their promise to serve as instruments of inclusion for low-literate people, rural residents, indigenous communities, and people with visual impairments. However, like any other social platform, voice forums come with their own pitfalls. They end up reflecting the existing sociocultural norms and values of the society, including its shortcomings and biases. For example, evaluation of several voice forums revealed extremely low participation of women despite that these services are designed to be inclusive, accessible, and usable for everyone: CGNet Swara and Sangeet Swara in India have only 12% and 6% female contributors, respectively; Baang and Polly in Pakistan have only 10% and 11% female contributors, respectively; and Ila Dhageyso, a voice forum to connect government officials and tribal people in Somaliland, has only 15% female users. Table 22.2 shows the usage of Sangeet Swara and Baang by male and female users.

To examine why the participation of women is almost non-existent on social media voice forums that are designed to be inclusive, accessible, and usable for everyone, Vashistha et al. examined Sangeet Swara and Baang, two widely popular social media voice forums in India and Pakistan, respectively (Vashistha et al., 2019a, b). Using a mixed methods approach spanning quantitative analysis of usage logs, content analysis of 10,361 posts containing 140 h of audio data, and qualitative analysis of 50 surveys and interviews, they investigated how men and women interacted with each other on these services, what content they posted and voted, and what factors affected their participation.

The analysis found that female users of Sangeet Swara and Baang faced systemic discrimination and harassment in the form of abusive, threatening, and flirty posts directed at them. Most women lacked agency to retaliate due to deep-rooted patriarchal values that discourage them from arguing and questioning others. They were worried of backlash, on the service from predatory men and in real life from family members, if they record threatening responses or responded to flirty posts. On

Table 22.2 Usage statistics by gender for Sangeet Swara and Baang for random 5000 posts

Voice forum	Gender	Total posts	Unique users	Likes	Dislikes	Shares	Comments	Reports
Sangeet Swara	Male	4764	419	21,630	58,644	189	Not applicable	
	Female	275	31	270	2636	15		
Baang	Male	4142	376	8181	5541	778	1942	2061
	Female	325	31	508	253	2	25	25

the other hand, many male users perceived women as objects of desire, reinforcing patriarchy in these digital social spaces. Male users who behaved inappropriately ganged up on those who criticized their behavior. Although only a fraction of male users recorded objectionable comments, most other male users condoned the unruly behavior of other men and disapproved of objectionable messages less strongly than women did.

A large fraction of women in low-income communities only have access to shared mobile devices where usage is directed by male family members. Social media voice forums like Sangeet Swara and Baang have been only marginally successful in reaching to some women in these communities; however, they are still a long way from providing a welcoming, vibrant, safe, and enriching environment to women. The experience of women users of Sangeet Swara and Baang demonstrates that access is just a first step toward actual and meaningful social inclusion, and more is required to address secondary barriers to women's digital inclusion beyond the basic hurdles of literacy, connectivity, and poverty. There is an urgent need to re-think the design of these services and use participatory design approaches to ensure that we provide equitable and inclusive social platforms for women. We also emphasize the importance of using the intersectionality lens while designing technologies aiming for inclusion to ensure that they do not widen existing economic, racial, cultural, and societal inequalities.

Like mainstream social media platforms, voice forums also face grand challenges when tackling misinformation and disinformation, especially since the posts are audio recordings in local languages that are unsupported by the advances in natural language processing. Our ongoing investigations have uncovered the presence of a significant amount of false posts, misinformation, and hoaxes related to the COVID-19 pandemic on Baang. It is important to note that voice forums like Baang differ greatly from mainstream social media platforms like Facebook in terms of scale, features, interfaces, supported languages, and target users. Consequently, solutions to tackle these challenges on a platform like Facebook might be ineffective for voice forums and vice versa. This presents interesting research challenges of understanding the composition of misinformation on voice forums, measuring diffusion properties of false posts, examining interaction of low-literate users with suspicious posts, understanding strategies users employ to verify information, and designing new tools and techniques to combat misinformation. The HCI4D community needs to tackle grand challenges like harassment, abuse, and misinformation to make these services truly diverse, inclusive, and impactful.

10 Summary

While Internet services have transformed how people participate in the information ecology and digital economy, these services have discounted the needs and wants of billions of people who experience literacy, language, socioeconomic and connectivity barriers. To address the information and instrumental needs of these people, several HCI4D practitioners and researchers have designed voice forums that enable users to interact with others via ordinary phone calls in local languages. Although voice forums have had a demonstrated impact on marginalized communities, most forums operate at a pilot scale because of known challenges in training and retaining users, managing local language content and costs of voice calls, difficulties in building and deploying these services, and measuring impact. This chapter discussed the innovation, implementation, and adaption of voice forums along with several approaches used to scale, sustain, and replicate them.

Discussion Questions

1. Why is speech-over-simple phones a viable strategy to provide information and connectivity to low-literate, non-tech-savvy, visually impaired, and marginalized segments of the society? How does this strategy compare with other modalities like television, radio, newspapers, websites, and smartphone apps?
2. Using the case studies discussed in the chapter as an inspiration, identify the features of voice forums that can help attain the following objectives:

   (a) The service should spread virally among users.
   (b) The service should promote user engagement.
   (c) The interface should promote improvements in usage patterns of the users.
   (d) Users should keep returning to the service for a long time (several weeks, months, or years).
   (e) Users should retain the information delivered by the service.

3. How does viral (person-to-person) spread compare with broadcast spread? Assume that you need to spread a voice forum among people in a rural developing region. You have two options: schedule automated calls (robocalls) to every phone number in the village or create a viral service and recruit a handful of initial users to seed the spread. Discuss the benefits and shortcomings of each of these approaches.
4. If you have a high accuracy speech recognizer for languages of developing regions, would speech input be preferable over touch-tone (key press) input in voice forums? Discuss the benefits and shortcomings.
5. In addition to speech transcription, which other tasks are suitable for voice forum users to gain airtime and earn money?
6. Would voice forums become obsolete when underserved people gain access to low-cost smartphones and affordable Internet access?
7. How could voice forums be designed to be more inclusive, safe, and welcoming to women?
8. What lessons could be used from this chapter to design a voice-based social media platform for people in developed regions of the world?

References

Gulaid, M., & Vashistha, A. (2013, December). Ila Dhageyso: An interactive voice forum to foster transparent governance in Somaliland. In *Proceedings of the sixth international conference on information and communications technologies and development: Notes-volume 2* (pp. 41–44).

Koradia, Z., Balachandran, C., Dadheech, K., Shivam, M., & Seth, A. (2012, March). Experiences of deploying and commercializing a community radio automation system in India. In *Proceedings of the 2nd ACM symposium on computing for development* (pp. 1–10).

Koradia, Z., & Seth, A. (2012). PhonePeti: Exploring the role of an answering machine system in a community radio station in India. In *Proceedings of the fifth international conference on information and communication technologies and development*.

Mudliar, P., Donner, J., & Thies, W. (2012, March). Emergent practices around CGNet Swara, voice forum for citizen journalism in rural India. In *Proceedings of the fifth international conference on information and communication technologies and development* (pp. 159–168).

Moitra, A., Das, V., Vaani, G., Kumar, A., & Seth, A. (2016, June). Design lessons from creating a mobile-based community media platform in rural India. In *Proceedings of the eighth international conference on information and communication technologies and development* (pp. 1–11).

Naseem, M., Saleem, B., St-Onge Ahmad, S., Chen, J., & Raza, A. A. (2020, April). An empirical comparison of technologically mediated advertising in under-connected populations. In *Proceedings of the 2020 CHI conference on human factors in computing systems* (pp. 1–13).

Patel, N., Chittamuru, D., Jain, A., Dave, P., & Parikh, T. S. (2010, April). Avaaj otalo: A field study of an interactive voice forum for small farmers in rural India. In *Proceedings of the SIGCHI conference on human factors in computing systems* (pp. 733–742).

Patel, N., Shah, K., Savani, K., Klemmer, S. R., Dave, P., & Parikh, T. S. (2012, March). Power to the peers: Authority of source effects for a voice-based agricultural information service in rural India. In *Proceedings of the fifth international conference on information and communication technologies and development* (pp. 169–178).

Patel, N., Agarwal, S., Rajput, N., Nanavati, A., Dave, P., & Parikh, T. S. (2009, April). A comparative study of speech and dialed input voice interfaces in rural India. In *Proceedings of the SIGCHI conference on human factors in computing systems* (pp. 51–54).

Smyth, T. N., Kumar, S., Medhi, I., & Toyama, K. (2010, April). Where there's a will there's a way: Mobile media sharing in urban India. In *Proceedings of the SIGCHI conference on human factors in computing systems* (pp. 753–762).

Raza, A. A., Ul Haq, F., Tariq, Z., Pervaiz, M., Razaq, S., Saif, U., & Rosenfeld, R. (2013b, April). Job opportunities through entertainment: Virally spread speech-based services for low-literate users. In *Proceedings of the SIGCHI conference on human factors in computing systems* (pp. 2803–2812).

Raza, A. A., Rosenfeld, R., Haq, F. U., Tariq, Z., & Saif, U. (2013a, January). Spread and sustainability: The geography and economics of speech-based services. In *Proceedings of the 3rd ACM symposium on computing for development* (pp. 1–2).

Raza, A. A., Kulshreshtha, R., Gella, S., Blagsvedt, S., Chandrasekaran, M., Raj, B., & Rosenfeld R. (2016, June). Viral spread via entertainment and voice-messaging among telephone users in India. In *Proceedings of the eighth international conference on information and communication technologies and development* (pp. 1–10).

Raza, A. A., Saleem, B., Randhawa, S., Tariq, Z., Athar, A., Saif, U., & Rosenfeld, R. (2018, April). Baang: A viral speech-based social platform for under-connected populations. In *Proceeding of the 2018 CHI conference on human factors in computing systems* (pp. 1–12).

Raza, A. A., Tariq, Z., Randhawa, S., Saleem, B., Athar, A., Saif, U., & Rosenfeld, R. (2019, May). Voice-based quizzes for measuring knowledge retention in under-connected populations. In *Proceedings of the 2019 CHI conference on human factors in computing systems* (pp. 1–14).

Swaminathan, S., Medhi Thies, I., Mehta, D., Cutrell, E., Sharma, A., & Thies, W. (2019). Learn2Earn: Using mobile airtime incentives to bolster public awareness campaigns. In *Proceedings of the ACM on human-computer interaction, 3(CSCW)* (pp. 1–20).

Sherwani, J., Ali, N., Mirza, S., Fatma, A., Memon, Y., Karim, M., Tongia, R., & Rosenfeld, R. (2007, December). Healthline: Speech-based access to health information by low-literate users. In *2007 international conference on information and communication technologies and development* (pp. 1–9). IEEE.

Vashistha, A., & Thies, W. (2012). {IVR} junction: Building scalable and distributed voice forums in the developing world. In *Presented as part of the 6th USENIX/ACM workshop on networked systems for developing regions*.

Vashistha, A., Garg, A., & Anderson, R. (2019a, May). ReCall: Crowdsourcing on basic phones to financially sustain voice forums. In *Proceedings of the 2019 CHI conference on human factors in computing systems* (pp. 1–13).

Vashistha, A., Garg, A., Anderson, R., & Raza, A. A. (2019b, May). Threats, abuses, flirting, and blackmail: Gender inequity in social media voice forums. In *Proceedings of the 2019 CHI conference on human factors in computing systems* (pp. 1–13).

Vashistha, A., Sethi, P., & Anderson, R. (2017, May). Respeak: A voice-based, crowd-powered speech transcription system. In *Proceedings of the 2017 CHI conference on human factors in computing systems* (pp. 1855–1866).

Vashistha, A., Sethi, P., & Anderson, R. (2018, April). BSpeak: An accessible voice-based crowdsourcing marketplace for low-income blind people. In *Proceedings of the 2018 CHI conference on human factors in computing systems* (pp. 1–13).

Vashistha, A., Cutrell, E., Borriello, G., & Thies, W. (2015a, April). Sangeet swara: A community-moderated voice forum in rural India. In *Proceedings of the 33rd annual ACM conference on human factors in computing systems* (pp. 417–426).

Vashistha, A., Cutrell, E., Dell, N., & Anderson, R. (2015b, October). Social media platforms for low-income blind people in India. In *Proceedings of the 17th international ACM SIGACCESS conference on computers & accessibility* (pp. 259–272).

Wolfe, N., Hong, J., Raza, A. A., Raj, B., & Rosenfeld, R. (2015, September). Rapid development of public health education systems in low-literacy multilingual environments: Combating Ebola through voice messaging. In *SLaTE* (pp. 131–136).

Chapter 23
The Open Data Kit Project

Waylon Brunette and Carl Hartung

1 Introduction

Advances in information and communication technologies (ICTs) have transformed the way people create, retrieve, and update information. Despite the digital evolution during the "Information Age," much of the world's population was not directly benefiting from ICTs in the 2000s. At the Millennium Summit in 2000, the United Nations established the Millennium Development Goals (MDGs) (Jensen 2010) to combat global issues such as poverty, disease, environmental degradation, and illiteracy. Many organizations, ranging from non-profits to governments, began collaborating on a variety of projects that aimed to help achieve the MDGs. The success of data-driven decision-making during the "Information Age" led development agencies and donors to advocate for using evidence-based development to inform future interventions. However, the scarcity of digital technologies designed to operate in environments with limited infrastructure slowed international development organizations' progress in gathering timely data for evidence-based decision-making. Additionally, the unavailability of suitable software tools for mobile data management was a limiting factor to the types of services organizations could provide in low-resource contexts.

Even though modern computing devices were transforming data collection workflows, many global development organizations continued to use paper solutions because mobile digital solutions were designed with assumptions that were problematic for remote locations. Examples of problematic design assumptions include available Internet connectivity, available resources, and the existence of infrastructure. The expansion of cellular telephone infrastructure, combined with

W. Brunette (✉) · C. Hartung
Paul G Allen School of Computer Science & Engineering, University of Washington, Seattle, WA, USA
e-mail: wrb@cs.washington.edu

© The Author(s) 2023
. Madon et al. (eds.), *Introduction to Development Engineering*,
https://doi.org/10.1007/978-3-030-86065-3_23

the decline of mobile devices' cost, presented organizations with an opportunity to leverage mobile devices to replace paper data collection. In 2008, the Open Data Kit (ODK) project was established to empower global development organizations working in resourced-constrained contexts to build information services. ODK provides a suite of scalable data collection tools by leveraging commercially available mobile devices and cloud platforms. Using mobile devices to digitize data in the field has been shown to decrease latency, improve data accuracy, and enable richer data types to be collected than with paper, such as pictures, video, or GPS traces.

2 Development Challenge and Initial Design Decisions

A challenge for many global development organizations was finding "appropriate" technology that could operate effectively in a variety of locations where vulnerable populations require assistance. Mobile devices are among the few technologies that can operate in most locations because of their lower power requirements and multiple networking options to connect to the Internet. Even in places with abundant connectivity, infrastructure can be damaged or destroyed during natural disasters, thus demonstrating the need for technology to function in disconnected environments. The lack of suitable software tools to create mobile information services in resource-constrained contexts disrupted global humanitarian organizations' efforts to leverage ICTs to improve services provided to beneficiaries. The ODK project started with an investigation into "why information services technology was not being leveraged in international development?" The problem was that many information technologies were built for high-resource locations, leading to flawed design assumptions about connectivity, available resources, technical expertise, and infrastructure constraints. An "appropriate" technology provides the necessary functionality to meet organizational requirements and is capable of operating within the constraints of the deployment context.

For technology to assist international development organizations, the technology should be usable by minimally-trained users, should be configurable by people with average computer skills (e.g., not software developers), and should operate robustly despite intermittent power and connectivity. Dr. Gaetano Borriello, the Open Data Kit project's visionary leader, wanted to design ICTs that focused on magnifying human resources by leveraging technology to address global development problems. Dr. Borriello and his team at the University of Washington's Department of Computer Science & Engineering (UW-CSE) perceived an opportunity to accomplish this using Google's open-source Android operating system (OS) for mobile devices. The challenge was to create a reusable data collection system that was adaptable to any context so that international development organizations could use the technology wherever they operated. The three design decisions that shaped the ODK project were: (1) perform co-design with field organizations to est assumptions, (2) follow technology trends to leverage market forces and avoid early obsolescence, and (3) use modular design with open source and standards to encourage community.

To ensure that ODK would be an "appropriate" technology for global development, ODK was designed and implemented using an iterative design process. Researchers gathered feedback from stakeholders at multiple stages for validation and refinement. The iterative process began by listening to organizations about the problems they experienced when attempting to use technology in their deployment context. After identifying the technology issues, researchers would build a prototype and then travel to the deployment context to perform co-design with the field partners. While the researchers were in situ with field workers, they would work side by side with field workers to make iterative changes. These co-design stages would often last for weeks with multiple iterative changes tested in situ.

The ODK project leveraged technology trends to avoid premature obsolescence and take advantage of global market forces. Dr. Borriello believed that by using "appropriate" cutting-edge technology to solve global development problems, it would be possible to leverage global technology trends to drive innovation and keep costs low. Therefore, the ODK project focused on building a data collection platform using commercially available mobile devices and cloud services to simplify an organization's ability to scale its information systems. In 2008, Dr. Borriello decided to ignore many global development experts' advice and chose Android devices as an "appropriate" cutting-edge technology for low-resource contexts. He identified logic inconsistency by global development experts who had a relentless focus on "low-cost" solutions without examining how technology becomes a "low-cost" solution. The ODK project took a different approach and chose Android because was a new cutting-edge, open-source software platform that was encouraging various companies to manufacture a variety of mobile devices that would run the Android OS. The goal was that companies would design different device models with different form factors and prices for different local markets. The hypothesis was that global market forces would be applied to solve problems for international development organizations as companies sought to adapt Android devices to various local markets. The companies would use their resource to localize the technology with appropriate language and price points for communities. Different companies created multiple form factors that targeted different use cases, including high-cost devices that incorporated more processing power, improved cameras, and more accurate GPS systems. The variety of Android device form factors enabled global development organizations to find appropriate hardware for various interventions, contexts, and requirements. Leveraging "appropriate technology" to enable economies of scale from the global market was one of the unconventional decisions that contributed to the ODK project's success.

Dr. Borriello believed the diversity of requirements from the variety of global development organizations necessitated a modular design to encourage a community of practice to form. The ODK project's goal was to build an open-source community that would enable organizations to contribute both their experiences and software code to expand and refine the project. With many different world contexts, having an open and modular design was key to the ODK project's goal of engaging people to contribute their ideas. Therefore, to encourage the creation of multiple modules, the ODK project uses a permissive, open-source software license. The modular design and open-interface standards promote the creation of modules that either complement or compete with the existing modules. Additionally, by being open and modular, the barriers to contribution are lower as developers can reuse most of the ODK frameworks' functionality, thus saving time and resources. The ODK project uses a free and open-source software (FOSS) model to allow organizations with limited resources to leverage the project's software with minimal or no cost. In 2008, the free and open reuse model focused on a community-building philosophy was a different operational approach for many global development organizations. Many organizations were using custom software built by companies, and when the intervention's funding was depleted, the software was often abandoned. In contrast, the ODK project focused on creating a community that would build reusable software for many different subject domains that organizations could leverage without purchasing the software.

3 Implementation Context

Organizations working in developing regions rely on data to determine what projects to implement and to evaluate their program's effectiveness. For a long time, the most common data collection approach was to send workers into the field to collect data on paper forms. While paper is almost universally available, inexpensive and requires very little training for use, it comes with several drawbacks. The first drawback is that paper forms suffer from a lag between the time the data is collected and the time the data is actionable by an organization. With field worker often working in remote regions, the time between when the data is collected and when it is delivered to a centralized location to be processed can be months. Paper records can be lost or destroyed during transport, costing an organization month of progress. Once gathered, more time is needed to digitize the data so that it can be analyzed. Another issue is that paper forms can be error-prone since use input cannot be constrained. Data entry clerks must often interpret handwriting or determine the meaning and can introduce their own typographic errors. Lastly, data in paper format is hard to query, retrieve, or aggregate if not carefully organized.

As portable digital devices became more available, organizations seized the opportunity to digitize many of these data collection workflows from end to end. These digital solutions allowed organizations to address many of the paper issues. Digitally captured data is more accurate due to standardized inputs and input constraints. Digital information can be instantly transferred, copied, or backed up wirelessly. Additionally, digital data collection enables a richer set of data types to be collected than what is possible with paper. For example, a smartphone allows the collection of digital data such as pictures, video recordings, or GPS traces. Finally, it is much easier to categorize, sort, and aggregate digital data to spot trends, outliers, fake data, and problems that need immediate attention.

3.1 History

The first mobile data collection solutions were implemented on Personal Digital Assistants (PDAs) such as the Compaq iPAQ or Palm OS. These devices were around the same size as a modern smartphone but had minimal computational power and storage, and most lacked wireless communication. Many of the initial systems developed by organizations targeted very specific use cases such as collecting data about Malaria patients, providing decision support, or paramedics working in remote regions. Eventually, platforms emerged that could be used for more general data collection, such as EpiHandy and Pendragon Forms. However, many of these initial systems implemented proprietary data protocols to communicate between mobile devices and data storage servers, which resulted in tightly coupled, siloed systems. Updating seemingly small items like the data type of a field in a form often required hiring software developers to make changes to the code. As PDAs started to become obsolete, this monolithic architectural approach caused entire systems to be scrapped. Organizations found it was not worth the cost or effort to implement new data collection clients from scratch on new platforms in order to be compatible with the existing data storage components of deployed systems.

As mobile phones started to proliferate, new opportunities presented themselves for using wireless connectivity like SMS to send data. The first generation of these "dumb" phones did not allow external applications, so solutions were limited to call-and-response SMS solutions or sending of one-way unstructured data. The second generation of phones, dubbed "feature phones," provided limited application development capabilities. One of the first data collection applications for feature phones, JavaRosa, allowed organizations to design forms using the XForms specification, eliminating the need to hire programmers to make updates to forms.

Noting the success of the iPhone, Dr. Borriello and his team at UW-CSE saw an opportunity with the pending release of Android, an open-source operating system for smartphones. The goal was to try to address many of the shortcomings noted in previous mobile data collection systems. In late 2008, coinciding with the initial release of Android, the idea for the ODK project was created in collaboration with UW-CSE, the Android team at Google, and Google.org. ODK's design focused on making an open and modular approach for creating information services that could be deployed to address a wide variety of issues for global development/humanitarian organizations. Android was chosen because it had multiple flexible inter-process communication methods that enabled ODK frameworks to leverage the existing apps for additional functionality (e.g., taking pictures, scanning barcodes, determining location), thus speeding development.

4 Innovation: The Open Data Kit Project

The Open Data Kit project aims to empower resource-constrained organizations to build information services in under-resourced contexts through the creation of an extensible suite of open-source tools. The ODK project supports the convergence of computing and mobility to create an information services platform that specifically targets global humanitarian organizations (Brunette et al. 2013b; Hartung et al. 2010). However, an organization's tasks can vary widely across subject domains, locations, and cultures. To enable broad use by a variety of organizations, ODK tools are designed without prior knowledge of the application domain or deployment conditions. To empower a variety of users, the ODK project focuses on creating interfaces that minimize the technical skills needed to build mobile data applications. Its modular design and open standards enable organizations to create information systems solutions using composable software frameworks. The ODK project leverages Android mobile devices and cloud platforms to simplify scaling interventions. Android compatible devices were chosen because of the variety of device form factors and price points make them a popular device in economically constrained environments.

Microsoft (MS) Office is an example of a suite of tools that are designed without prior knowledge of the data or subject domain. Global development organizations often use MS Office (e.g., MS Excel, MS Word) or other productivity software to digitize data because data can be customized by staff with little programming expertise. For example, MS Excel is commonly used to build tables of data by workers with intermediate technology skills. Additionally, desktop versions of MS Office operate offline, allowing workers to perform most of their work offline and share their files when connectivity is available. While mobile devices were well suited to operate in contexts with sporadic grid power and Internet connectivity, their software applications did not provide the diverse features of conventional PC productivity software (e.g., MS Office). Often mobile apps are designed for a single purpose; thus, users often use multiple mobile apps to perform complex tasks. These focused apps often have minimal customizability to other subject domains, thus making it difficult to create custom reusable templates that are common to MS Word or MS Excel.

The ODK project's development was guided by a few simple principles (Brunette et al. 2013b):

- *Modularity: Create composable components that can be easily mixed and matched and can be used separately or together. This allows organizations to take a "best-of-breed" approach to create information systems that meet their needs.*
- *Interoperability: Encourage the use of standard file formats and data transfer protocols to support customization and connection to other tools.*
- *Community: Foster the building of an open-source community that would continue to contribute experiences and code to expand and refine the software framework.*

- **Realism:** *deal with the realities of infrastructure and connectivity in the developing world and always support asynchronous operation and multiple modes of data transfer.*
- **Rich user interfaces:** *Focus on minimizing user training and supporting rich data types like GPS coordinates and photos.*
- **Follow technology trends:** *Use consumer devices to take advantage of multiple suppliers, falling device costs, and a growing pool of software developers.*

4.1   ODK 1 Tool Suite

The first set of ODK tools, "ODK 1" (Hartung et al. 2010), focused on replacing paper-based data collection with enhanced digital data collection. Released in 2009, ODK 1 has been used by a variety of organizations working in different global development domains. Field workers collected data via digital surveys on mobile devices, and the survey responses are then aggregated on a server for analysis. When first released, it consisted of three primary tools: ODK *Build*, ODK *Collect*, and ODK *Aggregate*. These tools provide the ability to design forms (*Build*), collect data on mobile devices (*Collect*), and organize data into a persistent store where it can be analyzed (*Aggregate*).

To enable organizations to customize the data being collected, the questionnaires in ODK 1 are specified using the JavaRosa variant of the W3C XForms standard defined by the OpenRosa Consortium (OpenRosa 2011). The JavaRosa XForm specifies the data types, navigation logic, and input constraints. While the XForm specification helped the ODK project separate the organization's deployment-specific configuration from a reusable domain-independent rendering framework, organizations felt it was too complicated for a user to configure the system. *Build* provides a graphical drag-and-drop interface for users to design a questionnaire that will automatically generate an XForm specification. To collect data, field workers use the *Collect* app to record data by navigating a variety of question prompts (e.g., text, pictures, numbers, location, barcodes) specified in the XForm Fig. 23.1.

Fig. 23.1 ODK *Collect* provides multiple data input options including text, GPS, selection, and sound

By default, *Collect* operates disconnected from the Internet to enable use in any environment.

Aggregate provided an easy-to-deploy server that stored and aggregated collected data to make it easy to export the data to a diverse set of data analysis and visualization tools. To export data for analysis, *Aggregate* provided interfaces to query and extract data in standard formats (e.g., CSVs, KML) and directly integrate with various web services, allowing *Aggregate* to act as a store and forward system to other software services. *Aggregate* was designed as a "container-agnostic" platform to enable organizations to deploy it on an "appropriate" hosting infrastructure for the intervention or project based on the type of data to be stored and other constraints. *Aggregate* could be deployed to a cloud hosting service to enable an organization to take advantage of the cloud infrastructure's highly-available and scalable services. However, some organizations collect sensitive data creating data hosting locality and security concerns. Organizations are often constrained by local policies or laws that require data to not leave the country of origin, require the prevention of hosting company employees from viewing the data, or prohibit the collection of high-risk data that contains sensitive information.

While *Build* was created to simplify the creation of XForms by enabling users to graphically compose surveys, some users felt the graphical interface was too burdensome for creating lengthy or complex questionnaires. Additionally, some users found that *Build* was inconvenient for sharing forms across their organization, especially when reusing the majority of the form and only needing to make minor adjustments for the deployment context. To solve this problem, "pyxforms" was created by Columbia University to give users in the ODK community an option of designing their form in an MS Excel spreadsheet. The spreadsheet was then automatically converted into an XForm by "pyxforms." After "pyxforms" popularity in the community grew, "pyxforms" was transferred to the ODK project and was renamed *XLSForm* to reflect the connection of an XLS file becoming an XForm. *XLSForm* became the most popular method of creating XForms for ODK. The creation of *XLSForm* is an example of how the ODK project's open standards and modular design enabled contribution to the project. The modular design also provides options to users as it enables users to choose an appropriate tool for their use case or skill level. Users can choose between *Build*'s graphical interface and *XLSForm*'s Excel spreadsheet interface to design their XForm.

4.2   Iterative Design with Users

Much of the success of ODK results from performing co-design in the field with the researchers working side by side with field workers who would later use the software. This interaction with field workers gave ODK researchers real-time, in situ feedback that would not have emerged or been considered in a typical western research university development environment. One of the most important types of feedback received in the co-design process concerned the cultural appropriatenes of the solution to the community that would be using the mobile devices. The rest of this section gives several examples illustrating the benefits of in-the-field iterative design with actual target users.

For the first trial deployment of ODK, two graduate student researchers went to Kampala, Uganda, to work with the Grameen Foundation to collect data about several SMS information services they were piloting in the country. In this pilot, the users were a group of rural farmers who owned "dumb" cell phones and used those phones to run secondary businesses as communication services for their villages. The farmers could all read and speak in their local languages. For the project, the farmers were provided an HTC G1 touch screen smartphone in order to fill out survey information whenever any of their customers used Grameen's SMS information services. During the training, the graduate student researchers encountered several unexpected issues. After entering the information, the farmers were given the instruction to "touch the button" to send the information. At this point, most of the room stared in confusion at the phones in their hands. Since it was the first time any of them had used a device with a touch screen, they were looking for a physical button instead of the virtual image of a button on the screen.

Later during the instruction, several farmers alerted the instructors that they felt the phones were broken because the phones would not respond to their touch. The instructors used the phone, and the phone worked as expected. Upon handing the phones back to the farmers, the same non-functional behavior was observed. It turned out that farmers often had very rough, calloused fingertips that did not interact well with the capacitive touch screens. However, the farmers could use the larger pad of their fingers to interact with the phone, which led the graduate students to significantly increase the size of all the buttons that very night.

Another example occurred when a developer who was working on ODK from Kenya emailed asking if he could swap the main color scheme in *Collect* from white text on a black background to black text on a white background. The reasoning was that it was easier to see black text when working in bright sunlight. Up to this point, much of the software development had occurred in Seattle, where bright sunlight is not often a problem. The default text and background colors were switched and have remained that way ever since.

4.3 Example ODK 1 Deployments

Two examples of ODK 1 deployments are:

HIV Treatment and Prevention in Kenya One of the first large-scale deployments of ODK was by the Academic Model Providing Access to Healthcare (AMPATH) in western Kenya. Their goal was to reach two million people in the AMPATH's catchment area to test and counsel all eligible individuals for HIV, identify pregnant women not in antenatal care, identify orphaned and vulnerable children, and identify people at high risk for tuberculosis. AMPATH stored their patient information in OpenMRS, an open-source medical record system. Because of ODK's and OpenMRS' open and modular designs, AMPATH was able to use *Collect* for their data collection client. They simply implemented a module for OpenMRS that converted the XForm answers into the OpenMRS concept format. This allowed them to switch from using PDAs to using Android smartphones with very little effort. During deployment, AMPATH also ran a study comparing the deployment cost of ODK to that of a previous tablet-based system and a paper-based system. They found that the per-patient cost of deploying the systems was $0.21 for paper, $0.15 using the PDAs, and $0.13 using *Collect* on Android smartphones. To date, AMPATH has reached more than one million patients through home-based counseling and testing, they have over 160,000 patients in active HIV treatment, and 90% of the HIV-positive people found via the program have enrolled in their care program.

Monitoring Forests in East Africa The Jane Goodall Institute (JGI) has been using ODK in Uganda and Tanzania to monitor forest reserves. Local community members are selected as Village Forest Monitors to patrol selected forest reserve areas. Monitors used *Collect* to gather information on deforestation using smartphone features, such as GPS and images, and can send that information for real-time plotting on maps. The data is used to inform conservation decisions relating to the health of forests and habitats for chimpanzees.

5 Evaluation and User Feedback

Millions of people have used the ODK 1 tool suite to perform data collection activities in the majority of the world's countries. Its modular design successfully enabled multiple organizations to deploy customized application-specific mobile data collection solutions. ODK 1 has been used in a diverse set of subject domains including disaster response, public health interventions, and election monitoring. Although the ODK 1 tool suite was experiencing success as a data collection platform, the UW-CSE research team wanted to understand how well it met generic mobile information system requirements. Thus a survey of 73 organizations was conducted in 2012 to determine the adequacy of current mobile data collection and management frameworks in meeting the needs of the organizations. The goal was to identify limitations that were preventing mobile information systems usage in the field. The respondents were from a broad range of organizations working in over 30 countries. From the survey responses and field observations, there were four focus areas that users thought should be improved (Brunette et al. 2013b):

- *Support data aggregation, cleansing, and analysis/visualization function directly on the mobile device by allowing users to view and edit collected data.*
- *Increase the ability to change the presentation of the applications and data so that the mobile app can be easily specialized to different situations without requiring a recompilation.*

- *Expand the types of information that can be collected from sensing devices, while maintaining usability by non-IT professionals.*
- *Incorporate cheaper technologies such as paper and SMS into the data collection pipeline.*

ODK 1's original focus on enabling resource-constrained organizations to replace paper data collection with enhanced mobile data collection meant that many design decisions were made to leave out unnecessary functionality in order to create a solution that was configurable by users with limited technical skills. This purposeful simplicity meant ODK 1 lacked features required for certain use cases. For example, ODK 1 focused on collecting survey data and uploading completed surveys to a server for aggregation and analysis. It did not provide functionality for distributing data back out to mobile devices for review and updating. However, feedback showed that many organizations need to access previously collected data on the device. Many organizations reported being unable to use ODK 1 for usage scenarios that rely on previously collected data. For example, organizations performing logistics management, public health interventions, and environment monitoring often have workflows that require workers to return to a location and reference previously collected data. The worker verifies whether the previously collected data is still accurate and updates the data to reflect the current state. In usage scenarios with follow-up data collections to track progress, organizations requested the ability to use data from previous surveys to complete new entries, as workers complained about having to re-enter information from the previous visit (e.g., patient demographics, refrigerator information).

After gathering feedback from users, it became clear that there were some feature deficiencies in ODK 1 for supporting data management use cases. Incorporating new tools to contribute the missing functionalities into the existing ODK suite was proposed. New tools would maintain the modular design and add new functionality. However, if too many features were added, there were concerns that a core strength of enabling users with limited technical knowledge to create mobile data collection systems could be lost. Additional features often lead to additional complexity for users as they have more choices and more controls to manage. Dr. Borriello decided that instead of jeopardizing the first tool suite's success, a second tool suite should be made that targeted different use cases. The second parallel tool suite would maintain the ODK project's goal of creating domain-independent tools that operate in disconnected environments but would address a different set of constraints and requirements. Organizations could then choose which of the parallel tool suites to use based on their constraints and requirements. Both tool suites would continue to target low-resource contexts that have limited availability of resources, power, and expertise. UW-CSE continued to improve and expand ODK 1 and began to explore the creation of a second tool suite to address use cases where ODK 1 functionality was insufficient.

5.1 Creation of the Second Tool Suite

ODK 1's focus on replacing and enhancing paper-based data collection led to a unidirectional data flow design (similar to paper). Research showed that many organizations wanted bidirectional synchronization of data to enable users to update previously collected data and conduct follow-up longitudinal studies. Additionally, bidirectional synchronization of data can provide field workers with continually updated data to make better decisions in the field and support more complex workflows based on previous data inputs. The feedback from organizations about missing features led to the creation of a second tool suite called "ODK 2." The new generation of tools would focus on *bidirectional data management* applications instead of *unidirectional data collection* (Brunette et al. 2017). ODK 2 was a response to the need of humanitarian organizations for a free, reusable platform capable of *bidirectional data management*. The four areas of desired improvements as well as other user research and feedback gave four design principles for ODK 2 (Brunette et al. 2013b):

- *When possible, user interface elements should be designed using a more widely understood runtime language instead of a compile-time language, thereby making it easier for individuals with limited programming experience to make customizations.*
- *The basic data structures should be easily expressible in a single row, and nested structures should be avoided when data is in display, transmission, or storage states.*
- *Data should be stored in a database that can be shared across devices and can be easily extractable to a variety of common data formats.*
- *New sensors, data input methods, and data types should be easy to incorporate into the data collection pipeline by individuals with limited technical experience*

After the initial evaluation in 2012, the UW-CSE research continued to perform iterative development and evaluation with global development organizations. This iterative development cycle led to the identification of additional feature requirements beyond the four basic ODK 2 design principles (Brunette et al. 2017). For example, dynamic value checking based on previous data was requested to improve data integrity. Improving data integrity was seen as a benefit of digitizing data collection; thus, organizations requested an expansion in data verification capabilities. An example use case that demonstrates the limitations of static constraint checks versus dynamic constraint checks is an agricultural longitudinal study. Generally, in these agricultural studies, workers travel to crop fields multiple times during a growing season. During these visits, the workers record the growing conditions and track the crop's progress over time by recording the crop height. By recording the growing conditions and crop heights of different crop locations research can be conducted to ascertain the effects of different conditions on crop yields. Organizations requested the crop height value inputted be validated to make sure the field worker entered a reasonable value to avoid typing mistakes or other data entry errors. Unfortunately, the ODK 1 tool suite uses static formulas for validation, so the values used during the data integrity check are the same value at the beginning and end of the growing season, as the check is a static absolute min or max crop height value. Organizations desired a data integrity check that would dynamically adjust over time based on the previously collected crop data. The min and max values used to verify the reasonableness of the crop height would be calculated based on previous crop height measurements taken earlier in the growing season. As with any ODK functionality, this dynamic data integrity check needs to be performed during disconnected operation. ODK 2's bidirectional data synchronization enables organizations to support this type of longitudinal study by enabling disconnected access to previous crop heights allowing the creation of dynamic data checks that can automatically adjust to catch data anomalies. Thus, ODK 2 expanded the ODK project's ability to create mobile information systems by having a *data collection platform* (ODK 1) and a *data management platform* (ODK 2). The following list describes the key design goals of the ODK 2 frameworks based on research findings (Brunette et al. 2017):

- *Workflow navigation should use intuitive procedural constructs that function independently of data validation and allow for user-directed navigation of the form.*
- *The presentation layer must be independent of the navigation and validation logic.*
- *The presentation layer must be customizable without recompiling Android apps via HTML, JavaScript, and CSS.*
- *Partial validation of collected data should be possible and the validation logic should be able to be dynamic.*
  Local storage should be robust and performant for data curation and for longitudinal survey workflows using a relational data model.
  Multiple data collection forms should be able to modify data within a single, shared data table.
  Foreground and background sensors should be supported for data collection. Should support adding new sensing and other methods of input.
  Disconnected operation should be assumed as data must be able to be collected, queried, and stored without a reliable Internet connection. When the Internet becomes available, the framework and cloud components should efficiently replicate data across all devices.
  User and group permissions are needed to limit data access.
  Cloud components should be able to fully configure the data management application remotely as well as preserve a change log of all collected data.

Multiple pilot deployments were conducted to investigate, test, and verify the requirements gathered for ODK 2. A diverse set of field partners working on global health interventions, disaster response management, and other areas were involved in validating the suitability of the ODK 2 frameworks in resource-constrained environments. These iterative pilots identified necessary features for ODK 2 that were not fulfilled by ODK 1. Table 23.1 presents a list of case studies and the

Table 23.1 ODK 2 case study feature requirement summary (Brunette et al. 2017)

FEATURE	Childhood Pneumonia	Chimpanzee Behavior Tracking	HIV Clinical Trial	Disaster Response	Mosquito Infection Tracking	Tuberculosis	Patient Records
Complex / Non-Linear Workflows	X	X	X	X	X		
Link Longitudinal Data to Collected Data	X		X	X	X	X	
Data Security and User Permissions	X		X	X	X	X	
Reuse of Data Fields Across Forms			X	X			
Bidirectional Synchronization	X		X	X	X	X	
Customizable Form Presentation	X		X	X			
Custom JavaScript Apps		X	X	X	X	X	
Sensor Integration	X						
Paper Digitization							X
Custom Data Types Update Multiple Fields in a Single User Action	X	X		X			

corresponding ODK 2 features that were necessary to create a data management solution for the use case.

5.2 Multi-Perspective Design

One of the ODK project's challenges was accounting for the many different "users" of the ODK frameworks. During the evaluation of the ODK project, there was a recognition that the classic roles of software where the "software developers" role creates a software application for the "users" role did not adequately explain the various people involved in configuring reusable frameworks like ODK. Since the ODK project seeks to design configurable software frameworks that enable organizations to adapt to their specific subject-domain needs, the classic two-role model did not fit properly. Instead, the "software developers" are creating frameworks for an intermediate set of persons who configure the frameworks to fulfill an organization's data collection workflow. The specified data collection workflow defines the interface the field workers use to collect the data. This is in contrast to standard software development, when "software developers" make assumptions when designing the system that determines and limits the functionality for "end-users." Unfortunately, when making these decisions, the "software developer" may not fully understand the implications on how future software "users" may need to adapt the software for usage in different contexts with varying connectivity, laws, data policies, budgets, etc. During the evaluation process, the research team identified multiple roles of various ODK actors that create, configure, extend, deploy, and customize mobile data management applications. Identifying the various actors was needed to ensure that the proper features and interfaces were being created to support the various roles. The establishment of roles came from a multi-perspective analysis (Brunette et al. 2015) performed to create appropriate framework abstractions. ODK 2's interfaces and abstractions are designed to target the skill levels of the following four roles (Brunette 2020):

- **End users**—*Often field workers who have been deployed by an organization to perform remote tasks such as providing services to beneficiaries.*
- **Deployment Architects**—*These are often employees of the organization who are subject-domain experts who customize the technology to satisfy meet organizational requirements. They are often non-programmers who are responsible for adapting an ensemble of off-the-shelf software to meet an organization's information management needs and impose constraints derived from deployment considerations.*
- **Programmers**—*Skilled programmers who have been employed by organizations to customize the look, feel, or workflow of the ODK 2 frameworks to meet an organization's deployment requirements. Example tasks include adding new sensors, data input methods, custom prompts, and custom data types.*
- **ODK Framework Developers**—*Those who create the core reusable ODK 2 frameworks that are behind reusable abstractions targeted at the other three roles.*

The focus on four distinct roles shaped the design of ODK 2's framework abstractions (e.g., communication resources, complex workflows). It is difficult to design a single abstraction that is usable by a wide range of technical skills. The four roles have different expectations for their technical skills. The ODK 2 design splits the traditional "user" role into three roles *end users, deployment architects,* and *programmers* to recognize that there are multiple "users" with different roles. To avoid confusion between the "software developers" who build the ODK frameworks and "software developers" who use the ODK frameworks' interfaces to customize an organization's workflow, two developer roles were established *ODK framework developers* and *programmers.* Both *programmers* and *ODK framework developers* are expected to know how to program; however, the *programmer* role is not expected to understand how ODK 2 frameworks are implemented, instead simply how to use the ODK programming abstractions for customization. Both the *end users* and *deployment architects* are assumed to have no software programming skills. However, *deployment architects* are assumed to be computer literate so that they can enter enough information about the deployment to customize ODK frameworks using high-level abstractions.

Adaptation: ODK 2 and Extensibility Exploration

Based on the evaluation, researchers at UW-CSE began an exploration of extending the functionality of the ODK project by creating additional mobile frameworks. After years of research focused on creating information systems for global development organizations, UW-CSE released part of the ODK 2 tool suite in 2017 as a parallel effort to the ODK 1 tool suite. The 8 years between the initial release of ODK 1 and the initial release of ODK 2 were spent performing iterative development with multiple cycles of innovation, adaptation, and evaluation of both ODK 1 and ODK 2. Key aspects of ODK 2 are (Brunette et al. 2017): *(1) a modular design that enables individual software frameworks designed for particular tasks to be used together to achieve complex workflows, (2) data and configuration management that enables data protection, storage, and sharing of data across mobile devices and cloud components, (3) a synchronization protocol that is designed to operate in challenged networking conditions, and (4) a services-based architecture that abstracts common functionalities behind a consolidated unifying services layer.*

A usability problem is often created by having a single interface that targets multiple types of users. Users with basic computer skills often want something simple and can feel overwhelmed with too many options; conversely, a user with programming skills can feel frustrated when functionality is not exposed. Instead of one size fits all for the interfaces, ODK 2 frameworks were created with multiple interfaces and abstractions that can perform similar functionalities to help users with different skill levels configure their system. These multiple interfaces follow some of the ideas of design scaffolding (as demonstrated in Anderson et al. (2012); Quintana et al. (2004)) by providing an interface for users to learn about the ODK 2 frameworks without having to understand the full details of the entire system. The goal was to ramp up a user's ODK 2 knowledge through the use of simple interfaces that are less confusing or intimidating than more advanced interfaces that provide additional functionality. For example, in ODK 2 there are multiple ways for a *deployment architect* to create a database table. One method of database creation completely hides the database details from a *deployment architect*. Based on the success of the XLSForm in ODK 1, Excel was chosen again as a method to configure the ODK 2 frameworks, although with different syntax and structures to account for the new functionality. ODK 2 will take a questionnaire designed in an XLSX workbook by a *deployment architect* and will automatically generate all needed database tables. This automation removes the need for a *deployment architect* to understand databases. Alternatively, if the *deployment architect* wants to customize the structure of the database, a spreadsheet can be added to the XLSX workbook to manually define the database table by supplying column names and column types. The *deployment architect* has a third option to create the database table pragmatically by using ODK 2's JavaScript functions.

6.1 ODK 2 Mobile Frameworks

ODK 2 frameworks provide a diverse feature set that enables organizations to create customized data management applications. Since ODK 2 was designed to handle be use cases with additional complexity, there was concern that a single

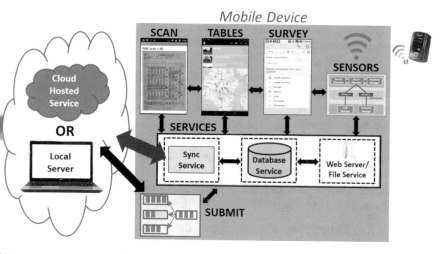

Fig. 23.2 This architecture diagram shows how six ODK 2 mobile frameworks work together to create what can appear to be a single customized mobile data management application (Brunette et al. 2017). The ODK 2 tool suite mobile apps are *Scan*, *Tables*, *Survey*, *Sensors*, *Services*, and *Submit*

app could have overwhelming complexity if it provided abstractions for every usage scenario. Therefore, to simplify customization interfaces, ODK 2 was built as multiple reusable frameworks that focus on specific areas of functionality. To give an *end user* the look and feel of a unified application, the multiple frameworks are designed to smoothly transition between each other. The six Android apps shown in Fig. 23.2) each contribute configurable software frameworks that provide specific functionality to ODK 2. The frameworks isolate the *deployment architect's* configurable interfaces from the reusable system components to create abstractions that are flexible enough to support varying types of workflows from different subject domains.

Tables - *Tables* (Brunette et al. 2013a) is a data management framework that enables users to view, update, and curate the entire data set with optional user permission enforcement. *Tables* is designed to render custom user workflows for use cases where previously collected data is viewed and updated (e.g., logistics management, medical care). Data can be viewed in either tabular or graphical format depending on the context, and views are customizable by modifying or extending the HTML, CSS, and JavaScript files. Since *Tables* uses runtime web rendering for its user interface, the user interface can be customized without recompiling the entire framework. The flexibility provided by *Tables* framework greatly increases the usability of the ODK 2 tool suite, as it enables organizations to encode complex workflows and decision logic outside the question workflow of *Survey*. It is also the creation of workflows for use cases where users often return to locations and need to reference multiple pieces of previously collected data.

Survey - *Survey* expands the possibilities of data collection through its dynamic question-rendering and constraint-verification framework that allows for interactive, non-linear navigation that enables organizations to customize workflow complexity. *Survey* collects data through the use of questions similar to *Collect*. However, *Survey* uses a runtime web rendering framework for its user interface. Since runtime languages (e.g., JavaScript) are used to define the prompt widgets, rendering logic, and event handling, the entire user interface and flow of *Survey* can be customized by an organization. The runtime customization permits a wider variety of customization possibilities than XForm's linear navigation flow used by ODK 1. The runtime flexibility is another example of providing multiple interfaces to *deployment architects*, as *Survey's* XLSX definition structure provides a simpler customization method that can allow someone to ramp up their web scripting knowledge as they require more advanced features.

Submit - Experiences from the field demonstrate that not all the data that organizations collect are considered equally valuable. Data has different priorities, and the system may need to be flexible when selecting the method of data transmission by accounting for inherent data properties, contextual data properties, and network properties (Brunette et al. 2015). Existing networking paradigms often assume that inherent data properties (e.g., size, type) are sufficient. However, data is not uniform and has both inherent data qualities and contextual data qualities. The inherent data properties are independent of the application domain, while the contextual data properties (e.g., priority, importance, deadline) are determined by the application usage scenario and deployment context. An important contextual data property is data public or private (e.g., private medical records vs. sports scores). Local laws and an organization's data policies often dictate how data can be stored and transmitted depending on if the data is public or private. *Submit* (Brunette et al 2015) is an experimental communication framework built to enhance performance in disconnected environments. It allows organizations to adapt their applications to share data in various network conditions to match their data communication needs. *Submit* also provides a modified peer-to-peer synchronization protocol.

Services - *Services* (Brunette et al. 2017) provides the common functionality to other ODK 2 mobile apps to maintain a modular design and avoid re-implementing core functionality within each tool, thus creating a services-based architecture. For example, *Services* abstracts the database access to a single-shared service interface that enforces consistent data semantics and provides data-access restrictions. Other examples of the core functionality provided by *Services* include the data synchronization protocol, a web server, framework preferences, and user authentication.

Sensors - *Sensors* (Brunett et al. 2012) is a device-connection framework designed for Android devices. *Sensors* simplifies connecting and integrating external sensor into an organization's data collection workflow by providing reusable functionality for sensor connection and processing. *Sensors* is a modular framework that simplifies both the application and sensor-driver development by creating abstraction that separate functionalities between the sensor framework, device-driver, and user application.

Scan - Many organizations still rely on paper forms, as the cost of providing a mobile device for every field worker can be problematic for organizations with limited financial resources. During the evaluation, some organizations requested that the ODK project provide the ability to use a hybrid approach that integrates paper forms with mobile devices. The idea was that field workers could use paper forms to collect data and then use a few mobile devices in the field to digitize the data. *Scan* is a paper digitization framework that explored how to combine paper and digital data collection (Dell et al. 2012, 2015). *Scan* enables organizations to digitize paper forms in disconnected environments by adding *Scan*-compatible data input components to their forms. Mobile devices have the computing resources necessary to locally process QR codes, check boxes, multiple-choice bubbles, and structured handwritten number boxes. *Scan* digitizes paper forms using the Android's camera and computer vision algorithms that run on the device into snippets corresponding to the individual data components (Dell et al. 2012, 2015). Additional information can be collected using handwritten text boxes that *Scan* cannot digitize programmatically. These handwritten text boxes are presented to the field worker as an image through an automatically generated *Survey* form to enable the field worker to manually transcribe the data by looking at an image and manually transcribing the handwritten text.

5.2   Example ODK 2 Deployments

Two examples of ODK 2 deployments that at the time of publication have ongoing field evaluations and iterations to improve the "data management platform" are:

Humanitarian Disaster Response The International Federation of Red Cross and Red Crescent (IFRC) partnered with UW-CSE to create an innovative humanitarian relief platform called "RC^2 Relief" (Fig. 23.3). RC^2 Relief leverages ODK 2 frameworks to simplify distributing relief items or providing cash assistance in response to disasters or humanitarian crises. The IFRC had previously leveraged ODK 1 to collect information without Internet connectivity during disaster responses. However, the IFRC found ODK 1's unidirectional dataflow limited their ability to use ODK as a humanitarian information platform. The issue was that ODK 1 could not provide the field worker with updated data to perform relief tasks. This limitation created a worker efficiency issue as field workers had to use both (1) a mobile device with ODK 1 for data entry and (2) either a paper or a laptop to obtain information about relief assistance. Using ODK 2's mobile information management frameworks, field workers can view what relief assistance should be delivered as well as the beneficiary's complete history. Since ODK 2 is a customizable framework, it was a natural fit for the Red Cross movement allowing the reuse of core technology and training materials between National Societies while

RC² Relief may provide a
**TIME SAVING OF
UP TO 90%**
in the collection of information.
This means more time is
available to focus on delivery of
better quality, more efficient,
and personalized support to the
communities that require it.

Fig. 23.3 The IFRC brochure about RC² Relief (IFRC 2020)

also empowering National Societies to customize the system to their particular use cases. Since ODK 2 is a rendering framework, a volunteer's workflow can be easily updated to adapt to the dynamic conditions of a humanitarian relief operation.

Vaccine Cold Chain Inventory Immunization programs depend on a network of vaccine refrigerators and cold rooms to keep vaccines within a narrow temperature range to maintain the vaccine's potency. This network of temperature-controlled storage is called a vaccine cold chain. Management of the vaccine cold chain depends on having an accurate understanding of equipment's status at different facilities, including which vaccine refrigerators are broken and needing repair. Previous attempts to use ODK 1 to perform regular inventory updates were met with resistance because field workers objected to re-entering the same information about the refrigerator during each visit. Instead, workers simply wanted to make updates to the data that they previously entered. With ODK 2's data synchronization, remote field workers can view and update the most recent cold chain inventory data on their mobile devices (Brunette et al. 2020). Synchronizing data helps decrease work duplication between field workers as each field worker has the most current information when Internet connectivity is available.

7 Challenges and Lessons

Technology designed for reusability and flexibility can act as an innovation catalyst because of its applicability to a diverse set of use cases. By providing reusable frameworks, the ODK project seeks to enable global development organizations to leverage computing technologies to innovate and improve the organization's ability to deliver services to combat global issues such as poverty, disease, environmental degradation, and illiteracy. The ODK project focuses on creating information services that are configurable by non-programmers and deployable by resource-constrained organizations (including both financial and technical resource constraints) to address a broad variety of issues for global development organizations. ODK's modular design allows an organization to leverage different modules to create flexible information technology solutions that can be customized to satisfy an organization's contextual requirements. Designing a reusable system without prior knowledge of the type of data that will be stored requires a flexible design, as organizations are often constrained by local policies or laws that dictate how data can be stored and transmitted. The ODK project was designed to leverage "appropriate" cutting-edge technology to solve global development problems and to take advantage of the global technology trends that drive innovation. The project targets commercial cloud platforms and Android devices to enable organizations to take advantage of global market forces. The choice of Android smartphones and tablets in 2008 as an "appropriate" cutting-edge technology for low-resource contexts was because of Android's flexible and open architecture. Choosing Android led the majority of people in the global development space to ridicule the ODK project, with many global development experts stating that Android devices would never be sold in Africa or Asia. The critics insisted that an Android device would never have a price point below $400 USD, and therefore, Androids would only be sold in North America and Europe. At the time of publication of this textbook, Android is the dominant mobile device operating system in the world with devices being sold below $40 USD. A key lesson from the ODK project was identifying the logic inconsistency from global development experts that were exclusively focused on "low-cost" solutions while not examining how technology becomes a "low-cost" solution through market forces. The ODK project seeks to identify and leverage "appropriate" technology that can operate in low-resource contexts to exploit future research and development by the global marketplace.

The ODK project tries to avoid leveraging technology that is not supported by the global marketplace because lacking such support there will likely be minimal momentum to lower prices. By following technology trends, the market solved problems for global development organizations by localizing the technology to users, creating different types of mobile computing devices, and creating devices with prices appropriate to the context. To avoid early obsolescence, ODK used "appropriate" cutting-edge technology to leverage advancements by commercial companies, universities, and governments. Additionally, the timeline of scaling global development interventions often does not align with information technology's life cycles. The cheapest devices are often discontinued when newer, more capable technologies can be produced at a similar or lower cost. Unfortunately, technology obsolescence creates churn, as organizations have to find replacement technologies and often conduct new pilots to verify that the replacement technology will meet organizational requirements. Additionally, depending on how different the replacing technology is, organizations may need to produce new training material and resource mobilization strategies. Therefore, if the lowest-priced technology does not remain available for purchase through an organization's scaling timeline, it may not have the lowest overall cost. For organizations to scale technology to multiple countries, locations, and contexts, it often takes years, thus requiring the technology to remain available for purchase for 5 to 10 years to provide ample time for the technical solution to be adapted, tested, and deployed to the various locations.

While ODK has been successful, there have been many struggles with open-source adoption and management. From the beginning, it was obvious that a few people working together at a public research university would not scale to meet the needs of supporting hundreds of organizations and their field workers deployed around the world. To mitigate the problems of a small research team scaling, an open-source model was used to build an ecosystem of technology innovation. Three approaches were used to build a community of practice: (1) modularity—to make it easy for alternative solutions to be developed in parallel, (2) community building—establish an open-source community for users to contribute experiences and code, thereby establishing a process to expand and refine the software, and (3) open-source—to allow organizations with limited finical resources to benefit from innovations at little or no cost through a free and open-source software (FOSS) model. A permissive open-source software license was chosen to encourage commercial enterprises to build features and expand the core ODK technology. The idea was to embrace the concept of a competing ecosystem where many companies, consultants, and organizations could take FOSS technology and build products and services with it. Supporting the diversity of use cases in the field was something beyond the capability of a few people working at a public research university. The goal was to create a robust ecosystem that would give organizations choices on the types of scaling the technical expertise they needed to receive to meet the needs in a local context.

Many global development software projects, like ODK, rely on donor and grant funding for software development, expansion, and upkeep. Funding organizations often have a large influence on the type of features implemented as donor funding often targets-specific domains, such as health, causing certain use cases to be emphasized more because of the availability of funding. For a deploying orga nization, there are still costs associated with using free software. For example there are costs associated with procuring and maintaining computer resources. Additionally, there are costs to configure ODK for the deployment context and to train field workers. Furthermore, technology is not static; therefore, ODK needs to be continually upgraded to handle software maintenance tasks, such as library and operating system API changes, bug fixes, and security fixes. Software maintenance can be a difficult area to fund because most funding comes with requirements for innovation with evidence-based outcomes on beneficiaries.

Much of ODK's success came from the design principle of performing iterative development that included co-design with field organizations to test assumptions. Researchers working side by side in the deployment contexts with organizations gave researchers feedback that helped identify incorrect assumptions and expectations. Building real systems with organizations with challenging constraints helped reveal problematic design assumptions that were limiting functionality because of incorrect expectations about deployment contexts. Often small assumptions made by a developer can make a technology inappropriate for an organization's use in a specific context.

Summary

Since resource-constrained environments often experience a dearth of technical personnel capable of building and customizing information systems, the ODK project was established to provide a configurable set of open-source software tools for organizations to use in resource-constrained contexts. The ODK project's modular software frameworks target a variety of global development use cases. The ODK tools are designed to be configurable by global development organizations that have an understanding of the field conditions, the types of workflows and actions that are necessary, and which tasks are important to complete. Therefore, the configuration abstractions need to be simple enough for a non-developer to use but flexible enough to handle various constraints based on the deployment context. ODK tools are designed for flexibility because they are created without prior knowledge of the application domain, deployment conditions, or types of data that will be collected. Five decisions that led to the success of the ODK project were:

Maintaining functionality when operating disconnected from the Internet
Identifying "appropriate" technology and leveraging technology trends
Having partners who are local domain experts and perform iterative development in situ
Using a modular design with open standards
Establishing an open-source community of users and developers to create an ecosystem of support

After a decade, the Open Data Kit project continues to engage in innovation cycles that include requirements gathering, iterative development with field partners, evaluation, and adaption. Academic researchers continue to design, build, and evaluate an ensemble of software frameworks that can be used together or independently to construct mobile information systems. The impacts of the ODK project have been diverse in scope and scale. There are several levels of impact including: (1) the impact on the organizational-level users and whether it meets organizational goals of increased efficiency, reduced cost, and more accurate data; (2) the impact on the field workers and if it helps them to be more efficient; and (3) the impact on the quality and cost-effectiveness of the services the beneficiaries receive. Additionally, the cost-savings are multi-faceted in scope and each needs to be measured to understand the benefits of (1) the interventions themselves, (2) savings from shared software infrastructure, (3) employee productivity, and (4) reduced costs of data acquisition for evidence-based decision-making and metrics and evaluation.

9 Discussion Questions

1. What are five advantages and five disadvantages of creating reusable software frameworks that are designed to be used by different organizations trying to solve problems in multiple subject domains?
2. What makes software for international development different from other software projects?
3. What are some current technology trends that can be applied to resource-constrained contexts to improve global development outcomes?
4. What are some advantages and disadvantages of a free and open-source software model?
5. What challenges would you expect developing organizational human resource that are capable of deploying platforms like ODK and how would you address them?
6. What areas of international development could benefit from applying information technology solutions?
7. What technology trends will the ODK project likely need to adapt to over the next 5 years to remain a useful platform for data collection or data management for low-resource communities?

References

Anderson, R., Kolko, B., Schlenke, L., Brunette, W., Hope, A., Nathan, R., Gerard, W., Keh, J., & Kawooya, M. (2012). The midwife's assistant: designing integrated learning tools to scaffold ultrasound practice. In *Proceedings of the Fifth International Conference on Information and Communication Technologies and Development*, ICTD '12 (pp. 200–210). New York: ACM.

Brunette, W. (2020). *Open Data Kit 2: building mobile application frameworks for disconnected data management*. Seattle: University of Washington.

Brunette, W., Larson, C., Jain, S., Langford, A., Low, Y. Y., Siew, A., & Anderson, R. (2020). Global goods software for the immunization cold chain. In *Proceedings of the 3rd ACM SIGCAS Conference on Computing and Sustainable Societies*, COMPASS '20 (pp. 208–218). New York: Association for Computing Machinery.

Brunette, W., Sodt, R., Chaudhri, R., Goel, M., Falcone, M., Van Orden, J., & Borriello, G. (2012). Open Data Kit sensors: a sensor integration framework for android at the application-level.

In *Proceedings of the 10th International Conference on Mobile Systems, Applications, and Services*, MobiSys '12 (pp. 351–364). New York: ACM.

Brunette, W., Sudar, S., Sundt, M., Larson, C., Beorse, J., & Anderson, R. (2017). Open Data Kit 2.0: a services-based application framework for disconnected data management. In *Proceedings of the 15th Annual International Conference on Mobile Systems, Applications, and Services*, MobiSys 2017 (pp. 440–452). New York: ACM.

Brunette, W., Sudar, S., Worden, N., Price, D., Anderson, R., & Borriello, G. (2013a). ODK tables: building easily customizable information applications on android devices. In *Proceedings of the 3rd ACM Symposium on Computing for Development*, ACM DEV '13 (pp. 12:1–12:10). New York: ACM.

Brunette, W., Sundt, M., Dell, N., Chaudhri, R., Breit, N., & Borriello, G. (2013b). Open Data Kit 2.0: expanding and refining information services for developing regions. In *Proceedings of the 14th Workshop on Mobile Computing Systems and Applications*, HotMobile '13 (pp. 1–6).

Brunette, W., Vigil, M., Pervaiz, F., Levari, S., Borriello, G., & Anderson, R. (2015). Optimizing mobile application communication for challenged network environments. In *Proceedings of the 2015 Annual Symposium on Computing for Development*, DEV '15 (pp. 167–175). New York: ACM.

Dell, N., Breit, N., Chaluco, T., Crawford, J., & Borriello, G. (2012). Digitizing paper forms with mobile imaging technologies. In *Proceedings of the 2nd ACM Symposium on Computing for Development*, ACM DEV '12 (pp. 1–10). New York: ACM.

Dell, N., Perrier, T., Kumar, N., Lee, M., Powers, R., & Borriello, G. (2015). Paper-digital workflows in global development organizations. In *Proceedings of the 18th ACM Conference on Computer Supported Cooperative Work and Social Computing*, CSCW '15 (pp. 1659–1669, 2675145). New York: ACM.

Hartung, C., Anokwa, Y., Brunette, W., Lerer, A., Tseng, C., & Borriello, G. (2010). Open Data Kit: tools to build information services for developing regions. In *Proceedings of the 4th ACM/IEEE International Conference on Information and Communication Technologies and Development*, ICTD '10 (pp. 1–12).

Jensen, L. (2010). The millennium development goals report 2010. In *New York: United Nations Department of Economic and Social Affairs*.

OpenRosa API WorkGroup. (2011). OpenRosa WorkGroup API Page: OpenRosa 1.0 APIs. https://bitbucket.org/javarosa/javarosa/wiki/OpenRosaAPI/. Accessed September 2020.

Quintana, C., Reiser, B. J., Davis, E. A., Krajcik, J., Fretz, E., Duncan, R. G., Kyza, E., Edelson, D., & Soloway, E. (2004). A scaffolding design framework for software to support science inquiry. *Journal of the Learning Sciences, 13*(3), 337–386.

RC2 Relief. https://media.ifrc.org/ifrc/wp-content/uploads/sites/5/2019/12/RCR-Leaflet-V3-20191206.pdf. Accessed October 2020.

Epilogue

There is no simple solution to the complex challenges of sustainable human development. Just as international development banks are not the only instruments for financing economic development, "development engineering" (commonly abbreviated to DevEng) is not the only approach for sustainable innovation. Yet the iterative cycles of research described in this book have resulted in technologies benefiting millions of people, often with positive spillovers for the natural environment. These efforts aim to and can blunt the hard edge of inequality, while mitigating the impacts of global climate change.

The first two chapters of the textbook outlined the historical context of technology and human development and reviewed recent research on the role of technology through the lens of development economics. The third chapter presented the reader with a generalizable framework for the design and scaling of technologies that benefit both people and the planet. The framework defined four key activities – innovation, implementation, evaluation, and adaptation – that, when woven together in iterative cycles, can yield significant development impacts. The fourth chapter discussed the ethical considerations relevant when scientists and engineers engage in the design of technologies for communities facing resource constraints.

In each of the 19 chapters that follow the first 4, the authors detailed how a specific technology was designed and adopted, and in some cases scaled up, through an interdisciplinary and iterative approach. These case studies cover a wide range of sectors, from water treatment for affordable arsenic removal to information technology that improves the governance of public services (like electricity or social protection). Some authors demonstrated how an innovation achieved sustainability and impact; others have highlighted where their efforts failed or fell apart. The range in the scale of the case studies highlights how the framework of development engineering can be applied in small pilots as well as larger research studies. Overall, we think the textbook presents a versatile framework for designing, developing, and assessing technologies that can be broadly applied across resource-constrained settings.

© The Author(s) 2023
S. Madon et al. (eds.), *Introduction to Development Engineering*,
https://doi.org/10.1007/978-3-030-86065-3

The challenge of reducing poverty – and doing so in a sustainable manner– is fundamental to the future of human progress and requires a multi-pronged approach. The interdisciplinary nature of this text (integrating perspectives from development economics, engineering, and other fields) helps the reader explore how technology might help us in this complex task. While we aim to be versatile, the field of development engineering is just one approach among many seeking sustainable and equitable development. We do not claim DevEng will solve all problems that cause or maintain poverty, or DevEng relates to all forms of technological change. For example, ample scope remains for improving access to markets and making governments accountable. Meeting these challenges does not always require the design or adoption of new technologies, and in some cases introducing new technologies can even set us back.

Similarly, from an intellectual perspective, there are multiple frameworks for rigorous field-based research, stretching far beyond the approaches presented here. In defining this discipline, we have focused intensely on understanding the adoption and impacts of novel solutions. As a result, the textbook proposes a process of iterative discovery, prototyping, testing, and adaptation of technology-based solutions – with continuous cycles of movement between the laboratory and the field. As the case studies highlight, this process situates its interventions within local contexts, engaging directly with the market, institutional, and behavioral constraints facing individuals and communities living in resource-constrained settings. This intellectual approach differs in its objectives from disciplines like political science, critical development studies, macroeconomics, and global health (to name just a few allied fields). Broader macroeconomic trends in development, as well as structural factors and questions of political will and leadership, are not addressed in the textbook and are better understood and analyzed using other references and approaches.

We believe that technological change will play a fundamental role in achieving sustainable development. We also sense that the next generation of researchers, engineering practitioners, and students interested in sustainable development will need to draw on insights from multiple fields to successfully design, implement, and scale solutions to our global challenges – hence the effort to introduce development engineering to our readers.

Index

© The Author(s) 2023
S. Madon et al. (eds.), *Introduction to Development Engineering*,
https://doi.org/10.1007/978-3-030-86065-3

Printed in the United States
by Baker & Taylor Publisher Services